# MIDLATITUDE SYNOPTIC METEOROLOGY

GARY LACKMANN

# MIDLATITUDE SYNOPTIC METEOROLOGY
## DYNAMICS, ANALYSIS, AND FORECASTING

GARY LACKMANN

American Meteorological Society

Reprinted with corrections October 2022.

Reprinted with corrections July 2021.

Reprinted with corrections February 2015.

Reprinted with corrections August 2012.

Front cover photograph courtesy of the National Oceanic and Atmospheric Administration.

Figure 1.20 (a) from Hovmöller, E., 1949: *The trough-and-ridge diagram.* Tellus, 1. Reprinted with permission of Wiley-Blackwell. Figures 5.1, 7.9, and 7.10 from International Geophysics, Vol. 30: Adrian Gill, Atmosphere-Ocean Dynamics, p.549-593, © 1982. Reprinted with permission from Elsevier.

Published by the American Meteorological Society
45 Beacon Street, Boston, Massachusetts 02108

The mission of the American Meteorological Society is to advance the atmospheric and related sciences, technologies, applications, and services for the benefit of society. Founded in 1919, the AMS has a membership of more than 13,000 and represents the premier scientific and professional society serving the atmospheric and related sciences. Additional information regarding society activities and membership can be found at www.ametsoc.org.

For more AMS Books, see www.ametsoc.org/amsbookstore. Order online, or call (617) 226-3998.

Library of Congress Cataloging-in-Publication Data

Midlatitude synoptic meteorology : dynamics, analysis, and forecasting/
Gary Lackmann.

p. cm.
Includes bibliographical references and index.
ISBN 978-1-878220-10-3 (pbk.)

1. Synoptic meteorology—Textbooks. 2. Dynamic meteorology—Textbooks. I. Title.

QC869.L33 2011
551.5—dc23

2011029732

# CONTENTS

# PREFACE

Teaching synoptic analysis and forecasting consumes a large fraction of my time and attention each fall semester and has done so for over a decade. But I don't mind. In fact, teaching a course that is linked to exciting weather phenomena and current meteorological situations is as enjoyable as it is challenging. Each time I teach the class, I look for ways to improve it. Although my student course evaluations are strongly positive, many students have commented that a textbook that bears some semblance to the syllabus would be a valuable addition as a learning aide. In working with students during office hours as well as within the classroom, it is clear that this suggestion has merit; students need a comprehensive, up-to-date text that adequately reflects the material in the course. One motive for writing this text is to respond to these student requests; however, it is also clear that there are some areas of synoptic analysis and forecasting that are *not covered* in currently available texts, and texts are outdated for some other areas. My overall goal is to offer a comprehensive, contemporary text that is useful to the broader atmospheric science community in addition to students of synoptic meteorology.

Although the target audience for this book is any person with an interest in the atmospheric sciences, a college-level background in meteorology, math, and physics will greatly improve access to the material. Comments received from our alumni indicate that some students kept and utilized their old course notes as reference material when working as meteorologists. With this in mind, the current text represents a complete reference that may be of use to professionals as well as students.

Are there other books that cover the topic of synoptic analysis and forecasting? Yes, there are some excellent ones, but there is not (to my knowledge) a single text that covers the set of topics presented here with an emphasis on *application*. We begin with an applied review of dynamical tools such as quasigeostrophic theory, isentropic analysis, and potential vorticity (PV). This is followed by a phenomenological presentation of cyclones, fronts, cold-air damming, and winter storms. A short theoretical treatment of baroclinic instability is accompanied by a linkage to the real atmosphere via PV. In my experience, undergraduate students often receive scant information about numerical modeling, a topic covered here in chapter 10. Surprisingly, there are few textbooks that discuss the *human forecast process*, a need which is addressed by chapter 11. This chapter has benefitted from abundant input from many friends and colleagues who are current and former members of the professional forecasting community. Finally, a brief chapter on manual analysis is included.

Students sometimes comment that the synoptic and forecasting course "brings it all together" for them. At North Carolina State University (NCSU), students take the synoptic course as first-semester seniors, meaning that they have already taken semester-long courses in thermodynamics, physical meteorology, and atmospheric measurements and analysis and two semesters of dynamics, in addition to elective courses. However, synoptic meteorology is taught at various points in the curricula in different college and university programs. Accordingly, I have designed the text to be accessible to a spectrum of undergraduate abilities and experience, as well as students at the graduate level.

A primary objective in synoptic meteorology is to bridge the gap between theory and observation in the applications of weather analysis and forecasting. In keeping with this theme, I rely heavily upon real-time and case-study examples to illustrate various concepts. In my course, these examples are shared with students in the form of in-class exercises, take-home lab assignments, daily forecasting activities, weather briefings, and a semester-long term-paper project. Although each of these activities has a slightly different objective, they all work towards a common goal.

Students of the atmospheric sciences will be expected to exhibit increasing familiarity with numerical modeling systems and not just for use in weather analysis and forecasting. Chapter 10 is written with an emphasis on practical weather forecasting and how to best utilize models in a forecast process. My aim is to introduce the topic in a way that maintains focus on physical processes and weather phenomena such as winter storms, flooding rains, cyclones, jets, and fronts. Ensemble prediction and data assimilation have emerged as important subdisciplines of the atmospheric sciences. The introduction to these topics in chapter 10 is intended to serve as a starting point for advanced undergraduate students in understanding these topics.

I apologize in advance to my colleagues for any important citations I've omitted. Citations are especially emphasized for very recent results and findings deemed controversial or historical in nature. Looking back at the citation list, I see many of my own papers cited; this is mostly because of familiarity and because I've used many of my own research papers as examples in the text.

A long list of acknowledgements is in order. To the teachers, professors, advisors, mentors, and colleagues who have assisted me in my career, I am indebted not only for the knowledge they provided, but also for their encouragement, for challenging me, and for the ethical standards they emphasized. The content of this book reflects knowledge gained from the courses I've taken and textbooks that I've utilized throughout my ongoing education. I was fortunate to learn from the outstanding University of Washington Atmospheric Sciences faculty in the 1980s, including Cliff Mass and Mark Albright who taught my first forecasting class. During that time, I was also fortunate to have outstanding mentorship at NOAA's Pacific Marine Environmental Laboratory from Jim Overland, Nick Bond, Judy Wilson, and Allen Macklin. At the University at Albany, State University of New York, particularly influential advisors and teachers included Dan Keyser, Lance Bosart, John Molinari, and Arthur Loesch. I am indebted to my many graduate student colleagues both there and at the UW. My career has benefitted from enthusiastic support and encouragement from John Gyakum at McGill University. Steven Businger provided not only encouragement and resources, but also the invitation to undertake this endeavor in the perfect location: his home state of Hawaii. And a special thank you to Anantha Aiyyer, who taught a class that I normally teach during my sabbatical, enabling me to complete this project.

In working to improve the text, constructive and insightful suggestions were provided by many colleagues, including Dave Schultz, John Nielsen-Gammon, Rod Gonski, Gail Hartfield, Jonathan Blaes, Jason Millbrandt, Sukanta Basu, Brian Colle, Anantha Aiyyer, Kevin Tyle, Frank Alsheimer, Kermit Keeter, Brian Etherton, Steven Decker, Greg Fishel, Andrew Odins, Chris Nunalee, Darin Figurskey, Patrick Market, Steven Businger, David Novak, Scott Rochette, Phil Schumacher, Walt Robinson, Greg Hakim, Neil Stuart, and others. In addition to much of the "PV inspiration" evident in this text, Professor John Nielsen-Gammon of Texas A&M University also provided a thorough and thought-provoking set of suggestions, comments, and corrections on the first printing of the text, many of which have been incorporated in the second printing. Sarah Jane Shangraw, Beth Dayton, Ken Heideman, and others at the AMS have been helpful, supportive, and flexible; they were understanding of my desire to make this text accessible to students via AMS discounts. I would also like to acknowledge Lesley Williams and Roger Wood at AMS, for their copy editing of the book. Support in the form of a sabbatical leave provided by North Carolina State University made this effort possible. Graduate student Whitney Rushing assisted with the graphics and copyright permissions.

Those agencies granting copyright permission for use of their graphics are acknowledged, and a special thanks to Adrian Simmons of ECMWF for an updated version of Fig. 10.1. The COMET program generously offered their graphics for use in this text, which were especially useful in chapter 10. Graphics and examples from research projects undertaken in collaboration with several current and former students—Chris Bailey, Mike Brennan, Kelly Mahoney, and Wendy Moen, among others—are used in the text and these individuals are acknowledged for these contributions.

I am grateful to the students who have taken my synoptic courses over the years, both at NCSU and SUNY College at Brockport. These students often provided thoughtful suggestions and flattering compliments that helped to inspire me to develop this text.

The Unidata program is acknowledged for all they do to make data and visualization software available to the educational community. Many of the graphics shown are derived from data obtained via the Unidata Local Data Manager (LDM) and visualized with the Generalized Meteorological Package (GEMPAK) and Integrated Data Viewer (IDV). Many of the experiments presented in this text were conducted using sophisticated numerical models available to the atmospheric science community, such as the Weather Research and Forecasting (WRF) model. The availability of these models, along with helpful reference documentation, has played a major role in both teaching and research in the atmospheric sciences. The WRF model is made available through the National Center for Atmospheric Research (NCAR). Both Unidata and NCAR are funded by the National Science Foundation (NSF). Support of the scientists (including graduate students and faculty PIs) whose research contributed to this book came in the form of grants from several funding agencies, most notably NOAA and the NSF. Interactions with scientists at NOAA/NCEP/EMC (Brad Ferrier and Mike Ek), NCAR (Jimy Dudhia), COMET (Stephen Jascourt and Bill Bua), and NOAA/NCEP/NSSL (Jack Kain) strengthened the material in several chapters of this text.

Finally, I must acknowledge the support of my family, without which I could not have completed this task. My parents Fred and Jeanne and my sister Lisa have provided love and support over the years. Most importantly, my wife of 20 years, Jen, and my two patient daughters, Sandy and Grace, have tolerated my workaholic tendencies and helped to bring balance and inspiration to my life.

# CHAPTER 1

# Introduction, Background, and Basics

*The principal task of any meteorological institution of education and research must be to bridge the gap between the mathematician and practical man, that is, to make the weather man realize the value of a modest theoretical education, and to induce the theoretical man to take an occasional glance at the weather map.*

—C.-G. Rossby, "Comments on Meteorological Research" (1934)

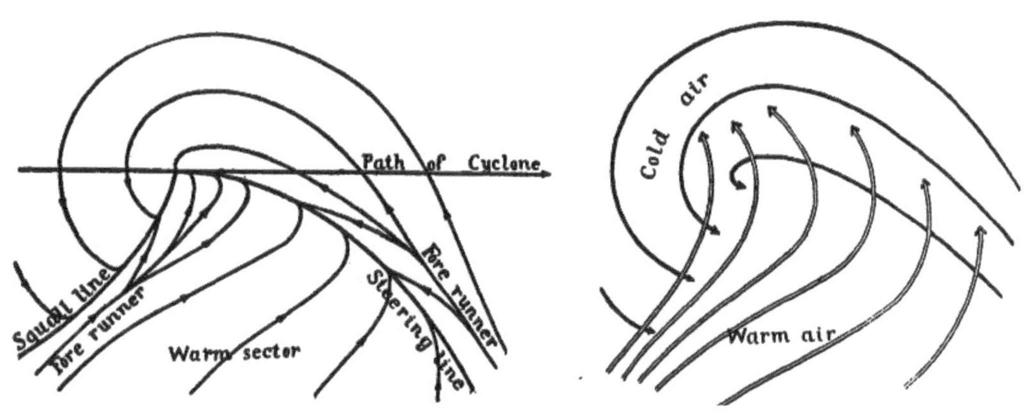


Near-surface streamlines (left panel), and the flow of warm and cold air in the lower troposphere (right panel) for a moving midlatitude cyclone. From J. Bjerknes (1919).

*Synoptic meteorology* traditionally involves the study of weather systems, such as extratropical high and low pressure systems, jet streams and associated waves, and fronts. The study of weather phenomena on somewhat smaller spatial and temporal scales, *mesoscale meteorology*, includes the study of convective storms, land–sea breezes, gap winds, and mountain waves. In recent years, advances in observing and computing technology have allowed the boundary between synoptic and mesoscale meteorology to blur as an increasing volume of high-resolution information has become available. As the resolution of analyses, forecasts, and observations increases, mesoscale weather systems begin to come into focus on the maps that were once the exclusive realm of synoptic-scale weather systems. Today's "synoptic meteorologist" benefits from in-depth knowledge of mesoscale processes and their interactions with synoptic-scale weather systems. The earth system exhibits a continuous spatial and temporal spectrum of motion that defies simple categorization.

Knowledge of numerical modeling and modern observational systems, in addition to a solid traditional foundation in the atmospheric sciences, is consistent with increasingly high expectations for today's scientists to solve complex problems, often of considerable societal import. Atmospheric scientists who are able to synthesize theoretical concepts, observations, and conceptual and numerical models in their work are best able to contribute to scientific advance. The advantages of the interplay between practice and theory were recognized by pioneers in the atmospheric sciences, such as Vilhelm Bjerknes, who actively encouraged such diverse activities. The quote at the beginning of this chapter from C.-G. Rossby, who at one point studied under Bjerknes, embodies this spirit; this philosophy remains as relevant today as it was in 1934.

*Weather forecasting* necessitates understanding a wide range of processes and phenomena acting on a variety of spatial and temporal scales. For example, forecasting for a coastal location requires information concerning the near-shore water temperature, the potential for land–sea breeze circulations, and the strength and orientation of the prevailing synoptic-scale wind flow. The prediction of precipitation type can benefit from knowledge of atmospheric thermodynamics and cloud physics. Other types of prediction, including air quality forecasting and seasonal climate prediction, also require knowledge that spans a broad spectrum of meteorological processes.

How will a single textbook comprehensively treat this variety of processes and phenomena? Although the focus of this text is on synoptic–dynamic meteorology, material concerning the full spectrum of relevant processes will be discussed in sufficient depth to provide the reader with an appreciation for the essential physics. In this chapter, a summary of basic physical processes, variables, and governing equations provides the prerequisite background for understanding the material presented in subsequent chapters, with an emphasis on examples from dynamics and thermodynamics.

A traditional starting point for synoptic meteorology is the *quasigeostrophic* (QG) *equations*, which are a simplified version of the full primitive equations, and are presented in chapter 2. Before discussing the QG system, it is necessary to review the governing equations and fundamental concepts with an eye toward physical interpretation and practical application; these are the goals of this chapter.

## 1.1. SCALES OF ATMOSPHERIC MOTIONS

The complete set of governing equations for the atmosphere is difficult to utilize and conceptually comprehend. For specific applications, the equations can be simplified through the elimination of terms that are unimportant to the situation in question. A procedure known as *scale analysis* is a systematic strategy to determine which terms in the equations, often associated with specific physical processes, are most important and which are negligible in a given meteorological setting. By characterizing the temporal and spatial scales associated with specific weather systems, we can systematically neglect "small" terms in the governing equations in the study of those systems. For example, for synoptic- and planetary-scale weather systems, such as cyclones and anticyclones, we know that in the midlatitudes, flow above ~1 km altitude tends to be fairly close to a state of geostrophic balance, whereas for mesoscale weather systems, such as thunderstorms, it does not, even in the same geographical location. We can use this knowledge of primary force balance to develop the QG equations. While these assumptions reduce accuracy, their use is justified by the physical insight obtained from the simplified equations.

Why is it important to derive customized equation sets and techniques for different classes of weather system, and why must students work through these derivations? Despite what some students may think, the equations are not designed as an instrument of torture for students of meteorology! In addition to the dynamical insight afforded by isolating the essential physics of a given weather system, one must know when to apply and when not to apply a given technique. Suppose that a particular weather forecasting technique is developed from simplified equations that are valid for synoptic-scale flows, and that this technique gains widespread acceptance in the forecasting community. If a forecaster were to apply this technique to a *mesoscale* weather system, the technique may fail, perhaps resulting in a poor weather forecast. This is one example of why students in the atmospheric sciences are required to derive equations, because it is important to know what assumptions were made in the development of a given technique, and this information is needed to deduce which tools are appropriate for which situations. In addition, the ability to apply a systematic approach to the governing

**Table 1.1.** Summary of basic scales, with example phenomena

| Scale | Length (km) | Time | Example phenomena |
|---|---|---|---|
| Microscale | <1 | <1 h | Turbulence, PBL |
| Mesoscale | 1–1,000 | 1 h–1 day | Thunderstorm, land–sea breeze |
| Synoptic | 1,000–6,000 | 1 day–1 week | Upper-level troughs, ridges |
| Planetary | >6,000 | >1 week | Polar front jet stream, trade winds |

equations allows atmospheric scientists to develop new equations and techniques to study unique problems.

In applying scale analysis to various weather systems, we must identify the characteristic horizontal length and time scales, and these are often related to one another. The length scale can be related to the size of a weather system, or how far an air parcel would travel within the system during a given time interval. The time scale can be related to how long it would take an air parcel to circulate within the system, or to traverse the characteristic length scale (implying a characteristic velocity scale). The values in Table 1.1 are typically used to define different scale regimes in atmospheric systems.

The values listed in Table 1.1 are approximate; there are not usually sharp distinctions or abrupt transitions in the dynamics at a certain specific scale, and there are exceptions. For example, the inner core of a tropical cyclone is characterized by mesoscale processes, but these systems can last a long time (>1 week). This demonstrates the need to be precise in our definitions of the terms *temporal scale* and *spatial scale*. The *advective time scale* is based on the time required for an air parcel to traverse the characteristic spatial scale when moving at a characteristic wind speed (velocity scale) for a given type of weather system. This can be independent of the time scale corresponding to the duration of the weather system.

## 1.2. VARIABLES, COORDINATE SYSTEMS, AND UNITS

Before introducing the governing equations and the simplifying relations that will facilitate their interpretation and application, we must first introduce the *dependent variables* in the system; the four *independent variables* are the three spatial directions and time, denoted as $x$, $y$, $z$, and $t$, respectively. The independent variables are fundamental to some definitions of dependent variables, as shown below.

We live on a spherical earth, but the governing equations would be simpler if they were applied to a flat, "Cartesian" earth! How do we account for the spherical complexity? For qualitative work, or if we study weather systems that are sufficiently small in spatial scale, we can neglect the distortion that results. However, for accurate quantitative calculations and numerical weather prediction, map scale factors are included in the equations to properly account for the distortion. Or, if we select a coordinate system that aligns with latitude and longitude lines, we can alternatively include "sphericity terms" in the governing equations to represent curvature effects. For now, we will forgo further discussion of these; however, bear in mind that the equations here are presented in Cartesian form, for simplicity of notation and interpretation.

Conventions for the spatial dimensions are presented in Fig. 1.1, which shows unit vectors in an orthogonal Cartesian coordinate system. For some applications, the $\hat{i}$ and $\hat{j}$ unit vectors may be *grid relative*, rather than *north relative*. In other words, the $x$ axis may be aligned with rows of points in a gridded analysis or numerical model domain; depending on the map projection used, the grid rows do not necessarily align directly with latitude circles. However, unless otherwise specified, the horizontal coordinate directions here will be taken to align with latitude and longitude lines as in Fig. 1.1.

In meteorology, a variety of vertical coordinate measures are used in place of the vertical distance $z$ to determine position along the vertical coordinate axis. Popular choices include *pressure*, *sigma* (a terrain-following ratio of pressure to surface pressure), *potential temperature*, or a hybrid among these. No matter what quantity is used to mark the distance along the vertical axis, notice that the

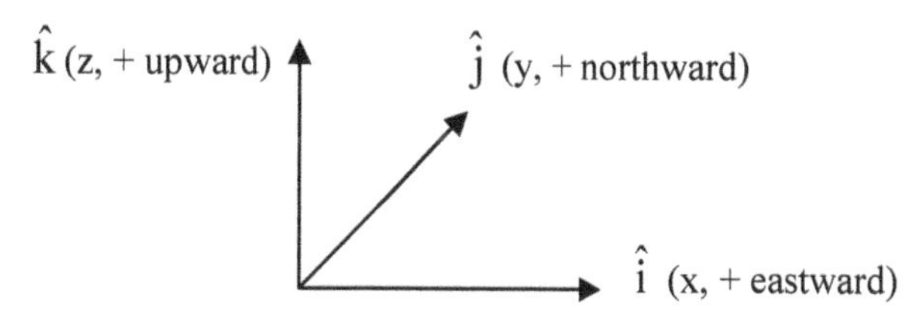


**Figure 1.1.** Coordinate axes and notation for the unit vectors and spatial axis directions.

$\hat{k}$ unit vector maintains right angles with the horizontal unit vectors.

There are significant advantages to using pressure as the vertical coordinate (known as *isobaric coordinates*). As we will see, some of the equations take on a simplified form in isobaric coordinates; however, an important practical advantage is that historically, it was easier to measure pressure than altitude for in situ measurements taken aloft. For instance, a rawinsonde can measure pressure with a baroswitch; however, it would be more difficult to measure the *altitude* above ground level of the sonde, as it would vary with changes in topography beneath, requiring nonlocal measurements to be made. Also, aircraft are designed to measure pressure and fly along surfaces of constant pressure, and modern meteorology has a strong historical linkage with the aviation industry, as discussed in section 5.4.

The wind velocity components are based on the time rate of change in the distance along the respective coordinate axes following the airflow, as defined below:

$$u = \frac{dx}{dt}, \quad v = \frac{dy}{dt}, \quad \text{and} \quad w = \frac{dz}{dt}, \tag{1.1}$$

where $d\,(\,)$ indicates a change in the variable in question. For example, $dx/dt$ indicates a change in zonal (east–west) distance, $dx$, for a given increment of time, $dt$, with positive values for motion toward the east. Similar conventions apply for the meridional (north–south) and vertical wind components. To be consistent with the expressions above, the vertical motion in isobaric coordinates, omega ($\omega$), is defined as

$$\omega = \frac{dp}{dt}. \tag{1.2}$$

The pressure-coordinate vertical velocity is negative for upward motion, because pressure decreases with height. Here, the notations for the 2D horizontal and 3D wind vectors are

$$\vec{V} = u\hat{i} + v\hat{j} \text{ and } \vec{U} = u\hat{i} + v\hat{j} + w\hat{k}, \text{ respectively.} \tag{1.3}$$

The *Coriolis parameter* is related to the projection of the earth's rotation onto the vertical ($\hat{k}$) axis and is given by $f = 2\Omega \sin \varphi$, where latitude is denoted $\varphi$ and $\Omega$ is the earth's rate of angular rotation, $2\pi$ radians day$^{-1}$, with a day here being the *sidereal day* (23 h 56 min). The numerical value of $\Omega$ is $7.292 \times 10^{-5}$ rad s$^{-1}$. Other basic variables include pressure ($p$), temperature ($T$), density ($\rho$), and specific volume ($\alpha = 1/\rho$). Recall that

**Table 1.2.** SI units of basic physical quantities

| Quantity | Unit | SI units and notation |
|---|---|---|
| Length | Meter | m |
| Time | Second | s |
| Mass | Kilogram | kg |
| Temperature | Kelvin | K |
| Velocity | Meter/second | m s$^{-1}$ |
| Force | Newton | N (kg m s$^{-2}$) |
| Pressure (force/area) | Pascal | Pa (N m$^{-2}$ or kg m$^{-1}$ s$^{-2}$) |
| Energy or work | Joule | J (N m or kg m$^{2}$ s$^{-2}$) |

density is defined as the mass of air $m$ per unit volume, a cubic meter. Measures of absolute water vapor content are dewpoint ($T_d$), specific humidity ($q$), and mixing ratio ($r$). The latter two quantities are similar in magnitude, they but have slightly different definitions: the specific humidity is the mass of vapor per unit mass (kilogram) of air, $m_v/m$, whereas the mixing ratio is given by the mass of vapor per unit mass of *dry* air, $m_v/m_d$. Values of the mixing ratio and specific humidity can be defined for saturated conditions, and these are denoted $q_s$ and $r_s$. A measure of the degree of saturation, the relative humidity (RH), is also a useful quantity; the RH is typically expressed as a percentage, for example, $100(q/q_s)$ or similar expressions involving the mixing ratio or vapor pressure ($e$). Those not familiar with the fundamental physical definitions of these quantities are referred to the readings at the end of this chapter for additional details.

Unless otherwise stated, the units to be used in this text are those of the System Internationale (SI), which is summarized in Table 1.2. The only frequent exception in this text is the use of the *millibar* (mb), equivalent to the hectopascal (hPa, or 100 Pa), as a convenient and commonly used unit for atmospheric pressure. Temperature will generally be expressed in degrees Celsius (°C), which are equivalent in size to kelvins (K); in a few instances, analyses and observations are shown with the temperature or dewpoint reported in degrees Fahrenheit, despite the advantages of the SI system.

## 1.3. THE GOVERNING EQUATIONS

The *ideal gas law*, or *equation of state*, provides a useful relation between the pressure, temperature, and density:

$$p = \rho R_d T_v \quad \text{or} \quad p\alpha = R_d T_v. \tag{1.4}$$

Here, $R_d$ is the dry-air gas constant, and $T_v$ is the *virtual temperature*, $T_v = T(1+0.61q)$. Atmospheric composition varies in space and time, mostly because of changes in water vapor content. Rather than introduce a *variable* gas coefficient, it is convenient to apply a modification to the temperature to account for variations in water vapor and retain the fixed value of $R_d$ = 287 J kg$^{-1}$ K$^{-1}$. Often when we speak of temperature, strictly speaking we are referring to the virtual temperature; the subscript "$d$" on the gas constant will hereafter be dropped for brevity.

A plethora of textbooks provide detailed and rigorous derivations of the equations of motion, or momentum equations, that stem from Newton's second law. There is no point in repeating these developments here. For the synoptic-scale weather systems of interest, it is acceptable to write the momentum equations in Cartesian form because the spatial scale of these systems is sufficiently small relative to the size of the earth. These equations are applied to a unit atmospheric mass of 1 kg:

$$\frac{du}{dt} = -\frac{1}{\rho}\frac{\partial p}{\partial x} + fv + F_x \tag{1.5}$$

and

$$\frac{dv}{dt} = -\frac{1}{\rho}\frac{\partial p}{\partial y} - fu + F_y. \tag{1.6}$$

The left-hand side of the zonal (east–west) momentum equation (1.5) represents the acceleration of the zonal wind component $u$; the first right-hand term denotes the zonal component of the pressure gradient force, followed by the (apparent) Coriolis force, and finally a term representing the zonal component of the frictional force. In the atmosphere, friction is mostly attributable to the transport of momentum by *turbulent eddies* in the planetary boundary layer (PBL), a process that is largely (but not entirely) confined to the lowest 1–2 km of the atmosphere. In vector form, the horizontal momentum equation is

$$\frac{d\vec{V}}{dt} = -\frac{1}{\rho}\nabla_h p - f\hat{k}\times\vec{V} + \vec{F}_r; \tag{1.7}$$

the horizontal and 3D forms of the *gradient operator* are

$$\vec{\nabla}_h = \frac{\partial}{\partial x}\hat{i} + \frac{\partial}{\partial y}\hat{j},\ \vec{\nabla} = \frac{\partial}{\partial x}\hat{i} + \frac{\partial}{\partial y}\hat{j} + \frac{\partial}{\partial z}\hat{k}; \tag{1.8}$$

and the Coriolis term is expressed in (1.7) using the vector cross product of the vertical unit vector with the horizontal velocity vector.

The derivatives marking the left side of (1.5)–(1.7) are *total* derivatives, meaning that they describe time rates of change *following moving air parcels*. By contrast, the time rate of change of a quantity "$a$" at a *fixed location*, the local derivative, is written as $\partial a/\partial t$. Notice that the total derivative is related to the local derivative in the following way:

$$\frac{d}{dt} = \frac{\partial}{\partial t} + u\frac{\partial}{\partial x} + v\frac{\partial}{\partial y} + w\frac{\partial}{\partial z}. \tag{1.9}$$

The total derivative is related to the local time rate of change (the first right-hand term) plus the contribution from *advective* changes, given by the three rightmost terms in (1.9). In vector form, (1.9) is

$$\frac{d}{dt} = \frac{\partial}{\partial t} + \vec{U}\cdot\nabla. \tag{1.10}$$

Using (1.10) to expand (1.7), we obtain

$$\frac{\partial\vec{V}}{\partial t} = -\vec{U}\cdot\nabla\vec{V} - \frac{1}{\rho}\nabla_h p - f\hat{k}\times\vec{V} + \vec{F}_r. \tag{1.11}$$

It is critically important to remain cognizant of the physical processes represented by these equations and to make sure that we can easily alternate between vector and scalar forms of the equations. If we take the scalar product (also known as the "dot product") of the zonal unit vector $\hat{i}$ with (1.11), we obtain the zonal momentum equation in Eulerian form:

$$\frac{\partial u}{\partial t} = -u\frac{\partial u}{\partial x} - v\frac{\partial u}{\partial y} - w\frac{\partial u}{\partial z} - \frac{1}{\rho}\frac{\partial p}{\partial x} + fv + F_x, \tag{1.12}$$

which is identical to (1.5) but with the total derivative expanded.

To summarize the processes represented in (1.12) verbally, this equation states that the local time rate of change of the zonal wind at a fixed location (the left side) is due to horizontal advection (the first two right-side terms), vertical advection, the pressure gradient force, the Coriolis torque acting on the north–south component of flow ($v$), and the zonal component of the frictional force. An interpretation of some of these terms for a simplified, idealized situation is presented in Fig. 1.2.

Given that the wind velocity in this hypothetical example is purely zonal, the second, third, and fifth right-hand terms in (1.12) are eliminated. Assuming that the frictional force is negligible (the sixth term), we are left with the first and fourth right-hand terms. We observe

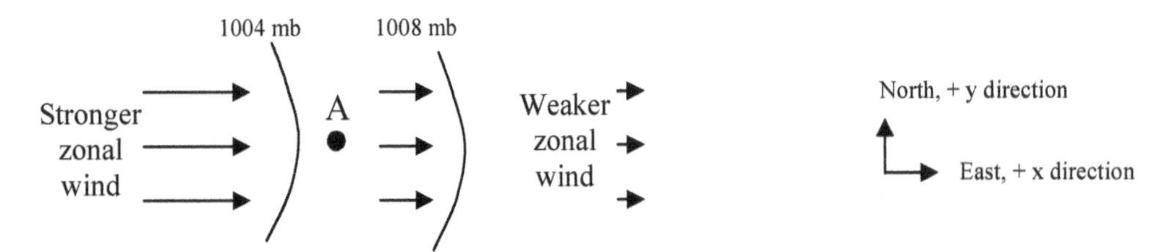


**Figure 1.2.** Hypothetical situation depicting horizontal wind velocity (solid arrows, length proportional to magnitude), isobars (solid contours, 4-mb interval), and a location at which we wish to evaluate the local tendency of zonal wind, denoted "A."

that the zonal component of the velocity decreases toward the positive $x$ direction ($\partial u/\partial x < 0$), but $u$ is positive throughout the region of interest. This, combined with the negative sign in front of the term, means that the quantity $-u\partial u/\partial x$ will be positive, leading to a local positive tendency of the zonal wind at location A because of advection. This makes intuitive sense, because air reaching point A originates from a region of larger positive zonal wind speed, consistent with a positive advective tendency. The pressure gradient force $-1/\rho(\partial p/\partial x)$ will yield a negative tendency at A because pressure is increasing in the positive $x$ direction, meaning that $\partial p/\partial x > 0$. With a negative sign leading the term, a negative value results, which should seem reasonable because the higher pressure to the east results in a net westward-directed horizontal pressure gradient force. Although this diagram presents a very simple and perhaps unphysical scenario, it illustrates the type of connections that we must make in linking mathematical equations to physical situations in the atmosphere using data from various sources.

For synoptic-scale motions, scale analysis reveals that the dominant terms in the horizontal momentum equation are the pressure gradient force and Coriolis terms. For the limiting case where these two terms are in exact balance, known as geostrophic balance, we can divide by $f$ (refer to Table 1.3 for definitions of parameters) in (1.5) and (1.6) to define the *geostrophic wind*, which is oriented parallel to isobars with lower values to the left (right) in the Northern (Southern) Hemisphere. Below are the vector and scalar expressions for flow that is in exact geostrophic balance:

$$\vec{V}_g = \hat{k} \times \frac{1}{\rho f}\nabla p;$$

$$u_g = -\frac{1}{\rho f}\frac{\partial p}{\partial y}; \text{ and } v_g = \frac{1}{\rho f}\frac{\partial p}{\partial x}. \qquad (1.13)$$

For flows that remain close to a state of geostrophic balance, the governing equations can be simplified considerably; these simplifications allow insight into the processes that govern the synoptic-scale flow and form the basis of traditional synoptic weather analysis and forecasting and QG theory as presented in chapter 2.

As mentioned previously, scale analysis allows for the determination of the relative importance of processes or terms in equations; Charney (1948) was one of the first to write about applying the scale analysis approach to atmospheric motions. We will now utilize this technique to define a useful measure of geostrophy (the extent to which the flow is in geostrophic balance).

If we assume that the relevant time scale is the advective time scale, then $T = L/U$, which is simply the time required for an air parcel to traverse the characteristic

**Table 1.3.** Scales of various parameters for midlatitude, synoptic-scale systems

| Symbol | Magnitude | Description |
|---|---|---|
| $W$ | $10^{-2}$ ms$^{-1}$ | Vertical velocity scale |
| $U$ | 10 ms$^{-1}$ | Horizontal velocity scale |
| $H$ | $10^4$ m | Depth of troposphere, vertical length scale |
| $H_{PBL}$ | $10^3$ m | Depth of PBL |
| $\Delta P_z$ | $10^5$ Pa | Vertical pressure change over troposphere |
| $L$ | $10^6$ m | Horizontal length scale |
| $\Delta P_h$ | $10^3$ Pa | Horizontal pressure change across synoptic system |
| $K_m$ | 10 m$^2$s$^{-1}$ | Turbulent eddy viscosity coefficient |
| $\rho$ | 1 kgm$^{-3}$ | Density scale for lower and middle troposphere |
| $f$ | $10^{-4}$ s$^{-1}$ | Coriolis parameter |

**Table 1.4.** Scale analysis of terms in the vertical momentum equation

| Term | $dw/dt$ | $-\frac{1}{\rho}\frac{\partial p}{\partial z}$ | $fu\cot\varphi$ | $-g$ | $F_z$ |
|---|---|---|---|---|---|
| Scale | $UW/L$ | $\Delta P_z/\rho H$ | $fU$ | $g$ | $K_m U^2/H_{PBL}^2$ |
| Numerical ($ms^{-2}$) | $10^{-7}$ | 10 | $10^{-3}$ | 10 | $10^{-4}$ |

length scale when traveling at the characteristic velocity scale. If we desire to isolate specific types of weather systems, then we can also define a characteristic frequency and phase speed corresponding to a given type of system (e.g., Rossby waves). If we restrict the analysis to systems in which the phase speed is similar to $U$, then fast moving but meteorologically unimportant tidal and sound waves are excluded. We will now go through some basic manipulations to reinforce these concepts and illustrate the utility of this approach.

Consider a synoptic-scale weather system located at 45°N latitude, characterized by a length scale of 1,000 km and an advective time scale of ~1 day.

1. What is an appropriate value for the order-of-magnitude velocity scale, $U$, in meters per second?
2. Using equation (1.7), consider the *acceleration term* $(d\vec{V}/dt)$. What are the *units* of this term? Using the given values for the velocity scale from question 1 and the time scale, what is the order-of-magnitude value of this term?
3. Now consider the *Coriolis term* in (1.7) and determine the appropriate synoptic-scale value of the term. Recall that $f = 2\Omega\sin\varphi$ and $\Omega = 7.292\times10^{-5}\,\mathrm{rad\,s^{-1}}$.
4. Consider the combined results of questions 2 and 3. What does this tell us about the degree of geostrophy characterizing synoptic-scale weather systems?
5. Finally, take the *ratio* of the acceleration term to the Coriolis term. Plug in the characteristic scales ($L$, $U$, and $f$) and simplify the expression, cancelling where possible. In general, the preceding example aside, what does it mean if the value of this ratio is very large, that is, much greater than 1? What if it is much smaller than 1?
6. Answers: The appropriate scale for $U$ in this example is $10\,\mathrm{m\,s^{-1}}$. The units of the acceleration term are $\mathrm{m\,s^{-1}/s}$, or $\mathrm{m\,s^{-2}}$, and the scale of this term is given by $\frac{U}{T}$ or $\frac{U}{L/U} = \frac{U^2}{L} \sim 10^{-4}\,\mathrm{ms^{-2}}$. The scale of the Coriolis term is $U$ multiplied by $f$; for a latitude of 45°N, this yields $10^{-3}\,\mathrm{m\,s^{-2}}$. Taking the ratio of the acceleration to the Coriolis term, we obtain $\frac{U^2/L}{fU} = \frac{U}{fL} \sim 0.1$. Thus, for synoptic-scale motions, the acceleration is an order of magnitude smaller than the Coriolis term, and the flow can be approximated by the geostrophic flow to within 10%.

In the preceding exercise, the ratio of the acceleration to the Coriolis term,

$$Ro = \frac{U}{fL}, \tag{1.14}$$

is known as the *Rossby number*; we will return to this useful parameter in chapter 2.

We now introduce other characteristic values for the dependent variables relevant to synoptic-scale systems to ascertain the magnitude of any term in the governing equations. As discussed by Charney (1948), the scale for $W$ cannot be assigned independently because it is constrained by our choices of $U$ and $L$. However, we can proceed by using observations of synoptic-scale systems to provide these values for the time being.

Using the values provided in Table 1.3[1], we can obtain order-of-magnitude estimates for the terms in the governing equations, thereby establishing which processes and terms dominate for synoptic-scale motions. The same principle could be applied by defining an analogous set of parameters for mesoscale or for planetary-scale flows.

Returning to the governing equations, the vertical equation of motion can be expressed as

$$\frac{dw}{dt} = -\frac{1}{\rho}\frac{\partial p}{\partial z} + fu\cot\varphi - g + F_z. \tag{1.15}$$

Scale analysis demonstrates that the vertical pressure gradient force and gravitational acceleration are far larger than the other terms in (1.15), as summarized in Table 1.4.

[1]Geophysical constants, such as the gravitational acceleration ($g$), are not included in Table 1.3 despite their appearance in the governing equations; the scale of $g$ for the troposphere is 10 m $s^{-2}$.

In this analysis, we find that the vertical pressure gradient force and gravitational acceleration are at least *four orders of magnitude* larger than the next largest term. Therefore, it is a convenient and highly accurate assumption to replace the vertical momentum equation with the *hydrostatic equation*, which expresses the balance between these two forces in the vertical:

$$\frac{\partial p}{\partial z} = -\rho g. \tag{1.16}$$

Although for synoptic-scale motions the atmosphere is very close to a state of hydrostatic balance, it is important to appreciate that the vertical motions which do arise (and are usually responsible for clouds and precipitation), are the result of small imbalances between these largely cancelling vertical forces. Utilization of the hydrostatic equation is *not* equivalent to ignoring vertical air motions but rather a reflection of the smallness of vertical accelerations relative to the powerful vertical pressure gradient and gravitational forces.

Some numerical weather prediction (NWP) models, such as the Global Forecast System (GFS) model in the United States, are currently *hydrostatic models*, meaning that they do not explicitly solve the vertical momentum equation. How then can these models predict clouds, precipitation, and rising air? In fact, once the model has obtained predictions for the *horizontal* wind velocity components, the vertical motion is obtained *diagnostically*, from the *continuity equation*, which is discussed below.

Conservation of mass forms the basis of the continuity equation, which can be expressed in a number of ways but is most simply stated in isobaric coordinates:

$$\frac{\partial u}{\partial x} + \frac{\partial v}{\partial y} + \frac{\partial \omega}{\partial p} = 0. \tag{1.17}$$

This relation implies that horizontal divergence (given by $\partial u/\partial x + \partial v/\partial y = \nabla \cdot \vec{V}$) is accompanied by changes in vertical motion with height (pressure). Utilizing the smallness of the vertical motion at the earth's surface, the continuity equation can be integrated with respect to pressure to determine the vertical motion above the surface, given the horizontal wind field. Horizontal convergence or divergence is related to a vertical expansion or contraction of air columns, and this will later be shown to represent an important dynamical process.

An often-overlooked aspect of the continuity equation is that *atmospheric* mass is *not* conserved. This is mostly due to water vapor phase changes: when water vapor condenses and falls to the surface as precipitation, the amount of atmospheric mass is reduced, and the resulting hydrostatic pressure decreases slightly. Likewise, evaporation results in a slight increase in atmospheric pressure. Thus, a more accurate form of the continuity equation (see Richardson 1922, Dutton 1986, chapter 8, or Lackmann and Yablonsky 2004 for discussion of this process) includes source and sink terms due to evaporation ($E$) and precipitation ($P$) on the right side:

$$\frac{\partial u}{\partial x} + \frac{\partial v}{\partial y} + \frac{\partial \omega}{\partial p} = E - P. \tag{1.18}$$

Scale analysis demonstrates that the evaporation and precipitation terms are very small relative to other terms in this equation. However, when this equation is integrated vertically to form a pressure tendency equation, the source/sink terms emerge as leading-order terms because of cancellation in the vertical integral of divergence and convergence, and because vertical motions are small near the surface and above the tropopause. The source and sink terms are most important during heavy precipitation, for example, during a tropical cyclone. The hydrostatic mass equivalent of 25.4 mm (1 in.) of rain is 2.5 hPa (2.5 mb); thus, heavy precipitation can result in significant mass removal and pressure changes.

A statement of conservation of energy, known as the *first law of thermodynamics* or the *thermodynamic energy equation* for an air parcel of unit mass, can be expressed as

$$dq = du + dw. \tag{1.19}$$

The notation here differs from that used earlier in the chapter; $dq$ is the heat added or subtracted from the mass, $du$ is the change in internal energy, and $dw$ is the increment of work done on or by the air parcel. This expression can also be written as

$$dq = C_p\, dT + g\, dz, \tag{1.20}$$

where $C_p$ is the specific heat of air at constant pressure, 1004 J $K^{-1}$ $kg^{-1}$, and other symbols have their previously defined meanings. Using the hydrostatic equation, (1.20) can be expressed as either

$$dq = C_p\, dT - \alpha\, dp \text{ or} \tag{1.21}$$

$$dq = C_v\, dT + p\, d\alpha, \tag{1.22}$$

where $C_v$ is the specific heat of air at constant volume, 717 J $K^{-1}$ $kg^{-1}$ and $C_p = C_v + R$, where R is the dry-air gas constant, ~287 J $K^{-1}$ $kg^{-1}$. In many situations, diabatic processes are dominated by adiabatic ones, leading to a cancellation between the work term and changes in the internal energy of air parcels. One example is the case of adiabatic (unsaturated) vertical air motion. Setting $dq = 0$ in (1.20) and dividing by $dz$, the *dry-adiabatic lapse rate* is obtained:

$$-\frac{dT}{dz} = \frac{g}{C_p} \equiv \Gamma_d \sim 0.0098\,\mathrm{K\,m^{-1}}. \qquad (1.23)$$

For approximate applications, the dry-adiabatic lapse rate is often taken to be 10 K $km^{-1}$. Again, for adiabatic motions, (1.21) can be expressed as $0 = C_p dT - \alpha dp$; using the ideal gas law (1.4) and dividing by $C_p$ and $T$, we obtain

$$\frac{dT}{T} = \frac{R}{C_p}\frac{dp}{p}.$$

Integrating from level $p_0$ (where the temperature is $\theta$) to the arbitrary level $p$ (where the temperature is $T$), and rearranging to solve for $\theta$, we obtain *Poisson's equation*:

$$\theta = T\left(\frac{p_0}{p}\right)^{\frac{R}{C_p}}. \qquad (1.24)$$

The potential temperature $\theta$ has a straightforward physical interpretation: It is the temperature that an air parcel would have if it were adiabatically compressed (or expanded) until the pressure was equal to $p_0$, which is usually taken to be 1000 mb. In weather analysis and forecasting, the vertical potential temperature profile is used to determine static stability, and examination of the horizontal $\theta$ distribution is useful in locating frontal boundaries. When diabatic processes are negligible, air parcels are constrained to travel along surfaces of constant $\theta$, known as *isentropic surfaces*. Chapter 3 further explores the utility of isentropic techniques in weather analysis and forecasting.

Consider the idealized vertical cross section depicted schematically in Fig. 1.3. The lowest 1 km is characterized by unsaturated turbulent mixing, meaning that air parcels are undergoing quasirandom periods of adiabatic ascent and descent. In this adiabatic mixing process, the temperature of rising and sinking air parcels follows the dry-adiabatic lapse rate, meaning that potential temperature is *conserved* for parcels moving in this layer. As a result of the adiabatic motions and the establishment of a "mixed layer," the vertical potential temperature profile appears qualitatively similar to that shown on the right side of the diagram. In the presence of solar heating, there may even be a slight *superadiabatic* layer immediately above the surface, indicated by a small zone where potential temperature decreases with height at the bottom of the right panel in Fig. 1.3.

Surface-based mixed layers usually correspond to the *planetary boundary layer.* The mixed layer depth has important implications for air pollution meteorology, as a deep mixed layer allows pollutants to disperse over a greater depth, reducing concentrations. Mixed layer identification is also useful in forecasting surface temperature and wind conditions. With strong winds aloft, the expected presence of a deep mixed layer can indicate that vertical momentum transport may produce strong surface winds, and the development of a mixed layer often means warmer surface temperatures because of

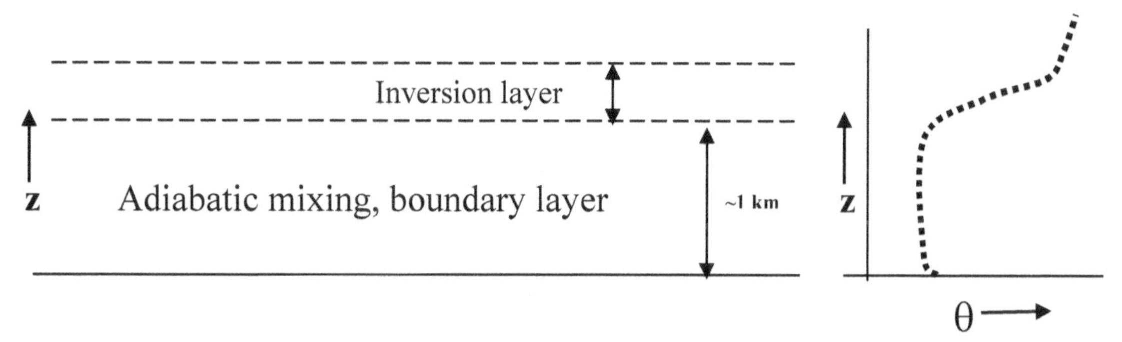


**Figure 1.3.** Idealized vertical cross section through a well-mixed PBL beneath a capping inversion layer. The inset at right shows the corresponding profile of potential temperature as a function of height.

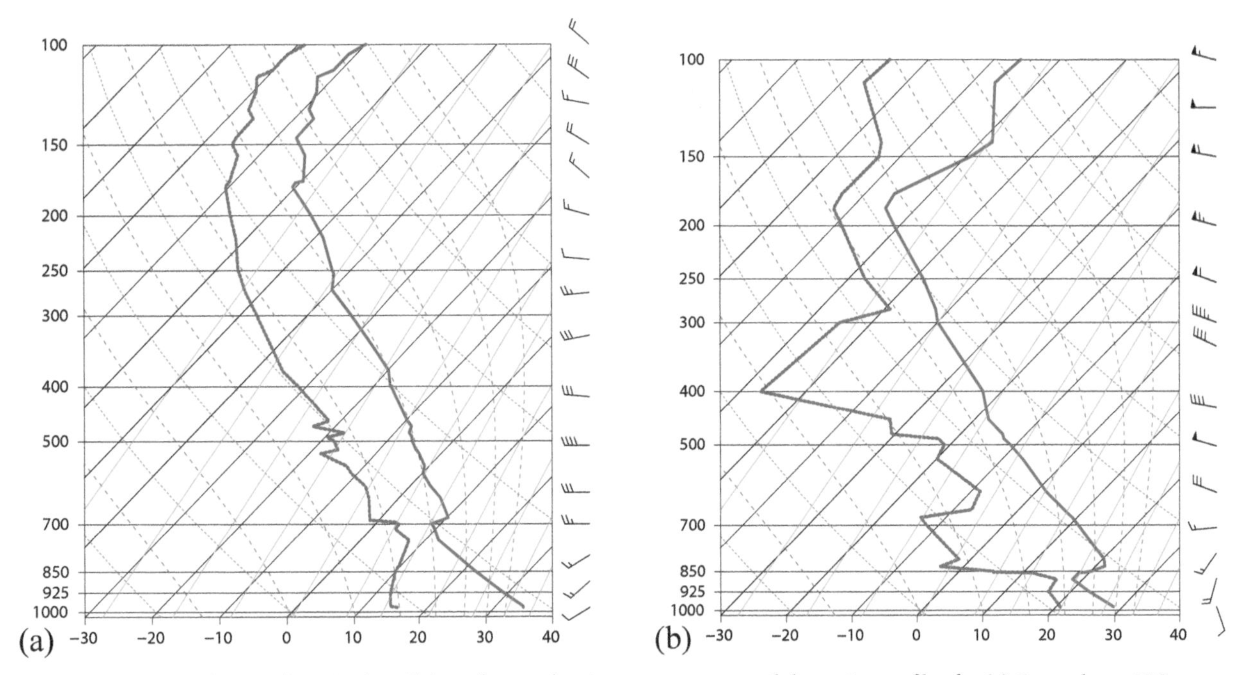


**Figure 1.4.** Example soundings in skew $T$–log$p$ format showing temperature and dewpoint profiles for (a) Greensboro, NC, at 0000 UTC 21 Aug 2007 and (b) Dallas–Fort Worth, TX, at 0000 UTC 19 Apr 2000.

compression of air with warmer potential temperature from aloft. At night, adiabatic mixing can prevent the development of a stable surface-based inversion layer, resulting in warmer minimum temperatures than would otherwise occur.

The examples provided in Fig. 1.4 exhibit adiabatic mixed layers that correspond to surface-based planetary boundary layers. In Fig. 1.4a, an unusually deep mixed layer extending to nearly the 700-mb level (~3 km altitude) has developed over central North Carolina on a day when record high temperatures were observed. The Fort Worth, Texas, sounding shown in Fig. 1.4b indicates a mixed layer extending to just above the 900-mb level, with a strong temperature inversion above.

When the environmental lapse rate is greater than the dry adiabatic value (9.8°C km$^{-1}$), the atmosphere is *absolutely unstable*, meaning that lifted parcels will become warmer than the surrounding environment, becoming less dense, and thus positively buoyant. This corresponds to the situation where the vertical derivative of potential temperature is negative (potential temperature is decreasing with height). Adiabatic mixed layers correspond to neutral stability, illustrated in the lower portions of Figs. 1.3 and 1.4. In the midlatitude troposphere, potential temperature typically increases with height. The more rapid the increase of $\theta$ with height, the more stable the atmospheric profile.

Examination of Fig. 1.4b to determine the vertical profile of static stability demonstrates the presence of a mixed layer of neutral stability extending to a pressure level of approximately 875 mb, capped by a strong, stable inversion layer that extends to roughly 825 mb. Between 825 and 175 mb, the troposphere exhibits weak stability; however, above the 175-mb level, very strong stability corresponding to the stratosphere is observed.

Some diabatic processes are relatively easy to quantify using thermodynamic equations or diagrams, such as a skew $T$–log$p$ chart. For example, condensational heating during the ascent of saturated air can be determined by knowledge of the condensation rate, which is tied to the rate of change of the saturation mixing ratio. For such *pseudoadiabatic motions* where the diabatic process is known (condensation during saturated ascent), we can compute thermodynamic quantities such as the *moist-adiabatic lapse rate* and the *equivalent potential temperature*, $\theta_e$:

$$\theta_e = \theta \exp\left(\frac{L_v r}{C_p T_{LCL}}\right) \tag{1.25}$$

Here, $T_{LCL}$ is the temperature that a lifted parcel would have at the *lifting condensation level*, which is the level at which lifted air parcels first reach saturation. A formula for $T_{LCL}$, useful in computation of (1.25), is obtained from Bolton (1980)[2]:

$$T_{LCL} = \left( \frac{1}{T-55} - \frac{\ln(RH/100)}{2840} \right)^{-1} + 55. \tag{1.26}$$

Physically, $\theta_e$ represents the temperature air would have if it were adiabatically adjusted to a reference pressure as for $\theta$ but additionally if all the vapor were condensed, adding latent heat to the parcel. For dry parcels, when $r$ is zero, (1.25) indicates that $\theta_e$ and $\theta$ are identical.

The equivalent potential temperature is conserved for a wide majority of atmospheric flows, and is related to stability. When $\theta_e$ decreases with height, the atmosphere is said to be *potentially* or *convectively unstable*. Physically, this means that if the layer in which $\theta_e$ decreases with height was lifted to saturation, then an unstable lapse rate would result. Physically, this situation is often found when potentially colder or drier air is located above warm, moist air. When lifting takes place, the moist lower part of the layer reaches saturation first, and then more condensational heating takes place there relative to the drier upper portion of the layer (which continues to cool at the larger dry-adiabatic lapse rate during ascent). Therefore, the layer experiences differential latent heating in the vertical, with the lower portion warmed relative to the upper, with instability being the eventual result.

Concerning the "parcel" versus "layer" approaches to analyzing static stability, it is a major—and often unrealistic—assumption to suppose that a single, cohesive parcel of air could be lifted undisturbed in an unchanging environment. Especially for turbulent, convecting motions, air parcels are constantly mixing and distorting. However, the parcel concept provides a useful thought experiment regarding what would happen if a parcel were somehow allowed to move undisturbed in that manner. The concept of a layer of air being lifted is more realistic, as broad airflows encountering mountain ranges or rising along sloping isentropic surfaces often exhibit some cohesiveness as layers during ascent.

[2]In (1.26), the temperature is in kelvins and RH is expressed as a percentage.

Differentiating the first law of thermodynamics (1.21) with respect to time, we obtain prognostic equations for the local tendency of temperature or potential temperature:

$$\frac{\partial T}{\partial t} = -\left( u\frac{\partial T}{\partial x} + v\frac{\partial T}{\partial y} \right) + \omega\left( \frac{R_d T_v}{C_p p} - \frac{\partial T}{\partial p} \right) + \frac{J}{C_p} \tag{1.27}$$

and
$$\frac{\partial \theta}{\partial t} = -\left( u\frac{\partial \theta}{\partial x} + v\frac{\partial \theta}{\partial y} \right) - \omega\left( \frac{\partial \theta}{\partial p} \right) + \frac{J}{C_p}\frac{\theta}{T}. \tag{1.28}$$

In (1.27) and (1.28), $J$ is the diabatic heating rate per unit mass.

Having introduced Cartesian forms of the basic governing equations, we should think carefully when considering whether all relevant physical processes have been accounted for in these equations. Even these equations, despite widespread use and acceptance in the atmospheric sciences, have approximations, and there are certain physical processes that are not included in them. For instance, acceleration of air due to momentum transferred by falling precipitation is not included in the momentum equation (1.7). Sensible heat transfer from falling precipitation is not typically included in (1.28), but it could be represented as a component of $J$. There are other processes that are not included as well, but one must consider whether it is worthwhile or necessary to include all of these terms for general application. If a very specialized problem is under investigation, then it is worthwhile to revisit the underlying assumptions and consider any neglected processes that may be important.

## 1.4. GEOPOTENTIAL, THICKNESS, AND THE THERMAL WIND

The shape of the earth's surface exhibits a small departure from that of a perfect sphere because of centrifugal effects associated with the earth's rotation, which result in a slight equatorial bulge. It is therefore convenient to define a slightly modified gravitational acceleration that combines the true gravitational force with the centrifugal force associated with the earth's rotation. Defined in this way, surfaces of constant geopotential are exactly aligned with the earth's oblate surface, and the effective gravitational force is given by the gradient of the geopotential:

$$\vec{g} \equiv -\nabla\Phi. \tag{1.29}$$

We can then define the earth's surface at mean sea level as the surface of zero geopotential, $\Phi$. Physically, the geopotential is related to the work required to lift a unit mass from sea level to a given altitude. The work in question is done against the effective gravitational force (hereafter "gravity," the magnitude of which at mean sea level is 9.80665 m s$^{-2}$). The geopotential at any level $z$ is given by the integral

$$\Phi(z) = \int_{\Phi_{(z=0)}}^{\Phi_{(z)}} d\Phi = \int_0^z g\, dz\,. \qquad (1.30)$$

This integral is complicated by the fact that the magnitude of the gravitational acceleration is a function of height, according to Newton's law of gravitation. Specifically, the magnitude of this force is proportional to the inverse square of the distance between the centers of mass in question. However, we can utilize the constant value of gravitational acceleration at mean sea level, $g_0 = 9.80665\ \mathrm{ms}^{-2}$, to define the *geopotential height Z*:

$$Z = \frac{\Phi(z)}{g_0} = \frac{1}{g_0}\int_0^z g\, dz\,. \qquad (1.31)$$

In the lower atmosphere, where $g$ is ~$g_0$, $Z$ is approximately equal to the geometric height $z$; the differences between $Z$ and $z$ are sufficiently small in the troposphere to justify their neglect. However, when the geopotential height is plotted on constant-pressure surfaces, as is frequently done, the values are actually geopotential—not geometric—height.

A small change in geopotential, $d\Phi$, is determined for a vertical distance sufficiently small to assume that the magnitude of the gravitational acceleration is constant. Using the hydrostatic equation, this can then be related to the density and pressure change as

$$d\Phi = g\, dz \approx -\alpha\, dp. \qquad (1.32)$$

The geostrophic wind defined by (1.13) in height coordinates can be written with pressure as the vertical coordinate, with the geopotential height gradient replacing the pressure gradient:

$$\vec{V}_g = \hat{k}\times\frac{g_0}{f}\nabla Z = \hat{k}\times\frac{1}{f}\nabla\Phi\,. \qquad (1.33)$$

From (1.32), the hydrostatic equation (1.16), and the equation of state (1.4), we obtain

$$d\Phi = -\frac{R_d T_v}{p}\, dp, \qquad (1.34)$$

which we can integrate between two pressure levels to yield

$$\Phi_{upper} - \Phi_{lower} = -R_d \int_{p_{low}}^{p_{up}} T_v\, d\ln p\,. \qquad (1.35)$$

Dividing by $g_0$ and reversing the limits of integration, we obtain a useful relation known as the *hypsometric equation:*

$$Z_{upper} - Z_{lower} = \frac{R_d}{g_0}\int_{p_{up}}^{p_{low}} T_v\, d\ln p. \qquad (1.36)$$

The right side of (1.36) is complicated because the virtual temperature typically exhibits complex structure with height (and pressure). However, a convenient simplification is to extract the *average* virtual temperature over the layer of integration, since this quantity is constant with pressure. We then see that the vertical distance between two pressure surfaces (also known as the "thickness") is proportional to the mean virtual temperature of the layer:

$$\Delta Z = \frac{R_d \overline{T}_v}{g_0}\ln\left(\frac{p_{low}}{p_{up}}\right). \qquad (1.37)$$

For two given pressure surfaces, for example, the 1000- and 500-mb levels, the colder the mean layer virtual temperature, the smaller the thickness. Put another way, we see that pressure decreases more rapidly with height in cold air relative to warm air.

The thickness thus relates the structure of geopotential height on constant-pressure surfaces to the temperature distribution. This provides insight into the structure of a variety of weather systems, including the midlatitude jet stream, and tropical and extratropical cyclones, among others. Critical values of thickness have also been shown to facilitate the prediction of precipitation type (e.g., snow versus rain or freezing rain) as discussed in section 9.3.

We will now utilize the thickness concept to understand one of the fundamental features of the midlatitude atmosphere: the jet stream. Observed pressure differences at the surface of the earth are relatively small compared to those on constant-height surfaces aloft. The same is true of geopotential height differences displayed on constant-pressure surfaces in the lower versus upper troposphere.

In the hypothetical situation represented in Fig. 1.5, the *surface pressure* at all points along a north–south-oriented vertical cross section is 1000 mb; in other words, the 1000-mb pressure surface is flat, with an altitude of 0 m everywhere. In the 1000–700-mb layer, the layer-average virtual temperature grows gradually colder from south to north. As a result, despite the fact that the 1000-mb pressure surface is flat and aligned with the surface, the 700-mb surface slopes downward toward the north, because, as evident from (1.37), the thickness of the layer

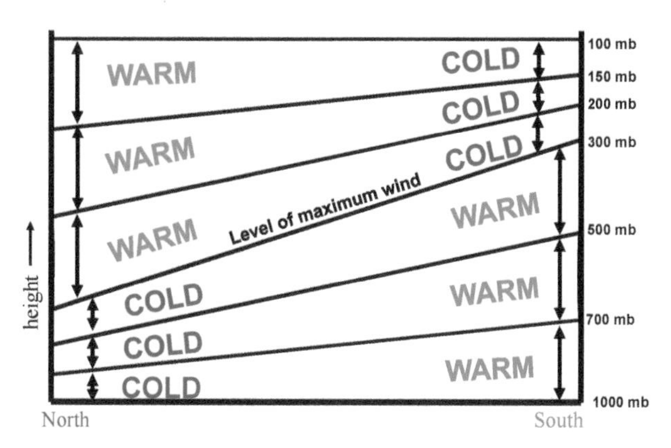


**Figure 1.5.** Idealized, hypothetical north–south cross section of temperature and constant-pressure surfaces. The designations "warm" and "cold" refer to relative horizontal temperature contrasts on a given pressure surface rather than to the absolute air temperature.

must decrease in the colder regions. If a similar temperature distribution is found in the 700–500- and 500–300-mb layers, then we find that the slope of each successive pressure surface increases as we ascend. This is analogous to "stacking doorstops" with the narrow ends all on the same side. For this reason, the standard contour interval when plotting geopotential height on constant-pressure surfaces increases with height; this interval is typically 30 m at the 700-mb level, 60 m at the 500-mb level, and 120 m at the 300-mb level, reflecting the increasing slope of these surfaces at higher altitudes.

As evident from (1.33), the geostrophic wind speed is proportional to the gradient of geopotential height on constant-pressure surfaces; thus, the geostrophic wind speed increases with height up to the 300-mb level in this idealized situation. Above the planetary boundary layer, the wind is fairly close to a state of geostrophic balance; thus, the expected result here is a westerly wind, increasing with height. Given that colder temperatures correspond to low pressure (or low geopotential height) aloft, a temperature configuration with colder air over the poles is consistent with westerly flow, increasing with height.

However, the temperature structure changes at the tropopause, which is lower in altitude over the poles relative to the tropics. Therefore, the north–south temperature gradient on a given pressure surface *reverses* for pressure surfaces that lie in the stratosphere in polar latitudes but remain in the troposphere closer to the equator. Typically at the 200-mb level and higher, temperatures are much colder in the tropics relative to the polar regions. This temperature configuration results in a leveling of the pressure surfaces above a certain altitude, and so we find that the speed of the westerly flow diminishes with height in the lower stratosphere (Fig. 1.5); the doorstops are now flipped around and stacked the other way, resulting in a reduction of the slope of pressure surfaces with height.

The similarity of the 500-mb *height* pattern to the 1000–500-mb *thickness* pattern is evident in the example shown in Fig. 1.6. This is a consequence of the 1000-mb pressure surface being relatively flat compared to the 500-mb surface. From Fig. 1.6b, it is also clear that the north–south temperature contrast differs from that shown in the idealized Fig. 1.5, which depicts a gradual cooling from south to north in the troposphere. Observations indicate that temperatures can be relatively uniform (horizontally) in the tropics and also in polar regions; the strongest temperature contrasts are often confined within relatively narrow zones. Such regions of temperature contrast correspond to *frontal zones*. The narrow zone of concentrated temperature gradient is associated with a correspondingly narrow band of enhanced horizontal height gradient aloft, which in turn corresponds to strong wind speed. For this reason, the *westerly jet stream* takes on a relatively narrow configuration. Frontal zones slope toward the cold air above the surface; nevertheless, a surface frontal analysis (Fig. 1.7) corresponding to the time of Fig. 1.6 demonstrates that the polar front has some linkage with the upper jet stream, as is evident from the eastern North Pacific Ocean across North America in this example.

### 1.4.1. The thermal wind relation

The geostrophic wind relation provides a link between the geostrophic wind field and the geopotential height, via (1.33). Therefore, the *vertical shear* of the geostrophic wind can be related to the *difference in height* (thickness) between two given pressure levels. The vertical shear of the geostrophic wind is known as the *thermal wind*, $\vec{V}_T$, and can be expressed as

$$\vec{V}_T = \vec{V}_{g\ upper} - \vec{V}_{g\ lower}. \tag{1.38}$$

The vector relation defining the thermal wind is illustrated in Fig. 1.8, which can easily be viewed in terms of vector subtraction. The thermal wind is not a true wind but a difference between the geostrophic wind at different levels, the vertical shear of the geostrophic wind.

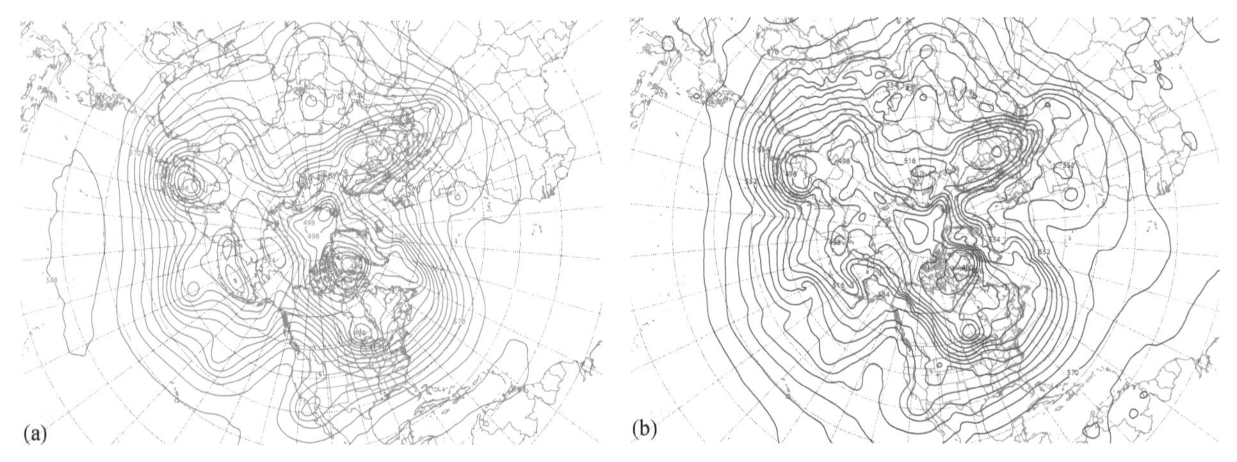


**Figure 1.6.** GFS analysis of (a) 500-mb geopotential height (dam), (b) 1000–500-mb thickness (dam) for 0600 UTC 22 Jan 2004.

A convenient coordinate system known as *natural coordinates* can be defined in which the horizontal coordinate axes are oriented with respect to the geostrophic wind. The $\hat{s}$ unit vector is everywhere parallel to the geostrophic flow (and parallel to geopotential height contours), and the $\hat{n}$ direction is oriented normal to height contours, pointing toward higher values. In this framework, the geostrophic wind relation can be written as

$$V_g = \frac{g_0}{f}\frac{\partial Z}{\partial n}. \tag{1.39}$$

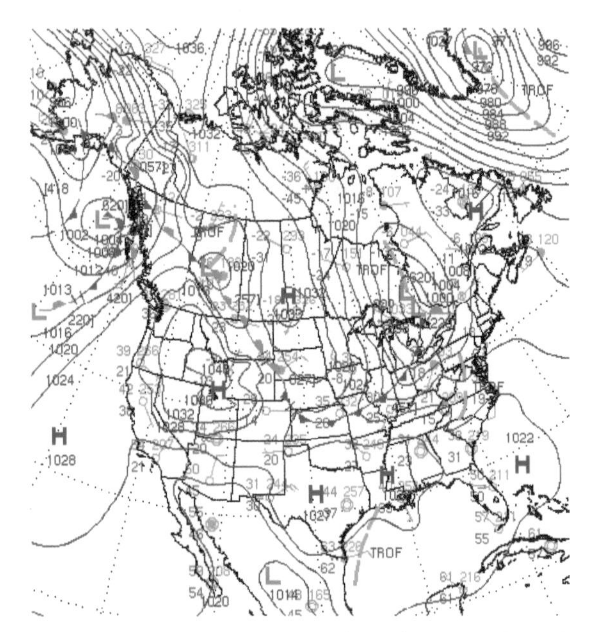

**Figure 1.7.** Surface analysis from the Hydrometeorological Prediction Center (HPC) valid 0600 UTC 22 Jan 2004. Sea level pressure (SLP, mb) and surface fronts are analyzed.

Applying the geostrophic wind relation at an upper pressure surface, denoted $U$, and at a lower level $L$,

$$V_{gU} = \frac{g_0}{f}\frac{\partial Z_U}{\partial n} \text{ and } V_{gL} = \frac{g_0}{f}\frac{\partial Z_L}{\partial n}, \tag{1.40}$$

we can then subtract the lower from the upper to obtain an expression for the thermal wind:

$$V_T = V_{gU} - V_{gL} = \frac{g_0}{f}\frac{\partial}{\partial n}\left[Z_U - Z_L\right]. \tag{1.41}$$

Recall from (1.37) that $Z_U - Z_L$ is the thickness ($\Delta Z$); substituting (1.37) in (1.41), it is evident that the thermal wind is related to the horizontal temperature gradient:

$$V_T = \frac{R_d}{f}\ln\left(\frac{p_L}{p_U}\right)\frac{\partial \bar{T}_v}{\partial n}. \tag{1.42}$$

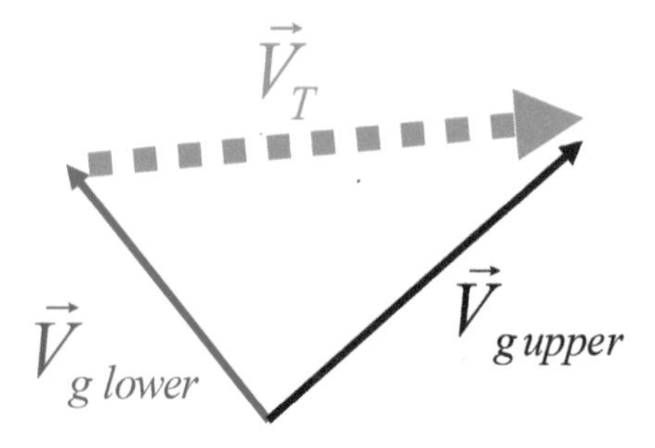


**Figure 1.8.** Schematic illustration of the thermal wind. Solid arrows depict geostrophic wind vectors at lower and upper levels; dashed red line represents the thermal wind.

Observations show that outside of the tropics, the atmosphere is generally close to a state of thermal wind balance, and that regions of strong horizontal temperature contrast are characterized by strong vertical wind shear. This finding has important practical applications in weather analysis and forecasting, some of which are listed below.

- Vertical wind shear is detrimental to tropical cyclones; therefore, frontal zones and regions of strong temperature contrast are not favorable for tropical cyclone activity.
- Convective storms forming in regions of strong shear are more likely to exhibit rotation and become severe relative to those forming in weak shear. Thus, convection taking place within zones of strong thermal contrast must be monitored carefully by operational forecasters.
- As will be demonstrated subsequently, *horizontal temperature advection* is also related to the character of the geostrophic shear, and this has useful forecasting applications that will become apparent when the QG equations are presented in chapter 2.

It is also convenient to express the thermal wind in Cartesian coordinates, in vector form, and by grouping constants (including the natural log of two fixed pressure surface values) into a constant "C":

$$\vec{V}_T = \vec{V}_{gU} - \vec{V}_{gL} = \frac{C}{f}\hat{k} \times \nabla \overline{T}_v, \tag{1.43}$$

where the gradient is evaluated on an isobaric surface. Splitting (1.43) into component form, we can relate vertical shear of the zonal or meridional geostrophic wind to the meridional and zonal temperature gradients, respectively:

$$u_{gU} - u_{gL} = -\frac{C}{f}\frac{\partial \overline{T}_v}{\partial y} \text{ and } v_{gU} - v_{gL} = \frac{C}{f}\frac{\partial \overline{T}_v}{\partial x}. \tag{1.44}$$

For later reference, we also write the thermal wind relation in vector form:

$$\frac{\partial \vec{V}_g}{\partial p} = \frac{1}{f_0}\hat{k} \times \nabla \frac{\partial \Phi}{\partial p}. \tag{1.45}$$

The situation depicted in Fig. 1.9 is completely consistent with Fig. 1.5, except that it features a narrow horizontal extent of the jet. The increase in wind speed with height

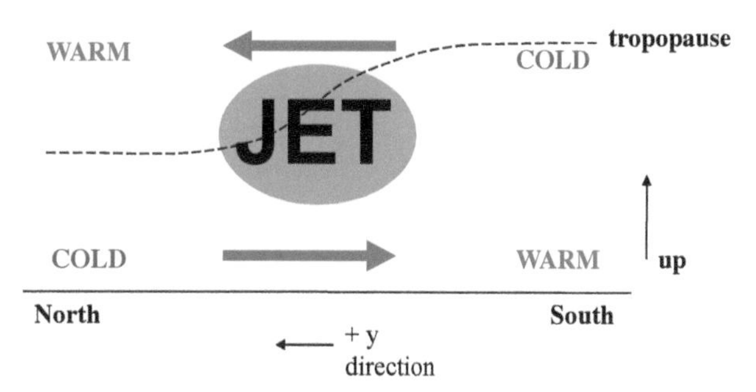


**Figure 1.9.** North–south cross section depicting the temperature anomalies relative to horizontal averages at a given altitude, the location of the tropopause (dashed line), the location of the jet stream core, and the meridional temperature gradient (red arrows) above and below the jet core.

within the troposphere is consistent with colder temperature values to the north (in the positive $y$ direction) as required by (1.44). Thus, we observe that the jet stream increases in strength with height up to the tropopause and then weakens with height in the lower stratosphere, where the sign of the meridional temperature gradient has reversed. At the level of the jet core, although winds are strongest there, the horizontal temperature gradient is weak at that level. This can be appreciated by noting that at the jet core, where the westerly wind is no longer increasing with height, (1.44) demands that the meridional temperature gradient be zero.

### 1.4.2. Temperature advection and the thermal wind

Consider Fig. 1.10, which displays an idealized pattern of 1000- (solid red contours) and 500-mb heights (dashed black contours). In reality, the magnitude of the 1000-mb height gradient is substantially weaker than that at the 500-mb level, in contrast to this hypothetical illustration. Based on the hypsometric equation, we draw lines of constant 1000–500-mb thickness (green dotted lines), by simply subtracting the value of the 1000-mb height from that of the 500-mb height at points of contour intersection. This method, known as *graphical subtraction*, also demonstrates that the thickness contours of a consistent interval can *only* cross 1000- and 500-mb height contours at points of intersection. In this example, regions where the 1000-mb height contours are crossing 500-mb height contours at a strong angle correspond to regions of strong horizontal temperature (thickness) gradient, as in the northern part of the domain. At the location denoted "A" in Fig. 1.10, we can utilize vectors corresponding to the 1000- and 500-mb

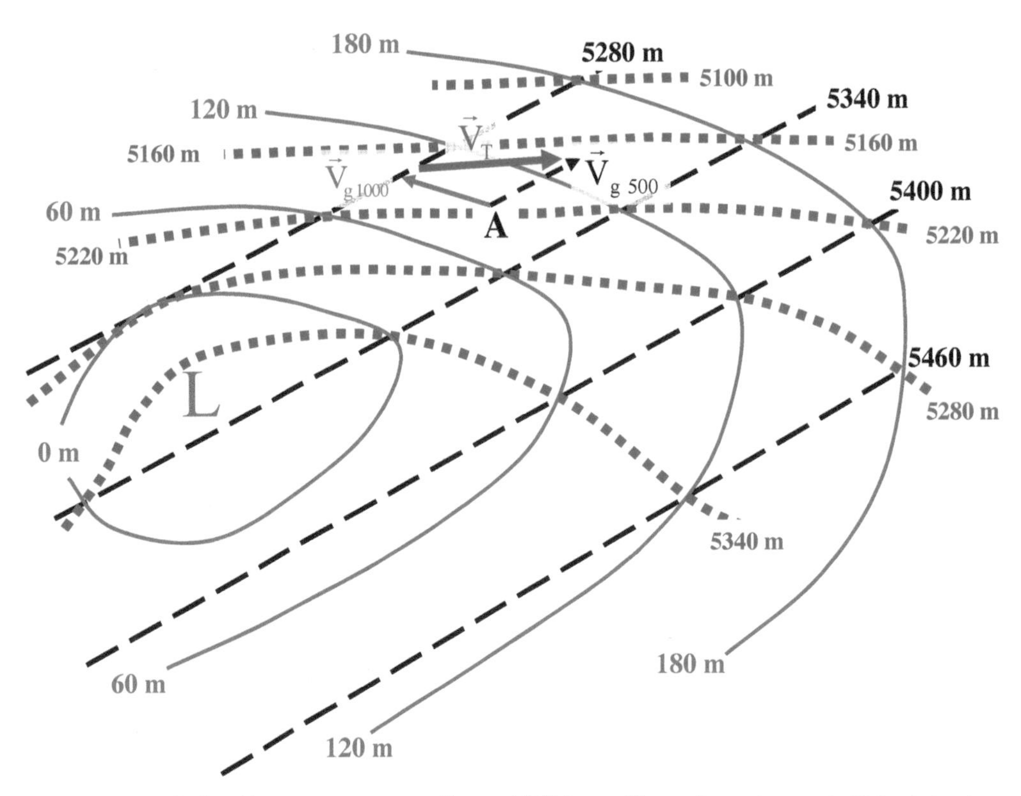


**Figure 1.10.** Idealized low-pressure center (denoted "L") located beneath southwesterly 500-mb-level flow in a Northern Hemisphere location. Solid red lines depict 1000-mb height, dashed black lines represent 500-mb height, and green dotted lines correspond to 1000–500-mb thickness. Vectors correspond to 1000- and 500-mb geostrophic winds at point A, and green vector corresponds to the thermal wind at this location.

geostrophic wind and perform a vector subtraction as in Fig. 1.8 to obtain the thermal wind vector.

Now, we are in a better position to understand why we refer to the geostrophic shear as the thermal wind: The relation between thickness contours and the thermal wind is exactly analogous to that between geopotential height contours and the geostrophic wind. In the Northern Hemisphere, the thermal wind is oriented parallel to thickness contours with lower (colder) values to the left.

Considering the thickness contours as layer-average isotherms, it is useful to evaluate the *temperature advection* within this layer. At point A, the southeasterly 1000-mb geostrophic wind is clearly associated with warm advection, that is, the geostrophic flow at this level is pointing from warmer toward colder thickness values. Similarly, the 500-mb geostrophic wind is also associated with warm advection. If we were to average the geostrophic wind at the top and bottom of this layer, then the resulting vector would fall between those at the top and bottom of the layer. There is clearly geostrophic warm advection taking place within the 1000–500-mb layer at point A.

The geostrophic wind profile in Fig. 1.10 turns clockwise with increasing height, which is known as a *veering* wind profile. In the Northern Hemisphere, veering geostrophic wind profiles are associated with warm advection. Conversely, *backing* geostrophic wind profiles (geostrophic wind turning counterclockwise with increasing height) in the Northern Hemisphere correspond to cold advection. We will exploit these relations to a much greater extent in chapter 2; however, at this point we can say that thermal advection is related to the forcing for vertical air motion through the QG equations, and that warm (cold) advection is often associated with forcing for ascent (descent). When examining soundings or overlays of geopotential height at different levels, one can thus obtain a qualitative sense of where ascending (or descending) vertical air motion is expected; this has

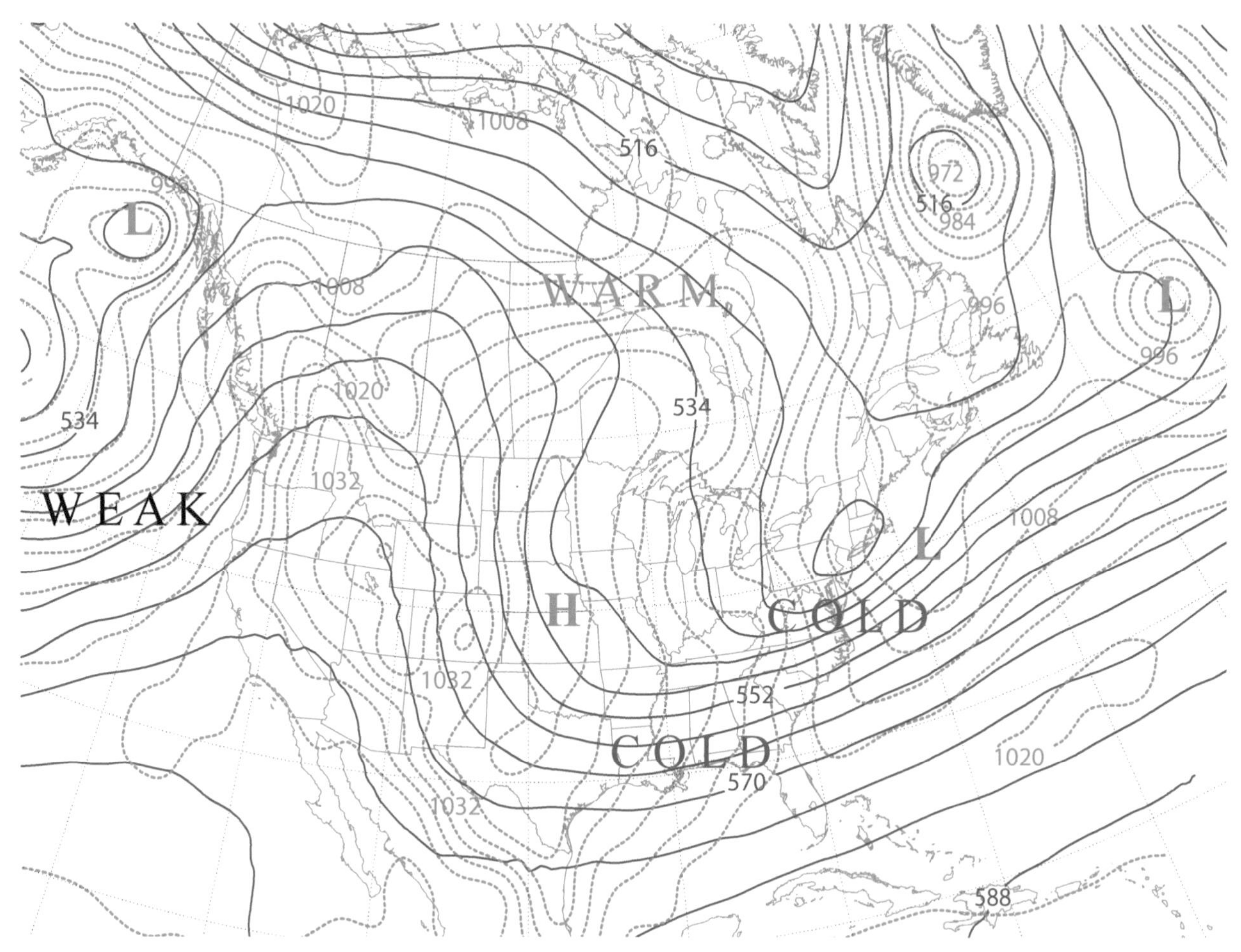


**Figure 1.11.** GFS analysis of SLP (dashed contours, interval 2 mb) and 500-mb geopotential height (solid contours, interval 6 dam) valid 1800 UTC 8 Jan 2010. High and low sea-level pressure centers are denoted "H" and "L," respectively. Selected areas of cold, weak, and warm advection are labeled.

value for weather forecasting because of the link between ascent, clouds, and precipitation.

How, physically, does the turning of the geostrophic wind with height specify the sense of the thermal advection in the layer? Because of the fixed orientation between the geostrophic wind and the height field via (1.33) or (1.39), and the link between the thickness and the layer-averaged virtual temperature (1.37), turning of the geostrophic wind with height requires flow across thickness contours, and thus temperature advection.

Given the similarity between the SLP field and the 1000-mb geopotential height, one can produce a useful overlay of SLP and 500-mb height to obtain a reasonable estimate of the thermal advection in the 1000–500-mb layer. In Fig. 1.11, pronounced backing of the geostrophic wind between that implied by the sea level isobars and that at the 500-mb level is evident over the southeastern United States and mid-Atlantic region; this is associated with cold advection in this layer. Conversely, geostrophic veering is evident across the Canadian provinces of Manitoba and Saskatchewan, coinciding with warm advection in the 1000–500-mb layer. The generally parallel alignment of SLP and 500-mb height contours near the low center over the North Pacific is indicative of weak thermal advection there, although veering and warm advection are apparent over western Washington, Oregon, and British Columbia.

The location of fronts can be identified via the layer-average temperature advection; for example, the cold advection over the southeastern United States in Fig. 1.11 is

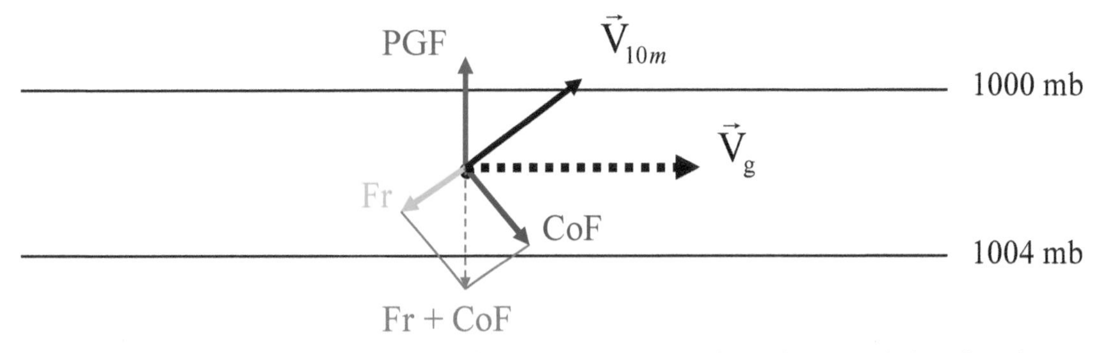


**Figure 1.12.** Plan view schematic of horizontal forces, along with geostrophic and geotriptic balanced wind vectors in idealized PBL. Isobars labeled every 4 mb.

taking place in the wake of a trough in the SLP field that corresponds to a cold front, extending from a low center in the western Atlantic southwestward into Florida.

### 1.4.3. Frictional veering

It is important to bear in mind that it is the veering or backing of the *geostrophic* wind that is related to temperature advection in the preceding development. Directional shear of the *ageostrophic* wind is not related to the sense of thermal advection. This is not to say that ageostrophic temperature advection is not important—just that the sense of its directional shear does not have the same relation to thermal advection as it does for the geostrophic flow.

We know that the influence of the frictional force results in a departure from geostrophic balance as indicated in Fig. 1.12. In this diagram, the wind vector at the 10-m level is drawn pointing toward lower pressure, consistent with the inclusion of friction in the force balance. Assume that the isobars, oriented east–west in this example, do not change orientation with height as we move from near the surface to the top of the boundary layer. The influence of friction will vanish above the boundary layer, and we expect the flow to be approximately geostrophic there. If we follow the 10-m wind vector aloft toward the region of geostrophic flow, then we see that the wind is veering. However, this is due to the vertical shear of the ageostrophic wind component and is not necessarily associated with any form of thermal advection; the geostrophic wind is not veering over this layer.

The measured winds in observed rawinsonde soundings often exhibit some degree of frictional veering. How can we separate this from the case where geostrophic veering is associated with warm advection? We must recognize that the frictional veering will be confined to the planetary boundary layer, generally 1–1.5 km in depth. Veering observed above that layer is much more likely to be associated with thermal advection. Recall also from section 1.3 that it is straightforward to identify the boundary layer depth in soundings by considering the depth of the surface-based mixed (adiabatic) layer.

## 1.5. VORTICITY AND THE VORTICITY EQUATION

Tornadoes, hurricanes, and extratropical cyclones and anticyclones are all characterized by varying degrees of wind rotation about a vertical axis. The motivation to understand the dynamics of these systems is one reason why atmospheric scientists often utilize the concept of *vorticity*, which is a local measure of rotation about a given coordinate axis. If we understand the mechanisms that lead to changes in vorticity, then we can better conceptualize the development and decay of important rotating weather systems. Our goal here is to briefly review and then apply the vorticity concept to several important atmospheric systems.

The *relative vorticity*, a vector representing a quantitative measure of rotation in a fluid, is given by the curl of the velocity:

$$\vec{\omega} = \nabla_3 \times \vec{U}. \tag{1.46}$$

The *absolute vorticity* takes into account rotation due to the spin of the earth on its axis, in addition to the relative vorticity. Expanding the vector cross product in (1.46), we obtain expressions for the relative vorticity component about each of the three directional axes:

$$\vec{\omega} = \left(\frac{\partial w}{\partial y} - \frac{\partial v}{\partial z}\right)\hat{i} + \left(-\frac{\partial w}{\partial x} + \frac{\partial u}{\partial z}\right)\hat{j} + \left(\frac{\partial v}{\partial x} - \frac{\partial u}{\partial y}\right)\hat{k}. \tag{1.47}$$

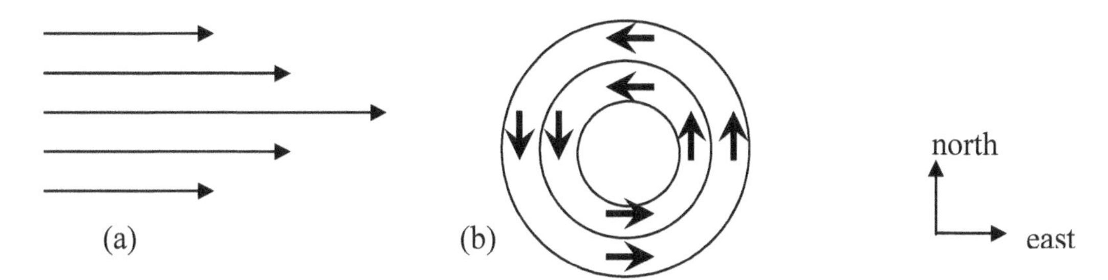


**Figure 1.13.** Idealized horizontal flows with (a) shear and (b) curvature vorticity.

Cyclonic values of vorticity are consistent with rotation in the same sense as the earth's rotation about its axis when viewed looking down on either pole [that is, counterclockwise (clockwise) rotation is cyclonic in the Northern (Southern) Hemisphere].

Meteorologists often focus on rotation about the vertical axis, the last term in (1.47). This is related to the desire to understand the development of rotation in weather systems, as mentioned above. For synoptic-scale motions (as will be discussed in chapter 2), the advection of the vertical component of vorticity is related to forcing for vertical air motion and is thus critical for weather forecasting. The relative and absolute vorticity about vertical axes can be respectively written as

$$\zeta = \hat{k} \cdot \vec{\omega} = \left( \frac{\partial v}{\partial x} - \frac{\partial u}{\partial y} \right), \quad \zeta_a = \left( \frac{\partial v}{\partial x} - \frac{\partial u}{\partial y} \right) + f, \qquad (1.48)$$

where $f = 2\Omega \sin\varphi$ is the vorticity due to the earth's rotation; this component is zero at the equator and is maximized at twice the earth's rotation rate at the poles. As with cyclonic relative vorticity, planetary vorticity is negative in the Southern Hemisphere.

A scale analysis of the $\hat{i}$ and $\hat{j}$ vorticity components in (1.47) using synoptic-scale values from Table 1.3 reveals that the horizontal gradients of vertical motion $(\partial w/\partial y, \partial w/\partial x)$ are approximately five orders of magnitude smaller than the vertical gradient of the horizontal wind and can be neglected. Note that these vertical shear terms, $\partial v/\partial z$ and $\partial u/\partial z$, also scale two orders of magnitude larger than the components of the vertical component terms $\partial v/\partial x$ and $\partial u/\partial y$, owing to the fact that the vertical distance scale $H$ is smaller than the horizontal length scale $L$ by that amount. This suggests that the potential for tilting of vorticity about a horizontal axis into the vertical could be an important mechanism for some types of weather system. Before delving further into the processes that dictate the vorticity tendency, we must first be comfortable recognizing and interpreting vorticity using standard analyses and data.

#### 1.5.1. Shear, curvature, and planetary vorticity

Given the practical importance and meteorological emphasis on the vertical component of vorticity, the term "vorticity" will refer only to the vertical component unless otherwise specified. Absolute vorticity is usually nonzero even with calm winds because of the presence of planetary vorticity, and relative vorticity can be strong even in the absence of flow curvature in the presence of *shear vorticity*. Figure 1.13a depicts a westerly flow characterized by lateral shear. From (1.48), we see that $\partial v/\partial x = 0$ in this example due to the absence of a meridional ($v$) wind component. However, $\partial u/\partial y$ is nonzero everywhere except along the axis of strongest winds, taking on either sign, with negative values north of the jet core. This is due to a positive westerly $u$ wind component decreasing in the $+y$ direction there; the negative sign multiplying this term in (1.48) means that this zone is characterized by positive values of vorticity, which is cyclonic in the Northern Hemisphere. Similarly, the zone south of the axis of strongest westerly flow exhibits anticyclonic (negative) relative vorticity in Fig. 1.13a.

For synoptic-scale systems, the value of $f$ is typically large enough to ensure that the absolute vorticity remains positive except in small regions of unusually strong anticyclonic shear or curvature. Regions of curved flow are characterized by perhaps more visually obvious vorticity because of clear evidence of curvature, whereas shear vorticity can be more subtle. Cyclonic *curvature* vorticity, characterized by flow rotating counterclockwise about a point, is illustrated in Fig. 1.13b. In this diagram, both $\partial v/\partial x$ and $-\partial u/\partial y$ contribute to positive values of relative vorticity at the center of circulation.

Suppose we wish to qualitatively estimate the geostrophic relative vorticity at the location denoted "A" in

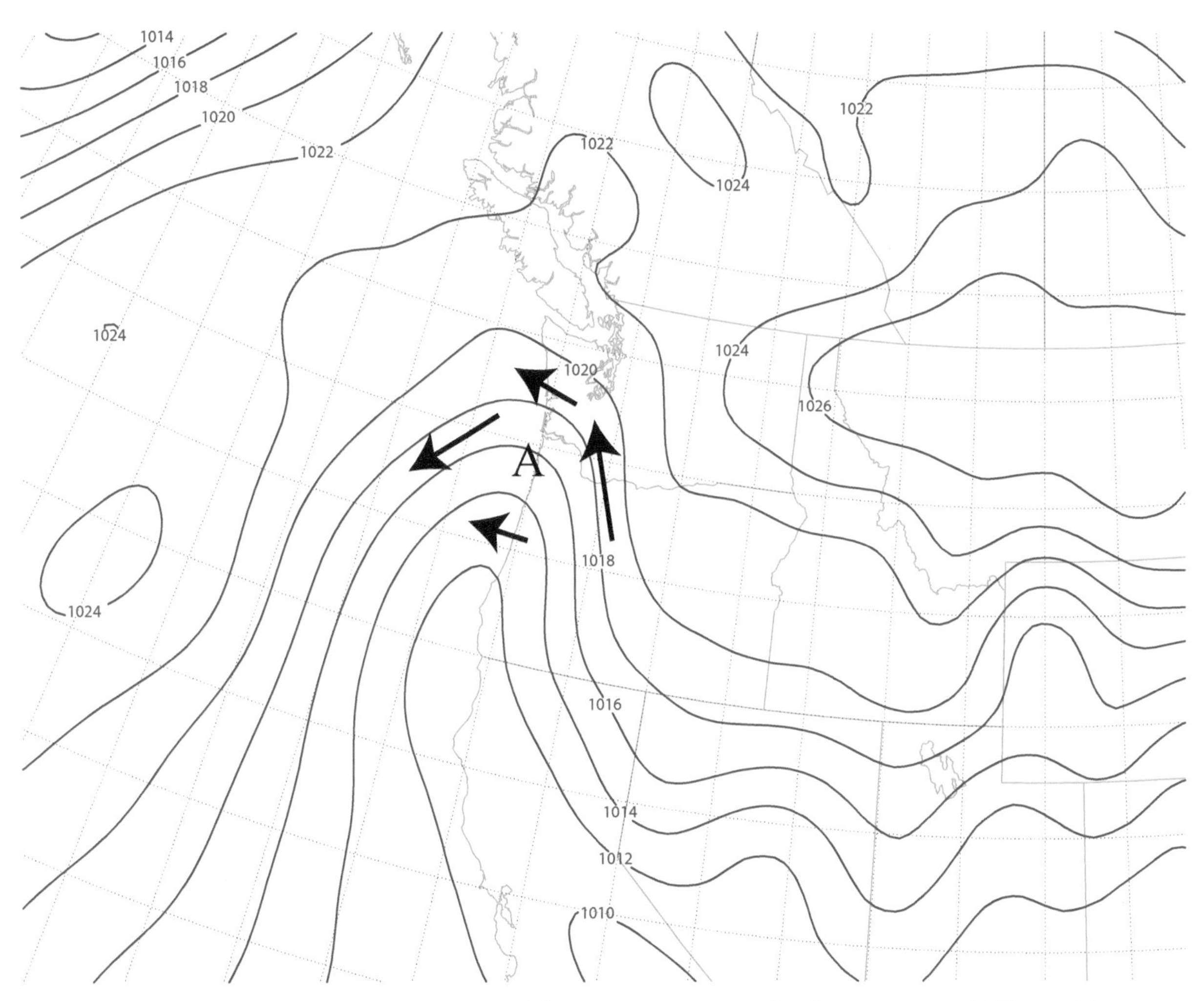


**Figure 1.14.** SLP analysis for 1200 UTC 1 Sep 2006. Solid contours represent SLP (2-mb contour interval); blue grid depicts latitude and longitude lines at 2° interval. Arrows represent geostrophic wind at selected locations.

Fig. 1.14. If the wind vectors were provided with more precision, then we could accurately compute this quantity using *finite difference* techniques. The geostrophic relative vorticity is given by $\varsigma_g = \partial v_g / \partial x - \partial u_g / \partial y$. The sign of $\partial v_g / \partial x$ is clearly positive at this location because we observe a southerly (positive) $v_g$ component to the east of point A, but a negative value to the west. The quantity $-\partial u_g / \partial y$ is relatively small, because $u_g$ is similar to the north and south of point A. Thus, we find that the $\partial v_g / \partial x$ term dominates, and that it is consistent with cyclonic vorticity, and a trough, in this example. Despite the ridge-like appearance of the isobars in the vicinity of point A, this is an *inverted trough*, as it represents a corridor of lower pressure relative to points on either side of the axis.

The relation between the geopotential height and vorticity for a typical wintertime situation is evident in Fig. 1.15. To highlight cyclonic vorticity, absolute vorticity, shaded for values greater than $10 \times 10^{-5}$ s$^{-1}$ (Fig. 1.15a), indicates values generally increasing toward the North Pole, partly because of the increase in planetary vorticity $f$ with latitude. Regions of both curvature and shear vorticity are evident, with the former showing prominently in the upper trough centered near the Great Lakes region of the United States. Cyclonic shear vorticity is evident along the northern flank of a strong jet over Northern Canada. A similar region is found over the eastern North Atlantic.

By displaying only relative vorticity contours in Fig. 1.15b, we highlight anticyclonic vorticity features

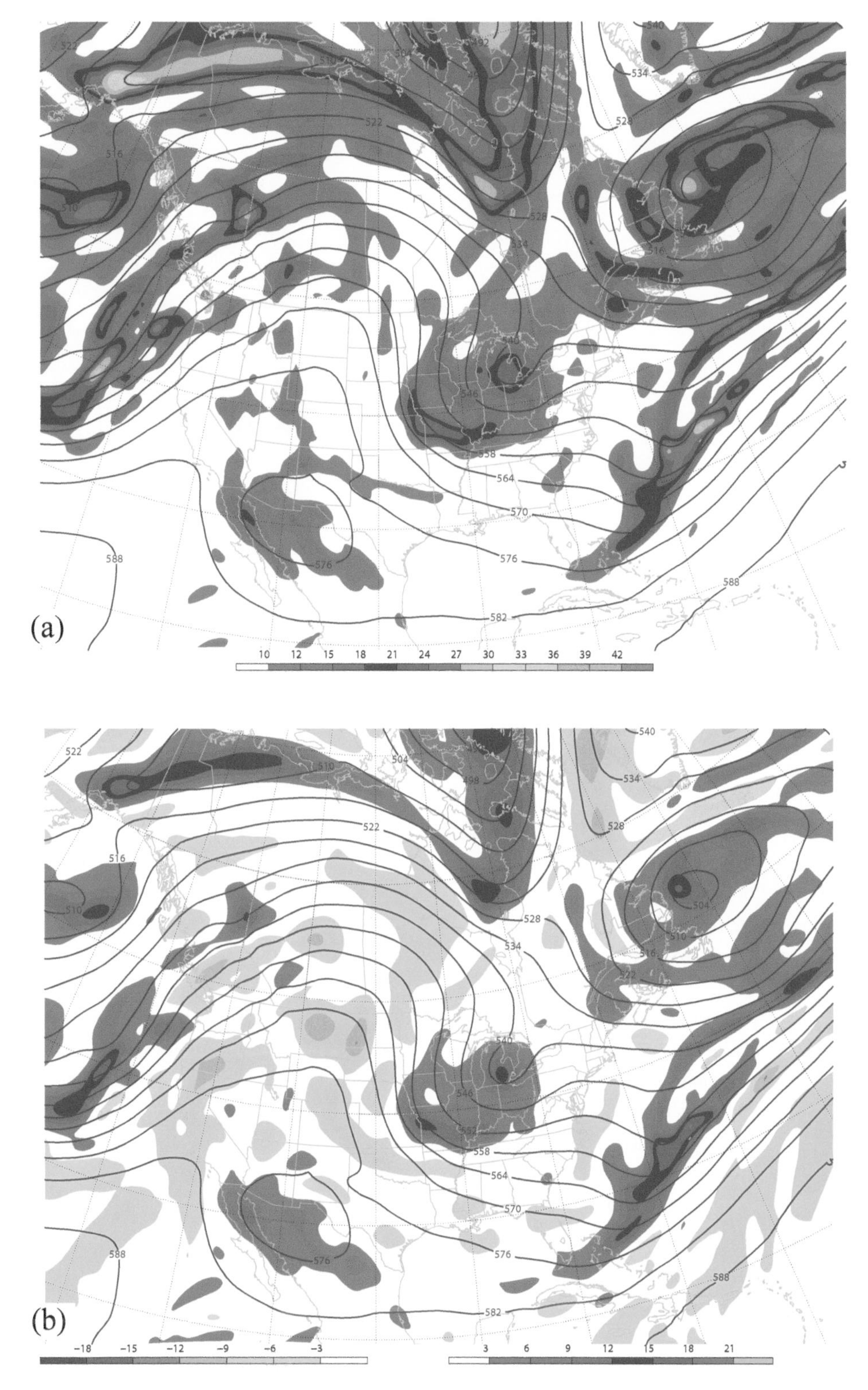


**Figure 1.15.** GFS analysis valid 1800 UTC 11 Jan 2010: geopotential height (solid, contour interval 6 dam) with (a) absolute vorticity (shaded above $10 \times 10^{-5}\ \mathrm{s}^{-1}$ as in legend at bottom) and (b) as in (a) except relative vorticity, shaded as in legend at bottom of panel.

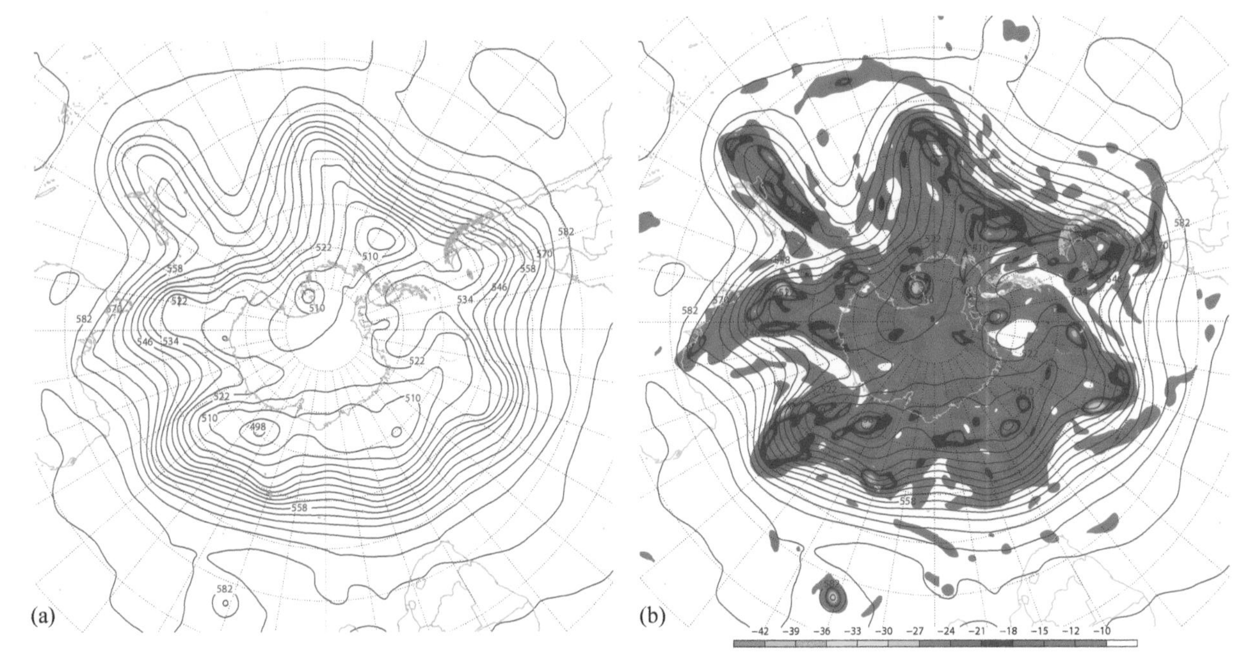


**Figure 1.16.** GFS analyses for the Southern Hemisphere valid 12 UTC 12 January 2010: (a) GFS 500-mb height analysis (solid contours, interval 6 dam). (b) As in (a), but with absolute vorticity superimposed (shaded less than $-10 \times 10^{-5}$ s$^{-1}$, as in legend at bottom of panel).

in the large upper ridge over western North America, and anticyclonic shear vorticity south of the axis of the North Pacific and North Atlantic jets. Even when viewing relative vorticity, there is a tendency for cyclonic values to be found on the poleward side of the jet, because of the presence of cyclonic shear vorticity there; the increases in absolute vorticity noted in Fig. 1.15a are thus not entirely due to increases in planetary vorticity with latitude.

The vorticity displayed in Fig. 1.15 is based on the full analyzed horizontal wind field and not only on the geostrophic flow. However, the clear correspondence between the full vorticity and the geopotential height field suggests that the geostrophic vorticity provides a reasonable approximation to that computed from the observed wind. It is a worthwhile exercise to plot the wind field along with the computed vorticity to gain a sense of the structure and character of the vorticity field.

To broaden perspectives, Fig. 1.16 displays an example from the Southern Hemisphere. Consistent with clockwise geostrophic flow around the broad region of lower geopotential height covering the South Polar region, a prevailing westerly jet stream is evident as for the Northern Hemisphere. Here, the absolute vorticity is generally negative, and we see that the maximum negative vorticity values (which are cyclonic in the Southern Hemisphere) are again found on the poleward (southern in this case) flank of the westerly jet stream.

### 1.5.2. The vorticity equation

The *vorticity equation* describes the mechanisms that give rise to changes in vorticity, and it is thus an excellent tool for understanding the evolution of atmospheric circulation systems. The vorticity equation is formed as a combination of the horizontal momentum equations; combining these two equations simplifies them into a form that more readily lends itself to physical interpretation.

The vertical component of the relative vorticity is given by $\zeta = \hat{k} \cdot \nabla \times \vec{U}$. To form an equation for the local time rate of change of this quantity, we must determine $\partial \zeta / \partial t = \hat{k} \cdot \nabla \times (\partial \vec{U} / \partial t)$; in other words, take $\hat{k} \cdot \nabla \times$(momentum equation). The complete derivation is included in many standard dynamics texts and is also suggested as an exercise at the end of the chapter for those wishing to review this important derivation.

Applications of the vorticity equation include conceptual and quantitative explanation of the formation of midlatitude cyclones, lee troughs, hurricanes, and tornadoes.

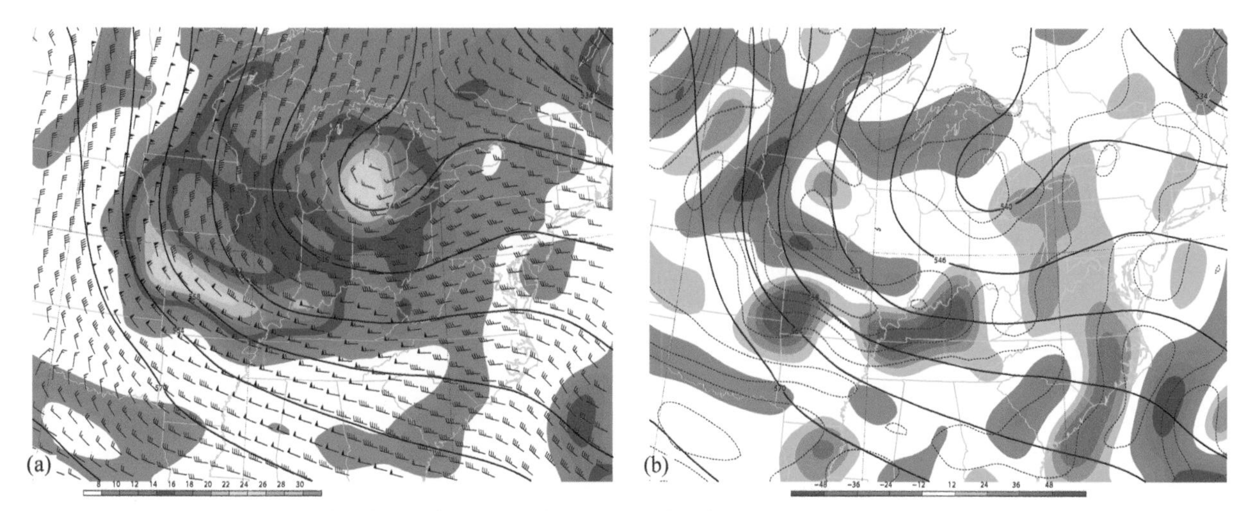

**Figure 1.17.** (a) As in Fig. 1.15, but for smaller geographical area and with 500-mb wind barbs added, and (b) 500-mb vorticity advection (shade interval $12 \times 10^{-10}$ s$^{-2}$, as in legend at bottom of panel). Vorticity and vorticity advection fields were smoothed by a weighted Gaussian filter to remove noise.

The "full" vorticity equation in isobaric, Cartesian coordinates can be written as

$$\underset{(a)}{\frac{\partial \zeta_a}{\partial t}} = \underset{(b)}{-\vec{V} \cdot \nabla \zeta_a} \underset{(c)}{- \omega \frac{\partial \zeta_a}{\partial p}} \underset{(d)}{- \left[ \frac{\partial \omega}{\partial x}\frac{\partial v}{\partial p} - \frac{\partial \omega}{\partial y}\frac{\partial u}{\partial p} \right]} \underset{(e)}{+ \zeta_a \frac{\partial \omega}{\partial p}} \underset{(f)}{+ \left[ \frac{\partial F_y}{\partial x} - \frac{\partial F_x}{\partial y} \right]}, \quad (1.49)$$

where $\zeta_a = \zeta + f$ is the absolute vorticity. The pressure gradient term has vanished in the cross-differentiation that is performed to obtain (1.49). The loss of information associated with the pressure field was noted by Sutcliffe (1947), and it suggests that there is a price to pay for the conceptual simplification afforded by this equation.

The local time rate of change of absolute vorticity is the same as the local tendency of relative vorticity, given that the Coriolis parameter $f$ is almost exactly constant with time at a fixed location. This equation states that the local vorticity tendency (term a) is due to horizontal vorticity advection (term b), vertical vorticity advection (term c), the tilting of vorticity about horizontal axes into the vertical (term d), vortex stretching (term e), and differential frictional processes (term f). These terms will be considered in turn below.

We will revisit the earlier example from Fig. 1.15 to illustrate *horizontal vorticity advection*, enlarged and with wind barbs added to highlight the Great Lakes trough (Fig. 1.17). Vorticity computations can be somewhat noisy because of the calculation of spatial derivatives, especially for high-resolution gridded datasets; the vorticity advection is also often quite noisy, as it involves still higher-order derivatives. Therefore, it is useful to apply a numerical filter to these quantities, as was done here. Two separate vorticity maxima are evident in Fig. 1.17a: one over southern Michigan, the other centered over Missouri. Over Kentucky, eastern Ohio, western Pennsylvania, and New York, winds are blowing from areas of larger vorticity toward smaller values, consistent with positive vorticity advection in these locations, as shown in Fig. 1.17b. Upstream of the trough, negative vorticity advection is evident. The strength of the horizontal advection depends on the strength of the wind vector component in the direction of the vorticity gradient, as well as on the magnitude of the vorticity gradient itself.

The vertical vorticity advection (term c in 1.49) is typically weaker than the horizontal advection because of the relative smallness of the vertical velocity. Both the tilting and vertical advection terms tend to be small near the earth's surface, but they can be large aloft. Tilting, in particular, can play an important role in upper-level frontal zones, which are characterized both by strong horizontal gradients of vertical motion as well as large vertical wind shear as suggested by the thermal wind relation. Large horizontal vorticity components in the presence of horizontal vertical velocity gradients can result in vorticity about a horizontal axis being tilted into the vertical.

The stretching term (e) in (1.49) is arguably the dominant vorticity production term, and this process will be explored in depth at several points in later chapters. In this term, the absolute vorticity multiplies the vertical derivative of omega, the pressure-coordinate vertical velocity. Using the isobaric continuity equation (1.17), and expanding the absolute vorticity, we can express this term as $-(\zeta+f)\nabla\cdot\vec{V}$. The spinup of vorticity is thus proportional to the convergence (related to the stretching of the vertical air column) as well as to the value of preexisting absolute vorticity. This important fact explains the rapid development of many strongly rotating cyclonic systems: for a given value of convergence, the rate of vorticity growth is proportional to the vorticity itself. Zones of cyclonic preexisting vorticity, such as fronts or troughs, are thus preferential sites for further growth of cyclonic vorticity. Conversely, in a ridge, where there is cancellation between the planetary and relative vorticity, it can be very difficult to generate cyclonic vorticity tendencies via stretching even with strong convergence because the multiplier (the absolute vorticity) is very small in that situation.

The frictional tendency generally acts in the sense of weakening a given vorticity feature and will not be discussed further here. However, for flow past topographic barriers or in the wake of mountains, frictional processes can lead to the development of cyclonic vorticity streamers that can be meteorologically important.

### 1.5.3. The barotropic vorticity equation and Rossby waves

Additional applications of the vorticity equation will be presented in subsequent chapters; however, at this point, we will explore one important application that stems from an approximated form of the vorticity equation. If we make the assumption of a frictionless, *barotropic* atmosphere, then we can greatly simplify the vorticity equation. Even in simplified form, the dynamical essence of an extremely important atmospheric phenomenon, Rossby waves, will be retained. Rossby waves play a major role in dictating weather on daily and longer time scales, and they are critically important to the meridional transport of heat, moisture, and momentum in the global energy balance. Many of the wavelike perturbations seen in the geopotential height contours of Figs. 1.15 and 1.16 are examples of Rossby waves.

The assumptions to be applied here to (1.49) are quite restrictive and are generally poor for the real atmosphere. What is the purpose of making such assumptions if we hope to understand the atmosphere? If we reduce a given equation to the simplest form that still retains the dynamical behavior of interest, then it serves to clarify the true underlying dynamics that are at work in the full unapproximated equations, and in the real atmosphere.

The *barotropic assumption* requires that the density depend only on pressure, which via the ideal gas law holds that the temperature is constant on an isobaric surface. Further, the geostrophic wind then must be independent of height, according to the thermal wind relation (1.43). We also assume frictionless flow, as would typically be found above the planetary boundary layer. For the flow to remain barotropic, it must also be nondivergent, meaning that all of the terms in (1.49) involving vertical motion and friction are neglected. With these assumptions, the vorticity equation (1.49) becomes

$$\frac{d(\zeta+f)}{dt}=0. \tag{1.50}$$

This is a statement of *conservation of vorticity*, meaning that following the flow, the absolute vorticity remains constant. If air parcels move poleward to a region of larger planetary vorticity, then there must be a compensating change in the relative vorticity to keep the sum constant, as illustrated in Fig. 1.18.

In a vorticity-conserving flow, once an air parcel is displaced from its latitude of origin, an oscillation comes about with alternating anticyclonic and cyclonic curvature as needed to allow the relative vorticity to compensate changes in planetary vorticity (Fig. 1.18). Note that the planetary vorticity $f$ is only a function of the meridional ($y$) direction, and so we can write

$$\frac{\partial(\zeta+f)}{\partial t}=-\vec{V}\cdot\nabla\zeta-v\frac{\partial f}{\partial y}. \tag{1.51}$$

The parameter $\beta\equiv\partial f/\partial y$ has a value of order $10^{-11}$ m$^{-1}$ s$^{-1}$ in midlatitude locations.

The behavior of Rossby waves will be illustrated here with a simple, idealized example. Assume a constant,

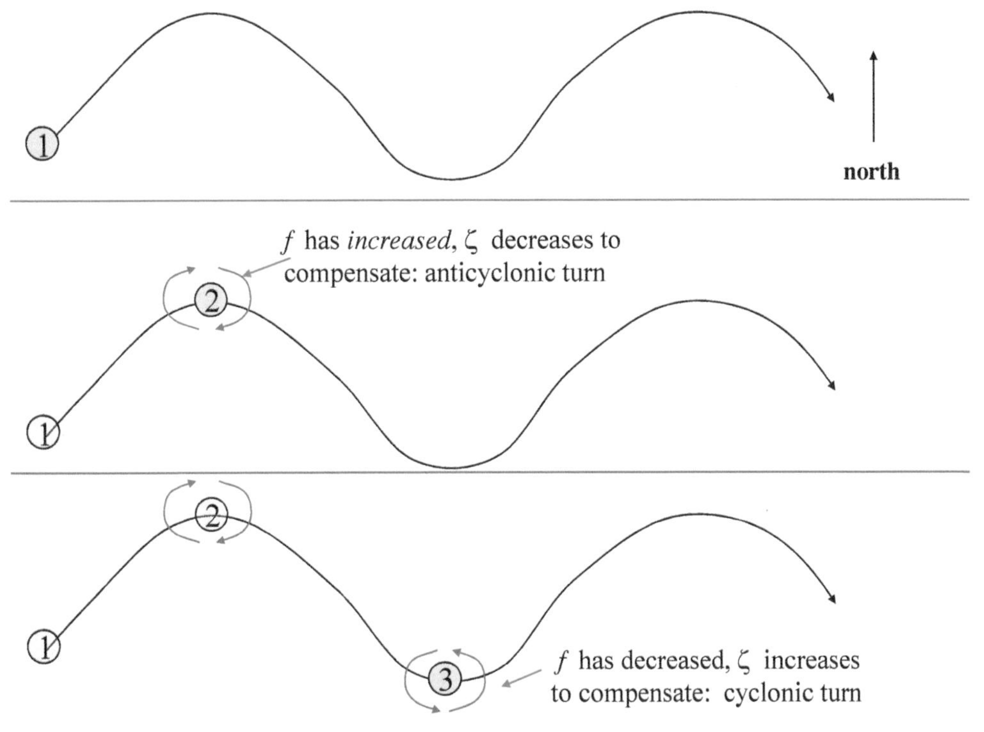


**Figure 1.18.** Air parcel moving on a constant absolute vorticity trajectory, shown at different locations along the flow in each panel.

uniform, time-independent westerly flow $U$, with a superimposed sinusoidally varying meridional wind component, defined as follows:

$$u = \frac{dx}{dt} = U \quad \text{and} \tag{1.52}$$

$$v = \frac{dy}{dt} = v_0 \cos\left[\frac{2\pi}{L}(x-ct)\right]. \tag{1.53}$$

Equation (1.53) describes a sinusoidal variation in the meridional wind of wavelength $L$, with the wave pattern moving at speed $c$, embedded in a uniform westerly current $U$ as defined by (1.52). The meridional wind component is maximized in magnitude at the "inflection points" between trough and ridge axes, as depicted in Fig. 1.19.

Given that we have complete knowledge of the wind field in this example, we can easily calculate the vorticity, which, because of the uniformity of the zonal wind component, is given entirely by $\partial v/\partial x$, obtained by taking $\partial/\partial x$ of (1.53):

$$\zeta = \frac{\partial v}{\partial x} - \frac{\partial u}{\partial y} = -\frac{2\pi}{L} v_0 \sin\left[\frac{2\pi}{L}(x-ct)\right]. \tag{1.54}$$

Noting that $df/dt = v\, \partial f/\partial y = v\beta$ and expanding the derivative in the barotropic vorticity equation (1.50), we obtain

$$\frac{d(\zeta + f)}{dt} = \frac{\partial \zeta}{\partial t} + U\frac{\partial \zeta}{\partial x} + v\beta = 0. \tag{1.55}$$

Using (1.54) we next evaluate the derivatives in (1.55) and substitute (1.53), then cancel common terms to obtain the following celebrated *Rossby wave phase speed equation*:

$$c = U - \frac{\beta L^2}{4\pi^2}. \tag{1.56}$$

Here, the speed at which the axes of troughs and ridges move in a barotropic atmosphere is given by the difference between the background zonal wind speed, $U$, and a term involving the square of the wavelength and the gradient of planetary vorticity. If we trace the origin of the right-hand terms in (1.56) back to the vorticity equation, we find that $U$ is associated with the advection of vorticity by the background zonal flow, and the $\beta$ term results from the advection of planetary vorticity by the meridional wind. The negative sign on the second term is consistent

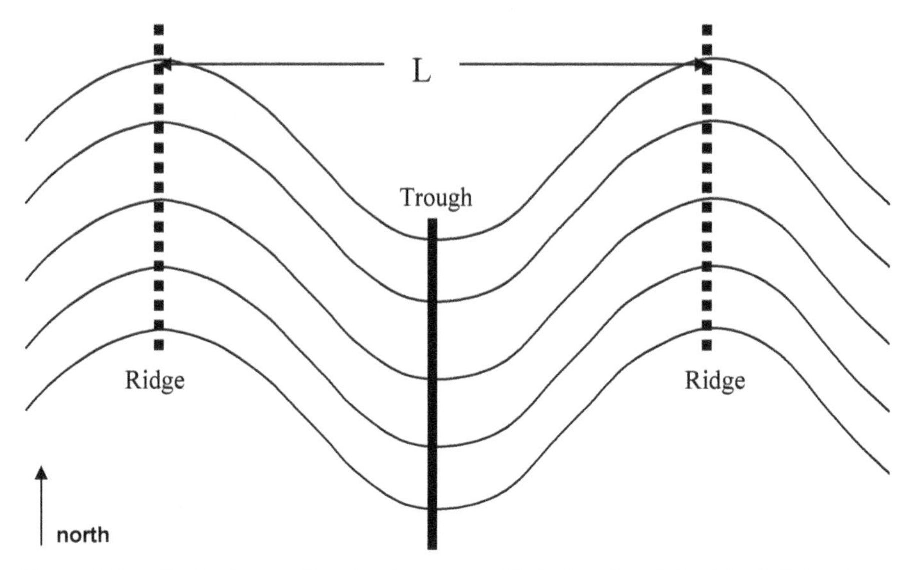


**Figure 1.19.** Idealized streamlines in a barotropic, frictionless flow defined by (1.52) and (1.53). Ridge and trough axes are shown with dotted and solid vertical lines, respectively.

with the effect of *propagation* of the waves against the tendency of the zonal flow to advect the waves eastward. For small wavelength $L$, or strong westerly flow $U$, the waves move eastward more rapidly because of a reduced cancellation between these terms.

The propagation of Rossby waves due to planetary vorticity advection can be understood by considering the influence of this term on the absolute vorticity tendency at a given location. For example, west of the trough axis in Fig. 1.19, the northerly wind component advects larger values of planetary vorticity southward, reinforcing the trough to the west of the axis. The opposite tendency is found east of the trough axis, where smaller planetary vorticity is being advected northward, building the downstream ridge westward.

Despite the drastic simplifications that went into the derivation of (1.56), it captures the dynamics of observed Rossby wave behavior quite well. Some additional questions may help to illustrate the insights available from this equation:

1. Given that $c$ is positive eastward, how does the east-west motion of troughs and ridges vary with the strength of $U$?
2. How does the movement of troughs and ridges vary with $L$?
3. Does there exist a certain wavelength for which $c = 0$? If so, what would be the implication for weather conditions in the vicinity?
4. How would the movement of troughs and ridges change if there was no gradient of planetary vorticity? For slowly rotating planets, such as Venus, should we expect Rossby waves to be important in midlatitude regions?

With some additional development, it can be shown that the speed at which Rossby wave *energy* travels differs from the speed at which individual troughs and ridges move. In fact, the energy of *Rossby wave packets* travels at the *group velocity*, which for Rossby waves is considerably faster than the *phase velocity* given by (1.56). This has important implications for weather prediction; large-amplitude wave events taking place in one part of the world frequently precede high-impact events downstream. Given the well-known dynamical mechanism of these waves, the possibility of accurate extended-range weather forecasting using Rossby wave ideas has enjoyed a recent resurgence. In chapter 2, we will explore these ideas further in the context of the QG energy equation.

The theoretical work of Swedish meteorologist C.-G. Rossby in the 1930s and 1940s was quickly tested with upper-air observations, as the upper-air radiosonde network was being developed beginning around that time. A short but illuminating paper by E. Hovmöller (1949) provides a clever depiction of Rossby wave phase and group velocity in observations, reproduced here in Fig. 1.20a. The movement of individual troughs and ridges can be

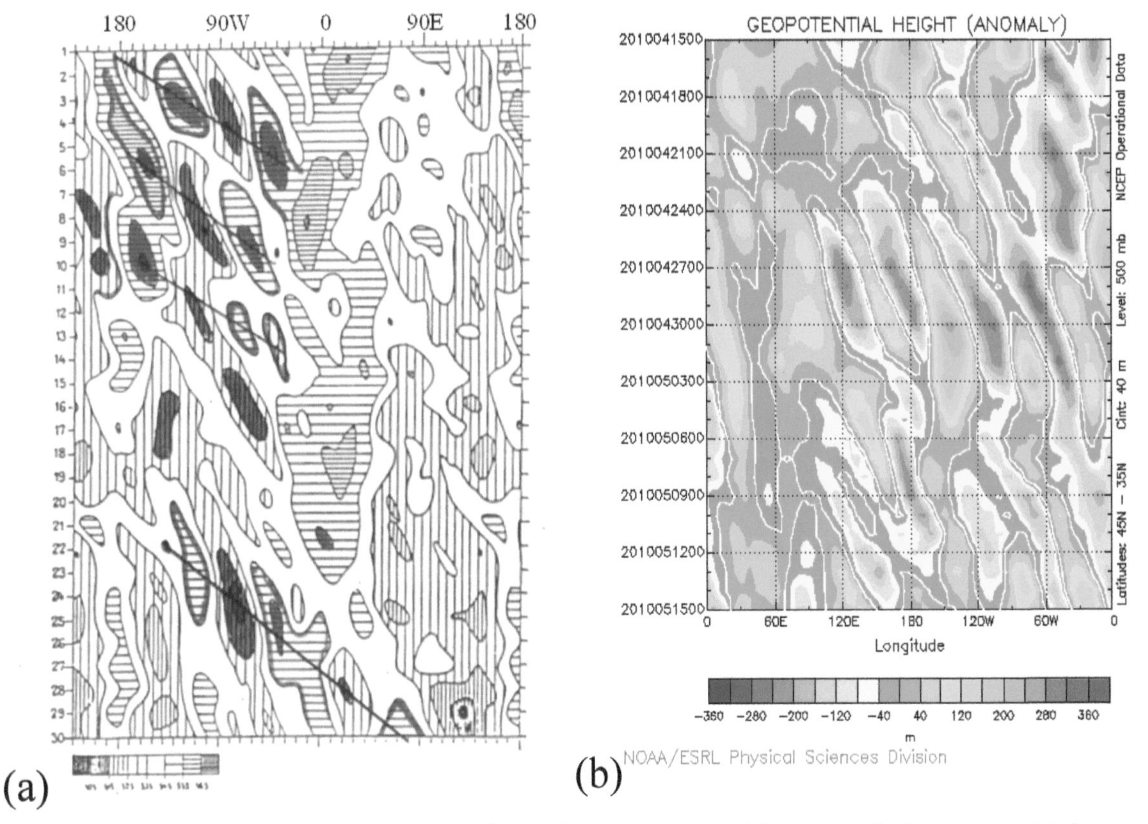


**Figure 1.20.** Time–longitude (Hovmöller) diagrams of 500-mb height anomaly: (a) for the month of November 1949 for a Northern Hemisphere latitude belt. The height anomalies are hatched according to the legend at the bottom of the figure. Phase velocity corresponds to the slope of the axis of individual troughs and ridges. Solid lines sloping downward to right indicate the Rossby wave group velocity (adapted from Hovmöller 1949); (b) As in (a), but for 15 Apr–15 May 2010, obtained from NOAA ESRL Web site (available online at www.esrl.noaa.gov/psd/map/time_plot/).

traced directly on this diagram, which is constructed with time progressing downward. It is also evident that longitudinal zones of large trough and ridge amplitude show progressive *downstream development*, as indicated by solid lines sloping downward to the right. These lines correspond to the group velocity of Rossby wave packets and exhibit an eastward component relative to the movement of trough and ridge axes. The group velocity is larger than the phase velocity in this diagram, as is typically the case. Plots of this type are now easily obtained using online tools, such as those provided by the National Oceanic and Atmospheric Administration (NOAA) Earth System Research Lab (ESRL); the behavior for the period 15 April to 15 May 2010, shown in Fig. 1.20b, exhibits similar characteristics to that seen in the original Hovmöller diagram, with a large-amplitude downstream development event evident between late April and early May.

## 1.6. MICROMETEOROLOGY

The primary focus of this text is on synoptic-scale weather systems. However, as will become evident in subsequent chapters, even these systems can be influenced in important ways by smaller-scale motions, including microscale processes taking place in the planetary boundary layer. In introducing the momentum equations earlier in this chapter [e.g., (1.5)–(1.7), (1.11), and (1.15)], a generic representation of "friction" was included. However, the frictional force due to molecular viscosity is negligible for most atmospheric applications. Yet, in our analysis of sounding data in section 1.3 (Figs. 1.3 and 1.4), we saw evidence of the pronounced impact of turbulent mixing on profiles of temperature, and we discussed *frictional veering* of the observed wind with height in section 1.4.3. What is the physical mechanism of friction in the atmosphere?

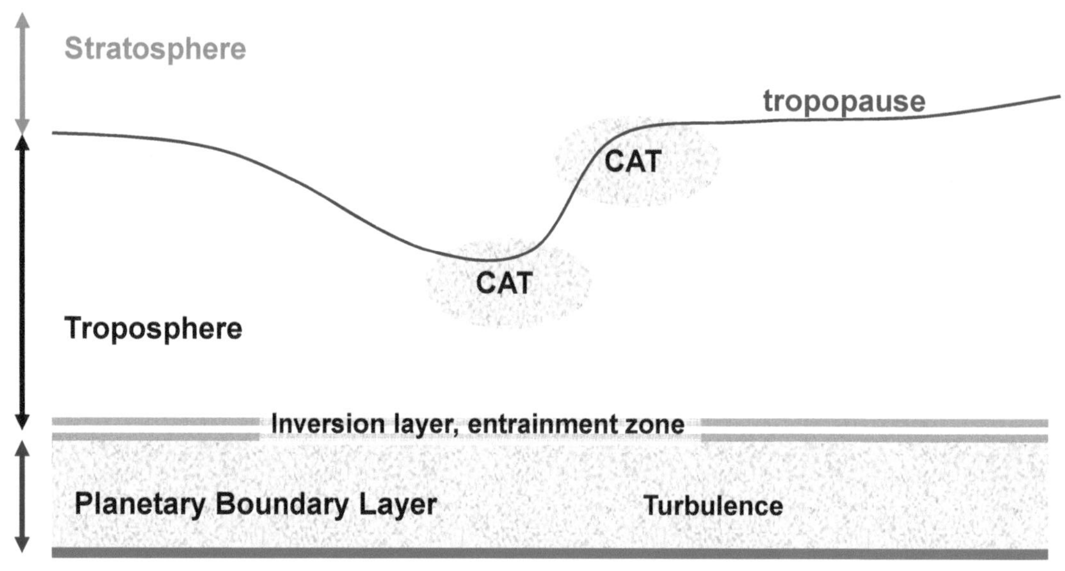


**Figure 1.21.** Idealized cross-sectional schematic indicating the typical presence of turbulence (gray shading) for a midlatitude location with a developed PBL and upper jet. The tropopause is indicated with a curved line, and the designation "CAT" refers to clear-air turbulence.

An important process that often operates near the earth's surface, and which serves to communicate surface influences to the atmosphere is *turbulence*, which alters the momentum, temperature, and moisture fields. Turbulence is most likely to be found in the lowest 1–2 km of the troposphere, in the *planetary boundary layer* (Fig. 1.21). Turbulence may also be found in regions of strong vertical wind shear above and below the jet stream core, and in the vicinity of convective storms. The boundary layer is not always turbulent; however, in situations characterized by strong winds or a warm lower surface, turbulence is typically present.

It is useful to partition the atmospheric flow into a turbulent part, which includes very high-frequency quasi-random fluctuations, and a more slowly varying field, such as would result from a 5- or 10-min time average of high-frequency measurements. In fact, standard surface observations report temperatures that have been averaged over 5 min to eliminate the high-frequency fluctuations that can be associated with turbulent flow. If we separate the basic atmospheric variables into a time-mean and perturbation values, we can better describe the influence of turbulence:

$$\theta = \overline{\theta} + \theta', \quad q = \overline{q} + q', \quad u = \overline{u} + u', \quad v = \overline{v} + v', \quad \text{and} \quad w = \overline{w} + w', \tag{1.57}$$

where the overbar denotes a time average (e.g., 5 or 10 min) and the prime designates the deviation from this average. By definition, the time average of a single-prime quantity is identically zero.

Suppose that high-frequency turbulence measurements are taken using rapid-response probes, as shown schematically in Fig. 1.22, along with some hypothetical sample data. Because of buoyancy, on average, air parcels ascending in turbulent eddies are more likely to possess a positive potential temperature perturbation, and vice versa for descending parcels. As a result, when the quantity $\overline{w'\theta'}$ is positive—that is, there is a positive correlation between vertical motion and potential temperature perturbations—there is an *upward turbulent heat flux*. In the presence of a vertical gradient of moisture, momentum, or any other variable, systematic correlations of this type can be expected. For example, over a water surface, an upward turbulent moisture flux is often found. Because wind speed generally increases with height, a downward turbulent momentum flux is often observed.

In a given layer, if the vertical turbulent flux at the top of the layer is equal to that at the bottom, then there would be no net tendency within the layer due to this process, because the turbulent transport out of the layer would match that coming in the bottom, and no net "accumulation" of the quantity would occur in the layer (Fig. 1.23a). If, however, the strength of the turbulent

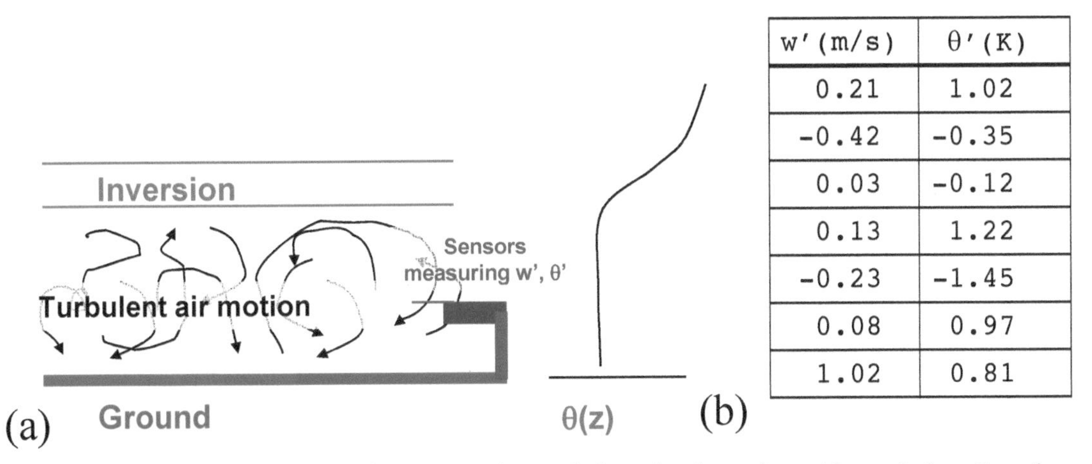


| w' (m/s) | θ' (K) |
|---|---|
| 0.21 | 1.02 |
| -0.42 | -0.35 |
| 0.03 | -0.12 |
| 0.13 | 1.22 |
| -0.23 | -1.45 |
| 0.08 | 0.97 |
| 1.02 | 0.81 |

**Figure 1.22.** Schematic of turbulent flow and sampling in the boundary layer, along with sample data. Curved black lines indicate hypothetical parcel trajectories in turbulent flow, and a typical profile of potential temperature is shown. Hypothetical sample data, as might be collected from rapid-response sensors, are provided.

flux changes with height, then a net tendency would result (Fig. 1.23b). At the top of the PBL, the flow transitions from turbulent to laminar, and the result is a zone of strong *turbulent flux convergence*, which imparts a net tendency there.

Based on these arguments, it is not the turbulent fluxes themselves that we would expect to appear in the governing equations but rather their vertical derivative, as indicated below for the potential temperature tendency equation:

$$\frac{\partial \overline{\theta}}{\partial t} = -\left(\overline{u}\frac{\partial \overline{\theta}}{\partial x} + \overline{v}\frac{\partial \overline{\theta}}{\partial y}\right) - \overline{w}\left(\frac{\partial \overline{\theta}}{\partial z}\right) + \frac{J}{C_p}\frac{\overline{\theta}}{\overline{T}} - \frac{\partial}{\partial z}(\overline{w'\theta'}). \tag{1.58}$$

However, adding terms resembling the rightmost in (1.58) creates a problem, because introducing new unknowns to the equations (and in this case, quantities that are very difficult to observe), without introducing additional governing equations, means that the set of equations is no longer *closed*. That is, by including the vertical flux divergence terms, the number of unknowns exceeds the number of equations, and the set can no longer be solved.

However, we know that the set of equations can be solved, because numerical weather prediction models generate forecasts many times each day! A scale analysis reveals that the turbulent flux terms are of leading order in equations such as (1.58), and so ignoring them is not an option if we wish to obtain accurate predictions. Instead, we must express these quantities in terms of the time-averaged variables, and their gradients, thereby accounting for them without introducing new variables to the equations. This approach, known as the *closure problem* in boundary layer meteorology, yields several different solutions of varying complexity and accuracy.

Logically, we expect that the strength of the turbulent flux of a given quantity may be proportional to the near-surface vertical gradient of the mean quantity, and the

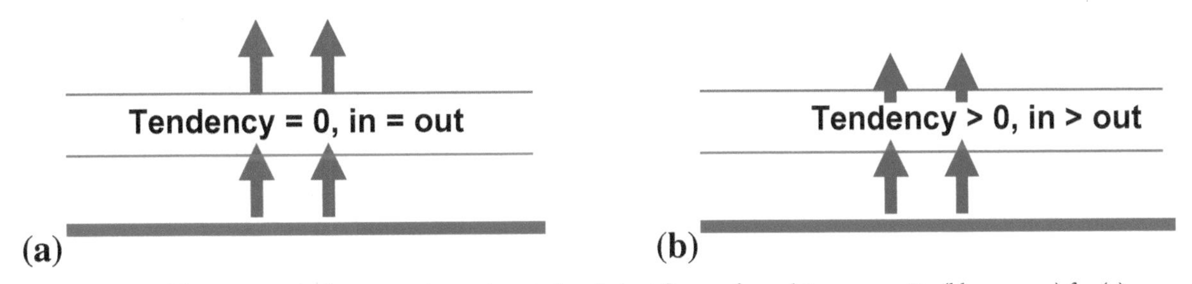


**Figure 1.23.** Schematic vertical cross sections of vertical turbulent fluxes of an arbitrary quantity (blue arrows) for (a) a layer with small vertical flux divergence and (b) a layer with vertical flux convergence (negative vertical flux divergence).

wind speed (which is related to the strength of mechanical turbulence production). To address this problem, field experiments featuring careful turbulence measurements have been undertaken and the functional dependence of the turbulent fluxes on mean quantities determined. The intuitive relations discussed above are consistent with the empirically defined *bulk aerodynamic* method of turbulence closure valid near the surface:

$$(\overline{w'\theta'}) \approx -C_H \left|\vec{V}\right| \left(\overline{\theta}_{10m} - \overline{\theta}_{sfc}\right), \qquad (1.59)$$

where $C_H$ is an observationally determined turbulent exchange coefficient, and similar forms are obtained for the turbulent moisture and momentum fluxes.

More complex (and accurate) parameterizations are available, some of which draw on analogies with molecular diffusion (flux gradient or $K$ theory). In some modern numerical weather prediction models, a predictive equation for the turbulent kinetic energy (TKE) is added to the governing equations and is solved at each grid cell in the model. In such a formulation, the strength of the turbulent fluxes is related to the predicted strength of the turbulence, among other things. Chapter 10 provides further discussion of how numerical weather prediction models represent turbulence (see section 10.4.1).

## REVIEW AND STUDY QUESTIONS

1. For a given pressure, temperature, and specific humidity value, which quantity has a larger numerical value, potential temperature, or equivalent potential temperature? Explain and justify your answer in terms of the *physical definition* of these two quantities.
2. Sketch two blank vertical cross sections extending from the surface into the lower stratosphere for a typical north–south-oriented transect that cuts through the midlatitude jet stream. Label the jet stream core with a large "J" centered near 10-km altitude. For one section, roughly sketch the distribution of *isotherms*, both beneath and above the jet core. For the other section, do the same for the distribution of *isentropes* (lines of constant potential temperature). Be sure that your sketches are consistent with thermal wind balance.
3. The lower-tropospheric thermal structure of a midlatitude cyclone is typically asymmetric, with cold air located west of the center and warmer air located to the east. It is also observed that height gradients aloft are generally much larger than those near the surface. Given these observations, what does the hypsometric equation suggest for the location of the region of lowest upper-tropospheric heights (trough axis) relative to a surface cyclone? Sketch an east–west-oriented vertical cross section passing through a midlatitude cyclone, indicating the location of the upper trough, surface cyclone, and locations of warmest and coldest air.
4. Suppose you are hiking with friends in a remote location and have no access to electronic communication devices of any type. How might observations of cloud motions at different altitudes be used to make a short-term weather forecast?
5. The Rossby wave phase speed equation (1.56) includes two terms on the right side. What physical processes do each of these terms represent, and do the processes work in concert or opposition to one another? What are the consequences for midlatitude weather conditions? Discuss and explain.
6. If the earth were rotating twice as fast on its axis as is currently observed, but the strength of westerly winds was similar to current values, would Rossby waves generally move more or less rapidly eastward? Explain.

## PROBLEMS

1. In our discussion of the continuity equation, it was stated that the hydrostatic pressure equivalent of 1" (25.4 mm) of rain is 2.5 mb. Given that pressure is defined as force per unit area, and that the density of liquid water is 1000 kg $m^{-3}$, reproduce this result.
2. Consider the following profile of potential temperature as a function of height, taken from a GFS model forecast for Greensboro, NC, valid 1800 UTC 21 Aug 2008.
   i. On a blank skew $T$–log$p$ diagram, such as the one provided, sketch the approximate corresponding **temperature** profile.
   ii. For this profile, what is the approximate pressure at the mixed layer top?
   iii. For the same profile, what is the approximate pressure at the tropopause?
   iv. Explain your reasoning for (ii) and (iii).
3. Suppose you are interested in an arbitrary quantity, $\mu$, given by the expression below. This expression has no

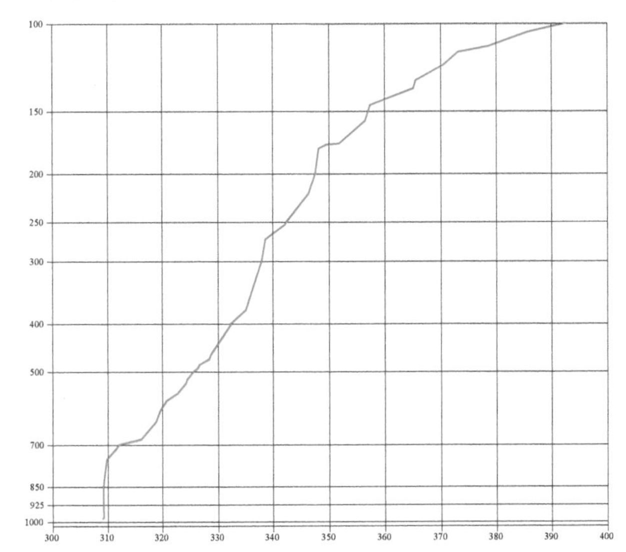


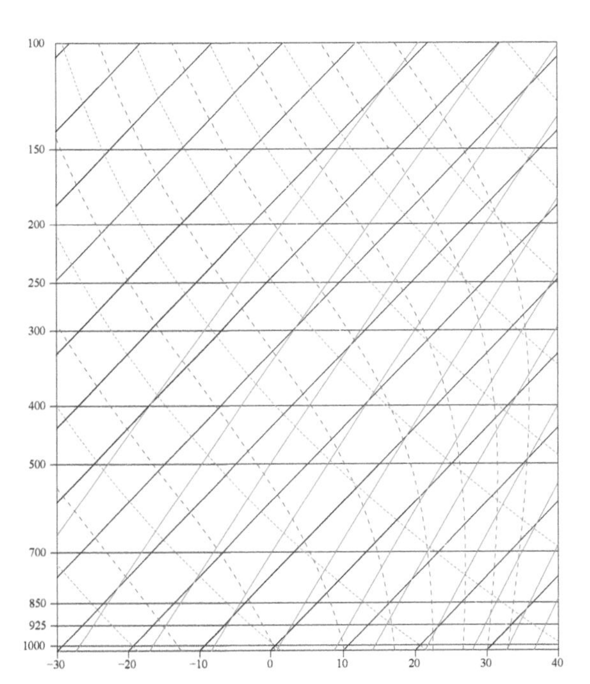


**Figure for Problem 2**

special meteorological relevance; however, it will provide practice with units and scale analysis,

$$\mu = g\,\frac{\partial v}{\partial x}\left(\frac{\partial u}{\partial y}\right).$$

i. What are the SI units (in meters, kilograms, seconds) of $\mu$?

ii. Use typical synoptic-scale, midlatitude values to determine an order-of-magnitude value for this quantity. Hint: the information in Table 1.3 will be useful.

4. Compute the *divergence* of equation (1.13), the geostrophic wind relation. Would the answer change if we replaced the variable $f$ with a constant $f_0$? This assumption is known as the $f$-plane assumption, and it will be used in defining the geostrophic wind in QG theory in chapter 2.
5. Beginning with equations (1.5) and (1.6), derive the vorticity equation using height ($z$) as the vertical coordinate. Hint: first expand the total derivatives on the left side of these equations.
6. Derive the Rossby wave phase-speed equation (1.56).
7. The plot below shows a 120-h forecast of 500-mb height (blue contours) with SLP (red contours) superimposed, valid 0600 UTC 11 Sep 2007. For each of the points indicated (A–D), determine whether the vertical geostrophic wind profile is **veering, backing, or weak** and indicate the corresponding sense (**warm, cold, or weak**) of the temperature advection (see Fig. for Problem 7).
8. Use the SLP forecast below (valid 1200 UTC 7 Sep 2007) to ***evaluate the sign*** of the *geostrophic relative vorticity* $\zeta_g = \partial v_g/\partial x - \partial u_g/\partial y$ at point A. Is the feature in the sea level isobars near this location a trough or a ridge? Explain (see Fig. for Problem 8).
9. Examine the cross section shown, derived from a GFS forecast valid 18 UTC 9 Oct 2009, depicting potential temperature in an east–west-oriented section along 50°N (the latitude/longitude values of the endpoints are shown at the bottom corners of the plot). The thermal wind relation for the meridional wind component can be written (proportionally) as $\frac{\partial v_g}{\partial z} \propto \frac{\partial \theta}{\partial x}$.

   i. Assuming that the near-surface geostrophic winds are light, use the thermal wind relation to sketch approximate ***section-normal isotachs*** ("section normal" means the component of the wind blowing into or out of the page). Draw negative values (if any) as thick dashed lines and positive values (if any) as solid lines. Assume that the absolute value of the strongest winds is ~125 kt and sketch approximate isotachs for positive and/or negative values of 50 and 100 kt.

   ii. Briefly explain and describe *what kind of weather system* is represented in this figure. In other words,

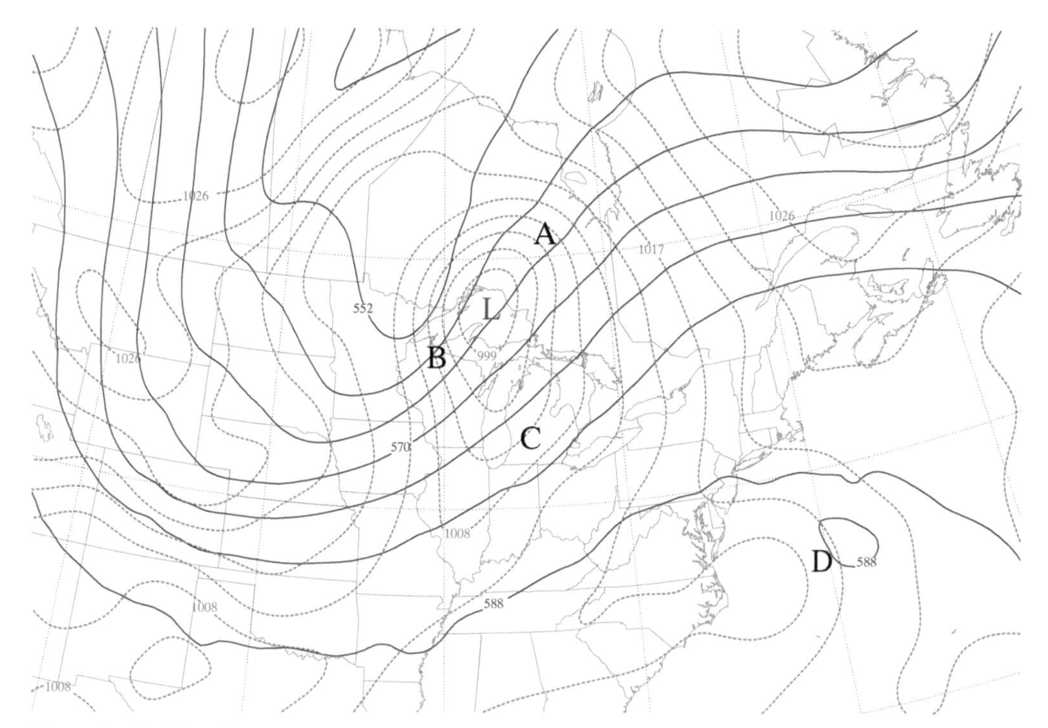


**Figure for Problem 7**

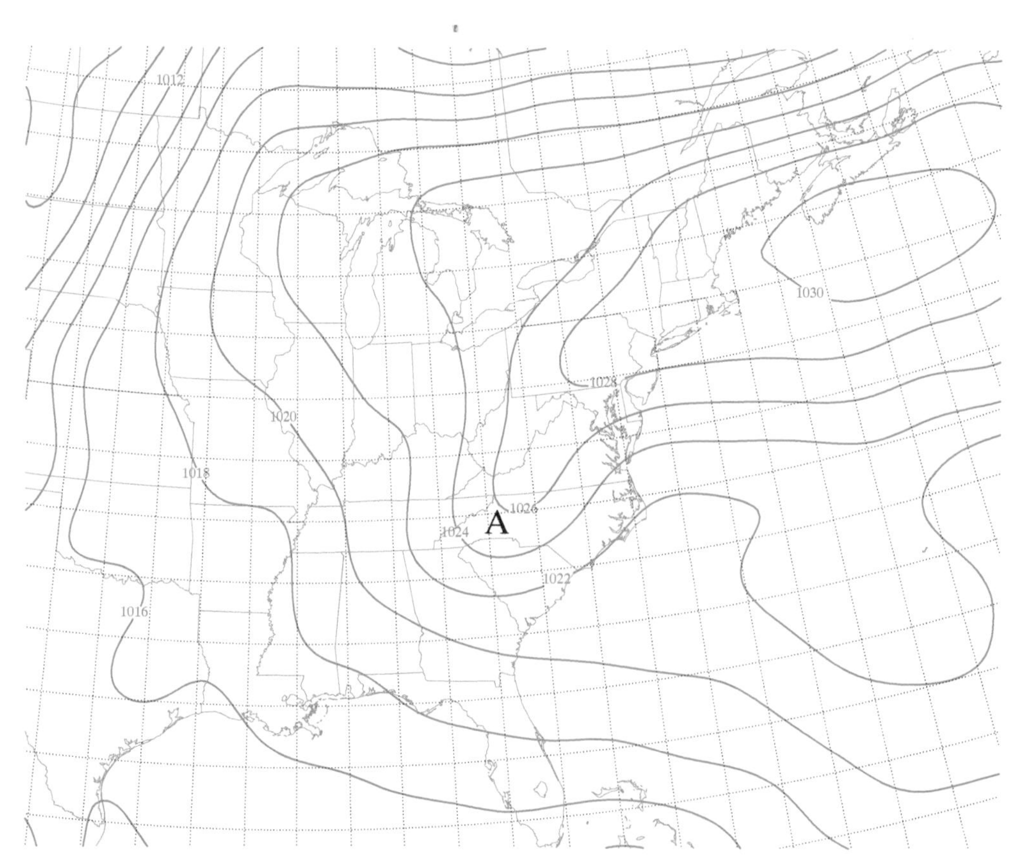


**Figure for Problem 8**

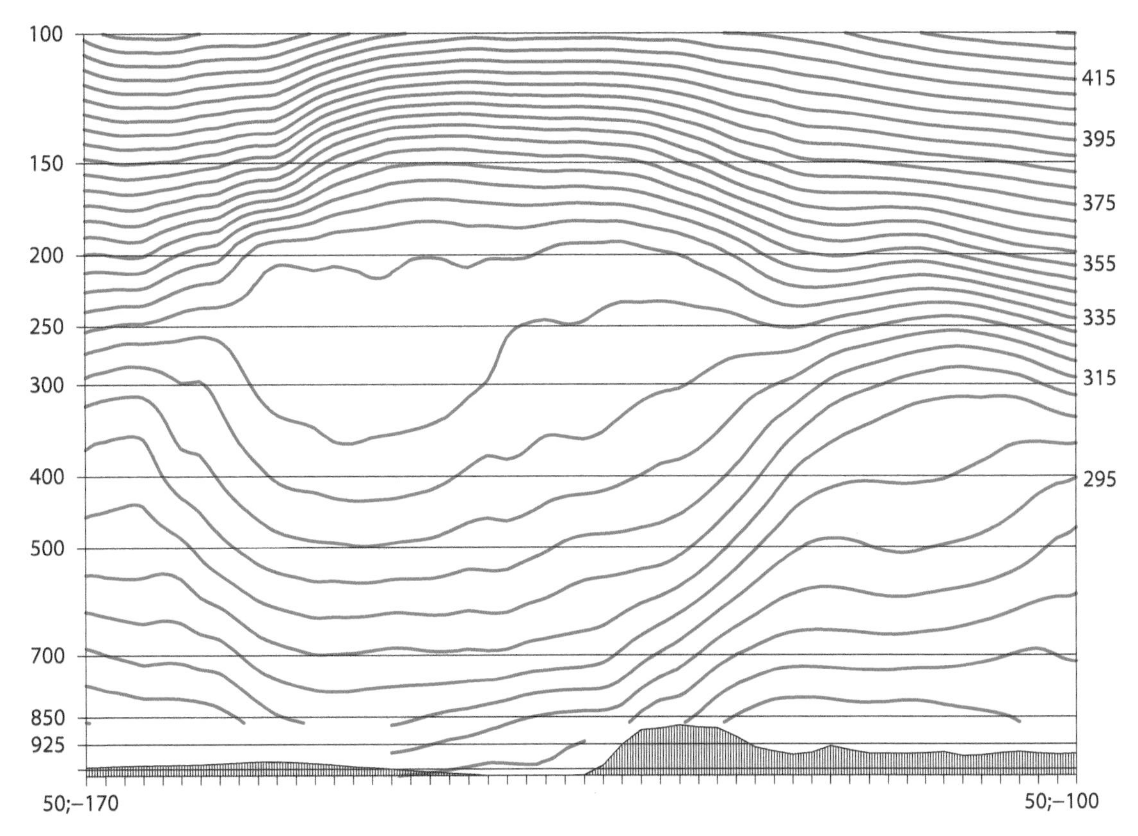


**Figure for Problem 9**

what feature or features does this cross section intersect that would be clearly evident on an upper-level isobaric map?

## REFERENCES


Bolton, D., 1980: The computation of equivalent potential temperature. *Mon. Wea. Rev.*, **108,** 1046–1053.

Charney, J. G., 1948: On the scale of atmospheric motions. *Geofys. Publ.*, **17,** 3–17.

Dutton, J. A., 1986: *The Ceaseless Wind: An Introduction to the Theory of Atmospheric Motion.* Dover, 617 pp.

Hovmöller, E., 1949: The trough-and-ridge diagram. *Tellus,* **1,** 62–66.

Lackmann, G. M., and R. M. Yablonsky, 2004: The importance of the precipitation mass sink in tropical cyclones and other heavily precipitating systems. *J. Atmos. Sci.*, **61,** 1674–1692.

Richardson, L. F., 1922: *Weather Prediction by Numerical Process.* Cambridge University Press, 1st ed.

Rossby, C.-G., 1934: Comments on meteorological research. *J. Aeronaut. Sci.*, **1,** 32–34.

Sutcliffe, R. C., 1947: A contribution to the problem of development. *Quart. J. Roy. Meteor. Soc.*, **73,** 370–383.


## FURTHER READING


Bluestein, H. B., 1992: *Principles of Kinematics and Dynamics.* Vol. 1, *Synoptic-Dynamic Meteorology in Midlatitudes,* Oxford University Press, 448 pp.

——, 1993: *Observations and Theory of Weather Systems.* Vol. 2, *Synoptic-Dynamic Meteorology in Midlatitudes,* Oxford University Press, 608 pp.

Charney, J. G., 1947: The dynamics of long waves in a baroclinic westerly current. *J. Meteor.*, **4,** 135–162.

Eady, E. T., 1949: Long waves and cyclone waves. *Tellus,* **1,** 33–52.

Holton, J. R., 2004: *An Introduction to Dynamic Meteorology.* Academic Press, 535 pp.

Martin, J. E., 2006: *Mid-Latitude Atmospheric Dynamics: A First Course.* Wiley, 324 pp.

Rogers, R. R., and M. K. Yau, 1989: *A Short Course in Cloud Physics.* 3rd ed. Pergamon Press, 293 pp.

Rossby, C.-G., 1945: On the propagation of frequencies and energy in certain types of oceanic and atmospheric waves. *J. Meteor.*, **2,** 187–204.

Wallace, J. M., and P. V. Hobbs, 2006: *Atmospheric Science: An Introductory Survey.* 2nd ed. Academic Press, 483 pp.

# CHAPTER 2

# Quasigeostrophic Theory

*Every day the operational meteorologist is asked to absorb a tremendous amount of weather data and make a forecast. The National Meteorological Center makes this job easier by providing a number of products, such as MOS (Model Output Statistics) guidance, which can be used to formulate the forecast. However, "good" forecasters do not rely exclusively on computer generated guidance when making a forecast. They examine the developing weather patterns and attempt to understand the factors which are likely to be responsible for the current and next day's weather.*

—D. R. Durran and L. W. Snellman, "The Diagnosis of Synoptic-Scale Vertical Motion in an Operational Environment" (1987)

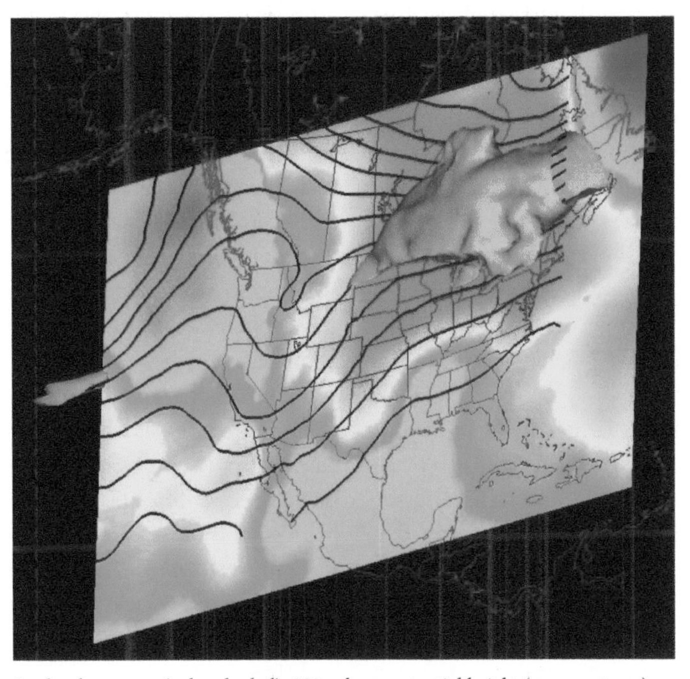

Sea level pressure (color shaded), 250-mb geopotential height (gray contours), and 60 m $s^{-1}$ isotach (shaded isosurface) from North American Mesoscale (NAM) model analysis valid 0000 UTC 16 Feb 2003.

The combination of practical applications and insights provided by the quasigeostrophic (QG) framework makes it a cornerstone of synoptic meteorology and forecasting. Rather than focus on the derivation of the QG equations, our emphasis here is upon the provision of examples and applications relevant to weather analysis and forecasting.

## 2.1. OVERVIEW

The governing equations that describe the behavior of the atmosphere (conservation of mass, momentum, and energy, along with other relationships discussed in chapter 1) are quite complex. Even when written in Cartesian form and with simplifying assumptions such as the hydrostatic approximation, it remains difficult to *conceptualize* the dynamical essence of synoptic-scale weather systems using the full set of equations. However, early pioneers in meteorology, including Reginald Sutcliffe, Carl-Gustaf Rossby, Jule Charney, and Arnt Eliassen, recognized that the equations could be greatly simplified by utilizing the observation that the flow in synoptic-scale weather systems is approximately geostrophic. Charney (1948) presents one of the earliest QG derivations; the motive of this work was, in part, to provide a useful set of equations for early numerical weather prediction efforts.

By using a carefully designed set of assumptions, the governing equations can be simplified and combined in ways that retain the fundamental dynamics of weather systems and yet are simple enough to comprehend. The simplified QG framework provides many tools for dynamic analysis of the atmosphere while suggesting specific applications for weather forecasting. This system allows us to understand and diagnose the processes leading to vertical air motion and weather system development or decay, along with explaining *why* weather takes place.

Although our aim is not to repeat the derivation of the QG equations, to understand the underlying assumptions we must revisit the derivations in sufficient depth to allow an appreciation for the limitations of this approach. The QG equations can be derived via an *asymptotic expansion* of the nondimensionalized governing equations in the Rossby number (e.g., Pedlosky 1979). Recall from chapter 1 that the smallness of the Rossby number is a measure of geostrophy, and that this dimensionless number was formed as the ratio of the acceleration to the Coriolis term in the horizontal momentum equations. An alternate approach is to utilize a scale analysis of the governing equations (e.g., Charney 1948; Holton 2004). Here, we simplify and combine the governing equations as presented in chapter 1 and reduce the system to two dependent variables that are useful in weather analysis and forecasting: the vertical velocity $\omega$, which is closely related to the formation of clouds and precipitation, and the geopotential tendency $\chi$, which is related to the development and movement of weather systems.

In this and subsequent chapters, we devote attention to the synoptic phenomena that are most important to midlatitude weather and forecasting, including extratropical cyclones, their attendant frontal systems, upper-level waves and vortices, and jet streaks. For example, the processes that lead to the formation, movement, and structural evolution of extratropical cyclones of the type pictured in Fig. 2.1 must be understood if we wish to predict day-to-day weather in the midlatitudes, or if we wish to diagnose the role that these systems play in the earth's climate system (see chapter 5).

## 2.2. THE BASIC EQUATIONS AND QG APPROXIMATIONS

The fundamental assumption underlying the QG equations is that the Rossby number $[R_o = U/(fL)]$ is small (on the order of 0.1). After partitioning the full velocity into geostrophic and ageostrophic components, the validity of this assumption allows us to neglect the ageostrophic velocity in some (but not all) terms. This does not mean that the flow must be exactly geostrophic for the equations to be valid. In fact, we will see that it is precisely the ageostrophic motions in the QG system of equations that give rise to much of what is considered "weather."

Several simplifications will be made in the following development; for example, our treatment of the synoptic-scale atmosphere can be limited to the portion of the troposphere that lies above the planetary boundary layer, allowing a further assumption of *frictionless* flow. This means that phenomena, such as frictional convergence (Ekman pumping), which can lead to significant vertical air motions in some circumstances, will not be explained by these equations. If we later wished to include frictional processes, then they can easily be included in the QG system; however, our goal is to simplify the system as much as possible while retaining the essential features of the midlatitude weather systems of interest; later, additional assumptions will be introduced, including *adiabatic* flow.

**Figure 2.1.** *Geostationary Operational Environmental Satellite-8 (GOES-11)* visible satellite image, 2200 UTC 8 Oct 2007.

The frictionless, horizontal momentum equations with pressure as the vertical coordinate can be written in a form similar to that shown in chapter 1, Eqs. (1.5)–(1.7), but for the omission of the frictional terms and the appearance of the gradient of geopotential in place of the pressure gradient. Expanding the total derivative, the scalar form of these equations is

$$\frac{\partial u}{\partial t} = -u\frac{\partial u}{\partial x} - v\frac{\partial u}{\partial y} - \omega\frac{\partial u}{\partial p} - \frac{\partial \Phi}{\partial x} + fv \text{ and} \quad (2.1)$$

$$\frac{\partial v}{\partial t} = -u\frac{\partial v}{\partial x} - v\frac{\partial v}{\partial y} - \omega\frac{\partial v}{\partial p} - \frac{\partial \Phi}{\partial y} - fu, \quad (2.2)$$

and the vector form is given by

$$\frac{\partial \vec{V}}{\partial t} = -\vec{V}\cdot\nabla_h\vec{V} - \omega\frac{\partial \vec{V}}{\partial p} - \nabla_h\Phi - f\hat{k}\times\vec{V}. \quad (2.3)$$

In (2.3), we have separated the horizontal and vertical advection to more clearly apply the QG assumption. The relation between the isobaric vertical motion used here, $\omega \equiv dp/dt$, to the height coordinate form, $w \equiv dz/dt$, is obtained by utilizing the chain rule of calculus and the hydrostatic equation

$$\omega = \frac{dp}{dt} = \frac{dp}{dz}\frac{dz}{dt} \approx -\rho g w. \quad (2.4)$$

If the synoptic-scale vertical motion is typically $10^{-2}$ m s$^{-1}$ (see Table 1.3), then the isobaric vertical motion is typically of order 0.1 Pa s$^{-1}$. Note that it is possible to have nonzero $\omega$ when $w$ is zero (e.g., with calm winds but locally falling pressure); however, to a good approximation, we can ignore these situations. Further, because pressure change is directly related to air parcel expansion and compression, $\omega$ is fundamentally important to adiabatic temperature change, along with the formation of cloud and precipitation; refer to section 1.2 for other advantages of isobaric coordinates.

Consider the advection of the horizontal velocity field by the horizontal wind, $-\vec{V}\cdot\nabla_h\vec{V}$ in (2.3). If we partition the horizontal wind into the geostrophic and ageostrophic components

$$\vec{V} = \vec{V}_g + \vec{V}_{ag}, \quad (2.5)$$

where the subscripts $g$ and $ag$ denote geostrophic and ageostrophic winds, respectively, then we can expand this advective term into four parts: (i) $-\vec{V}_{ag}\cdot\nabla\vec{V}_g$, (ii) $-\vec{V}_{ag}\cdot\nabla\vec{V}_{ag}$, (iii) $-\vec{V}_g\cdot\nabla\vec{V}_{ag}$, and (iv) $-\vec{V}_g\cdot\nabla\vec{V}_g$. We can again utilize the synoptic-scale values listed in Table 1.3 to perform a scale analysis of these terms, but now the velocity scale is replaced by the *geostrophic* velocity scale (10 m s$^{-1}$ as before) and a smaller *ageostrophic* velocity scale (1 m s$^{-1}$). Using these values, term (iv) is ~10$^{-4}$ m s$^{-2}$, an order of magnitude larger than terms (i) and (iii), and two orders of magnitude larger than term (ii). Furthermore, the vertical advection, using the synoptic-scale values listed in Table 1.3 and (2.4), is ~10$^{-6}$ m s$^{-2}$, justifying its neglect relative to the geostrophic advection. The QG assumptions amount to neglect of advection *by* the ageostrophic wind components and also advection *of* the ageostrophic wind components.

If we select a meridional location $y = 0$, with corresponding latitude $\varphi = \varphi_0$ about which to center our domain of interest, say, 45°N/S, then we can express the Coriolis parameter as a Taylor series expansion about this latitude:

$$f=f_0+\frac{\partial f}{\partial y}y+\frac{\partial^2 f}{\partial y^2}\frac{y^2}{2}+\cdots. \tag{2.6}$$

Scaling for synoptic-scale values demonstrates that $f_0$ (10$^{-4}$ s$^{-1}$) is an order of magnitude larger than $\partial f/\partial y(y)=\beta y$ (10$^{-5}$ s$^{-1}$). Thus, $f_0$ replaces $f$ in the geostrophic wind relation

$$\vec{V}_g=\frac{1}{f_0}\hat{k}\times\nabla\Phi. \tag{2.7}$$

Notation for the total derivative that applies to the QG equations, in which only the geostrophic advection is included, is expressed as

$$\frac{d}{dt_g}=\frac{\partial}{\partial t}+\vec{V}_g\cdot\nabla_h, \tag{2.8}$$

so that the QG momentum equation can be written as

$$\frac{d\vec{V}_g}{dt_g}=-\nabla\Phi-(f_0+\beta y)\hat{k}\times\vec{V}. \tag{2.9}$$

It is straightforward to demonstrate from (2.7) that

$$f_0\hat{k}\times\vec{V}_g=-\nabla\Phi. \tag{2.10}$$

Substituting (2.10) and (2.5) into (2.9), we obtain

$$\frac{d\vec{V}_g}{dt_g}=f_0(\hat{k}\times\vec{V}_g)-(f_0+\beta y)\hat{k}\times(\vec{V}_g+\vec{V}_{ag}); \tag{2.11}$$

expanding the right-hand term, we find that $f_0(\hat{k}\times\vec{V}_g)$ cancels on the right side, leaving

$$\frac{d\vec{V}_g}{dt_g}=-f_0\hat{k}\times\vec{V}_{ag}-\beta y\hat{k}\times\vec{V}_g-\beta y\hat{k}\times\vec{V}_{ag}. \tag{2.12}$$

Via the assumption of a small Rossby number, the third right-side term in (2.12) is an order of magnitude smaller than the second; therefore, the final form of the QG momentum equation is

$$\frac{d\vec{V}_g}{dt_g}=-f_0\hat{k}\times\vec{V}_{ag}-\beta y\hat{k}\times\vec{V}_g. \tag{2.13}$$

The scalar components of (2.13) are

$$\frac{du_g}{dt_g}=f_0v_{ag}+\beta yv_g \quad\text{and} \tag{2.14}$$

$$\frac{dv_g}{dt_g}=-f_0u_{ag}-\beta yu_g. \tag{2.15}$$

Physically, we can understand the first right-hand terms in (2.14) and (2.15) as the Coriolis torque acting at right angles to the ageostrophic wind, leading to the acceleration of the wind components perpendicular to the ageostrophic motions.

Other equations to be incorporated into the QG set include the hydrostatic equation (1.16), the equation of state or ideal gas law (1.4), which can be written as

$$\frac{\partial\Phi}{\partial p}=-\alpha=-\frac{RT}{p}, \tag{2.16}$$

and the continuity equation (neglecting the source/sink terms, consistent with an assumption of adiabatic flow), which can be expressed in the following ways:

$$\frac{\partial u}{\partial x}+\frac{\partial v}{\partial y}+\frac{\partial\omega}{\partial p}=0,\quad \nabla_h\cdot\vec{V}+\frac{\partial\omega}{\partial p}=0,\ \text{or}$$

$$\nabla_h\cdot\vec{V}_{ag}+\frac{\partial\omega}{\partial p}=0,\quad\text{because}\quad \nabla_h\cdot\vec{V}_g=0. \tag{2.17}$$

The QG form of the thermodynamic energy equation is

$$\frac{dT}{dt_g}=\frac{\sigma p}{R}\omega+\frac{J}{C_p}, \tag{2.18}$$

where $\sigma=-(RT/p)(d\ln\theta/dp)$ and $J$ is the diabatic heating rate (per unit mass) owing to processes such as radiation or phase changes of water. However, as stated earlier, the goal here is to maximize simplification initially; therefore, we assume adiabatic flow and neglect the final term in (2.18). The adiabatic assumption can be relaxed, and we will do so later in this chapter and also when considering the cyclone problem in chapter 5, given that latent heat release can exert important influences on cyclone dynamics. The manipulation and interpretation of the QG equations is further simplified by assuming that the stability parameter $\sigma$ is a function only of pressure, although this assumption is questionable in regions of horizontal stability gradients, such as may be found along coastlines with strong horizontal temperature gradients.

At this point, we have introduced a set of fairly restrictive assumptions, and we have applied them to the governing equations in isobaric coordinates. Summarizing these assumptions, they include, in approximate order from most restrictive to least, the following:

1. small Rossby number (the ageostrophic flow is assumed to be <~10% of the geostrophic flow),
2. adiabatic, frictionless flow (either of which can be relaxed),
3. horizontally uniform static stability ($\sigma$), and
4. hydrostatic balance.

Assumption (1) is ranked as the most restrictive because of the presence of a significant ageostrophic component in certain situations, for example, in the presence of highly curved flow, such as in an intense upper trough or midlatitude cyclone. Certainly, *tropical* storms are characterized by a highly ageostrophic flow because of the centrifugal force, which is an acceleration to the flow. Even for balanced flow, gradient wind balance in these situations gives rise to a large Rossby number. This situation is illustrated in Fig. 2.2 for Hurricane Ike in the Gulf of Mexico in September 2008. Because of the strong, highly curved flow, the geostrophic wind is a preposterous representation of the analyzed wind! As a result, the ageostrophic wind takes the form of an intense *anti*cyclone—for cyclonically curved flow, the actual wind is subgeostrophic, meaning that the ageostrophic wind will be oriented opposite to the geostrophic flow. Although this example illustrates the absurdity of using QG assumptions in the analysis of strong tropical cyclones, it may be acceptable to use QG reasoning in the diagnosis of interactions between tropical cyclones and the larger-scale environment, or after tropical systems have made landfall and weakened.

The QG thermodynamic energy equation, along with the hydrostatic equation, the geostrophic wind definition, the QG momentum equations, and the continuity equation, form a *closed set* of equations with the dependent variables being $\Phi$, $T$, $V_g$, $V_{ag}$, and $\omega$. Given this closed set, this system of equations can be solved for each dependent variable. Recall that our goal is to simplify the equation set and eventually solve to isolate processes that affect vertical air motion, and the geopotential tendency.

In chapter 1, we found that by forming an equation for the time tendency of the vertical component of vorticity, we were able to combine the momentum equations in a way that provided conceptual insight into processes leading to the development or decay of weather systems. We will follow an analogous procedure here but extend the combination to also include thermodynamic information (combining the vorticity and thermodynamic equations). In doing so, we will first obtain a useful diagnostic equation for the vertical motion, $\omega$, and then we form a *prognostic equation* for the geopotential height. In either case, our first task is to develop the QG form of the vorticity equation.

To derive the QG vorticity equation, we proceed as in chapter 1, taking the curl of the horizontal QG momentum equations [$\partial/\partial x$ of (2.15), the $y$ momentum equation; $-\partial/\partial y$ of (2.14), the $x$ momentum equation]. This exercise, left as a problem at chapter's end, is set up below in (2.19) and (2.20), with the resulting QG vorticity equation being (2.21):

$$\frac{\partial}{\partial x}\left[\frac{\partial v_g}{\partial t}+u_g\frac{\partial v_g}{\partial x}+v_g\frac{\partial v_g}{\partial y}+f_0u_{ag}+\beta y u_g\right]=0,\text{ and} \tag{2.19}$$

$$-\left\{\frac{\partial}{\partial y}\left[\frac{\partial u_g}{\partial t}+u_g\frac{\partial u_g}{\partial x}+v_g\frac{\partial u_g}{\partial y}-f_0v_{ag}-\beta y v_g\right]=0\right\} \tag{2.20}$$

$$\frac{\partial \zeta_g}{\partial t}=-\vec{V}_g\cdot\nabla(\zeta_g+f)+f_0\frac{\partial\omega}{\partial p}. \tag{2.21}$$

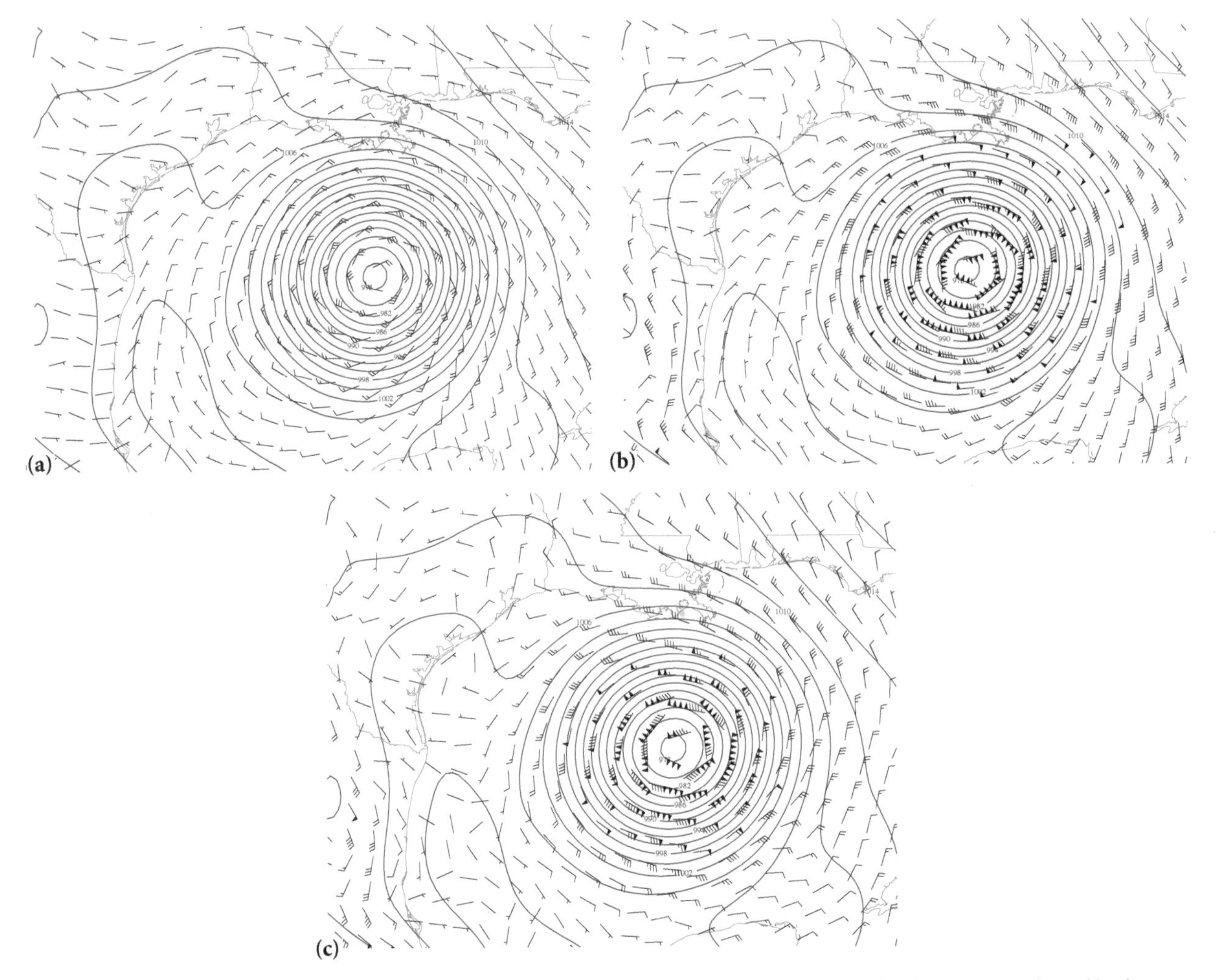

**Figure 2.2.** GFS forecast of Hurricane Ike valid 1800 UTC 11 Sep 2008. Solid contours are sea level pressure, and wind barbs are (a) 10-m winds, (b) geostrophic winds, and (c) ageostrophic winds.

The geostrophic relative vorticity $\zeta_g$ can be expressed as

$$\zeta_g = \frac{\partial v_g}{\partial x} - \frac{\partial u_g}{\partial y} = \frac{1}{f_0}\nabla^2\Phi = \frac{g_0}{f_0}\nabla^2 Z. \quad (2.22)$$

In (2.21), the $\beta v_g$ term is implicitly included in the advection because $\nabla f = \beta\hat{j}$. In keeping with the fundamental QG assumptions, there is no vertical advection, and this is evident in the QG vorticity equation (2.21) as well. This equation states that the local time rate of change of geostrophic relative vorticity is determined by the geostrophic advection of geostrophic relative and planetary vorticity, along with vortex stretching.

It is instructive to compare the QG vorticity equation (2.21) to the "full" vorticity equation (1.49), which is repeated below as (2.23) with slight notational changes to facilitate comparison:

$$\frac{\partial \zeta}{\partial t} = -\vec{V}\cdot\nabla(\zeta + f) - \omega\frac{\partial \zeta_a}{\partial p} - \left[\frac{\partial \omega}{\partial x}\frac{\partial v}{\partial p} - \frac{\partial \omega}{\partial y}\frac{\partial u}{\partial p}\right] + (\zeta + f)\frac{\partial \omega}{\partial p} + \left[\frac{\partial F_y}{\partial x} - \frac{\partial F_x}{\partial y}\right]. \quad (2.23)$$

The vertical advection, tilting, and frictional terms from (2.23) are absent in (2.21), consistent with the QG assumptions. The horizontal advection is now accomplished only by the geostrophic wind. The stretching term is present in both equations; however, the QG stretching term contains a simplification relative to that in (2.23), in that $f_0$ alone multiplies $\partial\omega/\partial p$ rather than $(\zeta+f)$ in the full vorticity equation. This is an important distinction. In the full equation, the magnitude of the stretching term is proportional to the vorticity itself, meaning that zones of large preexisting vorticity are sites for preferential vorticity growth, and also that an exponential feedback takes place. In the QG version, this effect is absent. We will return to this point when discussing cyclogenesis in chapter 5.

## 2.3. THE QG OMEGA EQUATION

In deriving the QG vorticity equation (2.21), we have a compact and conceptually useful equation with which we can understand synoptic-scale vorticity growth and decay. However, we wish to combine the governing equations in a more complete fashion to solve for processes leading to vertical motions or pressure and height tendencies. The first law of thermodynamics provides information relating to both the vertical motion and to the height tendency, and it can be combined with the vorticity equation to form a closed system in omega and geopotential tendency. The adiabatic form of the QG thermodynamic equation (2.18),

$$\left(\frac{\partial}{\partial t}+\vec{V}_g\cdot\nabla\right)T=\frac{\sigma p}{R}\omega, \tag{2.24}$$

can be modified using (2.16) to replace the temperature by the vertical derivative of the geopotential, which differs from the thickness only by the constant factor $g_0$. Dividing by $-p/R$, we obtain

$$\frac{\partial}{\partial t}\left(\frac{\partial\Phi}{\partial p}\right)+\vec{V}_g\cdot\nabla\left(\frac{\partial\Phi}{\partial p}\right)+\sigma\omega=0. \tag{2.25}$$

Next, we introduce $\chi$, which is exactly proportional to the local time tendency of the geopotential height $Z$:

$$\chi\equiv\frac{\partial\Phi}{\partial t}. \tag{2.26}$$

We will use the terms *geopotential tendency* and *height tendency* interchangeably. Substituting (2.26) in (2.25) and (2.21), and using (2.22) we have

$$\frac{\partial\chi}{\partial p}=-\vec{V}_g\cdot\nabla\left(\frac{\partial\Phi}{\partial p}\right)-\sigma\omega \quad \text{and} \tag{2.27}$$

$$\nabla^2\chi=-f_0\vec{V}_g\cdot\nabla\left(\frac{1}{f_0}\nabla^2\Phi+f\right)+f_0^2\frac{\partial\omega}{\partial p}. \tag{2.28}$$

Note that (2.27) and (2.28) are a system with two equations (the QG thermodynamic and vorticity equations) and two unknowns of interest, $\omega$ and $\chi$. We can now easily solve this system for either the height tendency or the vertical velocity. To gain insight into the processes that lead to the *time evolution* of a weather system, we can solve for $\chi$. In terms of weather forecasting, we benefit from under-standing the processes that lead to rising or sinking air motions by solving for $\omega$. The resulting "QG omega equation" provides a conceptual framework for understanding the fundamental causes of weather.

First, we will solve for omega by eliminating $\chi$ between these equations, which requires that we take $\nabla^2(2.27)-\partial/\partial p\ (2.28)$ and divide by $\sigma$:

$$\underbrace{\left(\nabla^2+\frac{f_0^2}{\sigma}\frac{\partial^2}{\partial p^2}\right)\omega}_{A}=\underbrace{\frac{f_0}{\sigma}\frac{\partial}{\partial p}\left[\vec{V}_g\cdot\nabla\left(\frac{1}{f_0}\nabla^2\Phi+f\right)\right]}_{B}+\underbrace{\frac{1}{\sigma}\nabla^2\left[\vec{V}_g\cdot\nabla\left(-\frac{\partial\Phi}{\partial p}\right)\right]}_{C}. \tag{2.29}$$

Although (2.29) may appear daunting at first glance, its interpretation is fairly straightforward. Term A, the left side of the equation, is essentially a 3D Laplacian acting on $\omega$. For sinusoidal patterns, this can be approximated as $-\omega$, with recognition that this is yet another simplifying assumption and not necessarily a valid one for many situations. To the extent that upper-level waves are sinusoidal, the second derivative of sine or cosine is simply $-$sine or $-$cosine. Given that omega is negative for upward motion, conveniently, $-\omega$ has the same sign as the height coordinate vertical velocity $w$.

Term B involves the vertical derivative of the absolute geostrophic vorticity advection by the geostrophic wind. This term is often referred to as the "differential vorticity advection" term. For cyclonic vorticity advection increasing with height, this term is positive, indicating forcing for ascent. Likewise, anticyclonic vorticity advection increasing with height represents forcing for descent, and other combinations are possible, for example, anticyclonic vorticity advection decreasing with height represents forcing for ascent.

Finally, term C is proportional to the Laplacian of the thickness advection and is often simply labeled the "thermal advection term." Local maxima of warm advection represent forcing for ascent, and local maxima of cold advection are associated with forcing for descent. The nature of the Laplacian operator affects the results; for example, a local minimum of warm advection also represents forcing for descent.

To actually solve for $\omega$, one must invert the operator on the left side, which is typically accomplished using a numerical procedure called "successive overrelaxation." This is a somewhat computationally intensive task, and it is not often done in operational forecasting environments. Rather than going through the trouble, it is common to interpret the forcing terms independently, with the understanding that there exists a somewhat loose connection between forcing and the actual QG vertical motion.

The QG omega equation states that forcing for QG vertical motion is proportional to differential vorticity advection and the Laplacian of temperature advection, where "advection" is by the geostrophic wind, and "vorticity" is the geostrophic absolute vorticity. The right-hand terms are best described as "forcing," and QG vertical motion as a response to this forcing. Forcing for ascent is associated with *cyclonic vorticity advection (CVA)* increasing with height and with local maxima of *warm advection (WA)*. Although less common, note that anticyclonic vorticity advection decreasing with height could also produce this result. Forcing for descent is associated with *anticyclonic vorticity advection (AVA)* increasing with height and local maxima of *cold advection (CA)*. Although less common, note that cyclonic vorticity advection decreasing with height could also produce this result.

In the Southern Hemisphere, the signs of the vorticity forcing terms are reversed, and so we have avoided the commonly used terms *positive vorticity advection* and *negative vorticity advection*, which are hemisphere dependent, in favor of the terms *cyclonic* and *anticyclonic* vorticity advection.

It must be borne in mind that QG forcing for a given sign of vertical motion does not guarantee that the sign of vertical motion will be observed; processes not included in the QG equations, such as orographic or frictional effects, can overpower the QG signal. In solving for omega, relaxation of the Laplacian operator smoothes and broadens the resulting omega field, providing a difference in the spatial structures of the forcing and response. Furthermore, rising air does not guarantee precipitation, or even clouds! These caveats lead to a list of additional considerations when using this equation in practical forecasting:

1. How intense is the forcing for ascent?
2. Do the two right-hand terms reinforce or cancel?
3. In forecasting clouds and precipitation, is there sufficient moisture available?
4. Are non-QG mechanisms at work that could obscure the QG signal?

Examples of the latter include land–sea breezes, orographic ascent or descent, or convection.

Several studies have demonstrated that the traditional form of the QG omega equation is not well suited for quantitative evaluation of forcing for ascent (e.g., Durran and Snellman 1987). But, our purpose here is not to obtain a *quantitative* estimate of omega, rather to arrive at a conceptual understanding of its cause, provided that we have a valid *qualitative* expression for the QG omega. A word of caution is needed: Often, the traditional form of the QG omega equation is used for operational weather forecasting. Provided that one remains cognizant of the underlying assumptions that went into its derivation, this is acceptable; however, there are other complications to be discussed below that suggest advantages for other methods in operational practice.

Despite complications in the use of the QG omega equation, we are now in a position to derive *physical understanding* of the causes for vertical air motions. Durran and Snellman (1987) offer a lucid physical interpretation of the QG vertical motion, and their example is followed here, utilizing an idealized situation in which the only forcing term active in (2.29) is term B, the differential vorticity advection.

Suppose that the 1000-mb geopotential height surface ($Z_{1000}$) equals 0 everywhere (i.e., the 1000-mb surface is flat). The 500-mb height contours indicate the presence of a westerly jet, with locally stronger wind speeds toward the east (Fig. 2.3). Isotachs indicate stronger wind speeds in the region of more closely spaced height contours.

Consider the local tendency of the zonal wind component at the 500-mb level at point A in Fig. 2.3. The advective tendency, $-u_g \partial u_g / \partial x$, is clearly negative in this location. Thus, the tendency for advection is to reduce the wind speed in the vicinity of the jet entrance region in this diagram. At the 1000-mb level, the wind remains

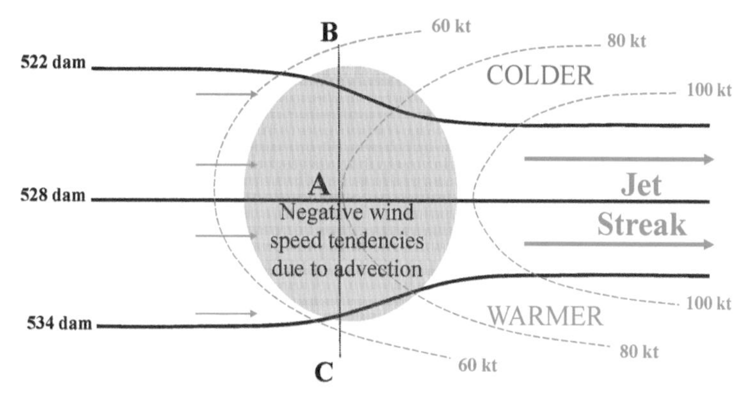


**Figure 2.3.** Idealized schematic of a jet entrance region in the Northern Hemisphere. Solid black lines represent 500-mb height and also 1000–500-mb thickness contours. Red arrows depict geostrophic wind vectors at select locations; red dashed lines represent isotachs. Dotted line B–C indicates the orientation of cross section in Fig. 2.4.

calm, and there is no advective tendency at point A. Next, consider the *geostrophic temperature (thickness) advection* at point A. In this example, the geostrophic thermal advection is exactly equal to zero, because the geostrophic wind in the layer is universally parallel to the thickness contours, which here are equivalent to geopotential height contours.

Finally, recall that thermal wind balance relates the strength of the geostrophic wind shear to the horizontal temperature gradient. At point A, where the geostrophic flow is purely zonal, we can utilize the scalar form of the thermal wind equation (1.44) for simplicity of interpretation: $u_{g\,U} - u_{g\,L} = -(C/f)\,\partial \overline{T}/\partial y$. Because of geostrophic advection in the jet entrance, the upper westerly flow is decreasing in strength while the tendency of the geostrophic flow at the bottom of the layer, the 1000-mb level, is zero. This means that the vertical wind shear is weakening with time. However, the geostrophic thermal advection is zero, as explained above. This state of affairs demonstrates that thermal wind balance is being disrupted in the jet entrance region—the vertical wind shear diminishes, yet the horizontal temperature gradient in the layer remains unchanged. Recalling that thermal wind balance is really a combination of geostrophic and hydrostatic balance, we see that even simple geostrophic flows can result in the disruption of balance. What are the consequences of this?

Given that the observed atmosphere is very close to a state of thermal wind (geostrophic) balance, which is in fact the observation upon which we have based the development of the QG system, something must be happening in nature to maintain a state of approximate balance. How might balance be restored in the particular situation described in Fig. 2.3 near point A? Clearly, either the shear must somehow be increased, or the meridional temperature gradient must decrease to mitigate the imbalance brought about by the geostrophic advections.

If the geostrophic advection in this example acts to disrupt thermal wind balance, then perhaps it is the ageostrophic circulation that serves to maintain it. In what sense must an ageostrophic circulation be configured to bring about the needed changes? Consider the north-south-oriented vertical cross section in Fig. 2.4, along the transect B–C indicated in Fig. 2.3. The perspective is from

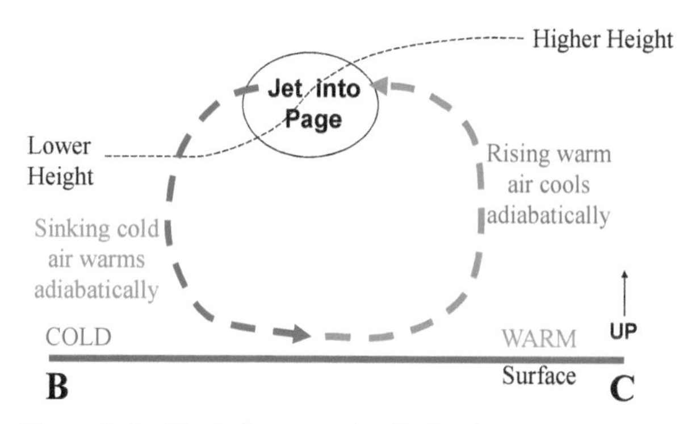


**Figure 2.4.** Vertical cross section B–C with orientation as indicated in Fig. 2.3. Dashed colored arrows indicate sense of ageostrophic circulation. Black dashed line represents a geopotential height contour, with exaggerated vertical displacement.

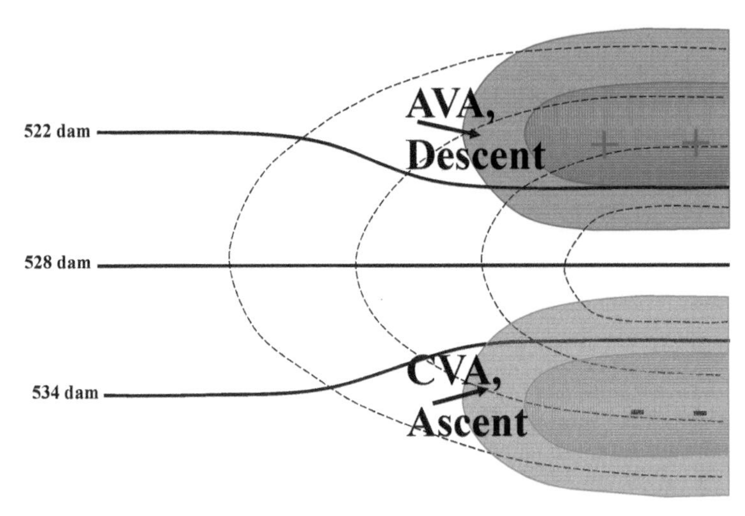


**Figure 2.5.** As in Fig. 2.3, but showing idealized geostrophic relative vorticity distribution in shaded colors, along with areas of significant differential vorticity advection and forcing for ascent and descent.

the west (i.e., from that of an observer standing with their back to the wind, looking eastward). If the north–south thermal gradient has become too strong for the weakening geostrophic shear, then it must weaken, requiring a cooling of the warm air and warming of the cool air. The mechanism to effectively accomplish these temperature changes is adiabatic expansion and compression, with warm air rising and becoming cooler and cold air sinking and becoming warmer (Fig. 2.4).

Is the vertical motion pattern shown in Fig. 2.4 consistent with the QG omega equation (2.29)? Given that thermal advection is exactly zero in this example, only term B, the differential vorticity advection, is at work as a forcing term. The geostrophic vorticity distribution that would accompany this situation is presented in Fig. 2.5. Regions of cyclonic and anticyclonic shear vorticity are found along the northern and southern flanks of the jet streak, respectively, as would be expected based on the discussion in section 1.5.1. In this example, the geostrophic vorticity advection is zero at the 1000-mb level, increasing in magnitude with height up to the 500-mb level. Therefore, whatever sign of vorticity advection is found at the 500-mb level is consistent with the sign of the *differential* vorticity advection in this layer. Thus, the region of ascent (descent) in the right (left) jet entrance region is consistent with forcing provided by cyclonic (anticyclonic) vorticity advection increasing with height.

The answer to the question of *why* differential vorticity advection leads to vertical motion is now clearer—the QG vertical motion is precisely that which acts in the sense needed to bring the atmosphere back toward a state of thermal wind balance. In the jet entrance, a thermally direct ageostrophic circulation develops, with warm air rising and cold air sinking. From the perspective of energy, a *thermally direct* circulation of this type leads to the conversion of potential energy to kinetic energy; this is also consistent with the expected acceleration of air parcels as they enter the core of the jet streak and gain kinetic energy.

In this example, the vertical shear becomes too weak for the thermal gradient because of the action of geostrophic advections; the adiabatic temperature changes accompanying the vertical motion therefore serve to weaken the horizontal temperature gradient. However, if the upper westerly flow could be accelerated or if lower-tropospheric easterly flow could be generated, then the geostrophic shear would increase, and this too would aid in thermal wind balance restoration. As air parcels enter the region of stronger height gradient aloft, the increasing northward-directed pressure gradient force leads to a northward acceleration. The resulting southerly ageostrophic flow is acted upon by the Coriolis force, accelerating the zonal ($u$) wind component, as described by the first right-hand term in (2.14).

At lower levels, the attendant mass convergence and divergence brought about by the upper ageostrophic flow lead to the development of higher pressure in the left entrance and lower pressure in the right entrance. This helps to bring about easterly geostrophic flow at low levels,

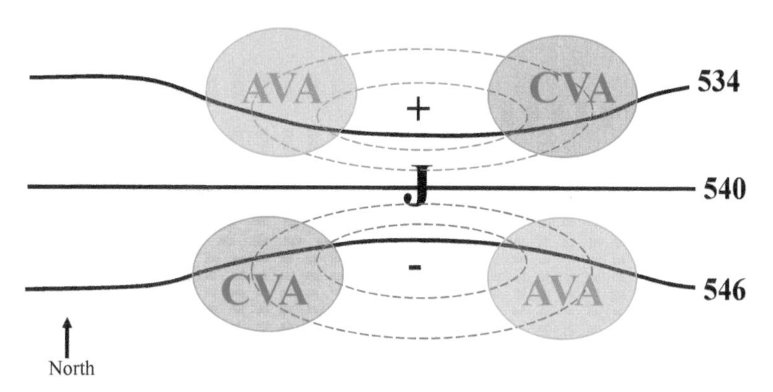


**Figure 2.6.** Idealized jet streak at the 500-mb level. Solid black lines are geopotential height contours, dashed red and blue lines represent relative vorticity isopleths, and the designations CVA and AVA indicate regions of cyclonic and anticyclonic vorticity advection, respectively.

further increasing the vertical shear of the geostrophic wind and also acting in the sense needed to restore thermal wind balance.

If we consider the situation in a jet *exit* region, as found on the right side of Fig. 2.6, then we expect that thermal wind balance disruption would take place in the opposite sense to that described in Figs. 2.3–2.5: geostrophic advection there would lead to increasing wind speeds aloft, leading to excess shear relative to the temperature gradient, and a *thermally indirect* circulation (warm air sinking and cold air rising) would be needed to bring the atmosphere back toward thermal wind balance.

The preceding interpretation is consistent with the right side of the QG omega equation representing "forcing" for ascent or decent. The following terms are thus related to thermal wind balance disruption: the right-hand terms consist of *geostrophic advections*, meaning that these terms describe thermal wind balance disruption by the primary, geostrophic flow. The left side of the equation, which describes the QG vertical motion, can be thus interpreted as a *secondary* circulation that arises in response to forcing from the primary circulation. The next time you observe rainy weather from a synoptic-scale weather system, do not let the gray skies dampen your spirits; you can take solace in the fact that the miserable weather is merely contributing what is needed to help restore thermal wind balance!

Turning now to real-data examples, while bearing in mind the limiting assumptions inherent in the QG approach, we can diagnose expected areas of ascent or descent associated with the differential vorticity advection or the thermal advection terms.

On 10 September 2008, a strong cyclonic vorticity maximum was centered over eastern Washington State, consistent with a pronounced trough in the 500-mb geopotential height field (Fig. 2.7a). Cyclonic winds were blowing through the vorticity center from northwest to southeast, leading us to expect cyclonic vorticity advection generally to the *east* of the trough axis, for example, at point C in Fig. 2.7a. At the center of the trough (point B), we expect the advection to be zero, with the contribution expected to become anticyclonic to the west of the trough axis, where the winds are advecting in smaller values of vorticity (e.g., point A). The computed *vorticity advection* (Fig. 2.7b) confirms our expectations regarding the locations of positive, negative, and zero advective vorticity tendencies relative to the trough axis. Inspection of the model-analyzed vertical motion (omega) field (Fig. 2.7c) reveals that the expected patterns of ascent and descent are generally consistent with the vorticity advection in this example, albeit somewhat noisy.

Several important comments and caveats are in order. First of all, we have not considered the thermal advection term, and thus our analysis of the right-side forcing terms in (2.29) is incomplete. Additionally, this is a region of complex terrain, and we are not considering topographic influences on the vertical motion. Furthermore, the QG omega equation demands that we consider the *differential* vorticity advection, not just that at the 500-mb level. And, the advection of the *geostrophic* vorticity by the *geostrophic* wind is what appears in (2.29), whereas Fig. 2.7 shows the full wind and vorticity fields. Despite all these caveats, in this example, the analyzed omega pattern shows excellent

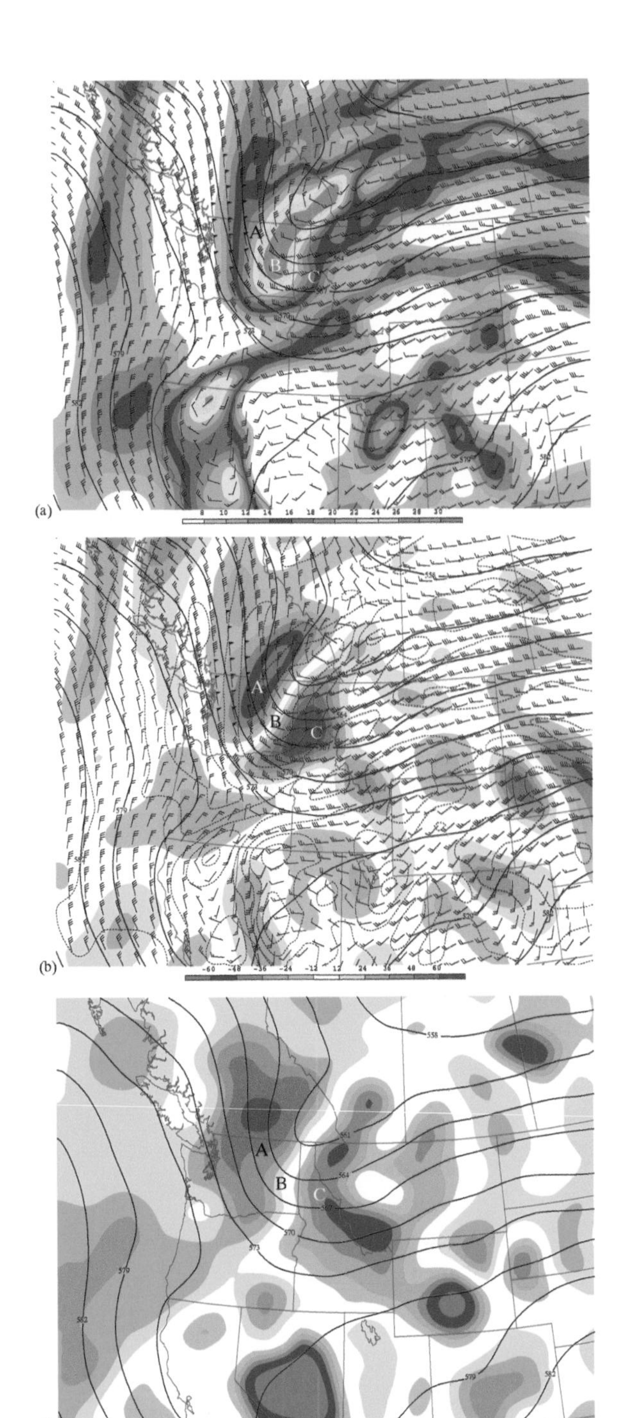

**Figure 2.7.** Analysis from the Rapid Update Cycle (RUC) model valid 0600 UTC 10 Sep 2008 for the U.S. Pacific Northwest: (a) 500-mb height (solid contours, interval is 6 dam), wind barbs, and absolute vorticity (shaded every $2\times10^{-5}\ s^{-1}$ beginning at $8\times10^{-5}\ s^{-1}$); (b) as in (a) except with 500-mb vorticity advection replacing vorticity (contoured every $12\times10^{-10}\ s^{-2}$); (c) as in (a) except with omega (shaded as in legend, interval is $2\times10^{-1}$ Pa $s^{-1}$) replacing vorticity.

qualitative agreement with the QG omega equation! This is consistent with the fact that the upper trough is strong and is characterized by a prominent vorticity center with a clear advection couplet bracketing it. Experience tells us that this is not an unusual example, but we must remain aware that we are taking some liberties in our interpretation and application.

By considering the 500-mb level in this example, we are looking sufficiently far aloft to avoid the strongest topographic forcing. Regarding the differential aspect of the vorticity advection, given that wind speeds and trough amplitudes typically strengthen with height up to (and above) the 500-mb level, the patterns of vorticity advection typically seen at this level are often consistent with the *differential* tendency as well, given the weaker flow (and advection) at lower altitudes. The geostrophic vorticity and flow are nearly always *qualitatively* consistent with that computed from the full wind, even if the former is often an overestimate in the base of a strong trough (because of the subgeostrophic flow found in regions of cyclonic flow curvature).

Now we will consider an example highlighting the thermal advection term in (2.29). Figure 2.8a indicates strong clockwise turning of the geostrophic wind with height over southern New York, eastern Pennsylvania, and New Jersey in a NAM forecast valid 1800 UTC 10 September 2009. Here, strong easterly geostrophic flow implied by the sea level isobars veers to southeasterly geostrophic flow, as indicated by the 500-mb height contours. Regions of backing geostrophic flow are found over Indiana and Ohio. The geostrophic veering and backing are related to the temperature advection, as discussed in section 1.4.2. The thermal advection obtained from inspection of the sea level isobars and 500-mb height field should be consistent with that within the surface-to-500-mb layer. By overlaying the 850-mb temperature advection (Fig. 2.8b), we do in fact observe a very high degree of consistency, both in regions of warm and cold advection.

The intense zone of warm advection is related to both a large angle of directional turning between the lower and upper geostrophic flow and the *magnitude* of the geostrophic flow at these two levels. In Fig. 2.8a, intense warm advection is present over eastern Pennsylvania. A model forecast sounding for Philadelphia, Pennsylvania, valid at this time, confirms the presence of a veering wind profile in the lower troposphere and indicates a deep layer of saturation, consistent with ascent (Fig. 2.8c). An inversion layer between the 850- and 925-mb levels in this sounding

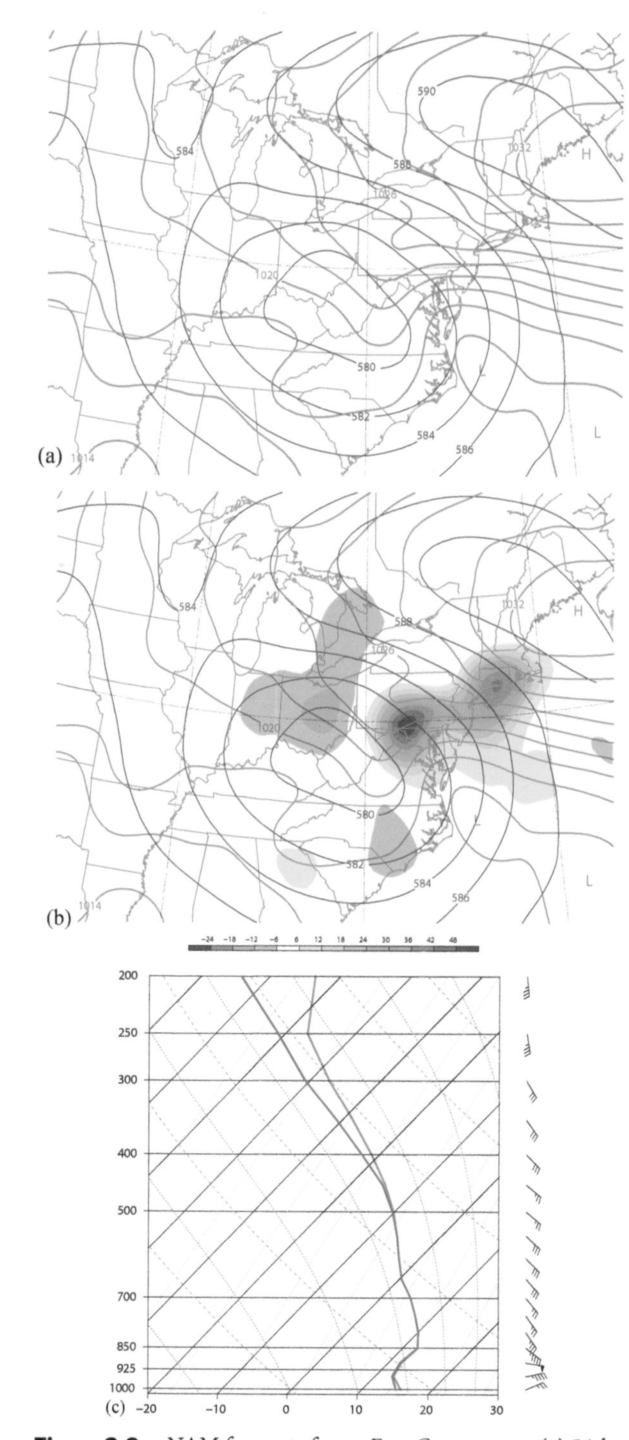


**Figure 2.8.** NAM forecasts for an East Coast storm: (a) 54-h forecast valid 1800 UTC 10 Sep 2009: 500-mb height (blue contours, interval is 20 m), sea level pressure (red contours, interval is 3 mb); (b) as in (a) but including 850-mb temperature advection [shaded, interval is $6 \times 10^{-5}$ K s$^{-1}$, warm (cool) colors indicate warm (cold) advection]; (c) 54-h model forecast sounding, skew *T*–log*p* format, for Philadelphia, PA.

is consistent with a warm-frontal structure and with the warm advection discussed above.

As with the previous example, we again emphasize that by considering only one term on the right side of the QG omega equation, we are not presenting a complete analysis. As is often the case when the forcing signal is strong, especially for intensive lower-tropospheric warm advection as shown in Fig. 2.8, it would be unlikely that differential vorticity advection could change the expected sign of the QG omega. However, as will be subsequently shown, other forms of the omega equation are better suited for operational use because of the tendency for cancellation between the forcing terms in the traditional form.

Before moving on to other forms of the QG omega equation, let us summarize what we have discussed so far. Equation (2.29) is useful for weather forecasting, and it helps to suggest useful quantities to plot to make weather forecasts. Perhaps more importantly, it provides a *physical interpretation* for the fundamental cause of the QG vertical motion. We also discovered that the QG omega is the vertical component of an ageostrophic circulation that arises in response to "QG forcing." The main points are as follows:

1. The atmosphere is constantly advecting itself out of thermal wind balance. Even advection by the geostrophic primary circulation can destroy balance.
2. An ageostrophic secondary circulation, including vertical air motion, arises as a response to thermal wind balance disruption by the primary flow in the necessary sense to bring the atmosphere back toward thermal wind balance.
3. The omega equation shows that differential vorticity advection and thermal advection (forcing terms arising from the primary flow) are related to forcing for vertical air motions.

The concept of thermal wind balance, primary and secondary circulations, and the purpose of deriving the QG omega equation (and of developing the simplified QG equation set in the first place) should now be clear.

In the previous two examples, there were strong QG signals, and several consistent features in various types of analysis fields, building our confidence in the veracity of the QG signal. In general, for situations in which a model precipitation forecast is consistent with the parameters suggested by QG analysis, such as favorable vorticity or thermal advection patterns as well as consistency with

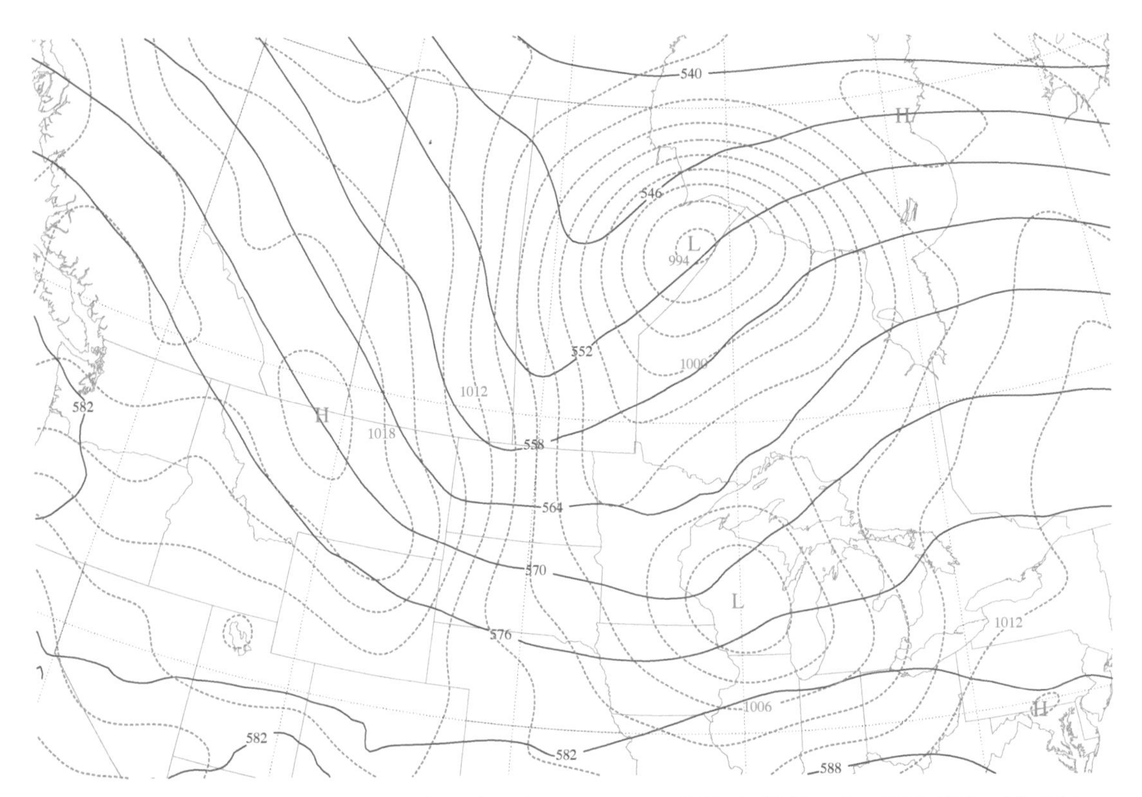


**Figure 2.9.** GFS forecast of 500-mb height and sea level pressure, valid 1800 UTC 13 Sep 2008: 500-mb height (solid blue lines, interval is 6 dam) and sea level pressure (dashed red contours, interval is 2 mb).

model forecast soundings, we can place more confidence in the signal. However, it is problematic to examine the QG forcing terms in isolation, because there may be cancellation between the two right-hand terms. An example of this situation is found over North and South Dakota at 1800 UTC 13 September 2008 (Fig. 2.9). Here, strong backing of the geostrophic flow with height is consistent with cold advection; however, a potent upper trough is located immediately to the west, suggesting cyclonic differential vorticity advection. Thus, barring an outright solution of the QG omega equation, the forecaster would be left to estimate which term is larger, an inexact and unsatisfactory task.

Fortunately, there are several ways around this issue. In fact, several aspects of the traditional QG omega equation have been mentioned that render this equation cumbersome for practical applications. To alleviate this issue, one can recast the right side of (2.29) to demonstrate that the majority of the forcing is related to the *vorticity advection by the thermal wind.* This is known as the *Trenberth approximation.* Another means of dealing with this issue is to recast the right side of (2.29) into the **Q**-vector form of the equation, in which the right side is related to the divergence of a vector field.

The derivation of the **Q**-vector form of the omega equation is not provided here, but the interested reader is encouraged to consult the original papers on this topic by Hoskins et al. (1978) and Hoskins and Pedder (1980) or dynamic meteorology texts that cover this topic. The **Q**-vector form of the QG omega equation is written as

$$\left(\nabla^2 + \frac{f_0^2}{\sigma}\frac{\partial^2}{\partial p^2}\right)\omega = -2\nabla\cdot\vec{Q} \tag{2.30}$$

where

$$\vec{Q} = -\frac{R}{\sigma p}\begin{bmatrix}\dfrac{\partial \vec{V}_g}{\partial x}\cdot\nabla\theta \\ \dfrac{\partial \vec{V}_g}{\partial y}\cdot\nabla\theta\end{bmatrix} = \begin{pmatrix}Q_1 \\ Q_2\end{pmatrix}$$

$$= -\frac{R}{\sigma p}\begin{bmatrix}\dfrac{\partial u_g}{\partial x}\dfrac{\partial \theta}{\partial x} + \dfrac{\partial v_g}{\partial x}\dfrac{\partial \theta}{\partial y} \\ \dfrac{\partial u_g}{\partial y}\dfrac{\partial \theta}{\partial x} + \dfrac{\partial v_g}{\partial y}\dfrac{\partial \theta}{\partial y}\end{bmatrix}. \tag{2.31}$$

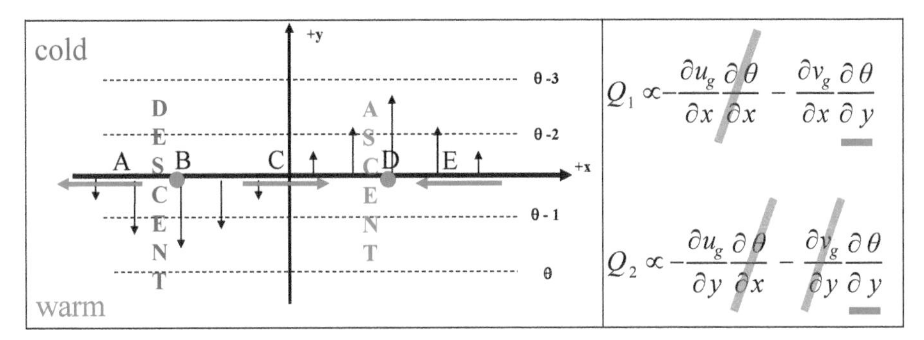


**Figure 2.10.** Idealized situation illustrating plan view of isentropes (dashed lines), geostrophic winds (solid black arrows), and coordinate axes. Red arrows represent **Q** vectors, and red dots indicate a zero-magnitude **Q** vector. At right, the component **Q**-vector terms are shown, with zero terms denoted by red slash.

The right side of (2.30) is *twice* the *divergence* of the **Q**-vector field. As was done with the traditional form of the equation, we can again consider the Laplacian on the left side of the equation as providing a negative sign; therefore, forcing for descent (positive $\omega$) is found where **Q** vectors diverge and forcing for ascent is found where **Q** vectors converge. Note that the convergence (or divergence) of **Q** vectors is **not** directly related to convergence (or divergence) of the wind; the **Q** vector is related to gradients of the wind and potential temperature, but the **Q** vector is not to be confused with the velocity vector! The two main advantages of the **Q**-vector form over the traditional form of the omega equation are (i) there is no cancellation problem and (ii) the **Q**-vector forcing can be computed at a single level (although it is best to consider several levels when using this technique). The second advantage allows one to infer forcing for QG vertical motions without having to consider the "differential" aspect, as was required with the traditional form of the equation (in the vorticity advection term).

Many meteorological software packages can automatically plot **Q** vectors and their divergence. Rather than plotting the **Q** vectors and eyeballing areas of convergence and divergence, a more accurate approach would be to invert the left-side Laplacian to solve for $\omega$; however, in practice we can simply plot **Q** vectors and overlay computations of the **Q**-vector divergence to get an idea of where there is forcing for vertical motion. However, when using high-resolution model output to compute the **Q** vectors, extremely noisy fields invariably result because of the products of spatial derivatives that make up the **Q** vector (2.31). To actually solve for $\omega$ provides a relatively smooth plot, because the numerical inversion process essentially acts as a low-pass filter on the recovered omega field. In practice, removing noise from model output fields with filters or spatial averages, or else using relatively coarse-resolution gridded data, are methods for isolating a meaningful synoptic signal using **Q** vectors.

Before examining actual data, we will consider an idealized example to demonstrate **Q** vectors in a relatively simple situation. A hypothetical case of northerly and southerly low-level jets in the presence of a meridional temperature gradient is depicted in Fig. 2.10. Let the $x$ axis be parallel to isentropes, with cold values to the north ($\partial\theta/\partial y < 0$), and note that $\partial\theta/\partial x = 0$ here. Assume that $v_g$ varies sinusoidally as indicated, and for notational simplicity, neglect the leading factor in (2.31) to facilitate the qualitative evaluation of the expressions. We will now evaluate **Q** at each of the points A–E indicated. Here, $\mathbf{Q}_1$ is the east–west component of the vector and $\mathbf{Q}_2$ is the north–south component.

In this example, the only nonzero term is the second part of the $\mathbf{Q}_1$ vector, and so we know *a priori* that all of the **Q** vectors in this example will be oriented in the zonal direction. Given that there is a negative sign in front of the remaining term, and that $\partial\theta/\partial y < 0$, the orientation of **Q** will be determined by the sign of $\partial v_g/\partial x$ in this example. Using the wind vectors provided to determine **Q** at the specified locations, we find that the **Q** vectors point toward areas of warm advection and away from areas of cold advection. Thus, regions of forcing for vertical motion in this example are exactly consistent with what one would expect from the traditional form of the QG omega equation—local maxima of warm (cold) advection are associated with forcing for ascent (descent).

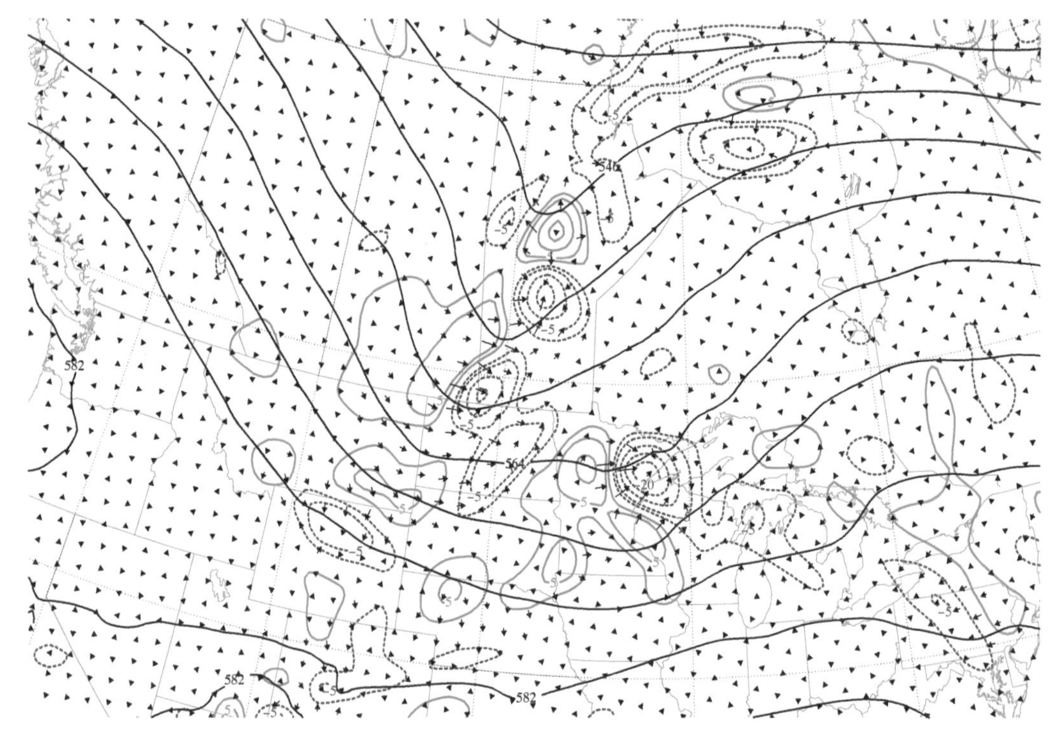


**Figure 2.11.** GFS 78-h forecast valid 1800 UTC 13 Sep 2008 (as in Fig. 2.9): 500-mb height (solid black lines, interval is 6 dam), **Q** vectors (arrows), and **Q**-vector convergence and divergence (dashed blue for convergence, solid red for divergence).

Figure 2.11 illustrates the 500-mb height and 400–700-hPa-layer **Q** vectors for the corresponding location and time to that in Fig. 2.9, where cancellation in the traditional forcing terms was subjectively implied over the Dakotas. As one might expect from the traditional form of the omega equation, the **Q** vectors exhibit convergence to the east of trough axes and divergence to the west. In this case, the **Q**-vector pattern indicates that in the regions of cancellation implied in Fig. 2.9, that **Q**-vector convergence is found over the Dakotas, suggesting that the differential vorticity advection is the dominant forcing term in this situation. However, **Q** vectors from the lower troposphere should also be examined to confirm that this is the case.

## 2.4. THE QG HEIGHT TENDENCY EQUATION

Recall from section 2.1.3 that we had consolidated the QG governing equations into the vorticity (2.28) and thermodynamic energy equations (2.27). These two equations contain two unknowns: $\omega$ and $\chi$ (the vertical motion and the geopotential tendency). We first eliminated $\chi$ between these equations and solved for $\omega$, providing the diagnostic omega equation (2.29), which is useful in weather forecasting. The physical interpretation of this equation revealed that the QG $\omega$ is the vertical component of an ageostrophic circulation that acts to bring the atmosphere back toward a state of thermal wind balance. Advections by the geostrophic primary circulation are capable of disrupting thermal wind balance, giving rise to secondary ageostrophic circulations that act to bring the atmosphere back toward a balanced state; what the general public often considers "weather" results from the ascending branch of this ageostrophic circulation.

Although the QG $\omega$ is useful in explaining patterns of clouds and precipitation, as well as in illustrating the concept of primary geostrophic forcing and a secondary ageostrophic response, it is purely a *diagnostic* equation, meaning that it does not contain information about the time evolution of weather systems. We can therefore obtain additional insight by working from (2.27) and (2.28) to instead eliminate $\omega$ and solve for $\chi$, the geopotential tendency. This will provide insight as to the processes responsible for the *development and decay* of weather systems. Recall that the geopotential and geopotential height differ by the constant $g_0$, allowing the terms *height tendency* and *geopotential tendency* to be used interchangeably.

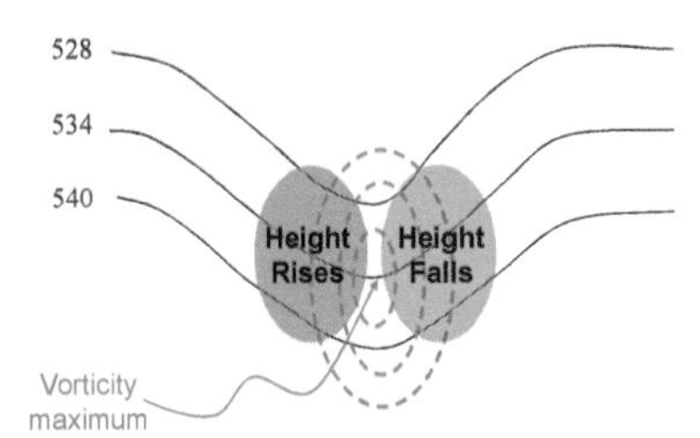


**Figure 2.12.** Idealized schematic representing the pattern of height rises and falls associated with the geostrophic vorticity advection term in the height-tendency equation. Solid contours represent geopotential height (interval is 6 dam); dashed lines represent relative vorticity. Shaded areas represent areas of height rise or fall as indicated.

We will proceed as in our development of the omega equation; only this time we will eliminate $\omega$ from (2.27) and (2.28), resulting in the QG height tendency equation

$$\underbrace{\left[\nabla^2+\frac{\partial}{\partial p}\left(\frac{f_0^2}{\sigma}\frac{\partial}{\partial p}\right)\right]\chi}_{A}=\underbrace{-f_0\vec{V}_g\cdot\nabla\left(\frac{1}{f_0}\nabla^2\Phi+f\right)}_{B}$$
$$\underbrace{-\frac{\partial}{\partial p}\left[-\frac{f_0^2}{\sigma}\vec{V}_g\cdot\nabla\left(-\frac{\partial\Phi}{\partial p}\right)\right]}_{C}. \qquad (2.32)$$

Note that the form of this equation is quite similar to that of the $\omega$ equation (2.29). Term A, as in the omega equation, includes an approximate 3D Laplacian operating on $\chi$, and we can treat this as before to contribute a negative sign to the term. Term B is a close counterpart to the vorticity advection term in the omega equation, except in that (2.32) the geostrophic vorticity advection appears without the vertical derivative; it is no longer the *differential* geostrophic vorticity advection. The vertical derivative operates on Term C in (2.32), the differential thickness advection.

The first right-hand term in (2.32) is the geostrophic absolute vorticity advection. Cyclonic (anticyclonic) geostrophic vorticity advection is associated with a tendency for height falls (rises), as illustrated in Fig. 2.12. At the center of a cyclonic vorticity center, the height tendency is zero, indicating that this term will often lead to a translational effect with an upper trough rather than amplification or decay. However, there are exceptions to this generalization; for example, when the vorticity distribution is asymmetric in an upper trough, this term can lead to significant amplification or decay.

When a wind speed asymmetry exists in the vicinity of an upper trough, as is the case for the examples shown in Fig. 2.13, the import or export of cyclonic shear vorticity can lead to a net height tendency in the base of the trough. In the case where a jet streak is located to the west of the trough axis, as in Fig. 2.13a, there is a net *import* of cyclonic vorticity into the base of the trough, and we expect the trough to amplify and dig equatorward toward lower latitudes. Such a trough configuration is commonly referred to as a "digging" trough. In contrast, a wind speed maximum on the downstream side of the trough leads to a net export of vorticity, leading to a weakening and poleward movement of the trough; this is known as a "lifting" trough (Fig. 2.13b).

Examples based on analyses from the Global Forecast System (GFS) model are shown in Figs. 2.14 and 2.15. At 1200 UTC 10 January 2010, a pronounced upper-level trough was located over the southeastern United States

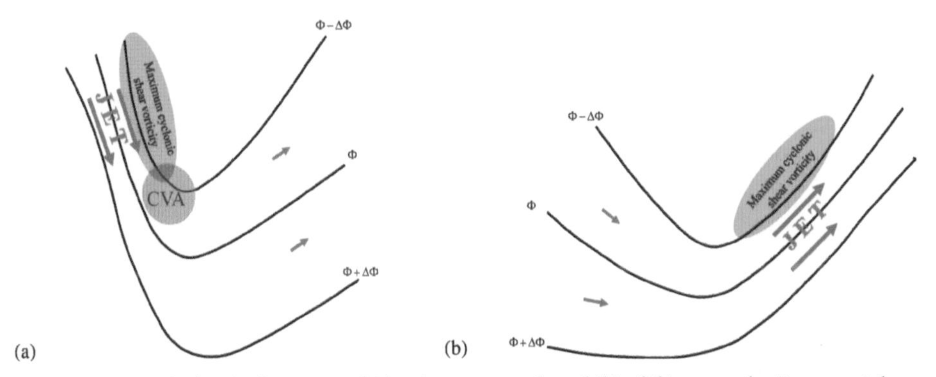


**Figure 2.13.** Idealized schematics of (a) a digging trough and (b) a lifting trough. Geopotential (solid contours), select wind vectors (arrows), and the locations of vorticity and vorticity advection maxima are labeled (from Bluestein 1993).

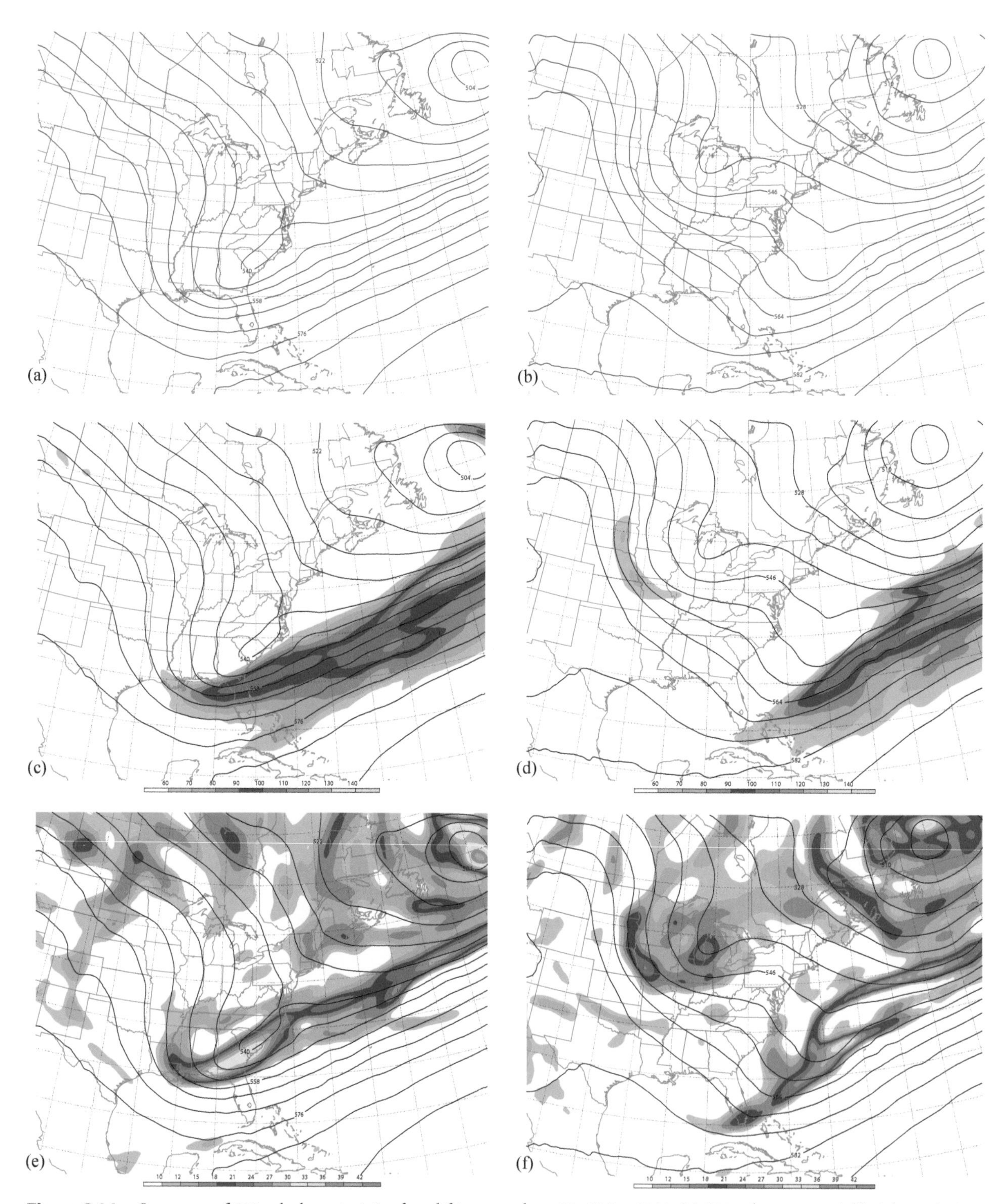


**Figure 2.14.** Summary of 500-mb characteristics for a lifting trough on 10–11 Jan 2010: (a) 500-mb geopotential height analysis valid 1200 UTC 10 Jan contour interval 6 dam; (c) as in (a) but including 500-mb isotachs (kt, shaded as in legend); (e) as in (a) but including 500-mb absolute vorticity (every $2 \times 10^{-5}$ s$^{-1}$, shaded as in legend); and (b),(d),(f) as in (a)–(c), but valid 1200 UTC 11 Jan 2010. Graphics are based on analyses from the GFS model.

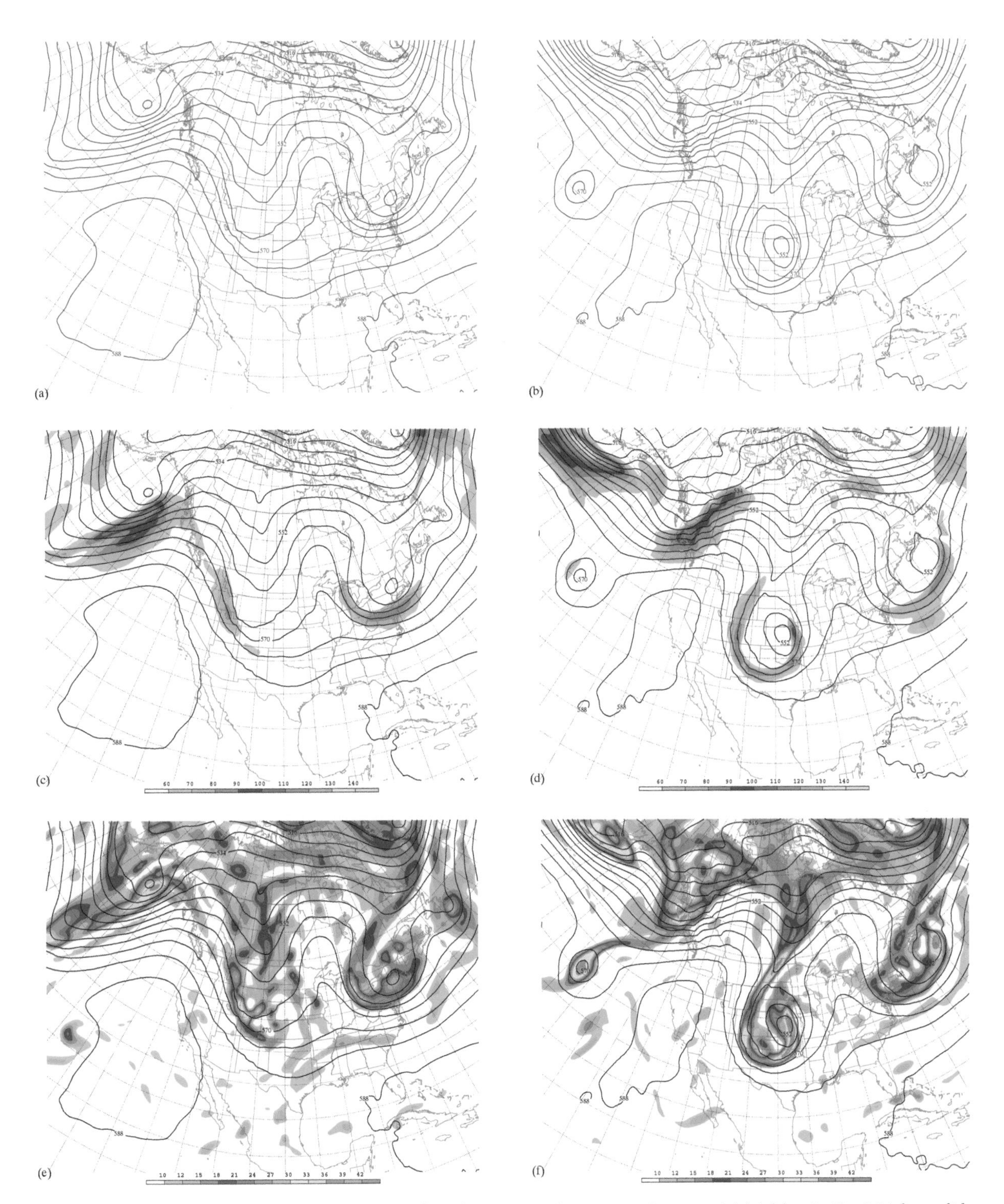

**Figure 2.15.** Summary of 500-mb characteristics for a digging trough on 22–23 Oct 2008: (a),(c),(e) as in Fig. 2.14, but valid 0000 UTC 22 Oct; (b),(d),(f) as in (a),(c),(e) but valid 0000 UTC 23 Oct.

(Figs. 2.14a,c,e). At this time, the closely spaced 500-mb height contours to the east of the trough axis relative to those to the west imply stronger geostrophic flow in the eastern portion of the trough, and a lifting trough structure (Fig. 2.14a); the stronger winds there are confirmed by the isotachs shown in Fig. 2.14c. The vorticity maximum associated with this trough exhibits a strong maximum at its base, with an elongated region of large cyclonic shear vorticity extending northeastward into the western North Atlantic Ocean (Fig. 2.14e). The situation 24 h later, as shown in Figs. 2.14b,d,f, is consistent with expectations for a lifting trough. The region of cyclonic curvature in height contours marking the base of the trough exhibits smaller amplitude relative to those at the earlier time (Figs. 2.14b,d), the vorticity in the base of the trough is considerably weaker as well, and the center of the trough and vorticity maximum have indeed lifted northeastward (Fig. 2.14f).

We have at this point considered only the vorticity advection term; in assessing the time evolution of troughs and ridges, the differential thermal advection must also be considered. Further, just because a trough is lifting does not mean that it cannot be associated with high-impact weather. The pronounced right entrance region associated with the jet in lifting-trough situations can be associated with strong forcing for ascent and heavy precipitation ahead of the trough axis.

An example of a digging trough is evident over western North America during an event that took place on 22–23 October 2008 (Fig. 2.15). In Fig. 2.15a, a lifting trough is located over the eastern North Pacific Ocean, but our focus here is on the trough centered near the states of Wyoming and Colorado in the western United States. Based on the close spacing of the 500-mb geopotential height contours to the west of the axis of this trough, in conjunction with more widely spaced contours east of the axis, we infer relatively strong geostrophic winds west of the trough axis and a digging trough configuration. The isotachs (Fig. 2.15c) and absolute vorticity (Fig. 2.15e) are consistent with this interpretation. The analysis 24 h later, at 0000 UTC 23 October, reveals that a large cut-off low has formed, centered over Kansas. The 500-mb heights have fallen in the center of this cyclonic system, and although other processes may have contributed to this, the import of cyclonic shear vorticity from the west and north is likely to have played a significant role. The values of vorticity in the trough have clearly amplified during this 24-h period (Figs. 2.15e,f).

The two preceding examples are meant to illustrate the importance of wind-speed maxima, or jet streaks, in the evolution of upper-level trough and ridge patterns. However, as mentioned earlier, oftentimes troughs do not exhibit a strongly digging or lifting signature, and in those cases the vorticity advection term leads primarily to translation of the system.

A process that is often strongly linked to the amplification or decay of upper-level troughs and ridges is represented by term C in (2.32). Note the similarity of this term with its counterpart in the QG omega equation (2.29). The term involves the thickness advection; however, unlike in the omega equation where the Laplacian of thermal advection was seen, here there is a vertical derivative acting on this quantity. This term is the *differential thermal advection*. The physical working of this term can be easily understood through consideration of the thickness tendency that would accompany a given sign of temperature advection within an atmospheric layer. Suppose a maximum of warm advection is located near the 700-mb level, as in the idealized schematic of Fig. 2.16a. If the column experiences net warming due to this process, the thickness of the layer must increase, consistent with the hypsometric equation (1.37). Now, it is clear why $\partial/\partial p$ operates on the thermal advection in the height tendency equation—the sign of the height tendency depends on whether the pressure surface in question lies above or below the level of maximum thermal advection. For a pressure surface located exactly at the level of maximum thermal advection, the height tendency would be zero.

A similar interpretation applies to a local maximum of *cold advection* in the lower troposphere (Fig. 2.16b); only the signs of the height tendency are reversed relative to those in the warm-advection case. Differential thermal advection of either sign is frequently observed in the vicinity of midlatitude cyclones. Warm advection to the east of such systems (ahead of a warm front) is often associated with pressure falls in the lower troposphere and a building of an upper ridge downstream of the system. Behind the surface cold front to the west of a cyclone, lower-tropospheric cold advection is consistent with rising surface pressure and falling upper-level geopotential height surfaces.

The 6-h GFS forecast valid 1200 UTC 26 October 2009 indicates a region of pronounced backing of the

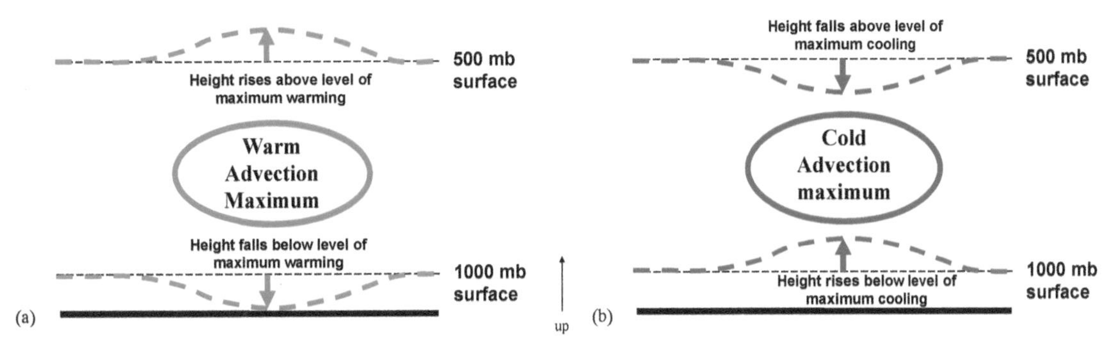


**Figure 2.16.** Idealized schematics depicting geopotential height changes associated with the differential thermal advection term in the QG height tendency equation (2.32).

geostrophic wind with height over the North Pacific Ocean to the west of the states of Oregon and Washington, corresponding to cold advection in the surface-to-500-mb layer (Fig. 2.17a). As this region of cold advection moves eastward, the upper trough amplifies, with a major trough forecasted for the western United States by 0600 UTC 27 October 2009 (Fig. 2.17b). Note that the 500-mb height pattern in the vicinity of the trough does not imply a digging configuration, indicating that the differential thermal advection term is likely responsible for this amplification. The 48-h forecast valid 0600 UTC 28 October 2009 indicates that the upper trough has further amplified and developed a closed circulation over the U.S. Southwest. Often, strong geostrophic backing is associated with maximum temperature advection in the lower troposphere, but it is best to examine a vertical cross section to confirm this, as shown in Fig. 2.17d. In this case, there is not only strong lower-tropospheric cold advection but a layer of warm advection is maximized aloft, above the 500-mb level. Thus, the combination of cold advection decreasing with height beneath warm advection increasing with height works consistently to produce dramatic height falls at the 500-mb level. Based on these figures, it is highly probable that the differential thermal advection process was responsible for the trough amplification observed in these analyses.

A process that has been eliminated from (2.32) is the geopotential height change brought about by adiabatic warming and cooling associated with vertical air motion. This is a consequence of our eliminating omega in the derivation of this equation, and that the QG vertical motion is the secondary response to forcing brought about by thermal wind balance disruption. In fact, the adiabatic temperature changes generally act to oppose the advective tendencies: warm advection leads to ascent and adiabatic cooling, while cold advection often leads to descent and adiabatic warming. This is consistent with forcing and response—this adiabatic compensation slows the rate of warming or cooling relative to what advection alone would provide.

## 2.5. ADDITION OF DIABATIC PROCESSES TO THE QG SYSTEM

Our goal in the development of the QG equations was to simplify the system to the maximum extent possible to gain conceptual insight into the processes that generate vertical motion and that dictate the amplification and decay of weather systems. One of our assumptions in this development was that of adiabatic flow. However, many of the weather systems of greatest importance and societal impact are characterized by heavy precipitation and thus strong condensational heating. Fortunately, it is straightforward to retain the diabatic term in the QG thermodynamic equation. In chapter 5 when we consider cyclone dynamics, the importance of diabatic processes will be demonstrated.

If the diabatic term is retained in progressing from (2.18) to (2.24), then the previous development is repeated to obtain the omega and height tendency equations in modified form, including the diabatic heating term

$$\left(\nabla^2 + \frac{f_0^2}{\sigma}\frac{\partial^2}{\partial p^2}\right)\omega = \frac{f_0}{\sigma}\frac{\partial}{\partial p}\left[\vec{V}_g \cdot \nabla\left(\frac{1}{f_0}\nabla^2\Phi + f\right)\right] + \frac{1}{\sigma}\nabla^2\left[\vec{V}_g \cdot \nabla\left(-\frac{\partial \Phi}{\partial p}\right)\right] - \frac{R}{C_p p \sigma}\nabla^2 J. \tag{2.33}$$

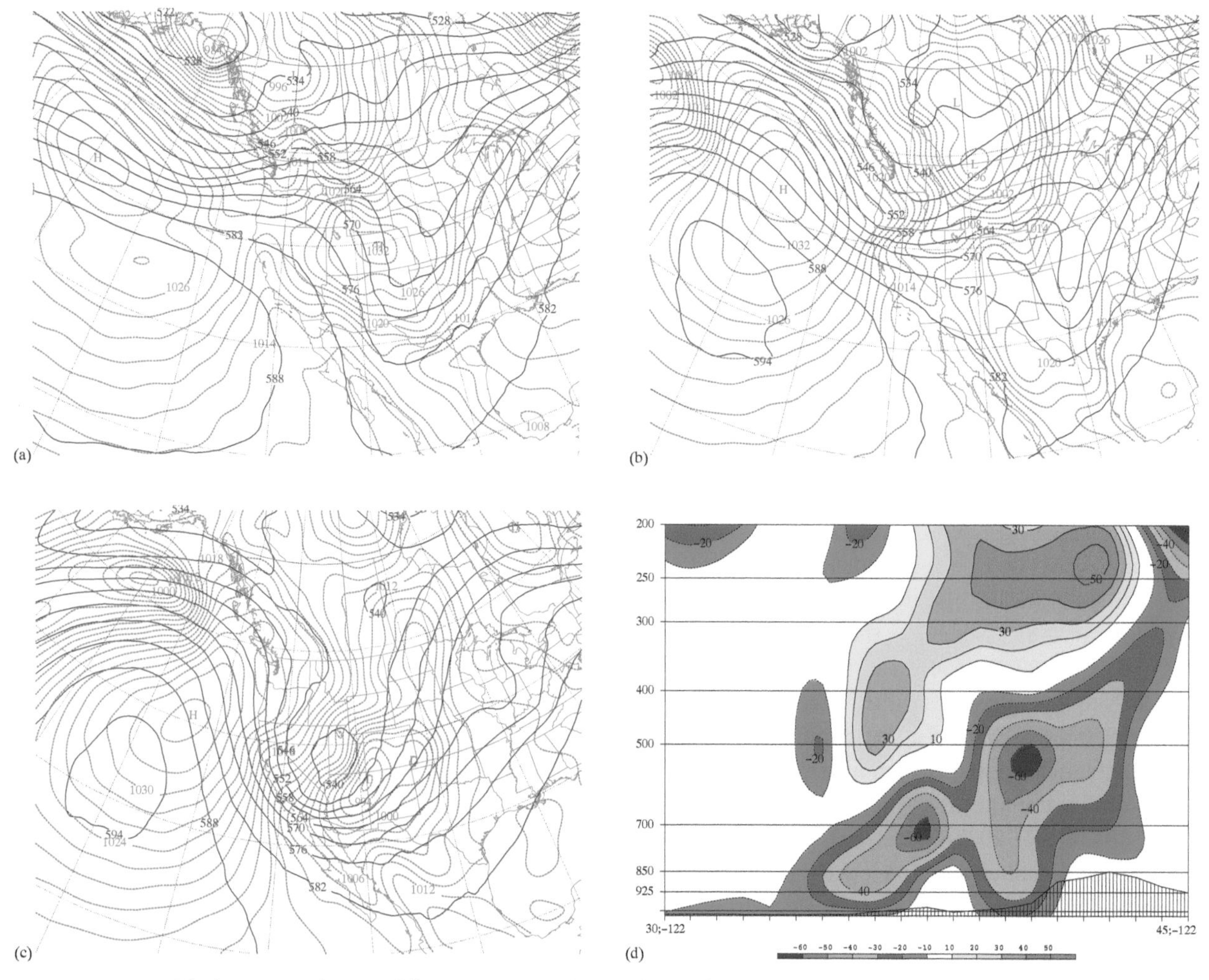

**Figure 2.17.** Cold advection and an amplifying upper trough: (a) 0-h GFS forecast valid 1200 UTC 26 Oct 2009; (b) 18-h forecast valid 0600 UTC 27 Oct 2009; (c) 42-h forecast valid 0600 UTC 28 Oct 2009. (d) North-south cross section of temperature advection (shaded every $10 \times 10^{-5}$ K s$^{-1}$ as in legend at bottom of panel) based on GFS analysis valid 12 UTC 27 Oct 2009.

The role of the diabatic term in (2.33) is similar to that of the thermal advection term, in that a local maximum of heating is consistent with forcing for ascent. Similarly, retaining the diabatic term in the development of the QG height tendency equation yields

$$\left[\nabla^2+\frac{\partial}{\partial p}\left(\frac{f_0^2}{\sigma}\frac{\partial}{\partial p}\right)\right]\chi=-f_0\vec{V}_g\cdot\nabla\left(\frac{1}{f_0}\nabla^2\Phi+f\right)-\frac{\partial}{\partial p}\left[-\frac{f_0^2}{\sigma}\vec{V}_g\cdot\nabla\left(-\frac{\partial\Phi}{\partial p}\right)\right]-\frac{f_0^2}{\sigma}\frac{\partial}{\partial p}\left(\frac{R}{C_p p}J\right). \quad (2.34)$$

We again see an analogous role for the diabatic term and the thermal advection term in the QG height tendency equation—here, it is the vertical derivative of the heating that determines the height tendency. The physical reasoning is also similar to that depicted in Fig. 2.16; net heating requires an increase in thickness, which is consistent with height rises above heating maxima and height falls beneath; this also requires that the vertical derivative enter the term.

There have been some spectacular storm systems in which latent heat release has led to dramatic model forecast failures and rapidly evolving high-impact weather situations. Perhaps the most famous example is the Presidents' Day blizzard of 19 February 1979. We will present a more detailed analysis of this event in chapter 5.

## 2.6. QG POTENTIAL VORTICITY AND THE HEIGHT TENDENCY EQUATION

Applying the chain rule to carry the vertical derivative through the differential thermal advection term (term C) in (2.32), we obtain the following two terms:

$$-\frac{\partial}{\partial p}\left[\frac{f_0^2}{\sigma}\vec{V}_g\cdot\nabla\left(\frac{\partial\Phi}{\partial p}\right)\right] = -\frac{\partial\vec{V}_g}{\partial p}\cdot\nabla\left(\frac{f_0^2}{\sigma}\frac{\partial\Phi}{\partial p}\right)$$
$$-\vec{V}_g\cdot\nabla\frac{\partial}{\partial p}\left(\frac{f_0^2}{\sigma}\frac{\partial\Phi}{\partial p}\right). \tag{2.35}$$

The thermal wind relation (1.45), repeated below, can be substituted in the first right-hand term in (2.35),

$$\frac{\partial\vec{V}_g}{\partial p} = \frac{1}{f_0}\hat{k}\times\nabla\frac{\partial\Phi}{\partial p}.$$

The result is the scalar product of two perpendicular vectors, and the term vanishes,

$$-\frac{\partial\vec{V}_g}{\partial p}\cdot\nabla\left(\frac{f_0^2}{\sigma}\frac{\partial\Phi}{\partial p}\right) = \frac{f_0}{\sigma}\hat{k}\times\nabla\frac{\partial\Phi}{\partial p}\cdot\nabla\frac{\partial\Phi}{\partial p} = 0. \tag{2.36}$$

Thus, returning to the height tendency equation (2.32), term C can be replaced with the rightmost term in (2.35). We can then divide (2.32) by $f_0$ and rearrange to obtain

$$\frac{\partial}{\partial t}\left(\frac{1}{f_0}\nabla^2\Phi + \frac{\partial}{\partial p}\left(\frac{f_0}{\sigma}\frac{\partial\Phi}{\partial p}\right) + f\right)$$
$$+\vec{V}_g\cdot\nabla\left(\frac{1}{f_0}\nabla^2\Phi + \frac{\partial}{\partial p}\left(\frac{f_0}{\sigma}\frac{\partial\Phi}{\partial p}\right) + f\right) = 0. \tag{2.37}$$

Inspection of (2.37) indicates that the quantity in parentheses is *conserved* for adiabatic, frictionless, geostrophic flow. This quantity is the *quasigeostrophic potential vorticity* (QGPV), denoted as

$$q = \frac{1}{f_0}\nabla^2\Phi + f + \frac{\partial}{\partial p}\left(\frac{f_0}{\sigma}\frac{\partial\Phi}{\partial p}\right). \tag{2.38}$$

As stated above, following adiabatic, frictionless, geostrophic flow,

$$\frac{dq}{dt_g} = 0. \tag{2.39}$$

Thus, the QG height tendency equation is simply a statement of QGPV conservation. Examination of (2.38) reveals that $q$ is the sum of the geostrophic absolute vorticity and a term involving the vertical derivative of the thickness, which is related to the static stability. There are two powerful properties of PV, each of which we will explore in more depth in chapter 4: (i) *conservation* under certain flow conditions (from 2.39) and (ii) *invertibility*. The invertibility property states that provided information about the QGPV distribution and boundary conditions, one can recover the geopotential field associated with any part of the QGPV field and from that recover the geostrophic wind and temperature (thickness) fields. One may wonder why it would be useful to recover the same quantities from which the QGPV was computed in the first place. The answer is that the QGPV can be divided into an arbitrary number of dynamically relevant pieces, for example, a local cyclonic PV anomaly associated with an upper-level trough. Then, the QGPV can be inverted in a *piecewise* manner, and the height field (and balanced flow) associated with a given QGPV feature can be uniquely identified. These two properties make PV one of the most useful and powerful dynamical tools available to meteorologists. In chapter 4, section 5.3.6, and again in chapter 7, we will explore PV applications in greater depth.

## 2.7. QG ENERGETICS

Another approach to atmospheric diagnosis is the study of mechanisms that control the exchange of *energy* between the environment and a given weather system. There are several different techniques for studying the mechanisms of energy transfer in the atmosphere. The study of downstream baroclinic development with Rossby wave packets can be analyzed via *local energetics*, where the instantaneous mechanisms of energy transfer are computed at specific locations. Other studies have computed volume averages around individual storm systems, or examined time-averaged contributions to energy tendency over weeks or longer. Here, we will utilize both local and volume-average approaches to understand how the configuration of upper-level trough and ridge patterns can inform us about evolving weather systems, again within the context of the QG equations.

### 2.7.1. Zonal averages and perturbations

A means of isolating synoptic-scale weather systems from the background flow in which they are embedded is to subtract the *zonal average* of relevant variables from the full instantaneous value. To illustrate this method, suppose that the following $u$ wind measurements are valid for a given set of longitude points taken along fixed latitude:

$u + 25\,\mathrm{m\,s^{-1}}$ $\quad +16\,\mathrm{m\,s^{-1}}$ $\quad +6\,\mathrm{m\,s^{-1}}$ $\quad +9\,\mathrm{m\,s^{-1}}$ $\quad +17\,\mathrm{m\,s^{-1}}$ $\quad +29\,\mathrm{m\,s^{-1}}$

$u' + 8\,\mathrm{m\,s^{-1}}$ $\quad -1\,\mathrm{m\,s^{-1}}$ $\quad -11\,\mathrm{m\,s^{-1}}$ $\quad -8\,\mathrm{m\,s^{-1}}$ $\quad 0\,\mathrm{m\,s^{-1}}$ $\quad +12\,\mathrm{m\,s^{-1}}$

These wind components can be represented as vectors, as indicated below the $u$ values listed. Let $\bar{u}$ be the zonal average of these wind observations, and $u'$ be the deviation from that average. At each point, $u = \bar{u} + u'$. For the data provided above, $\bar{u} = 17\ \mathrm{ms^{-1}}$, and $u'$ is computed by subtracting this value from the point value at each location, as indicated for the lower row of values.

This simple concept can be extended to zonal averages of any meteorological variable taken around latitude circles. We will refer to deviations from the zonal average of any variable as the "disturbance," "perturbation," or "eddy" values. The total horizontal perturbation velocity ($\vec{V}' = u'\hat{i} + v'\hat{j}$) can be taken to represent the perturbation flow associated with synoptic-scale systems, such as cyclones and anticyclones. The *kinetic energy per unit mass* of the perturbation flow, or *eddy kinetic energy* ($K_e$) for either the full ($K_e$) or geostrophic ($K_{eg}$) flow is defined as

$$K_e = \frac{(u'^2 + v'^2)}{2}, \quad K_{eg} = \frac{(u_g'^2 + v_g'^2)}{2}. \tag{2.40}$$

To further illustrate the process of decomposing atmospheric variables into zonal average and perturbation, Figs. 2.18 and 2.19 provide data from the GFS model for 30 September 2008. The full 500-mb geopotential height field (Fig. 2.18a) exhibits a wavy pattern of troughs and ridges; however, when the zonal mean height field (Fig. 2.18b) is subtracted, the resulting eddy height anomalies (Fig. 2.18c) appear as closed-off features. The perturbation geostrophic flow associated with these height anomalies can be computed directly from the height anomaly field, or one can simply subtract the zonal average of the horizontal wind components to determine the perturbation velocity field at each data point, as shown in Fig. 2.19. Again, in subtracting off the mean westerly flow, closed cyclonic and anticyclonic eddies become more evident.

The zero height anomaly contour or a certain kinetic energy contour can be used to define a spatial area over which to average the energy equation. Then, by integrating vertically with respect to pressure, the volume integral of geostrophic eddy kinetic energy $K_{eg}$ over an area or disturbance is denoted $[K_{eg}]$, which is given by

$$\left[K_{eg}\right] = \frac{1}{gA}\iint_{pA}\left(\frac{u_g'^2 + v_g'^2}{2}\right) dA\,dp. \tag{2.41}$$

The sequential downstream amplification and upstream decay of height anomalies of either sign are often associated with the eastward propagation of Rossby wave energy, as discussed in chapter 1. Viewing height anomalies, rather than the full height field, often reveals this process quite clearly.

### 2.7.2. The eddy kinetic energy equation

An equation for the time rate of change of $K_{eg}$ provides insight into the growth and decay mechanisms of weather systems in terms of energy transfer processes. Given that the kinetic energy per unit mass involves terms such as $u^2$ and $v^2$, how can we form an equation that gives us the time tendency of kinetic energy (starting from the $u$ and $v$ momentum equations)? Taking the scalar product of the horizontal velocity vector with the horizontal momentum equation provides an equation for the time tendency of the full kinetic energy. If we instead take the scalar product of the *perturbation velocity* with the perturbation momentum equations, rather than the full equations, the following eddy kinetic energy equation results. In the QG momentum equations (2.14) and (2.15), we substitute $u_g = \bar{u}_g + u_g'$ and $v_g = v_g'$, noting that $\bar{u}_g$ is a function only of the $y$ and $z$ directions, and then we take the scalar product with $\vec{V}_g'$ and rearrange to yield

$$\frac{\partial}{\partial t}\left(\frac{u_g'^2 + v_g'^2}{2}\right) + u_g\frac{\partial}{\partial x}\left(\frac{u_g'^2 + v_g'^2}{2}\right) + v_g\frac{\partial}{\partial y}\left(\frac{u_g'^2 + v_g'^2}{2}\right)$$
$$= f(u_g' v_{ag} - v_g' u_{ag}) - u_g' v_g' \frac{\partial \bar{u}_g}{\partial y}. \tag{2.42}$$

Combining the left-hand terms, substituting the perturbation geostrophic wind definitions $v_g' = (1/f)\partial\phi'/\partial x$ and $u_g' = -(1/f)\partial\phi'/\partial y$ in the rightmost term, and using the product rule of calculus, the equation

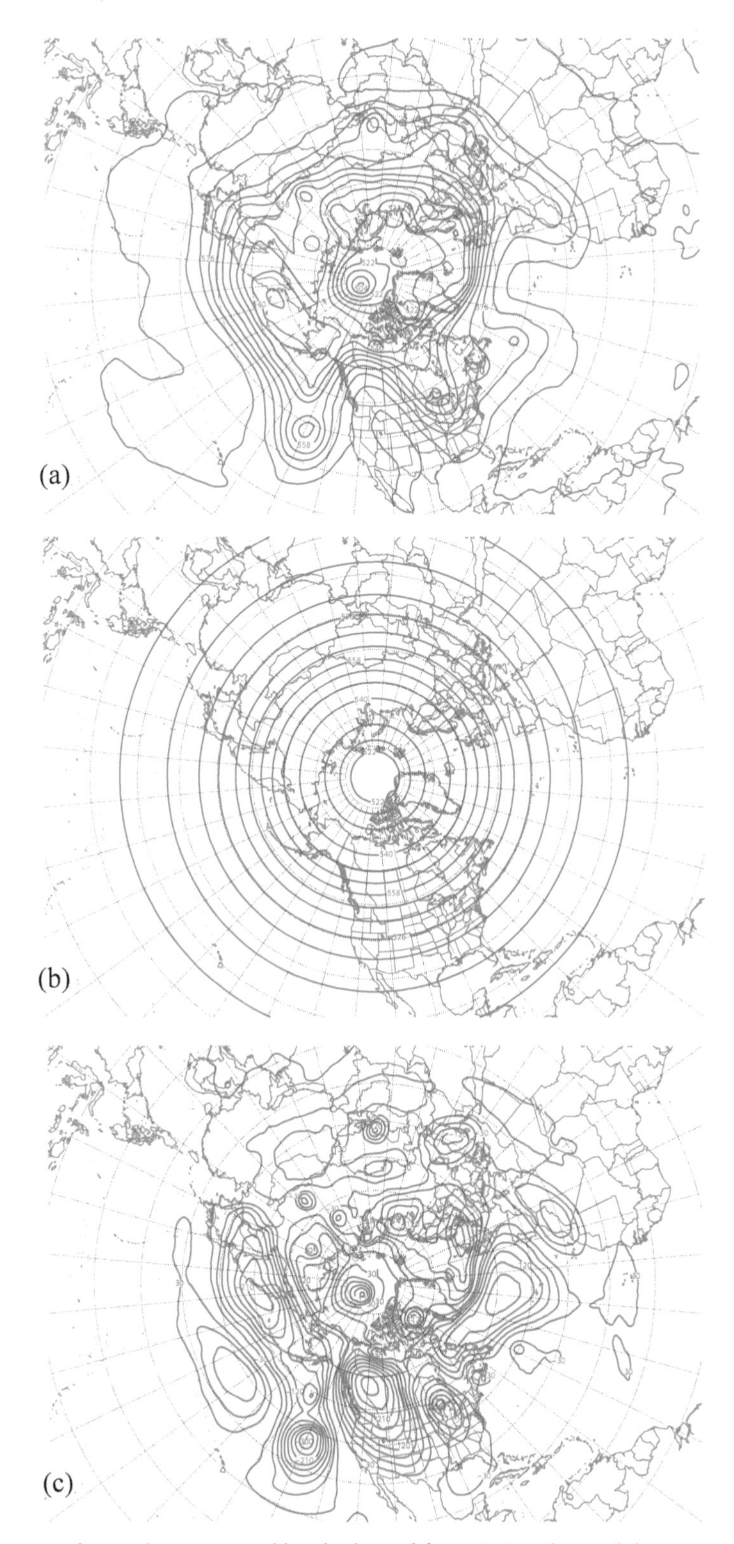

**Figure 2.18.** Decomposition of 500-mb geopotential height derived from GFS analysis valid 1200 UTC 30 Sep 2008: (a) full height field, contour interval is 6 dam; (b) zonal average height; and (c) height anomaly, interval 30 m, blue (red) lines represent negative (positive) anomaly values. The zero contour is omitted.

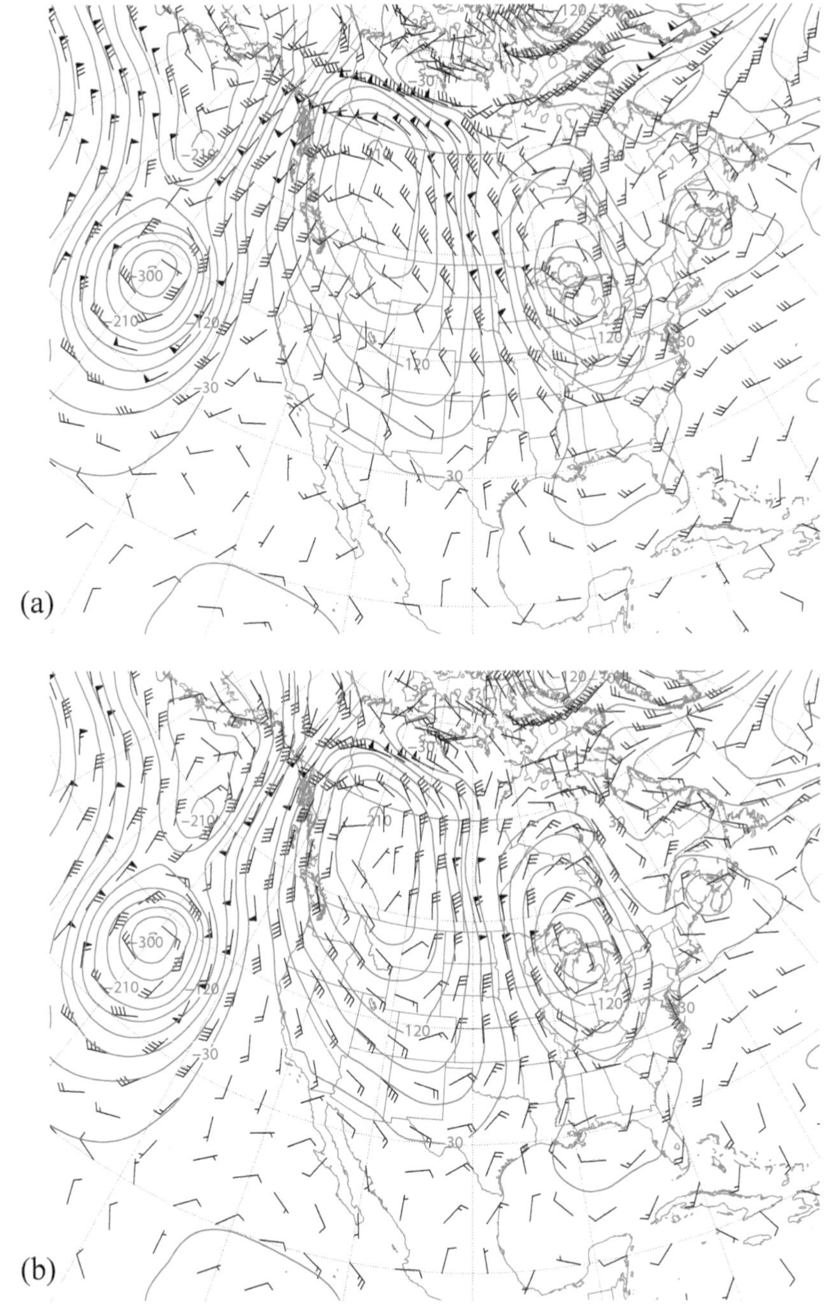


**Figure 2.19.** As in Fig. 2.18c, but adding (a) full wind velocity and (b) perturbation velocity. Winds are plotted as barbs using standard plotting convention.

takes the following form (the complete derivation is assigned as an end-of-chapter problem):

$$\frac{d}{dt_g}\left(\frac{u_g'^2+v_g'^2}{2}\right)=-\nabla\cdot(\phi'\vec{U}_{ag})-\frac{R_d}{p}\omega'T'-u_g'v_g'\frac{\partial\bar{u}_g}{\partial y}. \tag{2.43}$$

This equation is useful for the quantification of local energetics and can be applied in a variety of contexts. The left-hand side describes changes in $K_{eg}$ following the geostrophic flow, or we can split this into local and advective tendencies. The right-hand terms represent the *ageostrophic geopotential flux divergence*, *baroclinic conversion*,

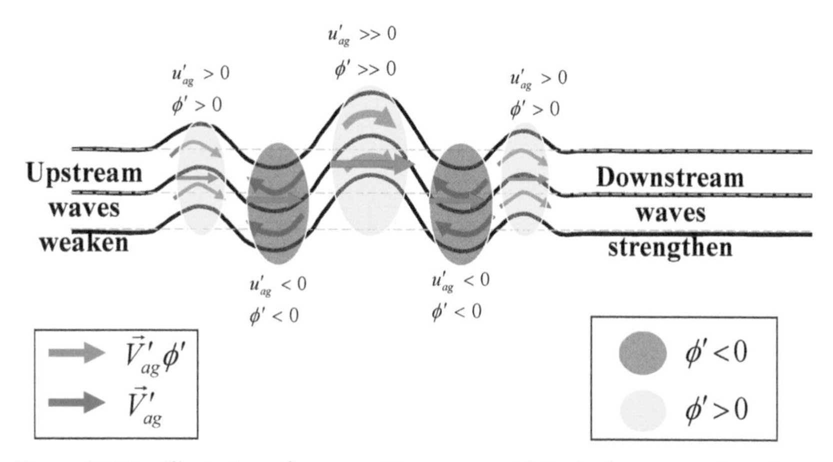


**Figure 2.20.** Illustration of ageostrophic geopotential flux in downstream baroclinic development. Solid black lines represent geopotential height contours; shaded yellow and blue ovals represent height anomalies of positive and negative sign, respectively; green curved arrows represent sense of horizontal ageostrophic flow; and red arrows depict the ageostrophic geopotential flux.

and *barotropic conversion*, respectively. We will explore these three terms in more detail below.

In chapter 1, we noted that the *group velocity* for Rossby wave energy exhibited a faster eastward movement than the speed of individual troughs and ridges, evident, for example, in Fig. 1.20. The eastward propagation of wave energy is described by the first right-hand term in (2.43), the divergence of the ageostrophic geopotential flux, as illustrated in Fig. 2.20. Because this term is related to the product of the ageostrophic velocity vector and the geopotential anomaly, the flux has an eastward-directed zonal component throughout a train of Rossby waves. This is because in locations where the flow is subgeostrophic (in troughs, providing a negative ageostrophic zonal wind component there), the geopotential anomaly is also negative, providing a positive (eastward)-directed flux; in ridges, both terms are positive. The result is an eastward-directed flux of Rossby wave energy. For a wave train without spatial variations in wave amplitude, the eastward flux is uniform across the troughs and ridges, and the flux divergence is zero. However, the atmosphere typically exhibits regions of highly amplified troughs and ridges, with regions of weak wave amplitude upstream and downstream. In this situation, similar to that shown in Fig. 2.20, there will be *energy flux convergence* downstream of the region of large-amplitude waves, and the wave amplitudes will tend to grow there. Meanwhile, the energy flux divergence upstream will result in weakening waves there. Orlanski and Katzfey (1991) and Orlanski and Chang (1993) have shown that the downstream energy flux associated with this process is a major factor in the amplification and decay of many upper-level disturbances.

An example of this, taken from a typical situation over North America, is provided in Fig. 2.21. Amplified upper waves are observed over the North Pacific and western North America at the initial time (Figs. 2.21a,b). These panels demonstrate the supergeostrophic (subgeostrophic) nature of the flow in the ridge (trough) and also the pattern of ageostrophic geopotential flux. Thus, Rossby wave energy is moving eastward, and we expect the upstream height anomalies to decay while the eastern anomalies amplify (including the trough over the central United States). The amplified trough over eastern North America evident in Fig. 2.21c is consistent with expectations based on the idealized schematic (Fig. 2.20).

Interpretation of the barotropic and baroclinic terms in (2.43) is facilitated by utilizing an average over a specified domain of interest. With suitable lateral boundary conditions, such as a periodic domain in the $x$ direction and a bounded domain in the $y$ direction, we can average over the depth of the troposphere and a horizontal area $A$ to obtain an integral form of the eddy kinetic energy equation

$$\underbrace{\frac{\partial [K_{eg}]}{\partial t}}_{A} = \underbrace{-\frac{1}{gA}\iint\limits_{p\,A} \overline{u'_g v'_g}\frac{\partial \bar{u}_g}{\partial y}\, dA\, dp}_{B} \\ \underbrace{+\frac{R_d}{A}[\rho]\iint\limits_{p\,A} \overline{w'T'}\, dA\, d\ln p.}_{C} \qquad (2.44)$$

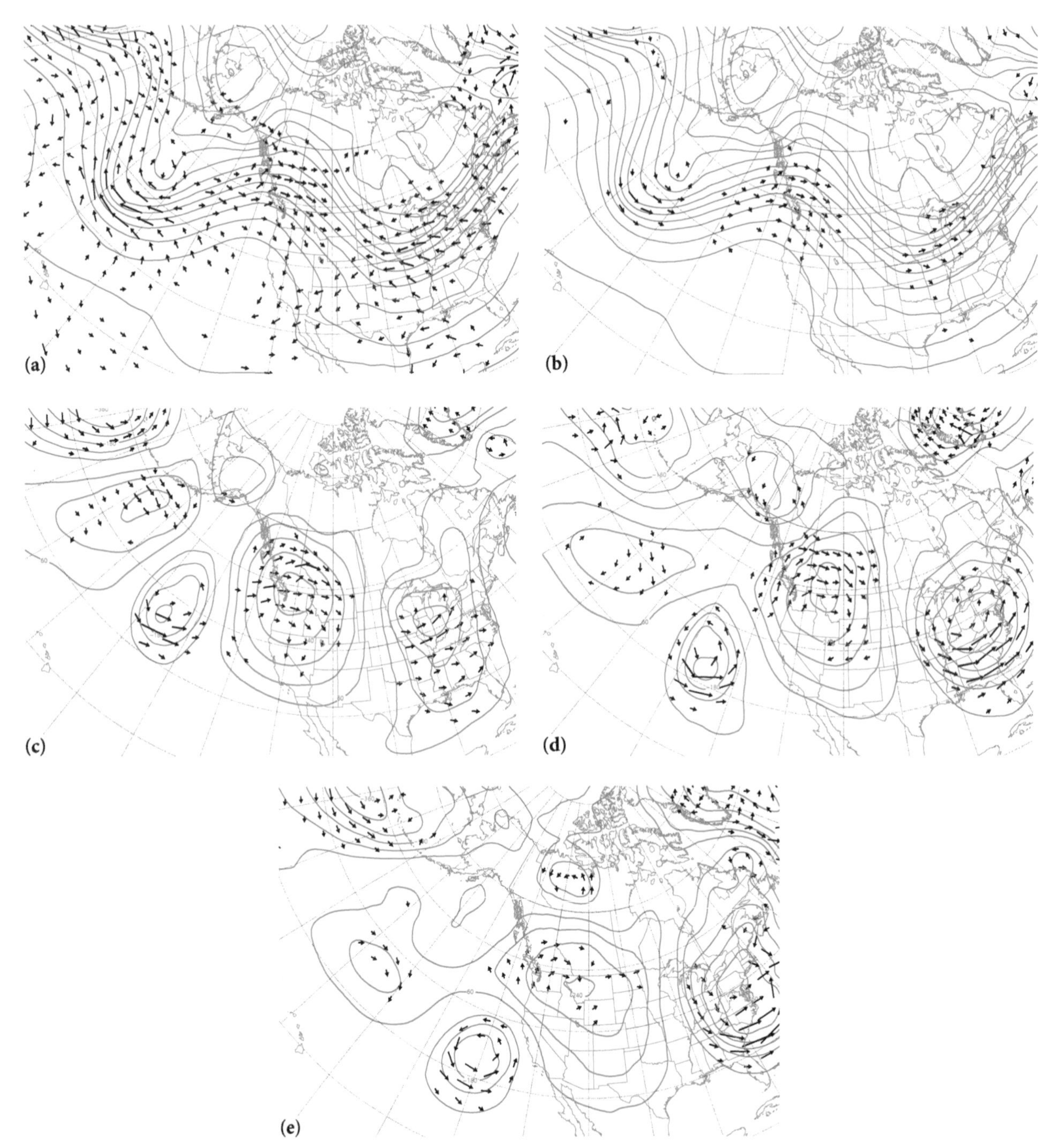


**Figure 2.21.** Example of downstream development for 0000 UTC 2 Feb 2009: (a) 500-mb height (interval 6 dam, blue contours) and ageostrophic wind vectors; (b) as in (a) but with ageostrophic geopotential flux (vectors); (c) 500-mb height anomaly (interval 60 m, zero contour omitted), positive (negative) anomalies red (blue) contours valid 0000 UTC 3 Feb 2009; (d) as in (c) except for 0000 UTC 4 Feb; (e) as in (c) except for 0000 UTC 5 Feb. Vectors in (c)–(e) are as in (b).

For example, the domain could be taken as a latitudinal belt encircling the globe, consistent with periodic boundary conditions in the zonal direction. Because the advection and ageostrophic geopotential flux redistribute eddy energy within such a domain, these terms do not appear in the volume-averaged version of the equation. Cyclones and anticyclones can be viewed as "eddies" in a background flow that is more or less zonal. Equation (2.44) tells us what processes and mechanisms determine whether the energy in such a disturbance will grow or diminish with time. The energetics framework can also be used in diagnosing the role that eddy disturbances play in the global energy balance. The left side of (2.44), term A, is simply the local time rate of change in eddy kinetic energy, averaged over the entire area and depth of a given domain or disturbance. This term will be positive for an eddy disturbance, such as a cyclone or anticyclone, that is growing in amplitude.

Inspection of the first right-hand term (term B) reveals a dependence on the meridional gradient of the zonal-mean flow, $\partial \overline{u}_g / \partial y$, and the spatial average of the perturbation velocity product $\overline{u'_g v'_g}$. The former will be largest in magnitude in regions where there is strong meridional shear of the mean flow, and it will be very small near the axis of the zonally averaged jet. The eddy product term $\overline{u'_g v'_g}$ requires more analysis. Considering this quantity for a perfectly circular eddy, as depicted in Fig. 2.22, we find that there is cancellation between this product in different parts of the eddy, and the average over the entire eddy is zero. The area-averaged eddy velocity product also equals zero for elongated eddies that do not have a pronounced axial tilt.

Consider now an asymmetric eddy with an axial tilt, as shown in Fig. 2.23. We see that when averaged around the circumference of the disturbance, there is a disproportionate region that is characterized by *negative*

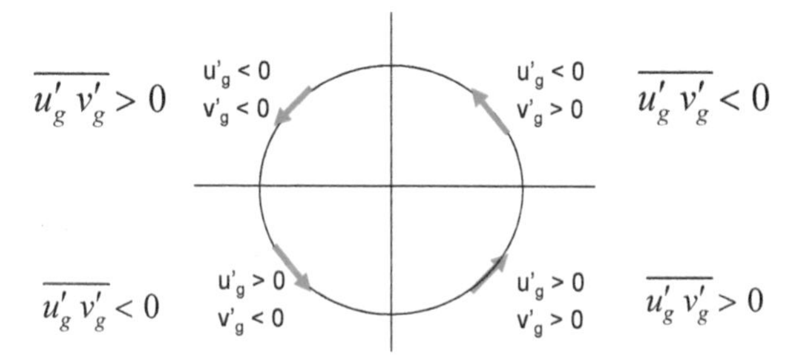


**Figure 2.22.** Eddy flow correlations for a circular eddy disturbance.

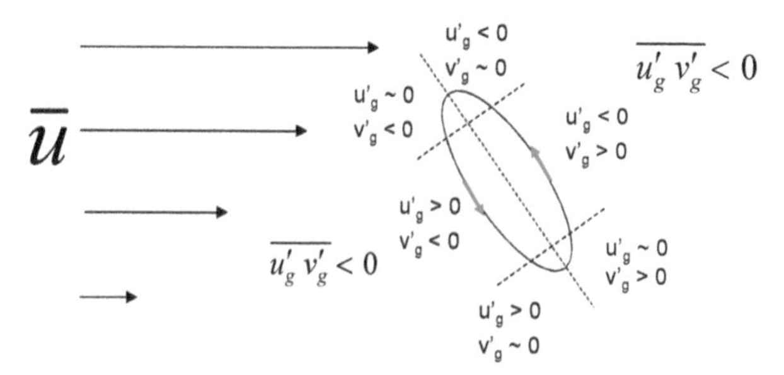


**Figure 2.23.** As in Fig. 2.22, but for an asymmetric eddy with a tilted axis. Arrows at left depict mean zonal flow.

values of $\overline{u'_g v'_g}$. For such "negatively tilted" eddies in the Northern Hemisphere, if located south of the region of strongest westerly $\overline{u}_g$ in the region where $\partial \overline{u}_g / \partial y > 0$, then a positive contribution to the eddy kinetic energy arises because of term B in (2.44). The physical process at work is related to the extraction of kinetic energy from the zonal mean flow, into the eddy, and it takes place with an axis oriented so that the eddy is *leaning against the background shear*. In many physical systems, this characteristic is associated with the growth of eddy energy at the expense of the reservoir of mean flow kinetic energy. To draw a loose analogy, imagine a stream or river with the strongest average flow in the center, weakening toward either bank. If one were to insert a barrier (e.g., a piece of plywood) into the flow oriented in a way that cut against the shear of the river current, then one can imagine a whirlpool (eddy) forming because of the extraction of energy from the mean current into the eddy. This is by no means an exact physical analog to what happens in the atmosphere, however, because the sense of the flow would only support an eddy with one sense of rotation for a given side of the river, whereas rotation in either sense is supported by term B.

Analysis of the barotropic energy conversion term provides useful information concerning the nature of trough and ridge structures and their potential for growth through this mechanism. For symmetric disturbances, or those centered near the mean jet core, this mechanism will not be important. In regions with strong shear of the mean zonal flow, however, it can be important for strongly tilted eddies. For Northern Hemisphere locations on the equatorward side of the mean jet core, a "negatively tilted trough" is a configuration favorable for barotropic energy growth, while positively tilted troughs lose energy to the mean flow through this mechanism. This terminology is

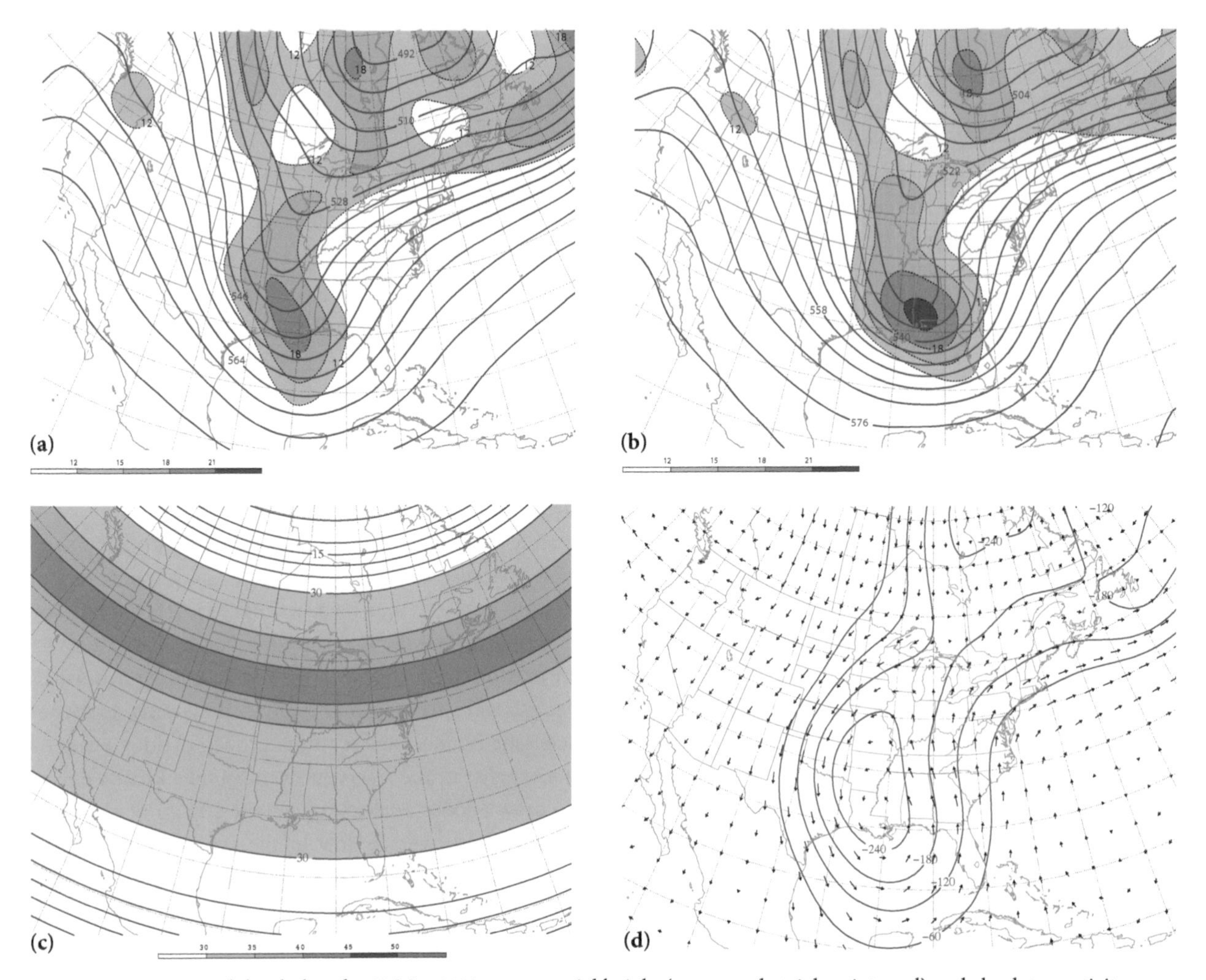


**Figure 2.24.** 500-mb level plots for 13 Mar 1993: geopotential height (contoured at 6 dam interval) and absolute vorticity (shaded as in legend) at (a) 0600 and (b) 1200 UTC; (c) zonal average of zonal wind contoured and shaded as in legend; and (d) 500-mb negative height anomaly (interval 60 m) and wind anomalies.

consistent with the sense of progression of the trough axis with increasing latitude; for example, a positively tilted trough exhibits an axis that is displaced in the positive $x$ direction with increasing $y$.

A classic example of a negatively tilted trough south of the mean jet core took place on 12–14 March 1993 in the so-called Storm of the Century in the eastern U.S. Several processes were at work to produce the highly amplified trough evident in Figs. 2.24a,b; however, one mechanism of potential importance was the barotropic kinetic energy conversion process. This trough is negatively tilted, and Fig. 2.24c confirms that $\partial \overline{u}_g / \partial y$ is positive in the region of negative trough axis tilt, ensuring that the sign of the barotropic term is positive for this system. It would require a careful computation of the entire energy budget to quantify the importance of this process in this storm; nevertheless, there is sufficient circumstantial evidence here for a forecaster to note this aspect and consider that this process would favor kinetic energy growth for the upper trough system.

Some words of caution are in order—First, not all negatively tilted troughs strengthen because of barotropic energy conversion! For instance, near the latitude of the mean jet core, where $\partial \overline{u}_g / \partial y \sim 0$, this term is negligible. Second, north of the jet core where $\partial \overline{u}_g / \partial y < 0$, a negatively tilted trough will result in the **weakening** of kinetic energy because of the barotropic conversion process. The need to compute $\partial \overline{u}_g / \partial y$ is somewhat cumbersome, but

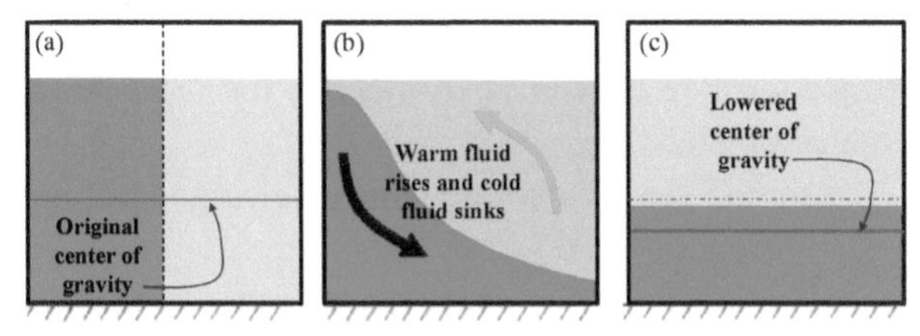


**Figure 2.25.** Schematic depiction of baroclinic energy conversion process. (a) System contains available potential energy; dashed line indicates a barrier separating dense (red) fluid and lighter (gray) fluid. (b) Once barrier is removed, dense fluid sinks and lighter fluid rises; kinetic energy is generated as potential energy is released. (c) In the final state, the center of gravity of system has been lowered, and a net loss of potential energy has occurred (adapted from Carlson 1998).

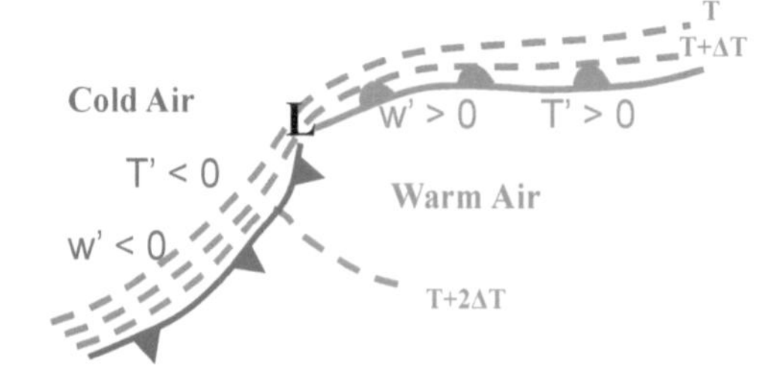


**Figure 2.26.** Schematic of baroclinic energy conversion in an idealized extratropical cyclone. Dashed lines represent isotherms, and the sign of temperature anomalies and vertical motions are indicated in select locations.

it is easily done using standard meteorological software packages. The main point of this analysis is that the structure of upper troughs and ridges is important to their energy growth or decay. Other aspects affected by these structural properties include the ability of such systems to transport momentum horizontally, as a part of the planetary-scale circulation.

The remaining right-hand term in (2.44), term C, is proportional to the correlation between the vertical motion and the temperature anomaly, averaged over the volume in question. This term represents the vertical heat flux. Unlike the barotropic conversion term, which describes the conversion of *existing* kinetic energy from the mean into the eddy flow, this term describes the *generation* of kinetic energy resulting from the conversion from potential energy, known as *baroclinic* energy conversion.

When, on average within a weather system, warm air rises and cold air sinks, this term contributes positively to the eddy kinetic energy through the conversion of potential to kinetic energy in the system. A visual analog is provided in Fig. 2.25, in which a container initially contains two fluids of different density, separated by a divider. At this initial time, because some of the dense fluid (left side in Fig. 2.25a) is located above less dense fluid (right side), there is *available potential energy* in the system that can be converted to kinetic energy. If the divider is removed, kinetic energy is generated as the dense fluid flows beneath the lighter fluid; potential energy is converted to kinetic energy at this stage (Fig. 2.25b). In the end after friction has eliminated the motion, the center of gravity of the system has lowered, and available potential energy has been converted to kinetic energy and then dissipated as thermal energy by friction (Fig. 2.25c).

If we carry this simple analogy over to a typical midlatitude cyclonic weather system, then it is clear that on average, air in regions of the system characterized by anomalous warmth (e.g., relative to a latitudinal average) is rising while anomalously cold air is sinking (Fig. 2.26). This is also related to the tendency for warm advection (cold advection) to be associated with forcing for ascent (descent), as evident from the QG omega equation; on average, there is conversion of potential to kinetic energy in a developing cyclone with this type of thermal pattern.

There are a number of useful insights offered by the consideration of baroclinic energy conversion. First, in general, the stronger the thermal contrast in the vicinity of a cyclonic system, the stronger the magnitude of the thermal advection pattern, and thus both factors in this term are maximized. Therefore, we expect that strongly developing cyclonic systems are characterized by large temperature contrasts, a finding that is entirely consistent with observation. Second, this explains the seasonality of cyclones—in the summer with generally reduced horizontal thermal contrasts, the baroclinic energy conversion term is weakened. In our discussion of cyclones in chapter 5, we will return to these arguments to explain some aspects of the observed cyclogenesis distribution.

## REVIEW AND STUDY QUESTIONS

1. Some of the assumptions used in the derivation of the QG omega and height tendency equations are questionable for midlatitude synoptic-scale cyclones. Discuss two of the most limiting assumptions, and

give an example of a meteorological situation in which these assumptions are violated.

2. If important QG assumptions are violated for some synoptic weather systems, then what is the point of presenting the QG analysis? What is the value provided by this development?
3. In the QG height tendency equation (2.32), the right-hand terms include differential thermal advection and vorticity advection. However, in a statically stable atmosphere, forced ascent can produce cooling over a deep layer of the atmosphere, potentially affecting the height field. Why is there no term on the right side of the height tendency equation related to vertical motion and adiabatic temperature change?
4. From the QG perspective, what is the ultimate cause of the synoptic-scale QG vertical motion (e.g., "weather")?
5. We often observe that cyclones form along preexisting frontal boundaries. Explain this observation within the context of the vorticity equation. Can this observation be explained using the QG form of the vorticity equation? Hint: consider the stretching term.

## PROBLEMS

1. The geostrophic wind, as defined by Eq. (2.7), is given by $\vec{V}_g = 1/f_0\, \hat{k} \times \nabla\Phi$, where $f_0$ is a constant (the $f$-plane approximation). An alternative definition is to include a *variable* Coriolis parameter $f$: $\vec{V}_g = 1/f\, \hat{k} \times \nabla\Phi$. Compute the *divergence* of the geostrophic wind using the variable $f$ definition, and compare that to the divergence of the $f$-plane definition. Discuss.
2. Using the geostrophic wind definition, show that the pressure gradient force can be expressed $-\nabla\Phi = f_0 \hat{k} \times \vec{V}_g$, equation (2.10).
3. Working from Eqs. (2.19) and (2.20), derive the QG vorticity equation (2.21). Be sure to show all of your work clearly and neatly.
4. Show that the geostrophic relative vorticity is given by $\zeta_g = 1/f_0 \nabla^2\Phi$, equation (2.22).
5. Starting from the QG thermodynamic and vorticity equations [(2.27) and (2.28)], derive the QG omega equation (2.29).
6. Derive the geostrophic eddy kinetic energy equation (2.43) from Eq. (2.42).

## REFERENCES

Bluestein, H. B., 1993: *Observations and Theory of Weather Systems.* Vol. 2, *Synoptic-Dynamic Meteorology in Midlatitudes,* Oxford University Press, 608 pp.

Carlson, T. N., 1998: *Mid-Latitude Weather Systems.* Amer. Meteor. Soc., 507 pp.

Charney, J. G., 1948: On the scale of atmospheric motions. *Geofys. Publ.,* **17,** 3–17.

Durran, D. R., and L. W. Snellman, 1987: The diagnosis of synoptic-scale vertical motion in an operational environment. *Wea. Forecasting,* **2,** 17–31.

Holton, J. R., 2004: *An Introduction to Dynamic Meteorology.* International Geophysics Series, Vol. 88, Academic Press, 535 pp.

Hoskins, B. J., and M. A. Pedder, 1980: The diagnosis of middle latitude synoptic development. *Quart. J. Roy. Meteor. Soc.,* **106,** 707–719.

——, I. Draghici, and H. C. Davies, 1978: A new look at the $\omega$-equation. *Quart. J. Roy. Meteor. Soc.,* **104,** 31–38.

Orlanski, I., and J. J. Katzfey, 1991: Simulation of an extratropical cyclone in the Southern Hemisphere: Model sensitivity. *J. Atmos. Sci.,* **48,** 2293–2312.

——, and E. K. M. Chang, 1993: Ageostrophic geopotential fluxes in downstream and upstream development of baroclinic waves. *J. Atmos. Sci.,* **50,** 212–225.

Pedlosky, J., 1979: *Geophysical Fluid Dynamics.* Springer-Verlag, 624 pp.

## FURTHER READING

Blackburn, M., 1985: Interpretation of ageostrohpic winds and implications for jet stream maintenance. *J. Atmos. Sci.,* **42,** 2604–2630.

Bluestein, H. B., 1992: *Principles of Kinematics and Dynamics.* Vol. 1, *Synoptic-Dynamic Meteorology in Midlatitudes,* Oxford University Press, 448 pp.

Sanders, F., and B. J. Hoskins, 1990: An easy method for estimation of Q-vectors from weather maps. *Wea. Forecasting,* **5,** 346–353.

# CHAPTER 3

# Isentropic Analysis

*The isentropic chart serves a twofold purpose as (1) a hydrodynamic, and (2) a thermodynamic chart—hydrodynamic in the sense that the paths of "tongues" of maximum water-vapor content serve as identifying indicators of the flow pattern and of lateral mixing; and thermodynamic with respect to indications of adiabatic changes in the air, including water vapor, as it flows along the sloping or broadly undulating isentropic surface.*

—H. R. Byers, "On the Thermodynamic Interpretation of Isentropic Charts" (1938)

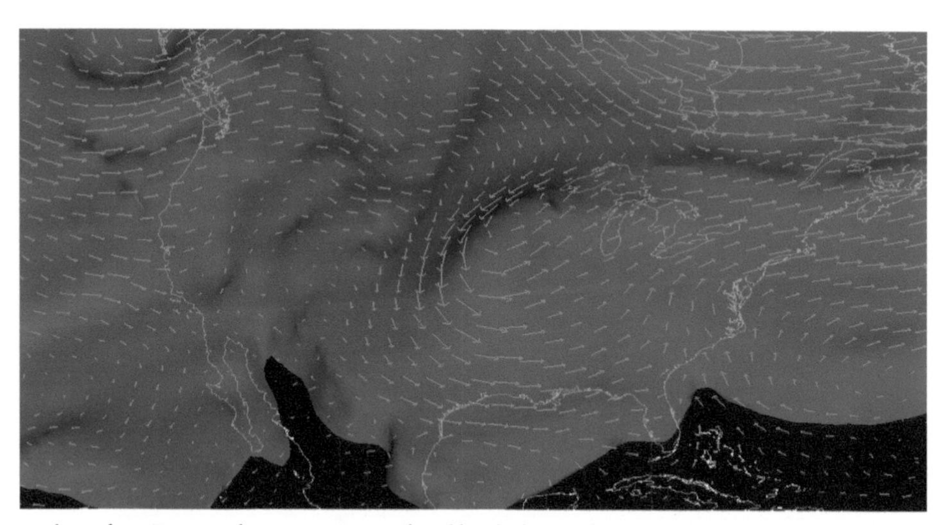

Analysis of 288-K potential temperature isosurface (blue shading) with winds (light blue vectors) valid 0000 UTC 3 Jan 2002.

In chapter 2, we developed and applied the quasigeostrophic (QG) system of equations, including the QG omega equation (2.29), which is widely used in synoptic weather forecasting. However, there are some drawbacks to the practical use of this equation, even when the more convenient **Q**-vector form (2.30) is used. The computation of Laplacians and evaluation of vertical derivatives yield noisy results unless either spatially coarse gridded data are used in the computations, or a filter is applied to the computed fields. Furthermore, to obtain the true QG omega field, one must invert the Laplacian on the left side of the equation, which is seldom done in operational practice. Finally, several restrictive assumptions were made in deriving the QG omega equation, such as the requirement of small Rossby number and horizontally uniform static stability. To include diabatic, frictional, orographic, and non-QG processes, the equation would become quite cumbersome to evaluate in an operational setting.

The limitations mentioned above should in no way diminish the importance of the understanding provided by the QG equations. That the QG vertical motion arises as an ageostrophic response to thermal wind balance disruption is an important conceptual interpretation. And despite the complications in applying the QG omega equation in weather forecasting, this framework suggests useful diagnostic quantities, such as differential vorticity advection and thermal advection, even if these fields are best used in a qualitative manner.

However, for a practical evaluation of the synoptic-scale vertical motion field, there are other options that offer some unique advantages. Some forecasters might opt to simply view the model-predicted vertical motion field. However, high-resolution model vertical motion fields are often noisy, and by simply viewing the model-output vertical velocity, much is lost in terms of understanding the associated physical processes.

The *isentropic analysis* framework is an alternate method that is conceptually simple and insightful, and is, in some respects, easier to use than the QG omega equation. Our objectives here are to present this method for diagnosis of vertical motion and moisture transport and to provide a conceptual framework for visualizing airflow in the vicinity of cyclones and fronts. The results obtained from the isentropic technique are generally consistent with those obtained from the QG omega equation for synoptic-scale motions; however, the method for visualizing and identifying regions of vertical motion differs, and several of the assumptions needed in derivation of the QG system are not required in isentropic analysis.

## 3.1. BASICS

For adiabatic motions, the first law of thermodynamics (1.21) can be expressed as

$$0 = C_p\, dT - \alpha dp. \tag{3.1}$$

Following vertical integration in pressure, we obtain Poisson's equation (1.24), repeated here as

$$\theta = T\left(\frac{p_0}{p}\right)^{\frac{R}{C_p}}. \tag{3.2}$$

Recall that the quantity *entropy* ($s$) is related to potential temperature by $s = C_p \ln(\theta) + const$; thus, if $\theta$ is conserved for a given process, then so is the entropy. This is why lines of constant potential temperature are called *isentropes* (lines of constant entropy).

As discussed in section 1.3, for a statically stable atmosphere, $\theta$ increases with height. For adiabatic motions, air parcels are *thermodynamically constrained* to move along isentropic surfaces (Fig. 3.1). In contrast, moving air parcels generally *do not* follow *isobaric* surfaces. This is a major advantage of the isentropic framework over others: by viewing flow on isentropic surfaces, one can trace the actual path of air parcels in three dimensions (provided that specific conditions, such as adiabatic flow, are met).

Consider a small fluid element of unit area with depth $\delta z$ bounded above and below by isentropes $\theta_1$ and $\theta_2$. The depth of the element can also be measured in isentropic coordinates as $\delta\theta$, as depicted in Fig. 3.2.

The mass contained in this volume is

$$M = \rho\, \delta A\, \delta z, \tag{3.3}$$

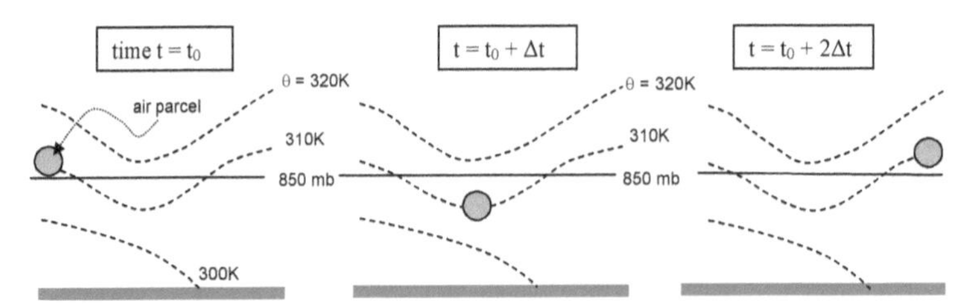


**Figure 3.1.** Idealized sequence of adiabatic air parcel motion on the 310-K isentropic surface at three times. Dashed lines represent isentropes, the solid horizontal black line represents the 850-mb pressure surface, and the thick gray lines represent the surface in each of the three panels.

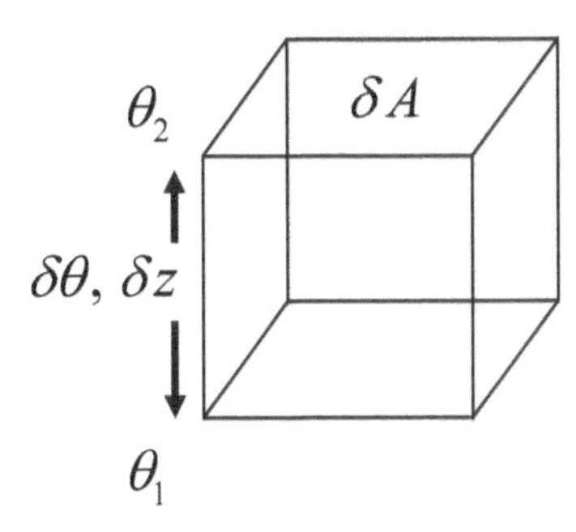


**Figure 3.2.** Idealized isentropic volume bounded by unit area $\delta A$ on top and bottom, of depth $\delta z$.

and for a small volume the hydrostatic approximation can be expressed as

$$\frac{\delta p}{\delta z} = -\rho g, \tag{3.4}$$

allowing (3.3) to be written as

$$M = -\delta A \frac{\delta p}{g}. \tag{3.5}$$

For a small volume, $\delta p = (\partial p / \partial \theta)\delta\theta$, so that we can express (3.5) as

$$M = \frac{\delta A}{g}\left(-\frac{\partial p}{\partial \theta}\right)\delta\theta. \tag{3.6}$$

Using the definition of density (mass divided by volume), we obtain an expression for the *isentropic density*, $\sigma$, of the volume depicted in Fig. 3.2 by dividing (3.6) by $\delta A \delta\theta$ to yield

$$\sigma = -\frac{1}{g}\frac{\partial p}{\partial \theta}. \tag{3.7}$$

Thus, we find that the isentropic density is inversely proportional to the static stability, $\partial\theta / \partial p$. This quantity will be utilized in the following chapter in deriving the isentropic potential vorticity.

## 3.2. CONSTRUCTION AND INTERPRETATION OF ISENTROPIC CHARTS

Pressure on an isentropic chart does not play the same role as geopotential height on an isobaric chart, although a quantity known as the *Montgomery streamfunction* ($\Psi_M$) can be used to define the geostrophic flow on an isentropic surface (Montgomery, 1937). However, the utility of isentropic charts lies in the diagnosis of 3D air trajectories and vertical motions, which can be shown using plots of pressure and wind on a given isentropic surface. Plots of pressure on an isentropic surface are essentially topographic maps of a given isentropic surface measured in pressure.

A rawinsonde sounding from Greensboro, North Carolina, at 0000 UTC 3 January 2002 is provided in Fig. 3.3. Recall that dry adiabats on a skew $T$–log$p$ diagram correspond to specific values of potential temperature, and these can easily be deduced by following the adiabats to the 1000-mb level; the dry bulb temperature of an adiabat at this level is equal to the potential temperature value, and these values are labeled in Fig. 3.3.

Suppose that we wish to construct an isentropic chart for the 296-K surface. From this diagram, we could estimate that the pressure corresponding to the 296-K isentrope at Greensboro is approximately 700 mb. By plotting the pressure values corresponding to the 296-K level for each rawinsonde site on a horizontal map, isobaric analysis of these values provides a complete view of the topography of this isentropic surface. Using a relatively coarse contour interval of 100 mb, the 296-K isobaric analysis in Fig. 3.4 reveals a region of lower pressure (higher altitude) for this surface centered over Missouri. We see that the pressure value from Greensboro, North Carolina, is 702 mb, close to our estimate from Fig. 3.3.

The plan-view analysis of the 296-K isentropic surface in Fig. 3.4 can also be visualized in three dimensions using the Unidata Integrated Data Viewer (IDV) to better illustrate the topography of this surface (Fig. 3.5). The 3D wind vectors are superimposed on this image to illustrate the sense of the flow, which is consistent with the rawinsonde wind measurements in Fig. 3.4, despite the fact that this graphic is based on the North American Mesoscale (NAM) analysis rather than solely on rawinsonde observations. The region of lowest isentropic pressure, centered over Arkansas in Fig. 3.4, appears as a region in which the isentropic surface bows upward in Fig. 3.5.

It is not the geopotential height but the Montgomery streamfunction that enters the definition of geostrophic wind on an isentropic surface. The Montgomery streamfunction is given by

$$\Psi_M = (C_p T + gZ)_\theta, \tag{3.8}$$

where $g$ is the gravitational acceleration and $Z$ is the geopotential height. The subscript $\theta$ indicates that the values are taken on a surface of constant potential temperature.

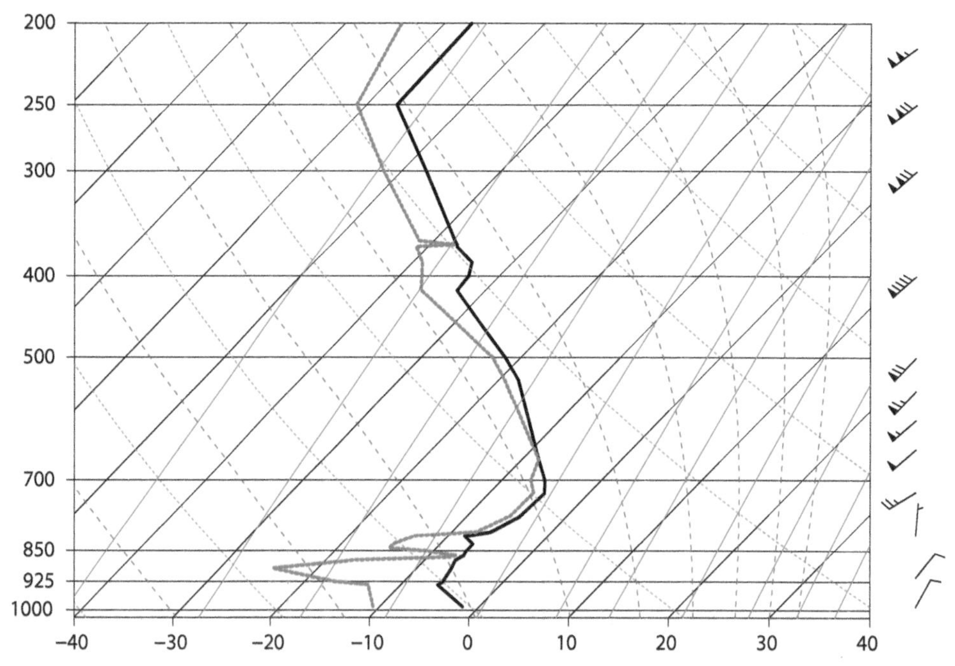


**Figure 3.3.** Skew *T*-log*p* diagram from Greensboro, NC, from 0000 UTC 3 Jan 2002.

In this case, the geostrophic wind components are given by

$$v_{g\theta} = \frac{1}{f_0}\frac{\partial \Psi_M}{\partial x} \text{ and } u_{g\theta} = -\frac{1}{f_0}\frac{\partial \Psi_M}{\partial y}. \tag{3.9}$$

The relation between the geostrophic wind on an isentropic surface and lines of constant Montgomery streamfunction is exactly analogous to the relation between the geostrophic wind on an isobaric surface and lines of constant geopotential height. The speed of the

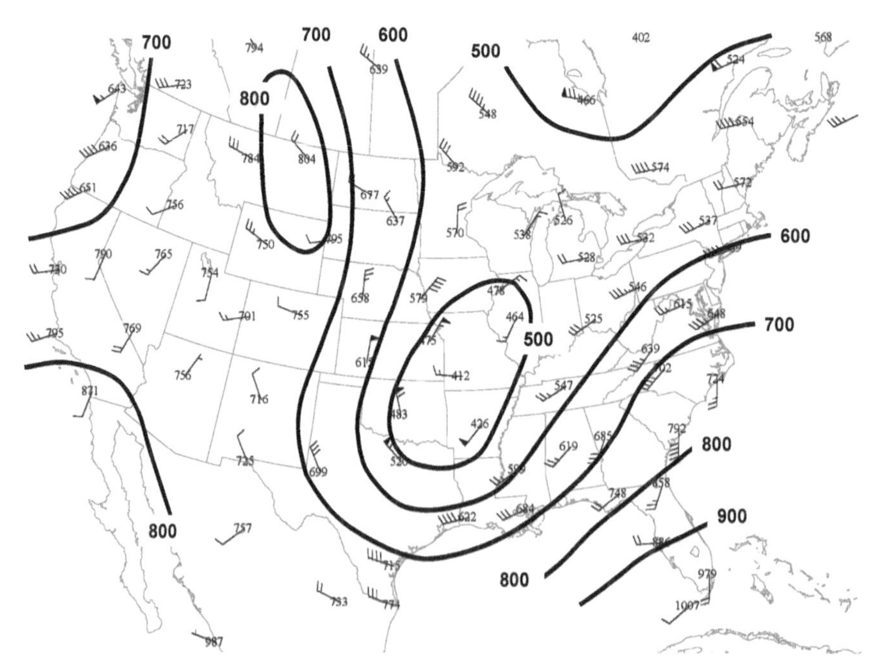


**Figure 3.4.** Pressure analysis (interval 100 hPa) on the 296-K isentropic surface for 0000 UTC 3 Jan 2002.

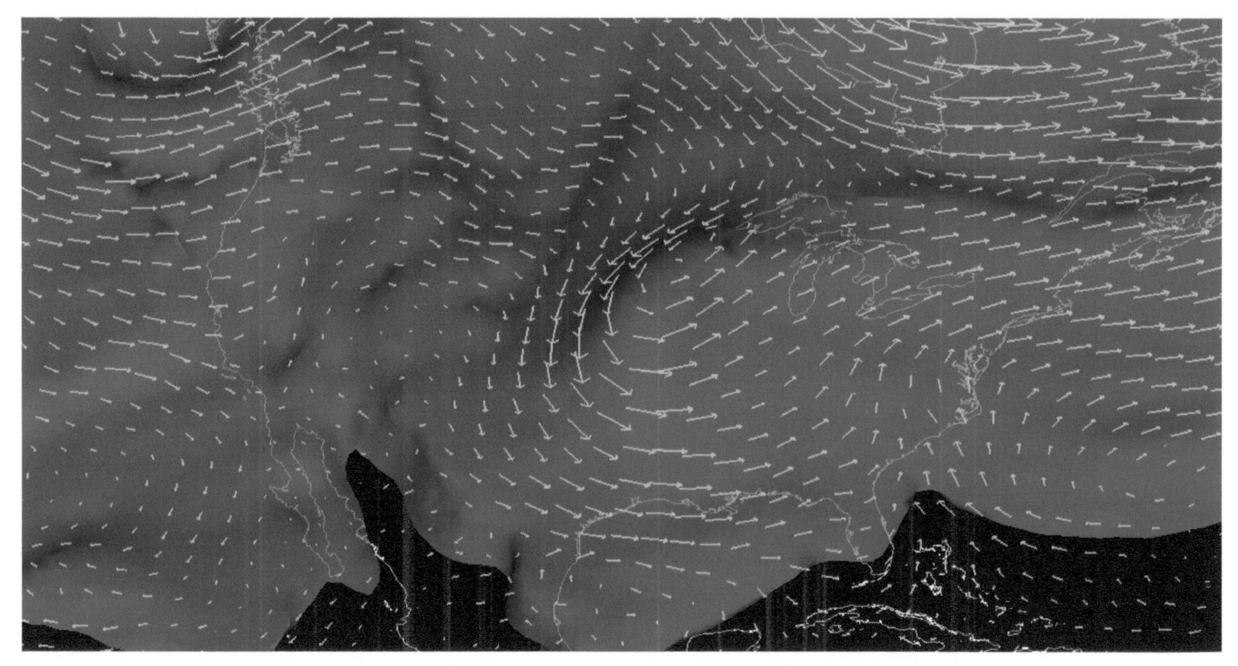

**Figure 3.5.** Three-dimensional rendering of 296-K potential temperature isosurface valid 0000 UTC 3 Jan 2002 (blue shading). Wind vectors have been interpolated to this surface (light blue vectors). Data source is NAM analysis.

geostrophic wind is directly proportional to the gradient of $\Psi_M$ and oriented parallel to $\Psi_M$ contours with lower values to the left in the Northern Hemisphere.

When computing the Montgomery streamfunc-tion from observational data, care must be taken when the expression (3.8) is used (Danielsen 1959; Bleck and Haagenson 1968). However, modern data assimilation systems used in the generation of gridded analyses provide sufficient accuracy to allow useful application of (3.8).

### 3.3. REPRESENTATION OF VERTICAL MOTION ON AN ISENTROPIC SURFACE

Using the definition of omega, and evaluating horizontal derivatives on isentropic surfaces, we can write an equation for the vertical air motion in isentropic coordinates:

$$\left(\frac{dp}{dt}\right)_{\theta=const} \equiv \omega = \underbrace{\left(\frac{\partial p}{\partial t}\right)_{\theta}}_{A} + \underbrace{\vec{V}\cdot\nabla_{\theta}p}_{B} + \underbrace{\frac{\partial p}{\partial \theta}\frac{d\theta}{dt}}_{C}. \quad (3.10)$$

The three right-hand terms in (3.10) can be interpreted as follows: Term A represents the local pressure tendency. This is the local derivative of pressure with respect to time and accounts for the vertical movement of an isentropic surface at a fixed location. This can arise from local pressure tendencies, for example.

Term B in (3.10) is the pressure advection, which is the dominant right-hand term in many situations. It follows from Poisson's equation that an isobar on an isentropic surface is also an isotherm; if both $p$ and $\theta$ are constant, then $T$ must be as well. So, the interpretation of the pressure advection term is related to isentropic temperature advection. The sign of this term can be evaluated conveniently by examining *the cross-isobar wind component* on an isentropic analysis. If the wind flow is directed from higher toward lower pressure on an isentropic surface, then rising motion is implied. Airflow from lower toward higher pressure implies subsidence.

Diabatic processes also change potential temperature, and term C represents these heating and cooling effects. Diabatic processes such as radiation or the absorption or release of latent heat can cause isentropic surfaces to move vertically (Fig. 3.6). Put another way, a nonzero term C means that air parcels are moving from one isentropic surface to another.

We can simplify Eq. (3.10) by eliminating term A through the use of an assumption dubbed the *frozen wave approximation* by T. N. Carlson (more specifically, Carlson credits his former student John Takacs for this apt terminology). If we assume that the local change in a

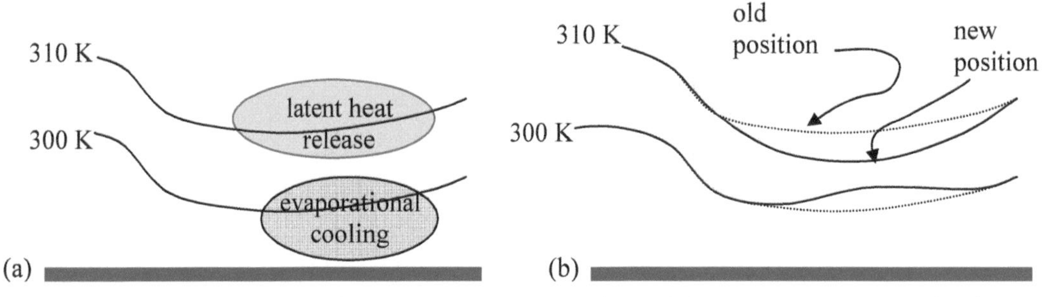

**Figure 3.6.** Schematic depiction of vertical isentropic motion due to diabatic processes at (a) initial and (b) later time. Locations of heating and cooling are shown for initial time only (after Moore 1993).

given variable is *due only to the motion of a synoptic-scale weather system*, and we assume that the system is not rapidly changing its speed of movement or shape with time, then we can write the expression for the change of any scalar variable ($S$) as

$$\left(\frac{dS}{dt}\right)_{system} = \frac{\partial S}{\partial t} + \vec{c}\cdot\nabla S \approx 0, \quad (3.11)$$

where $\vec{c}$ is the phase velocity of the weather system. The frozen wave approximation holds that the value of $S$ is constant *following the motion of the weather system*, as illustrated for the case of the value of geopotential height at the point indicated in Fig. 3.7. If we let $S$ = pressure ($p$), and solve for the local time tendency of pressure, then we can substitute from (3.11) into (3.10) to obtain

$$\omega \approx -\vec{c}\cdot\nabla_\theta p + \vec{V}\cdot\nabla_\theta p + \frac{\partial p}{\partial \theta}\frac{d\theta}{dt}, \text{ or}$$

$$\omega \approx (\vec{V} - \vec{c})\cdot\nabla_\theta p + \frac{\partial p}{\partial \theta}\frac{d\theta}{dt} \quad (3.12)$$

This equation relates the isentropic vertical motion to the *system-relative* horizontal air motion; as we will see, this allows us to identify storm-relative airstreams within a given weather system. We will apply this reasoning to both cyclones and fronts. At this point, we will explore the computation and interpretation of the storm-relative isentropic flow before addressing the diabatic term in (3.12).

In practice, there can be difficulty in accurately evaluating the system speed, $\vec{c}$. Sometimes, forecasters will neglect to evaluate the system motion and simply view the full rather than the storm-relative isentropic wind field; this is essentially equivalent to estimating $\omega$ based solely on term B in (3.10). For weather systems whose phase speed is much smaller in magnitude than the wind speed on a given isentropic surface, this approximation may be adequate; however, for rapidly moving systems, even the sign of $\omega$ may be misdiagnosed if the storm-relative flow is approximated by term B alone.

We will now extend our analysis of the example from Figs. 3.4 and 3.5 as a means of illustrating the relation between the QG methods developed previously and the isentropic technique presented here. Although the presentation will be limited to a single time from a single case, this will serve to demonstrate the strengths and weaknesses of the isentropic framework for the diagnosis and prediction of synoptic-scale vertical motion.

In early January 2002, a snowstorm affected parts of the southeastern United States. Useful QG-based graphics, including 500-mb height and vorticity (Fig. 3.8a), and

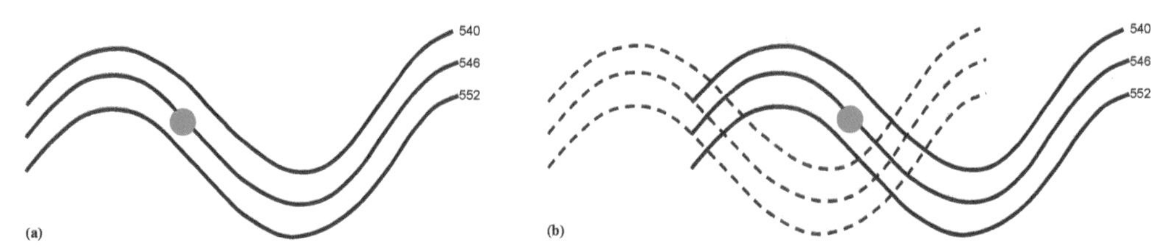

**Figure 3.7.** Idealized schematic illustrating the frozen wave approximation. A storm-relative location at an arbitrary point along a wave in the 500-mb height field is marked by the red circle, and the wave is moving from left to right with time: (a) initial time and (b) later time, with the outline of the wave at the earlier time shown in dashed contours.

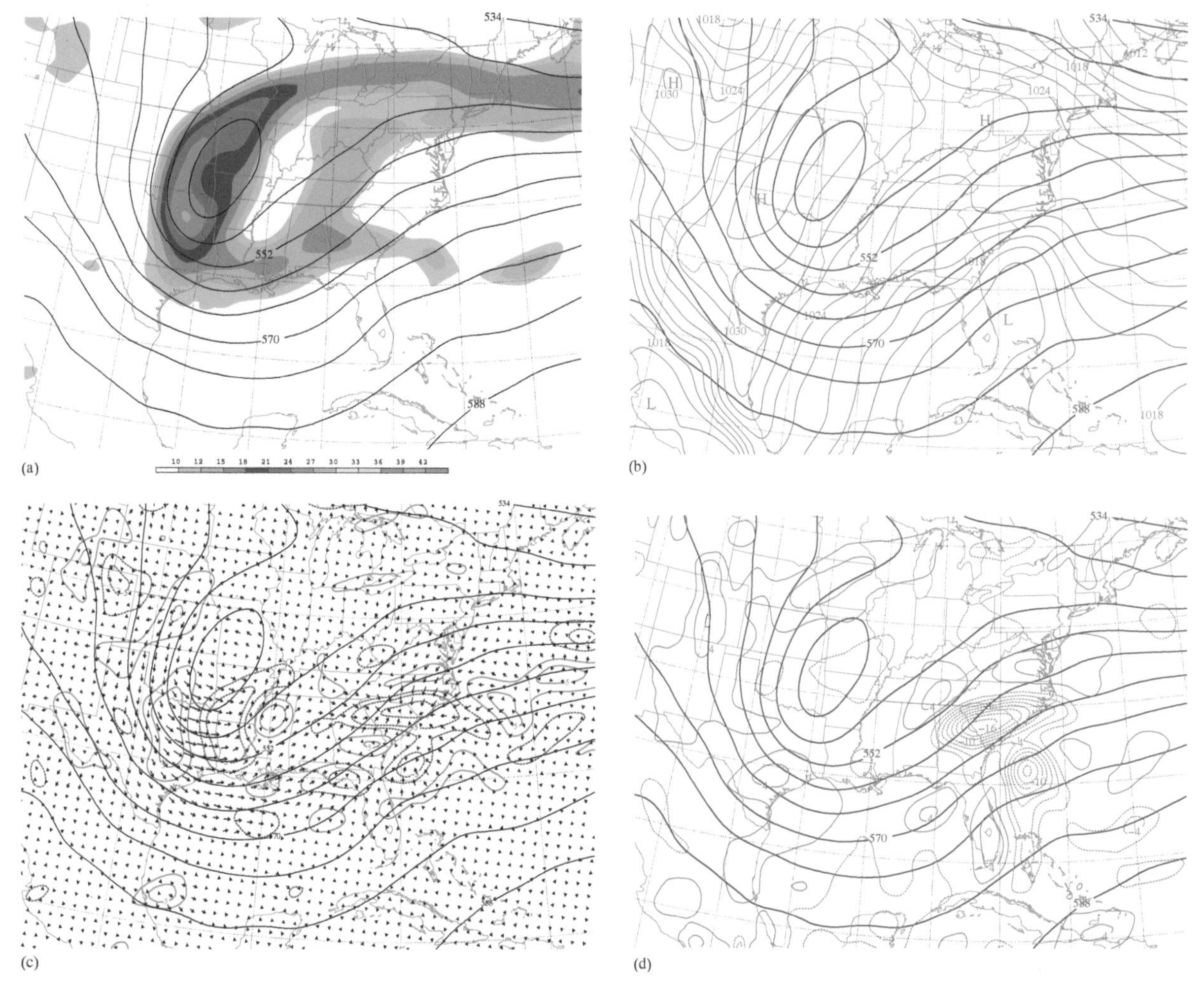


**Figure 3.8.** Summary for 0000 UTC 3 Jan 2002: (a) 500-mb height (contours) and absolute vorticity (shaded as in legend); (b) as in (a) but replacing vorticity with sea level pressure; (c) 500-mb height (contours), **Q**-vectors (arrows), and **Q**-vector convergence and divergence (dashed blue for convergence, solid red for divergence); (d) 500-mb height (solid blue contours) and 700-mb omega (red contours, ascent dashed, descent solid).

an overlay of the 500-mb height and sea level pressure (Fig. 3.8b) exemplify graphics that could be used for diagnosis of the traditional form of the QG omega equation. A strong upper low, evident in 500-mb height contours, is centered over southern Missouri, with an intense cyclonic vorticity maximum southwest of the low center. A secondary elongated vorticity maximum extends eastward along the Gulf Coast states, crossing Alabama and extending into Georgia. These fields imply cyclonic vorticity advection over eastern Texas, western Louisiana, and southwestern Arkansas (Fig. 3.8a). Cyclonic vorticity advection is also implied over northern Georgia and South Carolina. If we assume that the advection is increasing with height up to this level, then QG forcing for ascent is found in these regions. However, Fig. 3.8b, which depicts sea level pressure superimposed on the 500-mb height, indicates strong geostrophic backing and cold advection over eastern Texas, implying cancellation between the traditional QG terms, rendering the QG vertical motion diagnosis ambiguous there. Farther east, warm advection, implied by geostrophic veering between sea level and the 500-mb level, is consistent with forcing for ascent in the coastal portions of the Carolinas.

As discussed in the previous section, the **Q**-vector form of the omega equation solves the issue of cancellation in the traditional form of the QG omega equation, which is evident in comparison of Figs. 3.8a–c. A region of strong **Q**-vector convergence is observed south of the upper trough, over southern Arkansas and eastern Louisiana, consistent

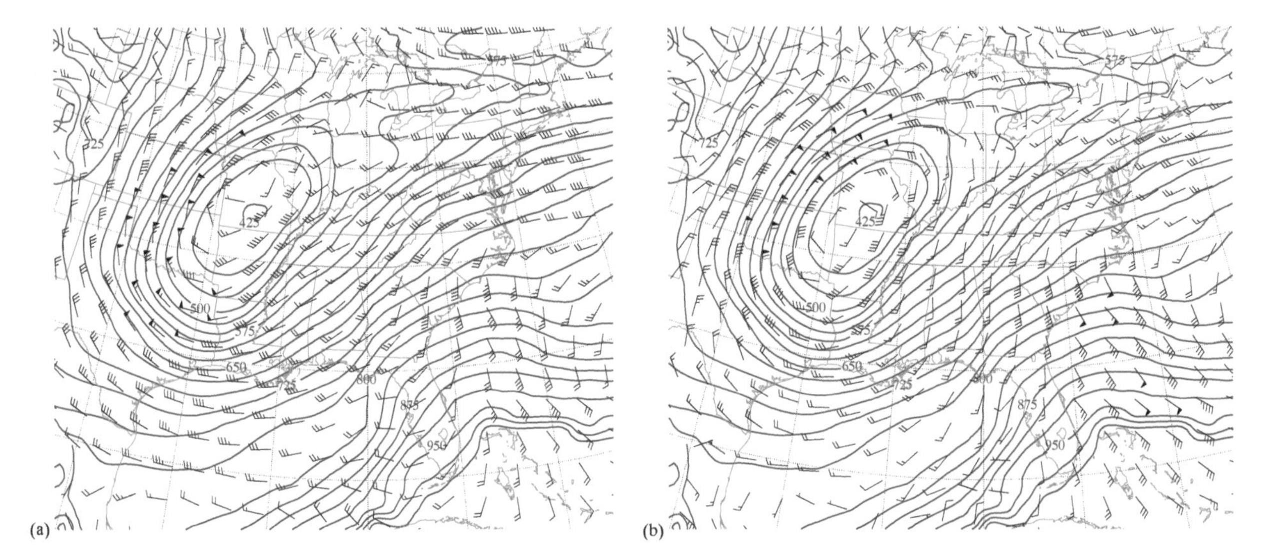


**Figure 3.9.** Pressure and wind on the 296-K isentropic surface, valid 0000 UTC 3 Jan 2002: (a) analyzed isentropic wind and (b) storm-relative isentropic wind.

with cyclonic vorticity advection (presumably increasing with height) identified in Fig. 3.8a. A second region of **Q**-vector convergence is evident over the western Carolinas, suggesting forcing for ascent there. Now, plotting the *analyzed* 700-mb vertical motion field for this time, we find strong ascent extending from central Georgia to eastern North Carolina (Fig. 3.8d). While there is weak ascent implied over eastern Louisiana northward to Arkansas, the magnitudes are far less than that over the Carolinas and Georgia.

Turning now to a plot of wind and pressure on the 296-K isentropic surface (Fig. 3.9a), derived from the same model analysis used to generate Fig. 3.8, a different picture emerges. There is strong ascent implied from a region east of Florida extending northward into Georgia, South Carolina, and southern North Carolina, nearly matching the pattern of strong ascent seen in Fig. 3.8d. The pattern of "isentropic upglide" corresponds closely to the region of warm advection evident in Fig. 3.8b.

As emphasized in the development of (3.12), it is the *storm-relative* isentropic flow that is most closely related to isentropic vertical air motion. While determining the storm-motion vector can be complex, closed centers or features on an isentropic surface can be tracked with time to establish the storm-motion vector. Here, a series of 12-hourly analyses was used to determine a system motion vector of 8.5 m/s toward the east-southeast. Subtracting this storm motion vector from the analyzed winds shown in Fig. 3.9a provides the *storm-relative isentropic flow* shown in Fig. 3.9b. By subtracting the system motion vector from the total wind to obtain the system-relative flow, the orientation of the storm-relative wind turns more strongly toward the region of lower isentropic pressure, consistent with the eastward motion of the system in this case. Thus, the pattern of storm-relative isentropic ascent expands and now includes weak ascent in eastern Mississippi and Alabama, along with strong ascent across Georgia, the Carolinas, and adjacent offshore locations.

Given that adiabatic airflow is largely constrained to follow isentropic surfaces, this framework is useful for the visualization of moisture transport. The plotting of *isentropic condensation pressure* in place of specific humidity has been advocated by Byers (1938) and others; however, plotting mixing ratio or specific humidity is more convenient, and it still provides a useful depiction of moist airstreams. In Fig. 3.10, which displays the isentropic mixing ratio on the 296-K surface, it is clear that abundant Atlantic moisture is feeding into the region of strong isentropic lift over the southeastern United States. The reduction in mixing ratio values following streamlines is consistent with ascent and vapor removal by condensation and precipitation. Moist airstreams, often accompanying a low-level jet (LLJ), have been referred to as the *warm conveyor belt*, or, more recently, as *atmospheric rivers*. In chapter 5, we will discuss the typical airstreams associated with midlatitude cyclones in more detail.

The lack of moisture entering the zone immediately ahead of the upper trough over Missouri in this example is consistent with a lack of precipitation there (Fig. 3.11). A close correspondence between areas of precipitation

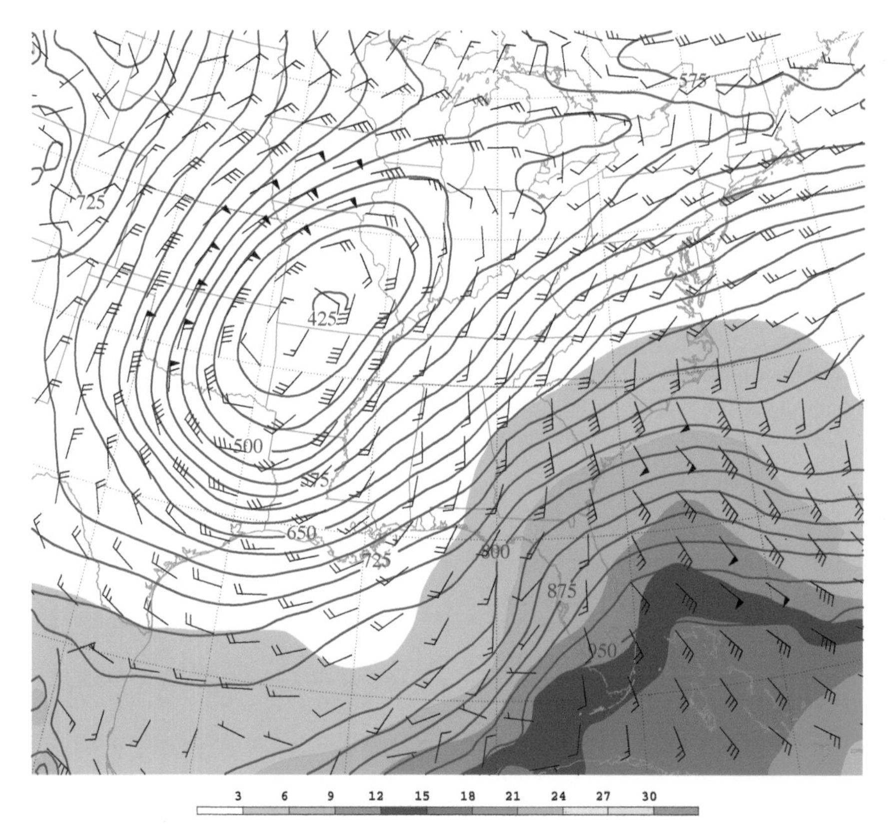


**Figure 3.10.** As in Fig. 3.9b, but with isentropic mixing ratio added to display, shaded beginning at 3 g kg$^{-1}$, as in legend.

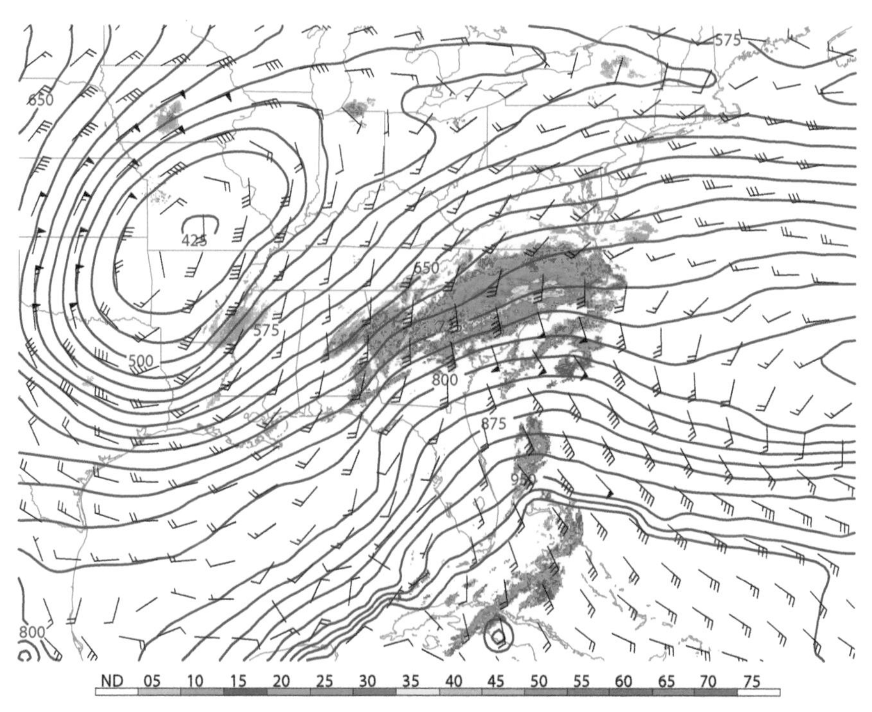


**Figure 3.11.** As in Fig. 3.9, but with composite radar reflectivity mosaic superimposed.

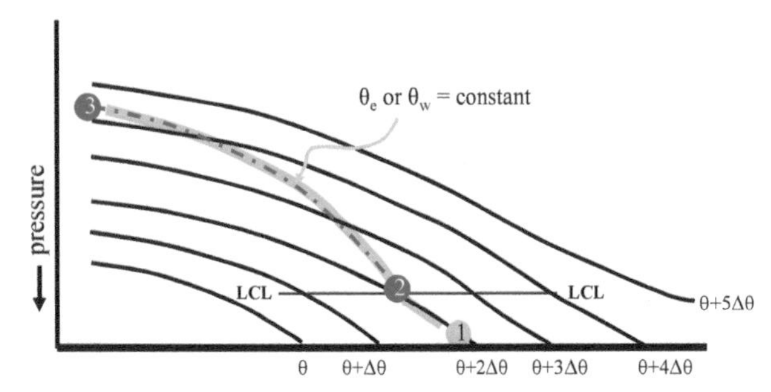


**Figure 3.12.** Idealized cross section showing isentropes (solid black contours) and three positions of a hypothetical air parcel, labeled in temporal order. The LCL is indicated, corresponding to position 2 of the air parcel [adapted from Fig. 12.4 of Carlson (1998)].

and isentropic lift is evident in this graphic, although the lack of radar reflectivity in the zone of strong isentropic lift over the Atlantic east of Florida is almost certainly due to the lack of radar coverage there.

At this point the reader may be wondering why we should care about systems with *adiabatic* airflow. In the weather systems that matter, condensation takes place and is typically strong. Does that not violate one of the fundamental assumptions in developing this technique? It is true that this technique is often applied in situations characterized by strong latent heat release, but the situation is not as bad as one would think, at least for qualitative purposes.

Suppose that an air parcel is rising along a sloping isentrope beginning at position 1 in Fig. 3.12. The parcel will move along an isentropic surface, following a dry adiabat, until reaching position 2, which corresponds to the lifting condensation level (LCL) for the parcel. At this point, further lifting results in the release of latent heat of condensation, and the parcel cools more slowly than the dry adiabatic rate and therefore no longer follows the isentrope it started out upon. Above the LCL, the parcel follows a moist adiabatic path given by a line of constant equivalent potential or wet-bulb potential temperature, as indicated. While the adiabatic assumption is violated, we see that the motion is not only still upward, but it is more strongly upward than one would infer based on dry isentropic lift alone. Thus, violation of the adiabatic assumption only results in an underestimation of the isentropic lift, and the qualitative result remains the same.

The isentropic framework for the diagnosis of vertical motion is visually appealing, convenient, and when properly applied, accurate. It is also consistent with the previously discussed QG techniques, although care must be taken to compare the same altitude range when undertaking a direct comparison of these techniques. Based on the example shown here, and everyday use in forecasting, it is easy to understand why many forecasters prefer the isentropic framework to traditional QG techniques. Some of the advantages of the isentropic framework are the following:

- clear, visual depiction of air parcel motion and three-dimensional airflow in weather systems, including vertical motions, and moisture transport;
- conceptual simplicity, an *explicit* representation of vertical motion on horizontal maps;
- that the adiabatic assumption is quite good much of the time and when it is not valid, the qualitative answer remains unchanged; and
- the QG assumptions of small Rossby number and horizontally uniform static stability are not needed.

Some disadvantages of isentropic analysis include:

- often in the areas we are most interested in, diabatic processes *are* important;
- sometimes $\theta$ does not increase with height, and the $\theta$ surfaces fold and intersect the ground, complicating practical application of the technique;
- computations must be performed to interpolate pressure, wind, and moisture data onto isentropic surfaces prior to analysis;
- the isentropic technique fails to provide *insightful dynamical interpretation* of cause and effect the way the QG framework does; and
- users of this technique must remain cognizant of the altitude of the isentropic surface being examined; the appropriate isentropic surface to examine depends on the season, latitude, and the phenomenon in question. One must typically look at several layers, and the relevant isentropic surfaces vary with season and location.

For the diagnosis of lower-tropospheric vertical motion, it is often most useful to select isentropic surfaces that fall between the 850 and 700-mb levels over the region of interest. This pressure range often corresponds to that in which precipitation formation is taking place and is sufficiently far above the surface to be free from the complicating influences of the planetary boundary layer and terrain. For elevated terrain, one would need to evaluate warmer isentropic surfaces, corresponding to higher altitudes. Some might consider the need to think carefully in the selection of an isentropic surface as an advantage rather than a disadvantage, as it demands some useful meteorological awareness!

For analysis of upper-tropospheric phenomena—such as tropopause folds, jet streams, and upper-level fronts—the selection of isentropic surfaces also requires care, and the choice varies with season and location. When isentropic analysis is used in conjunction with potential vorticity concepts, it is possible to diagnose the entire global circulation from an isentropic framework, as Hoskins (1991) demonstrates.

In summary, the isentropic vertical coordinate has a wide variety of uses, including in numerical models, in analysis of stratosphere–troposphere exchange, and even in diagnosing the global circulation (e.g., Bleck and Benjamin 1993; Hoerling et al. 1993; Hoskins 1991). Here, the presentation was limited to the diagnosis of isentropic synoptic-scale vertical motion as a convenient alternative to the QG methods introduced in chapter 2. However, the isentropic and QG vertical motion should not be viewed as corresponding to fundamentally different circulation systems. For example, in the case of a maximum of warm advection poleward of an advancing warm front, both the QG and isentropic methods would indicate ascent. In the QG system, the vertical motion is viewed as a secondary response to thermal wind balance disruption by thermal advection in the primary circulation; whereas in the isentropic framework, the ascent would be viewed as the vertical component of the primary circulation.

## REVIEW AND STUDY QUESTIONS

1. Suppose that you are analyzing a cyclonic weather system using isentropic analysis and notice an area of strong isentropic lift in the warm-frontal region (where the storm-relative winds are oriented toward lower pressure). If you were to instead have used the quasigeostrophic omega equation for this diagnosis, would you expect to have found similar results? Discuss.
2. During an event characterized by strong westerly winds, you are interested in forecasting the cloud cover for a location immediately east of the Rocky Mountains. Would you expect consistency between the QG and isentropic vertical motion in this setting? Why or why not?
3. A critical advantage of the isentropic framework is that air parcels undergoing adiabatic motion are constrained to move along isentropic surfaces. However, for many atmospheric systems of interest, the flow is not adiabatic. Is the isentropic framework invalid for these systems? Discuss.
4. The vertical motion in isentropic coordinates $d\theta/dt$ is zero for adiabatic conditions. Does this mean that only geometrically horizontal air motion is possible for adiabatic flow? Explain.
5. The equivalent potential temperature, $\theta_e$, is conserved for a wider range of conditions than the potential temperature $\theta$. Why would it often be impractical to use $\theta_e$ as a vertical coordinate?
6. What are the units of the isentropic density (3.7)? Explain, physically, why the units differ from more traditional measure of density, $\rho$.

## PROBLEMS

1. The plot below shows pressure and storm-relative wind on the 308-K isentropic surface for 0600 UTC 9 October 2007, based on a Global Forecast System (GFS) analysis. For the points A–E listed below, indicate the most likely vertical motion as rising, sinking, or weak.
2. The 308-K isentropic chart below shows pressure and total (not storm relative) wind. Assume the system is moving east at 20 kt, that diabatic effects are small, and that the frozen wave approximation is valid. Estimate the vertical motion at points A–C as ascent, descent, or weak.

## REFERENCES

Bleck, R., and P. L. Haagenson, 1968: Objective analysis on isentropic surfaces. NCAR Tech. Note NCAR-TN-39, 27 pp.

——, and S. G. Benjamin, 1993: Regional weather prediction with a model combining terrain-following and isentropic coordinates. Part I: Model description. *Mon. Wea. Rev.,* **121,** 1770–1785.

Byers, H., 1938: On the thermodynamic interpretation of isentropic charts. *Mon. Wea. Rev.,* **66,** 63–68.

Carlson, T. N., 1998: *Mid-Latitude Weather Systems.* American Meteorological Society, 507 pp.

Danielsen, E., 1959: The laminar structure of the atmosphere and its relation to the concept of a tropopause. *Arch. Meteor. Geophys. Bioklimatol.,* **11A,** 293–332.

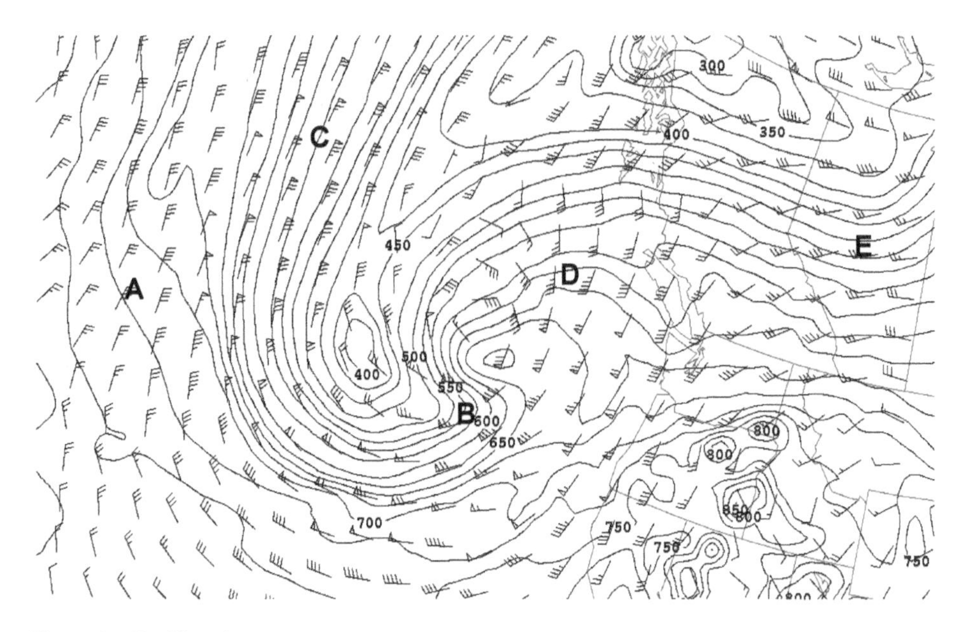


**Figure for Problem 1**


Hoerling, M. P., T. K. Schaack, and A. J. Lenzen, 1993: A global analysis of stratospheric–tropospheric exchange during northern winter. *Mon. Wea. Rev.,* **121,** 162–172.

Hoskins, B. J., 1991: Towards a PV-theta view of the general circulation. *Tellus,* **43,** 27–35.

Montgomery, R. B., 1937: A suggested method for representing gradient flow in isentropic surfaces. *Bull. Amer. Meteor. Soc.,* **18,** 210–212.


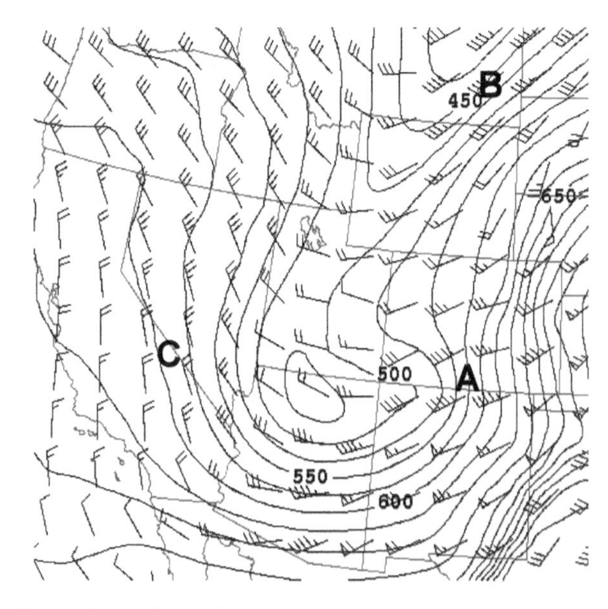


**Figure for Problem 2**


Moore, J. T., 1993: Isentropic analysis and interpretation: Operational application to synoptic and mesoscale forecast problems. NWS Training Center Manual, 99 pp.


## FURTHER READING


Browning, K. A., and T. W. Harrold, 1969: Air motion and precipitation growth in a wave depression. *Quart. J. Roy. Meteor. Soc.,* **95,** 288–309.

Carlson, T. N., 1980: Airflow through midlatitude cyclones and the comma cloud pattern. *Mon. Wea. Rev.,* **108,** 1498–1509.

Danielsen, E., 1968: Stratospheric-tropospheric exchange based on radioactivity, ozone and potential vorticity. *J. Atmos. Sci.,* **25,** 502–518.

Green, J. S. A., F. H. Ludlam, and J. F. R. McIlveen, 1966: Isentropic relative-flow analysis and the parcel theory. *Quart. J. Roy. Meteor. Soc.,* **92,** 201–219.

Market, P. S., J. T. Moore, and S. M. Rochette, 2000: On calculating vertical motions in isentropic coordinates. *Natl. Wea. Dig.,* **24,** 31–37.

——, and C. C. Buonanno, 1994: The eastern Missouri sleet storm of 29–30 December 1990: A diagnostic analysis. *Natl. Wea. Dig.,* **18,** 16–28.

Namias, J., 1956: Isentropic analysis. *Weather Analysis and Forecasting,* S. Petterssen, Ed., McGraw-Hill, 351–377.

Petersen, R. A., 1986: Detailed three-dimensional analysis using an objective cross-sectional approach. *Mon. Wea. Rev.,* **114,** 719–735.

Rossby, C.-G., 1937: Isentropic analysis. *Bull. Amer. Meteor. Soc.,* **18,** 201–209.

Saucier, W. J., 1955: Isentropic analysis. *Principles of Meteorological Analysis.* Dover, 250–260.

# CHAPTER 4

# The Potential Vorticity Framework

*This quantity, which may be called the potential vorticity, represents the vorticity the air column would have if it were brought, isopycnically or isentropically, to a standard latitude ($f_0$) and stretched or shrunk vertically to a standard depth $D_0$ or weight $\Delta_0$.*

—C.-G. Rossby,
"Planetary flow patterns in the atmosphere" (1940)

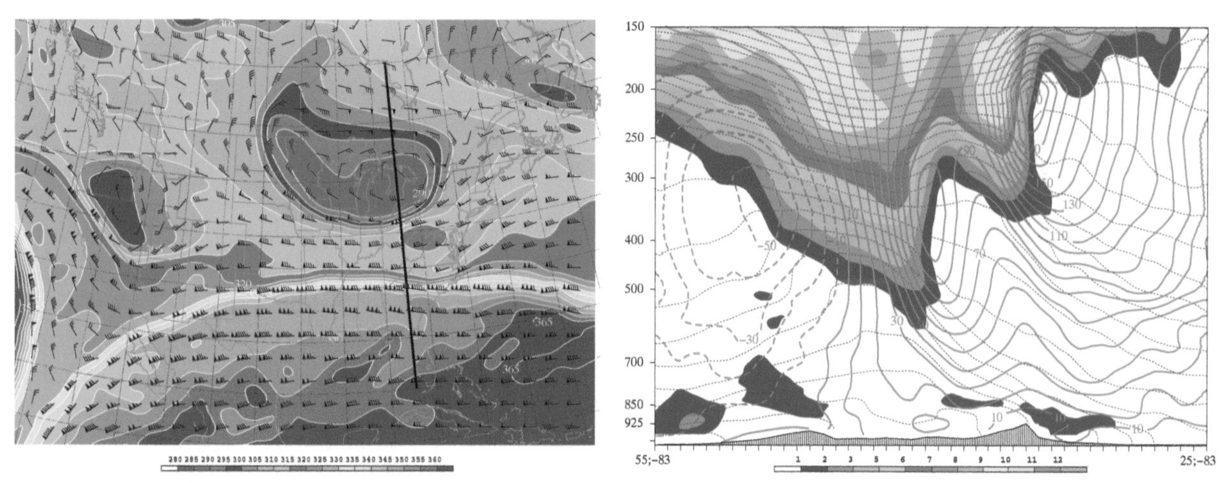


Plan view of potential temperature on the dynamic tropopause (left) with cross-section orientation indicated. (right) Cross section of potential vorticity (shaded), isentropes (blue contours), and section-normal isotachs (red contours, dashed negative). Data are taken from a GFS-model forecast valid 1200 UTC 10 Feb 2010.

There are many tools and frameworks that can be utilized in the analysis of atmospheric processes, and the popularity of different methods changes and evolves over time. On some occasions, a method may gain popularity because of an especially insightful paper or series of papers illustrating the benefits afforded by that technique. If a given analysis strategy provides unique advantages, then it is more likely to gain a foothold in the scientific community. In the case of potential vorticity (PV), it was introduced in the late 1930s, but its widespread use did not take place until the years following a remarkable paper by Hoskins et al. (1985). The Hoskins et al. article demonstrated the applicability of PV-based interpretation and techniques to a broad spectrum of geophysical problems, and it illustrated the succinct and powerful manner in which PV relates to both theoretical and observational aspects of the atmospheric sciences.

## 4.1. HISTORY AND DEFINITION

In developing the quasigeostrophic (QG) height-tendency equation in chapter 2, we utilized the thermal wind relation to express it as a statement of conservation of the variable $q$, the quasigeostrophic potential vorticity (QGPV). The QGPV, which is conserved following adiabatic, frictionless, geostrophic motion, was introduced by Charney and Eady in their famous baroclinic instability studies of the late 1940s. However, other forms of PV were introduced earlier and are more generally conserved, as will be shown in the following sections.

Early papers describing the conceptual advantages of PV were authored by Ertel (1942b,a,c) and Rossby (1940); the non-QG form of the PV is often referred to as the "Ertel" PV or "Rossby–Ertel" PV. Historical analysis reveals that Rossby and Ertel communicated prior to and during the period when these papers were published, and that they may have in some way collaborated on this work. Rossby (1936) described a shallow-water form of the PV in a study of ocean currents, and he later generalized this form to continuously stratified conditions, again for oceanic flows (Rossby 1938). However, the term *potential vorticity* was not introduced until 1940. Rossby (1940) is often cited as the original reference introducing PV; it is curious that in this paper, Rossby does not cite his own earlier oceanographic papers, in which fundamental preliminary concepts were introduced.

Meanwhile, Hans Ertel had visited Rossby at the Massachusetts Institute of Technology (MIT) in 1937 (Samelson 2003) and had worked on vortex dynamics for some years before that. Later, in a series of papers, Ertel (1942a,b,c) extended the PV ideas and presented a gen-eral form of the potential vorticity involving the potential temperature [Ertel 1942a, his Eq. (2.12)]. Conveniently, these early papers by Ertel have recently been translated into English by Schubert et al. (2004). It is interesting that Ertel's papers contain parenthetical reference to Rossby, but they do not cite his earlier works directly. It is clear that both Ertel and Rossby contributed substantially to the development of the PV concept, and that they may have collaborated directly (Samelson 2003).

A remarkable early application of PV concepts to atmospheric systems was carried out in the 1950s by E. Kleinschmidt (e.g., Eliassen and Kleinschmidt 1957). Kleinschmidt developed and applied the PV conservation and invertibility principles discussed in section 2.6; these ideas will be explored in greater depth later in this chapter. For further discussion regarding the history of potential vorticity, see Thorpe and Volkert (1997).

For the full governing equations, the PV is given by

$$P = \frac{1}{\rho}\vec{\eta}\cdot\nabla\theta, \tag{4.1}$$

where $P$ is the Rossby–Ertel potential vorticity, $\vec{\eta}$ is the 3D absolute vorticity vector, and the other symbols were defined earlier. In isentropic coordinates, and with the use of the hydrostatic assumption, the isentropic PV (4.1) can be written as

$$P = -g(\zeta_{a\theta})\frac{\partial\theta}{\partial p}, \tag{4.2}$$

where $\zeta_{a\theta}$ is the absolute vorticity evaluated on an isentropic surface. For adiabatic, frictionless motion, $P$ is conserved following the quasi-2D flow on isentropic surfaces.

As implied by its name, an analogy can be drawn between potential vorticity and potential temperature. Rossby (1940) stated that the potential vorticity represents the vorticity the air column would have if it were brought isentropically to standard latitude and stretched or shrunk vertically to a standard depth or weight (see quote at beginning of chapter). From (4.2), we would state that the PV represents the vorticity that the air would have if it were adiabatically adjusted to a reference latitude and static stability. Rossby went so far as to suggest reference latitude and isobaric depth values, but this aspect of PV is not often used. This concept also has implications for spatial scale and balanced flow. For adiabatic mixed layers, such as the planetary boundary layer, the PV is ~0, owing to the dependence on $\partial\theta/\partial p$; however, this obviously does not mean that there can be no *vorticity* in such a flow!

The purpose of this chapter is to outline the ways in which PV can be used as a means of diagnosing the importance of various physical processes in the atmosphere. A derivation of the PV and its tendency equation will be provided in section 4.3.

## 4.2. THE PV DISTRIBUTION AND THE DYNAMIC TROPOPAUSE

Given that the PV is related to the product of the absolute vorticity and static stability (4.2), in general, we expect to find large values of PV in the stratosphere and in polar regions. Given the conceptual interpretation of PV outlined above, along with the conservation property (see section 2.6), we would expect that if stratospheric air were adiabatically stretched to the point where the static stability was similar to that typically observed in the troposphere, then very large vorticity would result to compensate for the reduction in static stability. This line of reasoning led Kleinschmidt to consider the stratospheric PV reservoir to be the "producing mass" of cyclonic disturbances. For the troposphere, scale analysis of (4.2) yields $10^{-6}\,\mathrm{K\,kg^{-1}\,m^2\,s^{-1}}$, a value defined by Hoskins et al. (1985) as one *potential vorticity unit*, or 1 PVU. The large stability and PV in the stratosphere suggest that the tropopause can be identified as an isosurface of PV, given that this zone marks the transition between small tropospheric and large stratospheric values of PV (e.g., Danielsen 1968). The 1.5- or 2.0-PVU surface is often used to define the *dynamic tropopause*, a term referring to the definition of the tropopause using a PV isosurface.

The PV invertibility principle (see section 2.6) allows for the association between a given PV anomaly and the associated thermodynamic and wind fields. A *PV anomaly* can be defined with respect to a climatological PV average, or a shorter-term time or spatial average, for example. A cyclonic PV anomaly, if inverted, demonstrates that the associated cyclonic flow can extend well away from the location of the PV anomaly itself (Fig. 4.1). Similarly, anticyclonic flow is associated with anticyclonic anomalies. In the Northern (Southern) Hemisphere, cyclonic PV anomalies are positive (negative). The extension of flow associated with a given PV anomaly to spatial locations far removed from that of the anomaly itself can be viewed as analogous to the action-at-a-distance principle in electrostatics [see Bishop and Thorpe (1994) for a complete discussion of this analogy].

To link traditional isobaric and PV-based analyses, we now compare these quantities for a representative case study example. The Global Forecast System (GFS) 144-h forecast valid 1200 UTC 10 February 2010 indicates a closed upper low at the 500-mb level over the Great Lakes region (Fig. 4.2a). The shaded regions of this plot indicate that large values of upper-tropospheric potential vorticity characterize this trough, as well as a weaker trough over the southwestern United States. This association is not a coincidence; if these cyclonic PV anomalies were inverted, then the result would be the contribution of a negative height anomaly, consistent with the lower geopotential heights found there. The vertical and lateral extent of the negative height anomaly extends beyond the immediate vicinity of the cyclonic PV anomaly.

A vertical cross section, taken from north to south across the Great Lakes cyclone, demonstrates that in fact

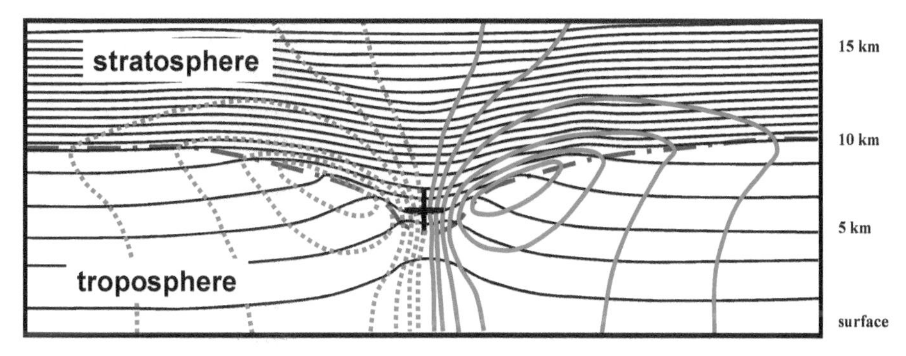


**Figure 4.1.** Cross section of idealized isentrope distribution (solid black contours) and section-normal wind speed (red contours, dotted for negative, solid positive) recovered from PV inversion. Brown dotted–dashed line indicates location of dynamic tropopause; the region of lower dynamic tropopause corresponds to a cyclonic PV anomaly, indicated by a "+" sign (adapted from Thorpe 1986).

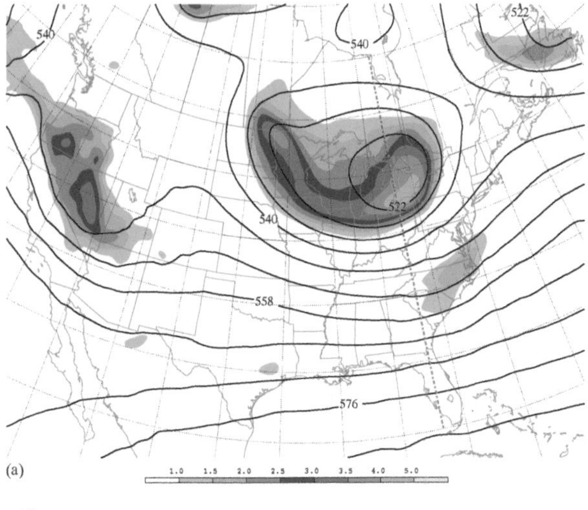


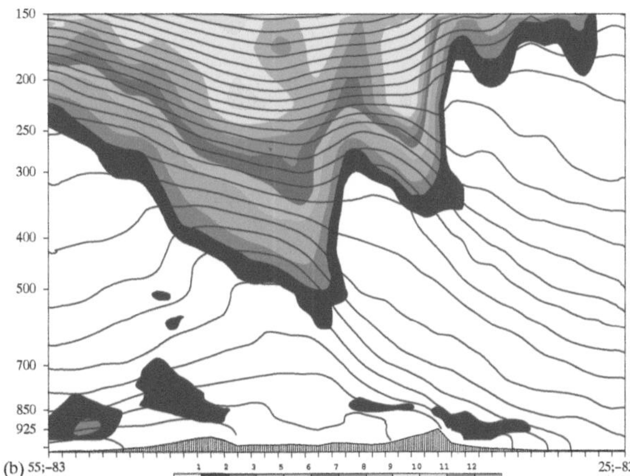


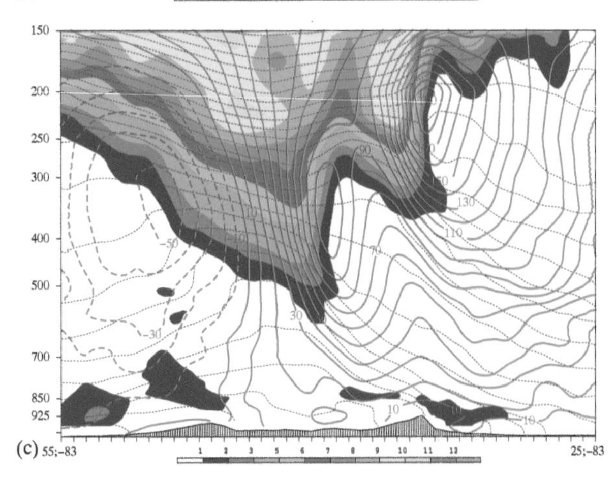


**Figure 4.2.** GFS 144-h forecast valid 1200 UTC 10 Feb 2010: (a) 500-mb height (solid contours) and potential vorticity in the 300–500-mb layer (shaded as in legend at bottom), dashed bold line indicates location of cross sections shown in (b),(c); (b) cross section of PV (shaded) and potential temperature (interval 5 K, solid contours); (c) as in (b) but section-normal wind component (10-kt interval, dashed for negative, solid for positive) added.

the PV maximum associated with this trough represents a downward extrusion of large PV from the stratosphere (Fig. 4.2b). The flow associated with the cyclonic PV maxima would also lead us to expect jet maxima in the regions of strong lateral PV gradient; the large cyclonic PV on the stratospheric side of the tropopause represents a cyclonic PV anomaly, with much lower PV on the tropospheric side. The double "stair step" PV pattern in this cross section suggests the presence of two distinct westerly jet maxima south of the center of the PV maximum marking the location of the upper trough, as well as an easterly jet to the north of the main PV maximum, consistent with the expected sense of the flow with a cyclonic PV anomaly to the south.

The isentropes in this cross section exhibit close vertical spacing, and thus large stability, in the sloping regions of large PV to the south of the main anomaly center. These zones are indicative of upper-level frontal zones. Based on the presence of two distinct zones of sloping isentropes (with large horizontal potential temperature gradients), the thermal wind relation also leads us to expect two isolated jet features, which is exactly what we inferred from the PV pattern. The inclusion of section-normal isotachs (Fig. 4.2c) confirms this consistency. The southernmost jet core, at higher altitude, is consistent with a subtropical jet stream, while the more northern of the two westerly jets would be classified as the "polar front" jet; the sloping frontal zone associated with the northern jet extends closer to the surface than does the southern zone. As will be shown in chapters 5 and 7, this has important implications for the kinds of cyclonic disturbances that can develop in association.

The data presented in this cross section suggest that *dynamic tropopause maps* would be a convenient way to represent all of the important upper-jet features on a single chart, given that the 1.5- or 2-PVU PV surfaces intersect the core of all tropopause-based jet streams. Plots of potential temperature on the dynamic tropopause ($\theta_{trop}$) are useful in that this quantity is conserved for adiabatic, frictionless flow and that it also describes the topography of a PV isosurface in isentropic coordinates (Fig. 4.3a). While not a conserved variable, pressure on the dynamic tropopause ($p_{trop}$) also gives a sense of the topography of the tropopause. Regions of higher $p_{trop}$ correspond to lower tropopause altitude, and gradients of $p_{trop}$ correspond to strongly sloped or even folded tropopause structures (Fig. 4.3b).

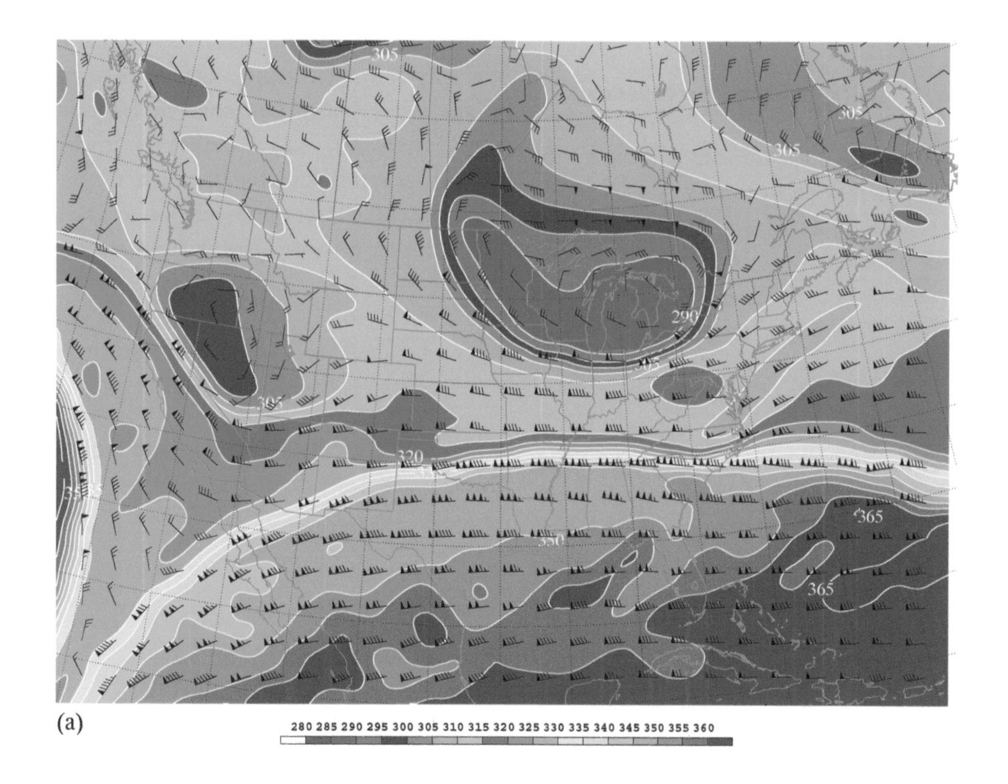


(a)

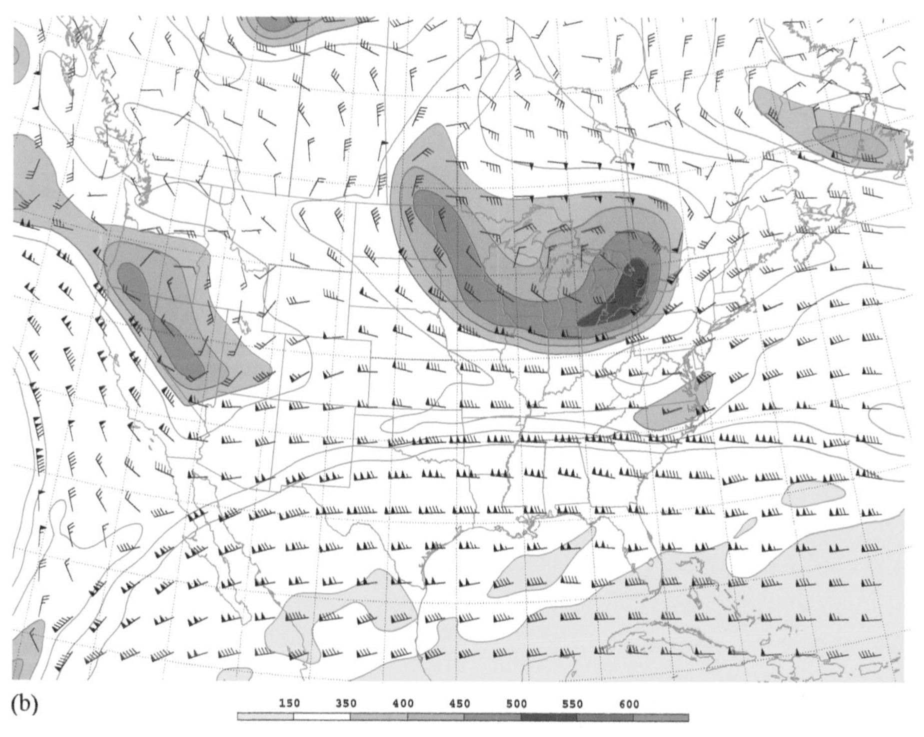


(b)

**Figure 4.3.** Example of dynamic tropopause analyses corresponding to Fig. 4.2, from GFS model 144-h forecast valid 1200 UTC 10 Feb 2010: (a) potential temperature (shaded, interval 5 K) and wind and (b) pressure (interval 50 mb, shaded as in legend) and wind.

## 4.3. DERIVATIONS

### 4.3.1. Potential vorticity and the PV-tendency equation

In isentropic coordinates, the horizontal momentum equations in scalar form may be written as

$$\left(\frac{\partial}{\partial t}+u\frac{\partial}{\partial x}+v\frac{\partial}{\partial y}+\dot{\theta}\frac{\partial}{\partial \theta}\right)u - fv + \frac{\partial M}{\partial x} - F_x = 0 \text{ and} \tag{4.3}$$

$$\left(\frac{\partial}{\partial t}+u\frac{\partial}{\partial x}+v\frac{\partial}{\partial y}+\dot{\theta}\frac{\partial}{\partial \theta}\right)v + fu + \frac{\partial M}{\partial y} - F_y = 0, \tag{4.4}$$

where $M$ is the Montgomery streamfunction from Eq. (3.3); $F_x$ and $F_y$ denote the frictional force components in the zonal and meridional directions, respectively; and $\dot{\theta} = d\theta/dt$. Cross differentiating (4.3) and (4.4) to obtain the isentropic vorticity equation, and then rewriting the result in flux form, this equation can be expressed as

$$\frac{\partial \zeta_{a\theta}}{\partial t} + \frac{\partial}{\partial x}\left[u\zeta_{a\theta} + \dot{\theta}\frac{\partial v}{\partial \theta} - F_y\right] + \frac{\partial}{\partial y}\left[v\zeta_{a\theta} - \dot{\theta}\frac{\partial u}{\partial \theta} + F_x\right] = 0. \tag{4.5}$$

In vector form, the isentropic vorticity equation is written as

$$\frac{\partial \zeta_{a\theta}}{\partial t} + \nabla\cdot(\vec{V}\zeta_{a\theta}) + \hat{k}\cdot\nabla_\theta\times\left(\vec{F} - \dot{\theta}\frac{\partial \vec{V}}{\partial \theta}\right) = 0. \tag{4.6}$$

Equation (4.6) can be rearranged to yield the following form:

$$\frac{\partial \zeta_{a\theta}}{\partial t} + \nabla\cdot\vec{j} = 0, \tag{4.7}$$

where

$$\vec{j} = (u\zeta_{a\theta}, v\zeta_{a\theta}, 0) + \left(\dot{\theta}\frac{\partial v}{\partial \theta}, -\dot{\theta}\frac{\partial u}{\partial \theta}, 0\right) + (-F_y, F_x, 0). \tag{4.8}$$

The purpose of expressing the vorticity equation in this form will become apparent below; however, note that the form of (4.7) specifies that local changes of absolute vorticity evaluated on an isentropic surface are due to the divergence of the vector $\vec{j}$, given by (4.8). In the absence of diabatic and frictional effects, there is no flux of isentropic vorticity across isentropic surfaces, a property described as the *impermeability theorem* by Haynes and McIntyre (1987); however, (4.7) describes the tendency of isentropic vorticity, not potential vorticity.

A general conservation relation for the mixing ratio of any arbitrary quantity $\chi$ can be written as

$$\frac{\partial(\sigma\chi)}{\partial t} + \nabla\cdot\vec{j} = \sigma S, \tag{4.9}$$

where $S$ represents all sources and sinks of $\chi$, $\sigma$ is the isentropic density defined in (3.7), and

$$\vec{j} = (u\sigma\chi, v\sigma\chi, \dot{\theta}\sigma\chi) + (-F_y, F_x, 0). \tag{4.10}$$

This flux vector can be subdivided into an advective flux, a diabatic flux, and a frictional flux should we desire to isolate contributions from different classes of physical processes. If we let $\chi$ represent the *dry air* mixing ratio (which is equal to 1)—assume that the frictional mass flux is small and neglect any sources or sinks of air (e.g., volcanoes)—then (4.9) becomes the isentropic continuity equation

$$\frac{\partial \sigma}{\partial t} + \nabla\cdot(\sigma\vec{V}) + \frac{\partial}{\partial \theta}(\sigma\dot{\theta}) = 0; \text{ or } \frac{1}{\sigma}\frac{d\sigma}{dt} + \nabla\cdot\vec{U} = 0. \tag{4.11}$$

Comparison of (4.9) and (4.10) with (4.7) and (4.8) suggests that we let

$$\chi = \frac{\zeta_{a\theta}}{\sigma}, \tag{4.12}$$

which is the density-weighted mixing ratio of absolute isentropic vorticity, or in other words, the amount of isentropic absolute vorticity per mass of dry air. However, from (3.7) we recognize that the isentropic density is related to the static stability; $\sigma^{-1} = -g(\partial\theta/\partial p)$, so that we can write

$$P = \frac{\zeta_{a\theta}}{\sigma} = -g\frac{\partial \theta}{\partial p}\zeta_{a\theta}, \tag{4.13}$$

where $P$ is the Rossby–Ertel potential vorticity. (In chapters 2 and 7, the symbol $q$ is used to designate the quasi-geostrophic form of the PV; in chapter 7, $Q$ denotes the basic-state PV). To obtain a tendency equation for $P$, we again utilize the conservation relation (4.9), this time letting $\chi = P$. This yields

$$\frac{\partial \sigma P}{\partial t} + \frac{\partial}{\partial x}\left[u\sigma P + \dot{\theta}\frac{\partial v}{\partial \theta} - F_y\right] + \frac{\partial}{\partial y}\left[v\sigma P - \dot{\theta}\frac{\partial u}{\partial \theta} + F_x\right] = 0. \tag{4.14}$$

For adiabatic, frictionless flow, (4.14) becomes

$$\sigma\left(\frac{\partial P}{\partial t} + u\frac{\partial P}{\partial x} + v\frac{\partial P}{\partial y}\right) + P\left(\frac{\partial \sigma}{\partial t} + u\frac{\partial \sigma}{\partial x} + v\frac{\partial \sigma}{\partial y} + \sigma\left(\frac{\partial u}{\partial x} + \frac{\partial v}{\partial y}\right)\right) = 0; \tag{4.15}$$

however, the second term in (4.15) is equal to zero because it is the adiabatic form of (4.11), the continuity equation, leaving

$$\frac{dP}{dt} = 0. \tag{4.16}$$

Equation (4.16) expresses the conservation principle of $P$. However, it is often the situations in which nonconservative processes are important that we can best exploit the PV conservation property, because it allows us to unambiguously identify the impact of diabatic processes or friction. To do this, the full PV-tendency equation is required, which is obtained from the flux form (4.14). The details of this derivation are straightforward and are left as an end-of-chapter problem; the resulting PV-tendency equation is

$$\frac{dP}{dt} = -g\frac{\partial \theta}{\partial p}\zeta_{a\theta}\frac{\partial \dot{\theta}}{\partial \theta} + g\frac{\partial \theta}{\partial p}\left[\frac{\partial \dot{\theta}}{\partial x}\frac{\partial v}{\partial \theta} + \frac{\partial F_y}{\partial x} - \frac{\partial \dot{\theta}}{\partial y}\frac{\partial u}{\partial \theta} + \frac{\partial F_x}{\partial y}\right], \tag{4.17}$$

or in vector form

$$\frac{dP}{dt} = -g\frac{\partial \dot{\theta}}{\partial p}\zeta_{a\theta} + g\frac{\partial \theta}{\partial p}\hat{k}\cdot\left(\nabla\dot{\theta}\times\frac{\partial \vec{V}}{\partial \theta}\right) + g\frac{\partial \theta}{\partial p}\hat{k}\cdot(\nabla\times\vec{F}). \tag{4.18}$$

The right-hand terms in (4.18) can be described as the *vertical diabatic*, *shear diabatic*, and *frictional* PV tendencies, respectively. Taken collectively, these terms represent the *nonadvective PV tendency*. We can consolidate the right-hand terms in (4.18) and use the definition of isentropic density (3.7) to write

$$\frac{dP}{dt} = \sigma^{-1}(\vec{\eta}\cdot\nabla\dot{\theta}) + \sigma^{-1}\hat{k}\cdot\nabla\times\vec{F}. \tag{4.19}$$

This form of the equation is appealing because the first right-hand term clarifies that the diabatic PV tendency is related to the projection of the heating gradient onto the absolute vorticity vector. In highly baroclinic environments, the horizontal vorticity components can be large, leading to a horizontal offset in the vertical redistribution of PV.

### 4.3.2. PV nonconservation

The fact that $P$ is conserved for adiabatic, frictionless flow should in no way imply that this is not a useful diagnostic quantity in situations characterized by strong diabatic or frictional processes. On the contrary, the signature of PV nonconservation is evident in the spontaneous appearance or disappearance of isolated PV maxima; such signatures tell us without question that these nonconservative processes are taking place. The PV anomalies created by such processes can then, in principle, be inverted to quantify the importance of those processes. This represents a convenient and rigorous method of determining cause and effect and assessing the dynamical impact of a given process.

Physically, how do diabatic processes change the PV? In the simplest case with no vertical isentropic shear, and neglecting frictional tendencies, the diabatic PV tendency is described by the first right-hand term in (4.18). Suppose that in such an idealized situation, a maximum of heating is centered in the lower troposphere, as pictured in Fig. 4.4a. Within the region of heating, isentropes will descend, as shown for the middle (330 K) isentrope in Fig. 4.4b. Though not shown on this schematic, recall that through the QG height tendency equation, height rises will occur above the level of maximum heating with height falls below, consistent with a thickness increase accompanying the heating.

As the height anomalies above and below the region of heating amplify, pressure gradient accelerations will lead to horizontal divergence above the heating maximum (with flow away from the region of higher pressure

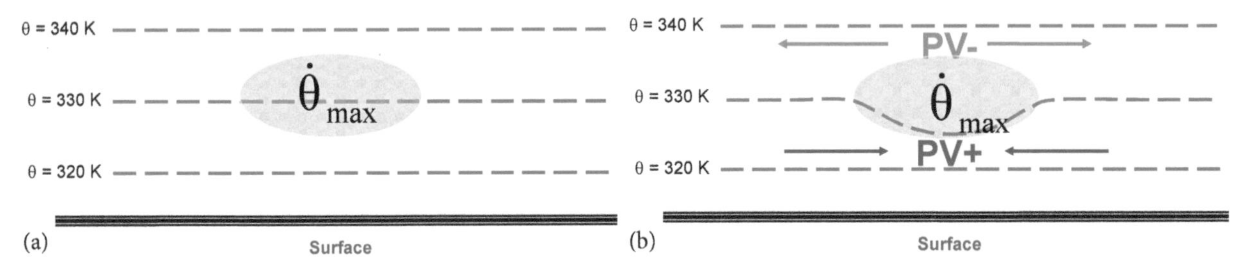


**Figure 4.4.** Diabatic PV redistribution in the presence of midlevel latent heat release: (a) heating region (shaded) superimposed with initially flat isentropes and (b) altered isentrope configuration, PV anomalies, and divergent flow vectors.

or height) with convergence below, as indicated by the arrows in Fig. 4.4b. Now, consider how these processes would change the PV. Above the heating maximum, the vertical spacing of the isentropes has increased, indicating a reduction of static stability there. Divergence has also resulted in a decrease in vorticity there (assuming a Northern Hemisphere location), and from (4.2) it is clear that PV cannot be conserved in this location—both of the PV components decrease in the region above the heating maximum. Conversely, below the heating maximum, both PV components have *increased*; stability has increased because of the vertical heating gradient (heating increasing with height), and vorticity has become more cyclonic because of convergence. Assuming the absolute vorticity remains positive, the relevant terms in (4.19) are $dP/dt = -g(\partial\dot{\theta}/\partial p)\zeta_{a\theta}$ and below the heating maximum (where $\partial\dot{\theta}/\partial p < 0$) a cyclonic PV tendency is found, with the opposite situation above the heating maximum.

Typically, stable ascent takes place in the presence of sloping isentropes and vertical wind shear; in that situation, a more complex diabatic PV redistribution pattern takes place according to the first two right-hand terms in (4.18), which are contained in the first right-hand term in (4.19). The diabatic PV tendency is related to the projection of the absolute vorticity vector onto the heating gradient (4.19). Recall that when vertical shear is present, the horizontal components of the absolute vorticity vector can be large, and the orientation of the vorticity vector can depart significantly from the vertical. This results in a horizontal displacement of the region of PV reduction aloft from the region of PV growth below the heating maximum, as depicted in Fig. 4.5.

Knowledge of the physical processes responsible for the observed PV distribution in a given weather system provides important information about the formation and evolution of the system. Consider the contrasting examples of PV structures accompanying three cyclonic systems shown in Fig. 4.6. For a strong extratropical cyclone, the main cyclonic PV feature is of stratospheric origin, with some contribution from diabatic processes in the lower troposphere (Fig. 4.6a). For a tropical cyclone (Fig. 4.6b), a diabatic PV tower is the dominant feature, consistent with a warm-core structure and an absence of stratospheric influence. Finally, some cyclonic systems are characterized by strong PV maxima of both stratospheric and diabatic origin; such hybrid systems include recurving tropical cyclones, extratropical cyclones accompanied by heavy precipitation, or subtropical low pressure systems (Fig. 4.6c). The representation of diabatic processes in numerical weather prediction (NWP) models is inherently difficult; therefore, recognition of a strong contribution of diabatically generated PV features diagnosed in NWP model forecasts provides information not only concerning the dynamical factors important to the system but also regarding the level of confidence one should place in the model prediction.

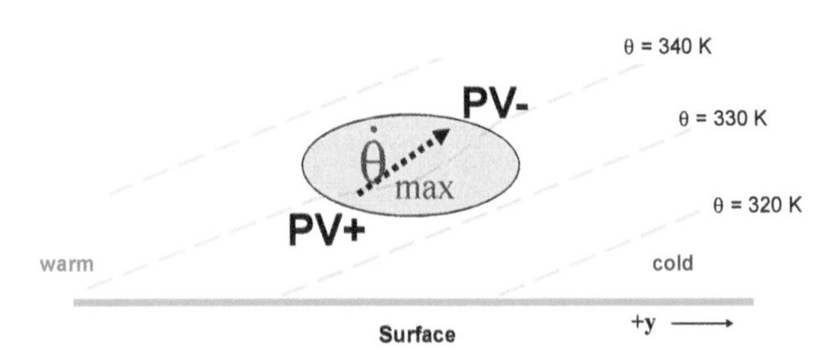


**Figure 4.5.** As in Fig. 4.4, but in an environment characterized by westerly shear. Dotted black arrow represents absolute vorticity vector; the upward orientation of the vorticity is due to an assumed cyclonic vertical absolute vorticity component. The northward tilt is consistent with westerly shear and colder potential temperature toward the north, as would typically be found in the Northern Hemisphere.

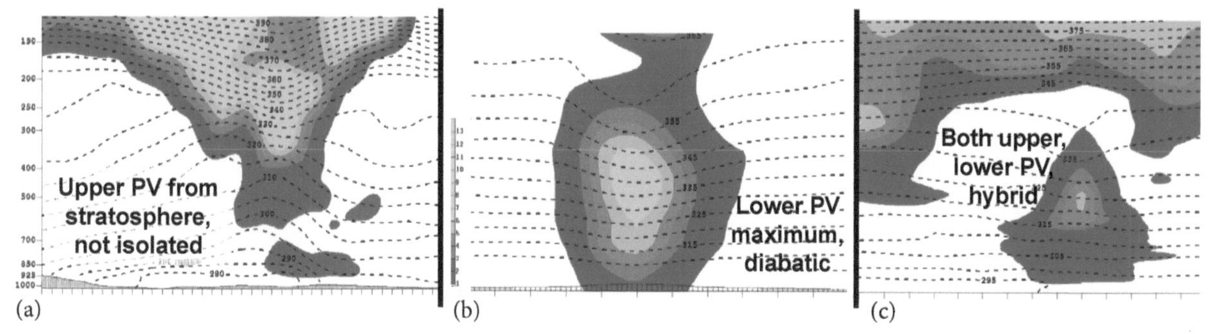


**Figure 4.6.** Potential vorticity (shaded above 1.0 PVU with shade interval 1 PVU) and isentropes (dashed lines, 5-K interval) for (a) a wintertime extratropical cyclone at 0000 UTC 25 Jan 2000, (b) a coarse-grain analysis of Hurricane Katrina at 0000 UTC 29 Aug 2005, and (c) a hybrid subtropical cyclone at 1800 UTC 24 Sep 2008.

Use of PV methods in weather analysis and forecasting offers the advantage of allowing one to quantify the importance of specific physical processes, such as latent heat release, and then use PV inversion to isolate the resulting dynamical behavior due to the modified PV distribution. For example, consider the major eastern U.S. snowstorm that took place on 24–25 January 2000. During this event, numerical forecast models did not accurately predict the inland extent of heavy precipitation associated with a coastal cyclonic storm system, and portions of North and South Carolina, Virginia, and the mid-Atlantic region were affected by heavy, unexpected snowfall (Fig. 4.7).

Retrospective analysis of the January 2000 snowstorm event identified an area of heavy rainfall that began early over Alabama and Georgia on the 24th but was not captured by numerical model predictions, even for short-term forecasts (Fig. 4.8). As expected based on the process of PV nonconservation, the diabatic PV production associated with this region of precipitation was grossly underestimated by the models, as was the associated cyclonic flow in the lower troposphere in the vicinity of this feature. Comparison of model forecasts with analyzed PV in the lower troposphere reveals the glaring absence of a strong PV maximum centered near the South Carolina–Georgia border (Fig. 4.9).

This analysis enabled a more detailed investigation into the origin of the initial precipitation feature shown in Fig. 4.9, and its role in the subsequent evolution of model error in this case. Another useful aspect of PV analysis

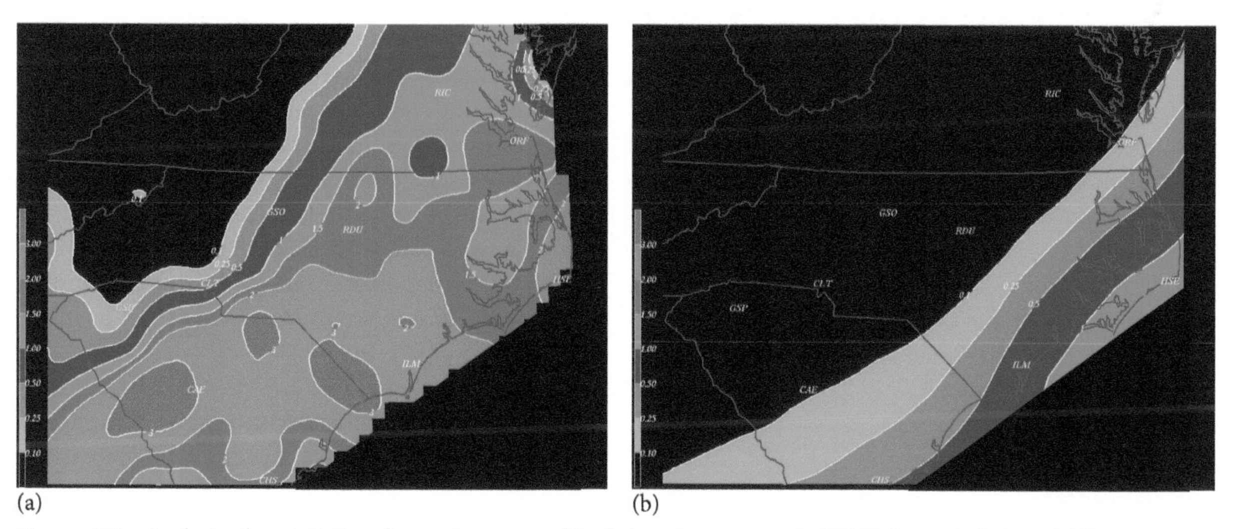


**Figure 4.7.** Analysis of precipitation observations versus North American mesoscale (NAM; formerly Eta) model forecast for 48-h period ending 0000 UTC 26 Jan 2000 (in inches; legend at left of panels): (a) objective precipitation analysis from the National Weather Service (NWS) co-op network and (b) Eta model 48-h forecast. Adapted from Brennan and Lackmann (2005).

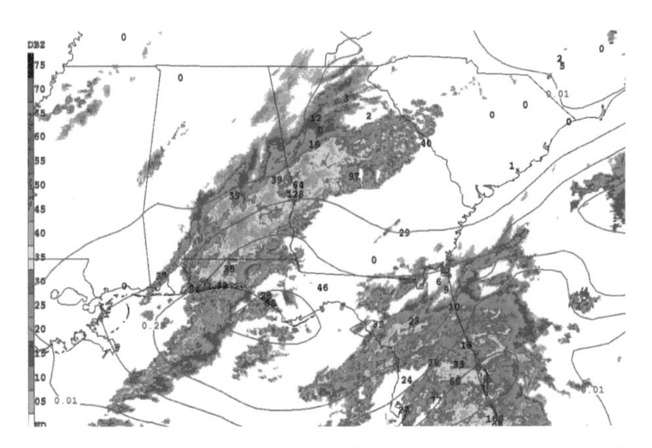

**Figure 4.8.** Composite radar, 6-h precipitation forecast, and observed gauge precipitation totals: Radar valid 0900 UTC 24 Jan 2000, and 6-h forecasted precipitation totals from Eta model (interval 0.01, 0.10, 0.25, and 0.5 in.) and observations (integer values in hundredths of inches), ending 1200 UTC 24 Jan 2000. From Brennan and Lackmann (2005).

is evident in comparing observational data, such as radar or satellite imagery, to short-term model forecasts. For situations in which the short-term model forecast misrepresents a precipitation system, a forecaster can gain a sense not only of the existence of a model error but the conceptual picture afforded by PV analysis may allow an informed, dynamically based adjustment to the forecast.

For the January 2000 event, research results led to the development of the schematic diagram shown in Fig. 4.10, indicating the role of moisture transport associated with the diabatic PV anomaly in question. Though not shown here, inversion of this PV feature quantified and confirmed its role in both driving isentropic ascent in the zone of easterly flow to the north of its center and also in contributing to onshore moisture transport from the Atlantic into the zone of heavy precipitation (Brennan and Lackmann, 2005).

The example provided here has direct implications for weather forecasting—and for the diagnosis of errors in NWP models. There are also important linkages between the PV and baroclinic instability theory, as will be discussed at length in chapter 7.

### 4.3.3. Integral constraints

The divergence theorem (Gauss's theorem) is

$$\int_V \nabla \cdot \vec{F}\, dV = \int_A \vec{F} \cdot \hat{n} dA \tag{4.20}$$

where $\hat{n}$ is the outward-directed unit vector perpendicular to the outer surface of volume $V$. For our PV applications, we can integrate (4.7) over an isentropic volume bounded laterally by a streamline on which friction and diabatic processes are negligible (or, alternately, we can integrate over a global isentropic layer), yielding

$$\int_V \frac{\partial(\sigma P)}{\partial t} dV + \int_V \nabla \cdot \vec{j}\, dV = 0. \tag{4.21}$$

The application of (4.20) yields

$$\int_V \frac{\partial(\sigma P)}{\partial t} dV + \int_A \vec{j} \cdot \hat{k}\, dA = 0; \tag{4.22}$$

but because of the nature of the flux vector $\vec{j}$,

$$\int_A \vec{j} \cdot \hat{k}\, dA = 0; \tag{4.23}$$

therefore,

$$\int_V \frac{\partial(\sigma P)}{\partial t} dV = 0, \quad \text{or} \quad \int_V \frac{\partial(\zeta_{a\theta})}{\partial t} dV = 0. \tag{4.24}$$

Equation (4.24) expresses the "PV impermeability theorem" of Haynes and McIntyre (1987), which was also evident from the isentropic vorticity equation. This is a very powerful constraint upon how the vorticity can change, and it suggests that it is better to view diabatic PV changes as the "redistribution" rather than the "creation" and "destruction" of PV. In fact, the *mass-weighted PV changes* due to a given area of diabatic heating or cooling will exactly cancel within a volume bounding the diabatic region. The interpretation provided earlier that the PV is related to the isentropic vorticity mixing ratio further suggests that one can think of diabatic PV increases or decreases as the "concentration" or "dilution" of *PV substance*, a term coined by Haynes and McIntyre (1987). However, it is important to recognize that PV substance is simply the isentropic absolute vorticity, not the potential vorticity.

From this perspective, any process that results in the movement of mass across an isentropic surface can alter the potential vorticity. This includes condensation and precipitation mass removal, as discussed in relation to Eq. (1.18) in chapter 1. Schubert et al. (2001) present the PV tendency equation for the equations of Ooyama (2001), which represent an unusually complete set of

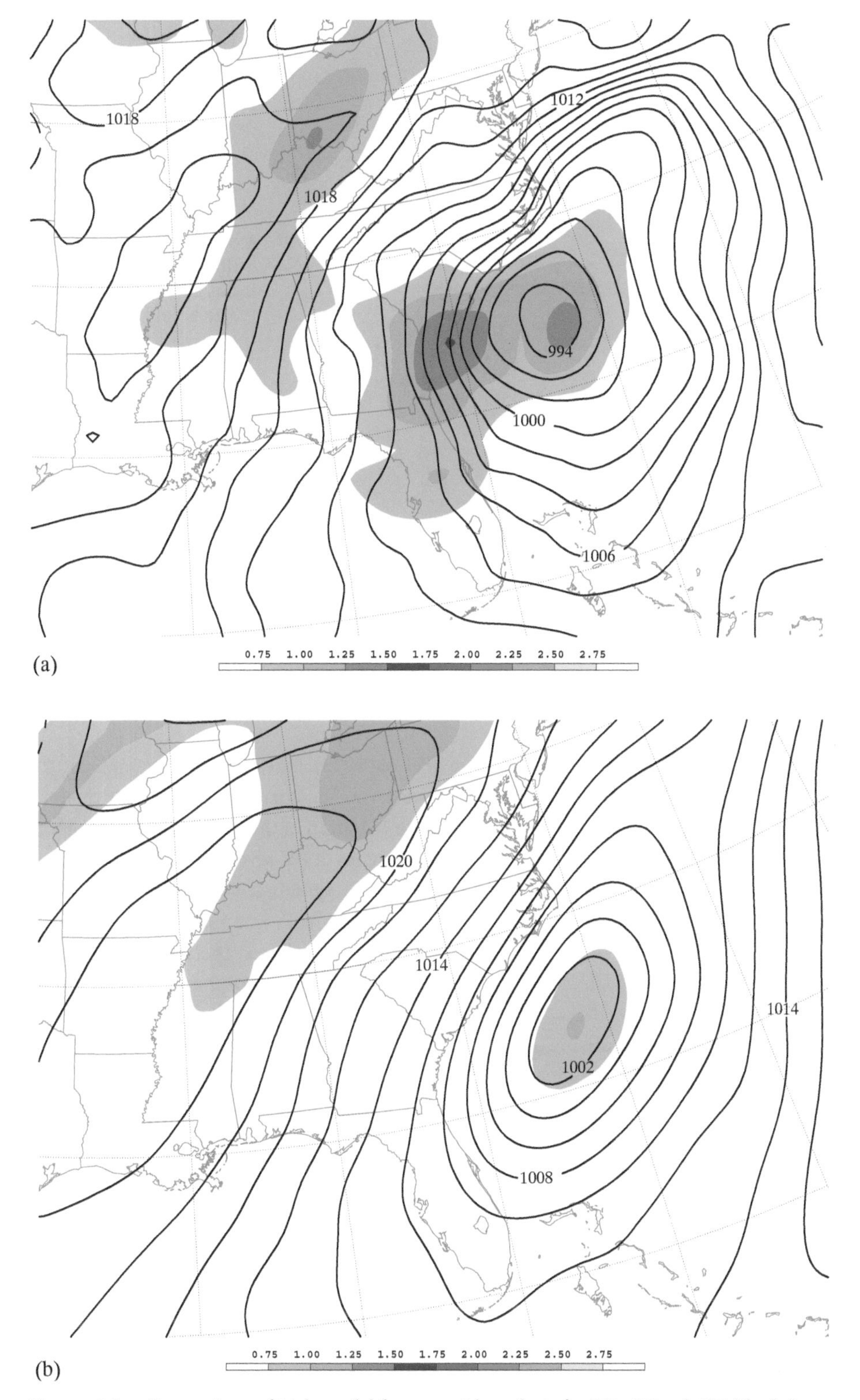


**Figure 4.9.** Comparison of 24-h model forecast with analysis for 900–700-mb PV (shaded as in legend at bottom of panels) and SLP (contour interval 2 mb): (a) Rapid Update Cycle (RUC) analysis valid 0000 UTC 25 Jan 2000 and (b) as in (a) but for Eta model 24-h forecast.

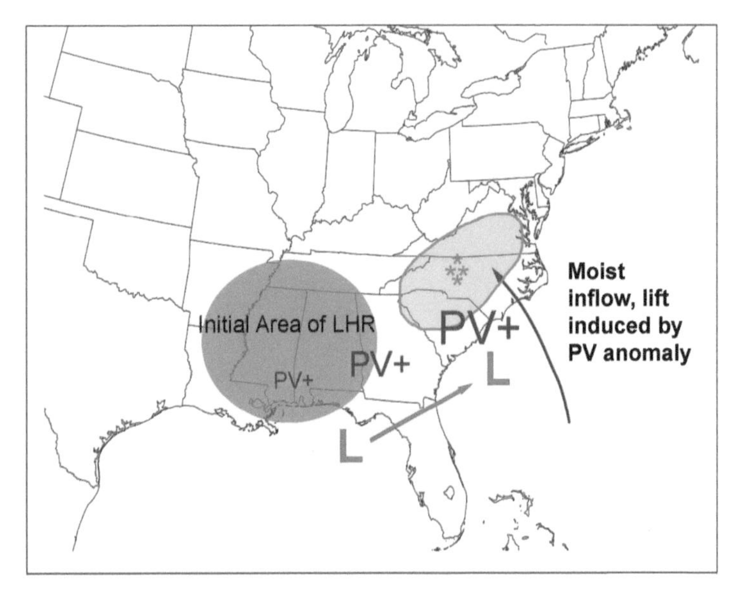


**Figure 4.10.** Schematic diagram summarizing the role of the diabatic lower-tropospheric cyclonic PV anomaly for the January 2000 snowstorm event. "L" marks the location of the surface cyclone center, "PV+" indicates the location of the center of the diabatically generated lower PV anomaly at approximately 12-h increments, with the size of the lettering being proportional to the strength of the anomaly (from Brennan and Lackmann 2005).

physical processes, including a PV tendency term arising from the precipitation mass sink. This form of the PV tendency equation is

$$\frac{dP}{dt} = \frac{1}{\rho}[(\nabla \times \vec{F}) \cdot \nabla\dot{\theta} + \vec{\eta} \cdot \nabla\dot{\theta} + P\nabla \cdot (\rho_r \vec{U}_r)], \quad (4.25)$$

where $\rho_r$ is the density of rain water and $\vec{U}_r$ is the fall velocity of precipitation relative to the air. This contribution is negligible for many systems; however, during heavy precipitation, such as that associated with a tropical cyclone, this effect can contribute to the cyclonic PV tendency in the lower troposphere, as shown for the case of Hurricane Lili (2002) by Lackmann and Yablonsky (2004).

## 4.4. PV INVERSION

In addition to the conservation principle for PV, a second advantage of the PV framework is the ability to *invert* the PV distribution to isolate the associated balanced wind, temperature, and pressure fields. Some investigators have gone so far as to isolate the contribution of water substance to the PV field, allowing the ability to isolate the role of water vapor in atmospheric dynamics (McTaggart-Cowan et al. 2003). The previous example of the January 2000 eastern U.S. storm included PV inversion to quantify the contribution of the diabatic lower PV anomaly. Additional examples of PV inversion are provided in chapters 5 and 7.

Given quantitative information about the PV distribution, along with boundary conditions and a balance relation, one can invert the PV field to recover the balanced part of other atmospheric field variables. The balance condition can range from the assumption of geostrophic balance to more accurate conditions, such as the nonlinear balance equations of Charney (1955).

The quasigeostrophic potential vorticity, obtained from the QG height tendency equation in chapter 2, can be expressed as

$$q = \frac{1}{f_0}\nabla^2\phi' + f + f_0\frac{\partial}{\partial p}\left(\frac{1}{\sigma_r}\frac{\partial \phi'}{\partial p}\right), \quad (4.26)$$

where $\sigma_r$ is the reference state static stability. Defining

$$q_* = q - f = \sum_{i=1}^{n} q_{*i}, \quad (4.27)$$

the PV field can be subdivided into as many pieces as desired. For example, if one wishes to determine the balanced wind field associated with a specific PV anomaly, it could be isolated and inverted in isolation from other anomalies. It is evident from (4.26) and (4.27) that there is a linear operator linking the $q$ and $\phi'$ fields:

$$\phi' = \ell^{-1}(q_{*_i}), \text{ where } \ell = \frac{1}{f_0}\nabla^2 + f_0 \frac{\partial}{\partial p}\left(\frac{1}{\sigma_r}\frac{\partial}{\partial p}\right). \quad (4.28)$$

Inversion of this operator is accomplished by iterative techniques, such as the successive over-relaxation (SOR) method. Owing to the linear nature of this operator, when the entire PV field is inverted and the resulting geopotential fields are summed, the result will almost exactly match the observed geopotential field.

Inversion of the full Rossby–Ertel PV is more accurate, but it is also more complex. Davis and Emanuel (1991) presented a case study analysis based on the inversion of the Ertel PV, and several subsequent papers present a comprehensive account of these methods. Use of the Charney (1955) nonlinear balance equations in this inversion process allows for a much more accurate recovery of the wind field than is possible using the QGPV for strongly curved flows. See Davis (1992b) for further discussion and comparison between QG and the Rossby–Ertel PV inversion.

In chapter 5, the application of PV concepts to extratropical cyclones is undertaken; in chapter 7, the utility of PV in understanding the conditions necessary for baroclinic instability will be discussed. In operational forecasting, PV has become widely used in Europe, but it has been slower to catch on in the United States (e.g., Brennan et al. 2008; Carroll and Hewson 2005).

## REVIEW AND STUDY QUESTIONS

1. In what sense is the *potential vorticity* analogous to the *potential temperature*?
2. For the idealized, large Northern Hemisphere convective system shown below, which of the PV distributions shown on the second row would most likely correspond? The notation "PV+" means larger PV than surrounding areas, and "PV−" means smaller.
3. Inspect the various forms of the PV tendency equation [(4.17)–(4.19)]. What *synoptic conditions* maximize the contribution of a given diabatic heating gradient to the PV tendency? Hint: consider the role of vorticity and feedbacks that could arise in the diabatic PV tendency.
4. Suppose that in a hypothetical worst-case scenario, a "subtropical storm" moves ashore and causes considerable damage and beach erosion at a U.S. coastal location. An investigation is launched into the watch and warning process for the storm. It is decided that an objective method must be established to determine whether a given storm falls under the jurisdiction of tropical weather forecast centers or not. How might potential vorticity be used in this circumstance to clarify the nature of such a system?
5. Consider a midlatitude cyclone characterized by heavy warm-frontal precipitation. Based on your knowledge of PV, sketch the lower-tropospheric PV distribution that might accompany such a system, including the location of the low pressure center and frontal boundaries. Sketch both cross-sectional and plan views of this distribution. Justify your reasoning.

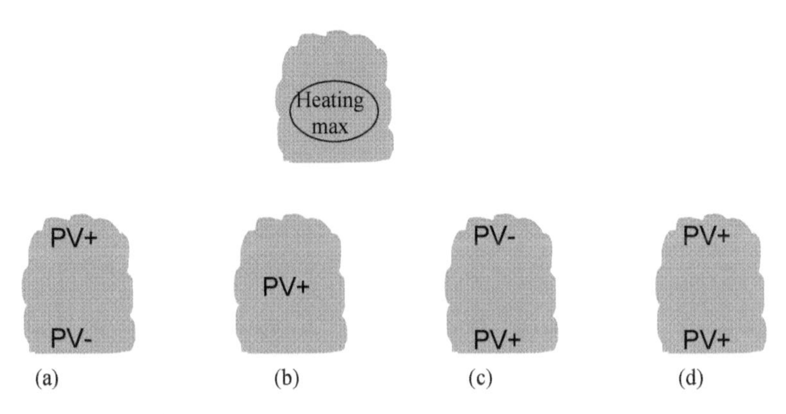


**Figure for Review and Study 2**

## PROBLEMS

1. Perform a scale analysis of the PV definition provided in Eq. (4.2) for typical midlatitude synoptic-scale conditions (see chapter 1).
2. Derive Eq. (4.17) by letting the conservation variable $\chi$ in Eq. (4.9) be *P*.
3. The image below shows the 200–400-mb layer-averaged potential vorticity (PVU), shaded as in the legend at right. Based on your knowledge of the relation between upper-level troughs and ridges, and the PV, sketch reasonable estimations of the 570- and 546-dam 500-mb height contours that might accompany this PV pattern. Hint: these contours roughly bracket the warm and cold side of the jet stream. At location A, do you expect that the *height of the tropopause* would be higher or lower in altitude relative to the surrounding areas to the east and west? Explain.
4. On next page, panels (a)–(c) depict a sequence of SLP (contours) and lower-tropospheric PV (in the 850–700-mb layer, shaded). Panel (d) shows SLP superimposed on a visible satellite image. Based on the evidence available, discuss what processes are likely to have contributed to the rapid development of this cyclone.
5. Briefly explain the reasoning behind your answer to question 4.

## REFERENCES


Bishop, C. H., and A. J. Thorpe, 1994: Potential vorticity and the electrostatics analogy: Quasi-geostrophic theory. *Quart. J. Roy. Meteor. Soc.,* **120,** 713–731.

Brennan, M. J., and G. M. Lackmann, 2005: The influence of incipient latent heat release on the precipitation distribution of the 24–25 January 2000 cyclone. *Mon. Wea. Rev.,* **133,** 1913–1937.

——, ——, and K. M. Mahoney, 2008: Potential vorticity (PV) thinking in operations: The utility of nonconservation. *Wea. Forecasting,* **23,** 168–182.

Carroll, E. B., and T. D. Hewson, 2005: NWP grid editing at the Met Office. *Wea. Forecasting,* **20,** 1021–1033.

Charney, J., 1955: The use of the primitive equations of motion in numerical prediction. *Tellus,* **7,** 22–26.

Danielsen, E. F., 1968: Stratosphere-troposphere exchange based on radioactivity, ozone, and potential vorticity. *J. Atmos. Sci.,* **25,** 502–518.

Davis, C. A., 1992b: Piecewise potential vorticity inversion. *J. Atmos. Sci.,* **49,** 1397–1411.

——, and K. A. Emanuel, 1991: Potential vorticity diagnostics of cyclogenesis. *Mon. Wea. Rev.,* **119,** 1929–1953.

Eliassen, A., and E. Kleinschmidt, 1957: Dynamic meteorology. *Encyclopedia of Physics,* S. Flugge, Ed., Springer-Verlag, 1–154.

Ertel, H., 1942a: Ein neuer hydrodynamischer Erhaltungssatz. *Naturwiss,* **30,** 543–544.

——, 1942b: Ein neuer hydrodynamischer Wirbelsatz. *Meteor. Z.,* **59,** 277–281.


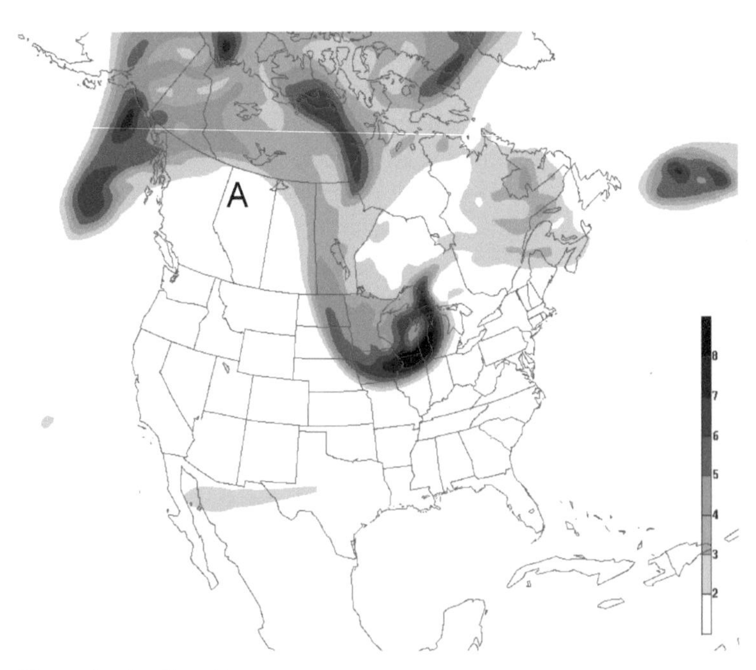


**Figure for Problem 3**

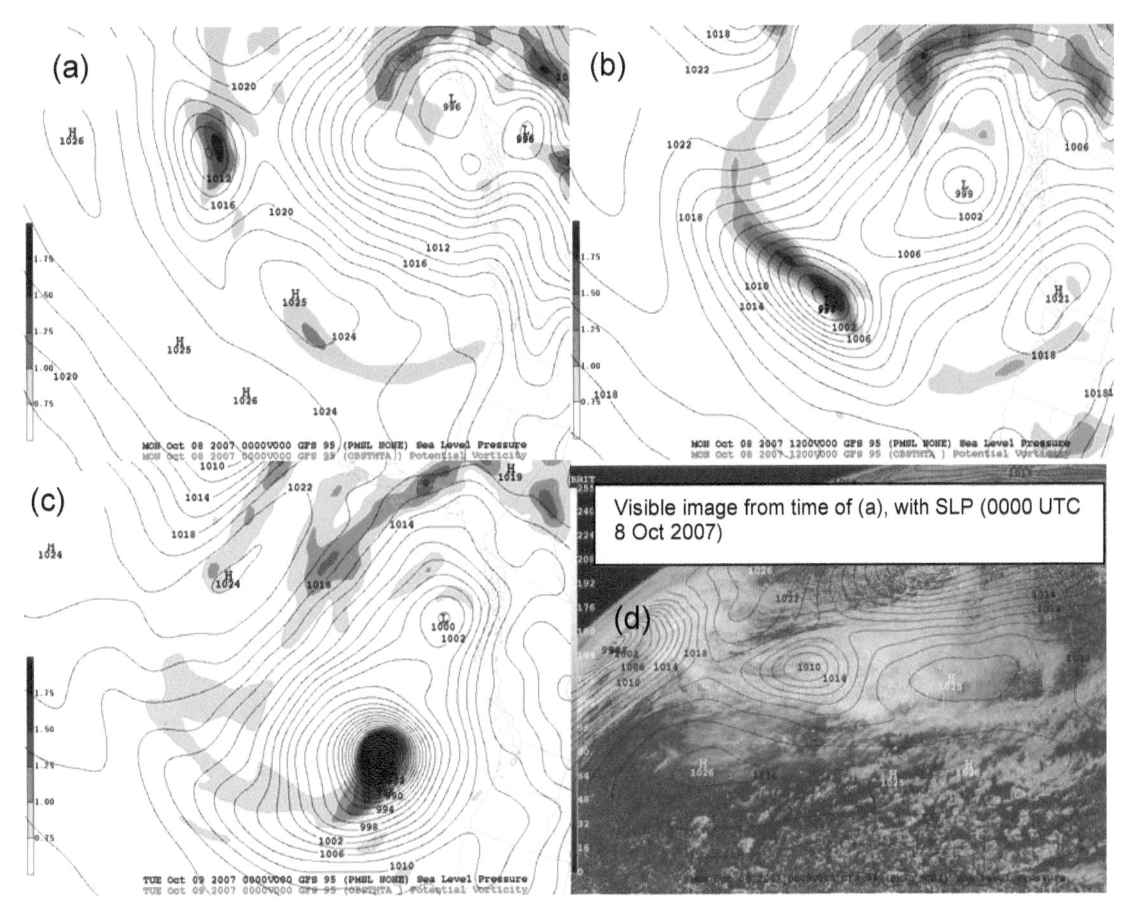


**Figure for Problem 4**

——, 1942c: Uber hydrodynamische Wirbelsatze. *Phys. Z.*, **43,** 526–529.

Haynes, P. H., and M. E. McIntyre, 1987: On the evolution of vorticity and potential vorticity in the presence of diabatic heating and frictional or other forces. *J. Atmos. Sci.*, **44,** 828–841.

Hoskins, B. J., M. E. McIntyre, and A. W. Robertson, 1985: On the use and significance of isentropic potential vorticity maps. *Quart. J. Roy. Meteor. Soc.*, **111,** 877–946.

Lackmann, G. M., and R. M. Yablonsky, 2004: The importance of the precipitation mass sink in tropical cyclones and other heavily precipitating systems. *J. Atmos. Sci.*, **61,** 1674–1692.

McTaggart-Cowan, R., J. R. Gyakum, and M. K. Yau, 2003: Moist component potential vorticity. *J. Atmos. Sci.*, **60,** 166–177.

Ooyama, K. V., 2001: A dynamic and thermodynamic foundation for modeling the moist atmosphere with parameterized microphysics. *J. Atmos. Sci.*, **58,** 2073–2102.

Rossby, C.-G., 1936: Dynamics of steady ocean currents in the light of experimental fluid dynamics. *Pap. Phys. Oceanogr. Meteor.*, **5,** 1–43.

——, 1938: On the mutual adjustment of pressure and velocity distributions in certain simple current systems, II. *J. Mar. Res.*, **1,** 239–263.

——, 1940: Planetary flow patterns i n the atmosphere. *Quart. J. Roy. Meteor. Soc.*, **66,** 68–87.

Samelson, R. M., 2003: Rossby, Ertel, and potential vorticity. http://www.aos.princeton.edu/WWWPUBLIC/gkv/history/RossbyErtelRMS.pdf

Schubert, W., S. A. Hausman, M. Garcia, K. V. Ooyama, and H.-C. Kuo, 2001: Potential vorticity in a moist atmosphere. *J. Atmos. Sci.*, **58,** 3148–3157.

——, E. Ruprecht, R. Hertenstein, R. Nieto Ferreira, R. Taft, C. Rozoff, P. Ciesielski, and H.-C. Kuo, 2004: English translations of twenty-one of Ertel's papers on geophysical fluid dynamics. *Meteor. Z.*, **13,** 527–576.

Thorpe, A. J., 1986: Synoptic disturbances with circular symmetry. *Mon. Wea. Rev.*, **114,** 1384–1389.

——, and H. Volkert, 1997: Potential vorticity: A short history of its definitions and uses. *Meteor. Z.*, **6,** 275–280.

## FURTHER READING

Davis, C. A., 1992a: A potential-vorticity diagnosis of the importance of initial structure and condensational heating in observed extratropical cyclogenesis. *Mon. Wea. Rev.*, **120,** 2409–2428.

——, M. T. Stoelinga, and Y.-H. Kuo, 1993: The integrated effect of condensation in numerical simulations of extratropical cyclogenesis. *Mon. Wea. Rev.*, **121,** 2309–2330.

Hakim, G. J., D. Keyser, and L. F. Bosart, 1996: The Ohio Valley wave-merger cyclogenesis event of 25–26 January 1978. Part II: Diagnosis using quasigeostrophic potential vorticity inversion. *Mon. Wea. Rev.*, **124,** 2176–2205.

Hoerling, M. P., T. K. Schaack, and A. J. Lenzen, 1991: Global objective tropopause analysis. *Mon. Wea. Rev.*, **119,** 1816–1831.

Lackmann, G. M., 2002: Cold-frontal potential vorticity maxima, the low-level jet, and moisture transport in extratropical cyclones. *Mon. Wea. Rev.*, **130,** 59–74.

——, and J. R. Gyakum 1999: Heavy cold-season precipitation in the northwestern United States: Synoptic climatology and an analysis of the flood of 17–18 January 1986. *Wea. Forecasting*, **14,** 687–700.

Morgan, M. C., and J. W. Nielsen-Gammon, 1998: Using tropopause maps to diagnose midlatitude weather systems. *Mon. Wea. Rev.*, **126,** 2555–2579.

Raymond, D. J., 1992: Nonlinear balance and potential-vorticity thinking at large Rossby number. *Quart. J. Roy. Meteor. Soc.*, **118,** 987–1015.

——, 1937: On the mutual adjustment of pressure and velocity distributions in certain simple current systems. *J. Mar. Res.*, **1,** 15–28.

Stoelinga, M. T., 1996: A potential vorticity-based study of the role of diabatic heating and friction in a numerically simulated baroclinic cyclone. *Mon. Wea. Rev.*, **124,** 849–874.

Thorpe, A. J., 1985: Diagnosis of balanced vortex structure using potential vorticity. *J. Atmos. Sci.*, **42,** 397–406.

Thorpe, A.J., 1994: An appreciation of the meteorological research of Ernst Kleinschmidt. *Meteor. Z.*, **2,** 3–12.

Uccellini, L. W., D. Keyser, K. F. Brill, and C. H. Wash, 1985: The Presidents' Day cyclone of 18–19 February 1979: Influence of upstream trough amplification and associated tropopause folding on rapid cyclogenesis. *Mon. Wea. Rev.*, **113,** 962–988.

Wang, X., and D.-L. Zhang, 2003: Potential vorticity diagnosis of a simulated hurricane. Part I: Formulation and quasi-balanced flow. *J. Atmos. Sci.*, **60,** 1593–1607.

Whitaker, J. S., L. W. Uccellini, and K. F. Brill, 1988: A model-based diagnostic study of the rapid development phase of the Presidents' Day cyclone. *Mon. Wea. Rev.*, **116,** 2337–2365.

Wu, C. C., and K. A. Emanuel, 1995: Potential vorticity diagnostics of hurricane movement. Part I: A case study of Hurricane Bob (1991). *Mon. Wea. Rev.*, **123,** 69–92.

# CHAPTER 5

# Extratropical Cyclones

*At every stage of development of a scientist or a science, the stock of knowledge already acquired, or the views of a dominating School or personality, will to some extent block the recognition, or even the observation, of certain otherwise obvious facts that do not fit in with this knowledge or view.*

—T. Bergeron, "Methods in Scientific Weather Analysis and Forecasting" (1959)

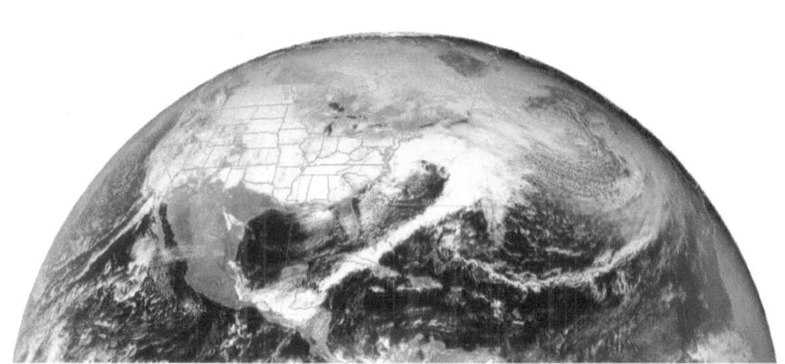

Visible geostationary satellite image during major East Coast snowstorm; and corresponding photo from Ellicott City, MD, 6 Feb 2010 (courtesy of William Mahoney).

Before the days of routine meteorological observations and surface map analyses, weather observers had noticed that changes in midlatitude weather often followed a similar sequence. Patterns of cloud, wind, and precipitation were tied to trends in atmospheric pressure. Written communication revealed that similar sequences of events were often observed at later times in more easterly locations. It was recognized that some sort of traveling disturbance, a minimum in the pressure field, influenced local weather conditions. Early meteorologists began to equip barometers with rudimentary weather forecast information, based on typical experiences. What were these travelling low pressure disturbances?

Today, extratropical cyclones are recognized not only for the important influence they exert on midlatitude weather conditions but also for their integral role in

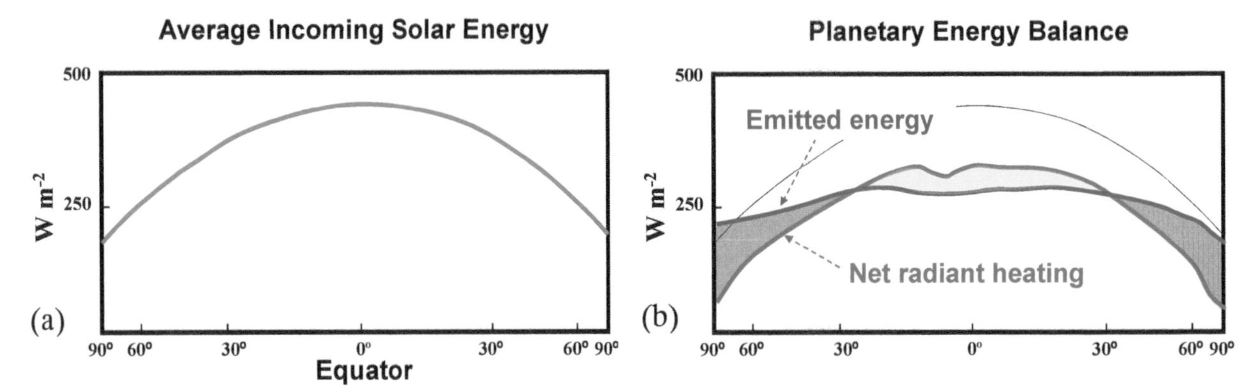


**Figure 5.1.** (a) Average flux of solar radiation reaching the top of the atmosphere. (b) Thin black curve as in (a), blue curve is average outgoing radiant energy flux, green curve is net radiant heating, and shaded areas represent the difference between emission and absorption [adapted from Gill (1982), originally from Winston et al. (1979)].

the earth's climate system. Nature provides periodic reminders of the societal impact of extratropical cyclones, as evident in the photograph shown at the beginning of this chapter, taken during the 6 February 2010 U.S. mid-Atlantic snowstorm. The accompanying satellite image valid for around that time reveals the presence of a cyclonic swirl of cloud in this region, accompanied by arcing bands of frontal cloud. In this chapter and in chapter 7, we will delve into the dynamical mechanism of extratropical cyclones.

In the preceding chapters, we have developed several frameworks with which to diagnose various types of synoptic weather systems. From the quasigeostrophic (QG) framework, we ascertain that ageostrophic motions, including ascent and descent, are part of a secondary circulation that arises in response to the disruption of thermal wind balance by the primary geostrophic flow. The *QG height tendency equation* provides insight into processes affecting the development or decay of weather systems, and this equation can be expressed as a statement of potential vorticity (PV) conservation. The PV framework as an analysis tool was developed in chapter 4; additional applications of PV will appear in this chapter and in chapter 7. Isentropic analysis, presented in chapter 3, can be used to determine patterns of cloud and precipitation in the vicinity of extratropical cyclones. Here, we apply these developed tools to the analysis of extratropical cyclones.

## 5.1. CYCLONES IN EARTH'S CLIMATE SYSTEM

Radiation measurements taken from satellites, averaged over time, provide an accurate account of incoming and outgoing radiative fluxes as a function of latitude (Fig. 5.1). Because of changes in albedo related to the climatological extent of cloud cover, as well as land surface characteristics—including ice, snow, and vegetation—the net radiant heating as a function of latitude follows the green curve in Fig. 5.1b. Comparison of the net radiant heating curve with the outgoing radiant emission curve demonstrates the well-known result that the tropics, equatorward of roughly 30° latitude in either hemisphere (corresponding to the yellow shading in Fig. 5.1b), receive a surplus of radiant energy, that is, these regions emit less radiant energy per unit area to space than they absorb from the sun. Conversely, polar regions experience a net deficit of radiant energy, shaded light blue in Fig. 5.1b.

Given that the long-term annual-mean temperature changes very slowly, we know that there must be a relatively rapid transfer of energy from regions of radiant energy excess to those of deficit, that is, from the tropics toward the poles. This poleward energy transport must be accomplished by the oceans and atmosphere. It is tempting to implicate a simple horizontal advective circulation to achieve the transport; however, in fact, the low-latitude Hadley circulation does not extend nearly far enough poleward to account for the observed energy transport.

Both oceanic and atmospheric heat transport are significant, and the availability of modern high-resolution reanalysis and observational datasets offer the ability to quantify these transports between ocean and atmosphere as a function of latitude. A recent study by Fasullo and Trenberth (2008) demonstrates that the atmosphere plays a dominant role in this transport relative to that of the oceans, especially in middle and higher latitudes (Fig. 5.2).

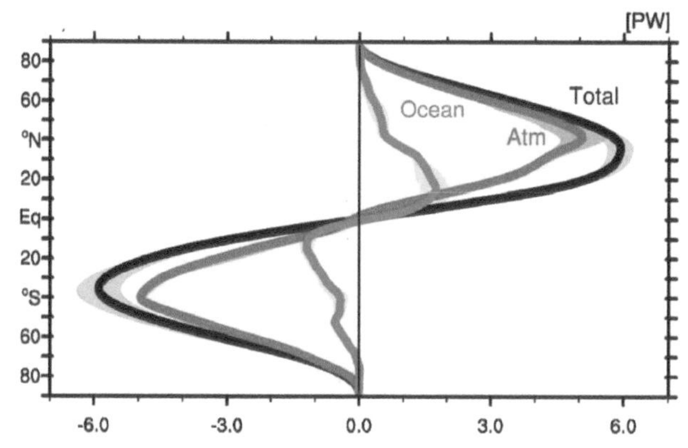


**Figure 5.2.** Annually and zonally averaged northward heat transport in petawatts (PW, $10^{15}$ W) to balance the net radiative imbalance; total transport shown by black curve, partitioned into atmospheric (red) and oceanic (blue) contributions. Uncertainty is shown as shading of the $2\sigma$ range. Negative values in the Southern Hemisphere are consistent with poleward transport there as well (from Fasullo and Trenberth 2008, their Fig. 7d).

The atmospheric contribution to the poleward energy transport involves both latent and sensible heat. Poleward transport of water vapor, and subsequent condensational heating as moist tropical airstreams are lifted, contributes significantly to the total heat transport. To those studying the planetary-scale circulation, extratropical cyclones, with their attendant fronts and airstreams, are essentially "synoptic-scale turbulence," disturbances acting to mix tropical and polar air and serving to reduce the overall equator-to-pole temperature gradient.

The importance of both tropical and extratropical cyclones in the climate system takes on additional significance when one considers *climate change*. Surface observations indicate that the strongest warming observed in recent decades has taken place in the arctic region, and future climate projections exhibit the same spatial tendency, a phenomenon known as *polar amplification* (Fig. 5.3). The role of transient disturbances, such as extratropical cyclones and anticyclones, is thought to be important to this process.

In a warming climate, it is expected that the lower troposphere will exhibit a 10%–20% increase in water vapor content. There follow several important implications, including the possibility of heavier precipitation, along with increased efficiency in latent heat transport associated with cyclones. Greater heat transport efficiency could alter the frequency, strength, and location of the midlatitude storm tracks. Increased precipitation in midlatitude cyclones could potentially result in a climatological change in their dynamical structure, with a tendency for stronger lower-tropospheric, diabatically generated cyclonic PV anomalies, although the possibility of such changes is the subject of current and future research.

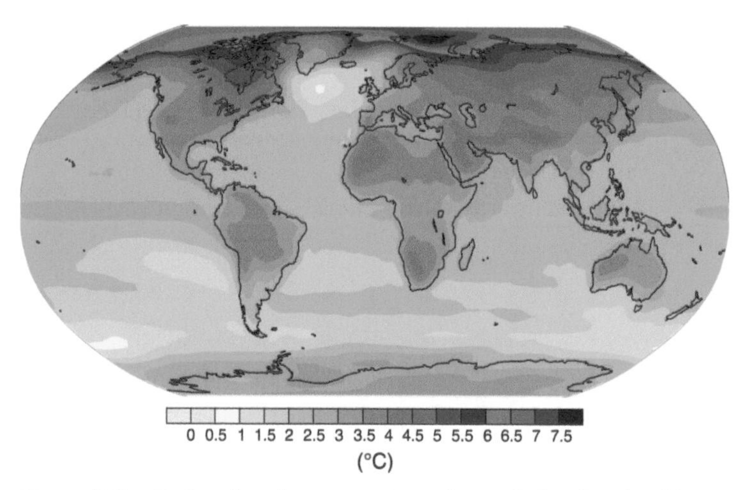


**Figure 5.3.** Projected surface temperature change for last decade of the twenty-first century (2090–99) relative to the 1980s, based on the ensemble mean of atmosphere–ocean general circulation models using the A1B emission scenario. [Source: Solomon et al. (2007), their Fig. 10.8.] Climate Change 2007: The Physical Science Basis. Working Group I Contribution to the Fourth Assessment Report of the Intergovernmental Panel on Climate Change, Figure 10.8. Cambridge University Press.

Another consequence of polar amplification that suggests possible changes in midlatitude storm tracks is the implied reduction of the equator-to-pole temperature gradient evident in Fig. 5.3, which, in isolation, suggests a weaker jet stream and a reduced baroclinic potential energy source for cyclones. However, both observations and climate models show an *increase* in the baroclinicity of the *upper* troposphere, thus complicating the picture.

The preceding discussion should clarify the important role cyclones play in the earth's climate system, as well as in day-to-day synoptic-scale weather conditions. Additionally, many mesoscale weather phenomena, including convective storms, are strongly sensitive to the synoptic-scale pattern in which they are embedded. In this chapter, we will explore the climatology, dynamics, life cycles, and prediction of midlatitudes cyclones. Additionally, given that the focus of several pioneering studies in meteorology was the analysis and prediction of cyclones, this chapter concludes with a brief account of the recent history of atmospheric science, with emphasis on research relating to extratropical cyclones.

## 5.2. CLIMATOLOGY OF CYCLONES

The midlatitude *storm tracks*, which can be measured through examination of the *variance* of the geopotential, meridional wind, or vertical motion fields, mark the most frequent paths of baroclinic, midlatitude disturbances. In the Northern Hemisphere, the most prominent storm tracks are found extending from eastern North America across the North Atlantic, and from near Japan extending across the North Pacific (Fig. 5.4). In contrast, the Southern Hemisphere exhibits a more continuous storm track, because of reduced influence from land masses and topography (not shown).

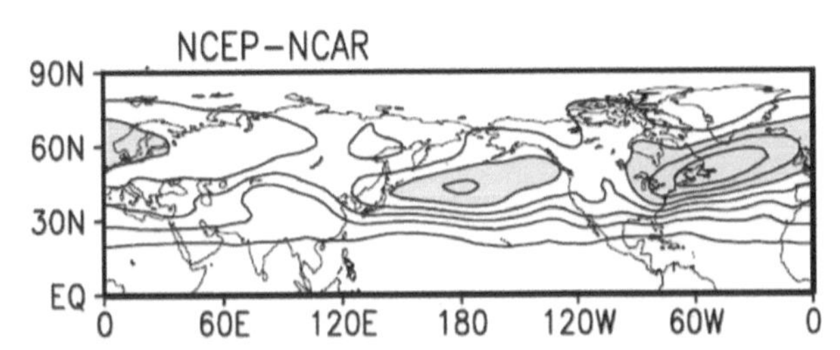


**Figure 5.4.** Northern Hemisphere storm track defined using the standard deviation of filtered 500-mb geopotential height (contour interval is 20 m, shaded greater than 100 m) from National Centers for Environmental Prediction–National Center for Atmospheric Research (NCEP–NCAR) reanalysis from January 1982–94 (from Chang 2009, his Fig. 2b).

The dynamics of the storm tracks, although outside the scope of this chapter, have been the subject of intensive study. It is known that extratropical cyclones convert potential to kinetic energy, and thus act to locally reduce baroclinicity in their immediate surroundings. Why, then, are storms frequently observed to follow one another in rapid succession? Hoskins and Valdes (1990) examined the maintenance of storm tracks, and they found that diabatic processes serve to replenish the baroclinic energy in storm tracks to a large extent. As cold continental air spills over the warm waters of the western oceanic boundary currents (the Gulf Stream and Kuroshio), rapid diabatic warming of the troposphere takes place; latent warming in the form of condensation is also important. Interestingly, Hoskins and Valdes also speculate that the ocean wind stress from the synoptic eddies themselves helps to drive the warm western oceanic boundary currents, and perhaps in this sense the storm tracks are self-maintaining. More recent studies have examined the influences of orography and diabatic heating on storm tracks. Chang (2009) identifies extratropical heating as exerting an important influence on the shape and intensity of the Atlantic storm track and as an explanation for why the Atlantic storm track is stronger than that in the Pacific, despite weaker baroclinicity.

Within the storm tracks themselves, baroclinic energy at the upstream end of the storm track leads to the development of strong disturbances; the energy from these Rossby waves then propagates downstream with the ageostrophic geopotential flux, as indicated in Fig. 5.5 and discussed in section 2.7.

Traditional cyclone climatologies were constructed by logging the location of sea level pressure minima on routine synoptic analyses. Recent reanalysis datasets have been used to objectively reconstruct cyclone climatologies and to study changes in the storm tracks over time. The basic patterns in the newer reanalysis-based climatologies are quite similar to the classic studies. Climatologies of cyclones and anticyclones for only the North American region are discussed below; however, global or hemispheric climatological cyclone studies based on reanalysis data, such as Sickmöller et al. (2000) and Zolina and Gulev (2002), are recommended reading.

The influence of the Rocky Mountains is evident in Figs. 5.6b,c for cyclones during the month of January over

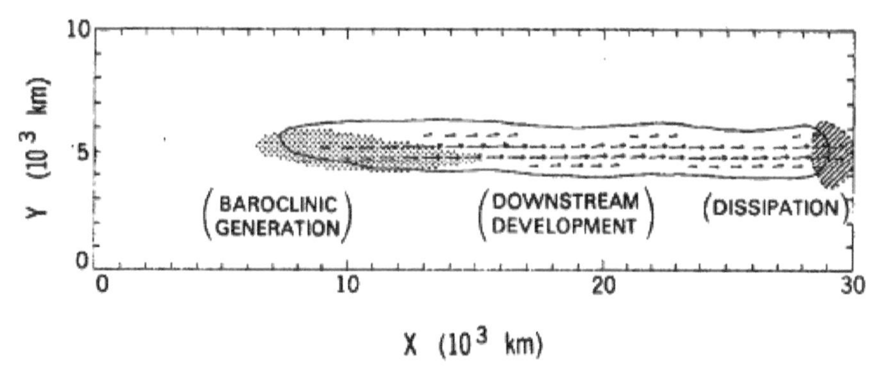


**Figure 5.5.** Horizontal schematic showing time-averaged storm-track diagnostics averaged from days 40 to 200 of a numerical model simulation. Solid curve corresponds to the outline of eddy kinetic energy maximum associated with a storm track. Dotted region indicates region of strong positive baroclinic energy conversion; hatched region indicates strong eddy kinetic energy dissipation. Vectors indicate ageostrophic geopotential flux above a given magnitude. From Chang and Orlanski (1993), their Fig. 11.

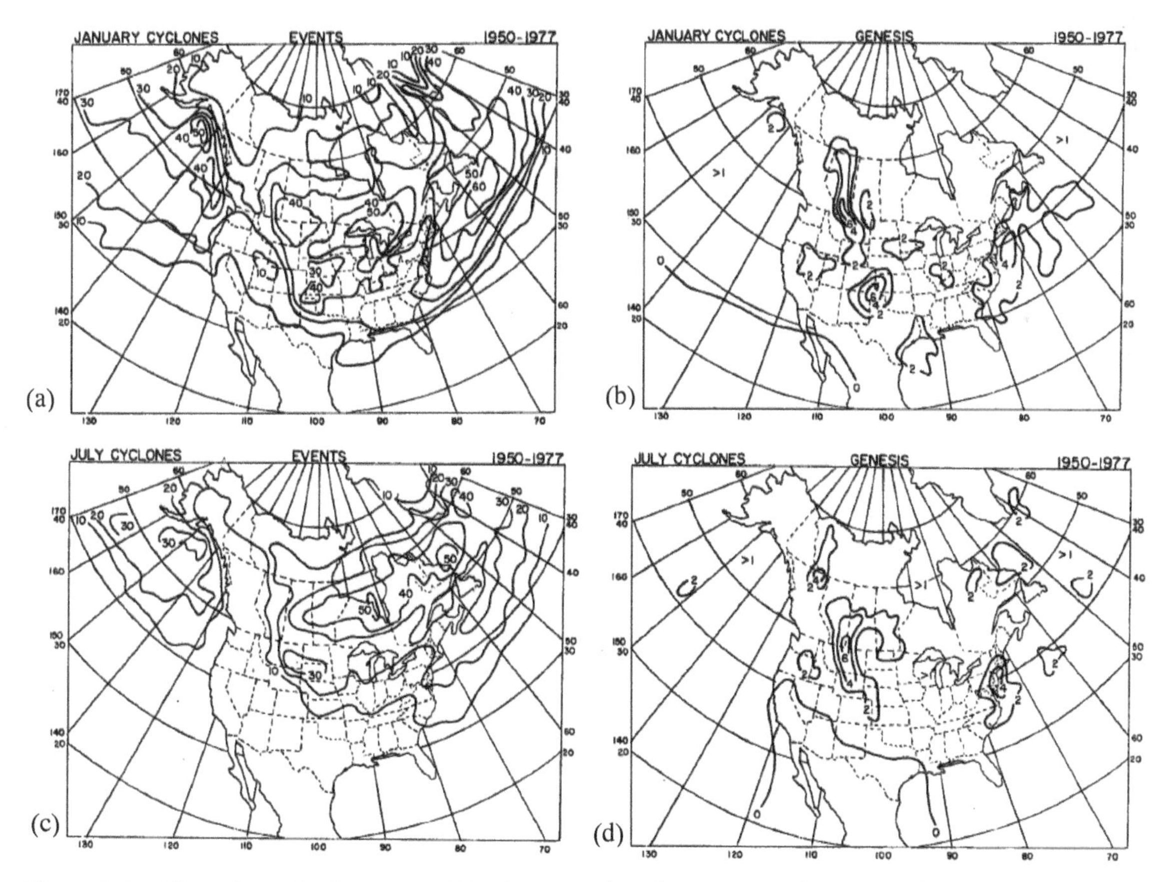


**Figure 5.6.** Climatology of cyclones around North America based on 28-yr samples: (a) areal frequency of events during January; (b) areal distribution of locations of cyclogenesis for January; (c) as in (a) but for July; (d) as in (b) but for July. Based on Figs. 2, 3 from Zishka and Smith (1980).

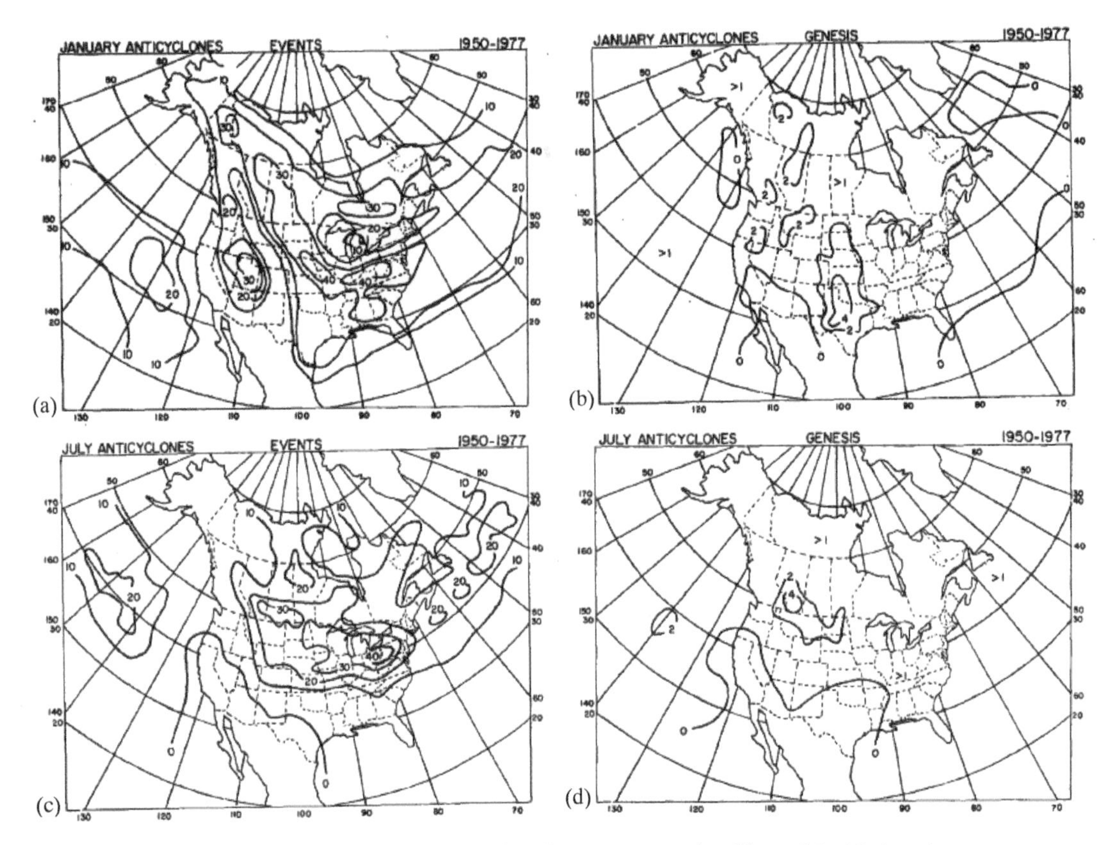


**Figure 5.7.** As in Fig. 5.6, but for anticyclones. Taken from Figs. 4, 5 of Zishka and Smith (1980).

a 28-yr sample analyzed by Zishka and Smith (1980). The formation of cyclones in the lee of the mountains is related to vortex stretching there, while the preferential lysis region along the windward slopes of the Rocky and Cascade Mountains in western North America is consistent with the compression of eastward-moving air columns. Also, the increased influence of friction and orographic channeling, which can enhance mass convergence into low pressure systems, may contribute to the lysis maximum along the mountainous coasts of western North America. The pattern of cyclogenesis in the month of July exhibits many of the same features seen in January, only shifted to the north, which is expected based on the northward retreat of the jet stream then.

The maximum in January cyclogenesis along the East Coast of the United States evident in Fig. 5.6b may have some orographic aspects, but it is undoubtedly also related to the strong baroclinicity present there due to the underlying contrast between the warm waters of the Gulf Stream and a much colder winter landmass to the west. The cyclogenesis process will be discussed in more detail in section 5.3.

Although *anti*cyclones are generally less likely to produce high-impact weather than cyclones, they are nevertheless an important part of the synoptic weather pattern and can be thought of as parts of the same baroclinic wave trains as their cyclonic counterparts. Examination of the climatological locations for their occurrence, formation, and lysis, as for cyclones, is provided by the classic Zishka and Smith (1980) study (Fig. 5.7). The contrasting influence of the Great Lakes is evident in the anticyclone frequency between January and July (Figs. 5.7a,d). During January, the lakes are relatively warm, and arctic anticyclones passing over them experience strong diabatic heating, leading to surface pressure falls, and thus a relative minimum in anticyclone frequency over the lakes in January. Conversely, the lakes are relatively cold and stable in July, favoring anticyclones at that time.

## 5.3. CYCLOGENESIS

The process of *cyclogenesis* (the formation of cyclones) has historically been a focus of cyclone research, often with specific emphasis on cyclones that form rapidly or unexpectedly. Theoretical, observational, and numerical modeling studies are all important research tools utilized by atmospheric scientists working on this problem, leading to the relatively complete understanding of cyclogenesis that exists today. Despite major advances in our understanding, there are still aspects of the problem that are not fully understood and research continues. There are several different, yet consistent, ways in which to interpret the physical mechanisms of cyclogenesis (or anticyclogenesis). Tools developed in the first four chapters that can shed light on this problem include the vorticity equation, the pressure tendency equation, the QG height tendency equation, potential vorticity applications, and QG energetics. These different tools and techniques each offer some unique interpretations, and therefore several perspectives will be presented.

### 5.3.1. Vorticity view

As discussed in section 1.5, weather disturbances of many types are accompanied by "cyclonic vorticity maxima," including tropical and extratropical cyclones, upper troughs, frontal systems, lee troughs, and jet streaks. Given that cyclones are identified by cyclonic vorticity maxima at the surface, the process of cyclogenesis may be explained through analysis of the vorticity equation (1.49), repeated here for convenience:

$$\frac{\partial \zeta_a}{\partial t} = -\vec{V}\cdot\nabla\zeta_a - \omega\frac{\partial \zeta_a}{\partial p} - \left[\frac{\partial \omega}{\partial x}\frac{\partial v}{\partial p} - \frac{\partial \omega}{\partial y}\frac{\partial u}{\partial p}\right] + \zeta_a\frac{\partial \omega}{\partial p} + \left[\frac{\partial F_y}{\partial x} - \frac{\partial F_x}{\partial y}\right]. \tag{5.1}$$

As mentioned above, an important characteristic of frontal zones, to be discussed in the following chapter, is that they are characterized by enhanced cyclonic vorticity. Cyclones are also observed to frequently form along preexisting frontal boundaries. The vorticity framework provides a straightforward means of understanding this observation. To simplify the interpretation for cyclogenesis, consider the change in vorticity near the surface, at the center of a band of cyclonic vorticity extending along a uniform frontal boundary. At the center of the band and near the surface, we can neglect the advection and tilting terms, leaving the stretching term in (5.1), $\zeta_a(\partial\omega/\partial p)$, as the main production term. Suppose that a mobile upper-tropospheric trough tracks toward the frontal boundary, as depicted in Fig. 5.8. Associated with the upper trough, there is QG forcing for ascent, resulting in lower-tropospheric stretching ahead of the trough axis (region of blue shading in Fig. 5.8).

In this situation, a low pressure center will typically not form until the upper-trough axis approaches the front, and then a cyclone, or frontal wave, will form along the frontal zone and subsequently amplify into a full-blown cyclonic system. Why are frontal zones so favorable for cyclogenesis? Examine the multiplier in the stretching term. To quantify this preference, suppose that surface divergence is $-1\times10^{-5}\,\mathrm{s}^{-1}$ (convergence) throughout the blue shaded area ahead of the upper trough, that $f$ corresponding to this latitude is ~$10^{-4}\,\mathrm{s}^{-1}$, and that the relative vorticity $\zeta$ is $10^{-4}\,\mathrm{s}^{-1}$ only in the hatched area near the frontal zone and is zero elsewhere.

Compare the rate of vorticity spinup due to stretching in the area ahead of the upper-level trough before it

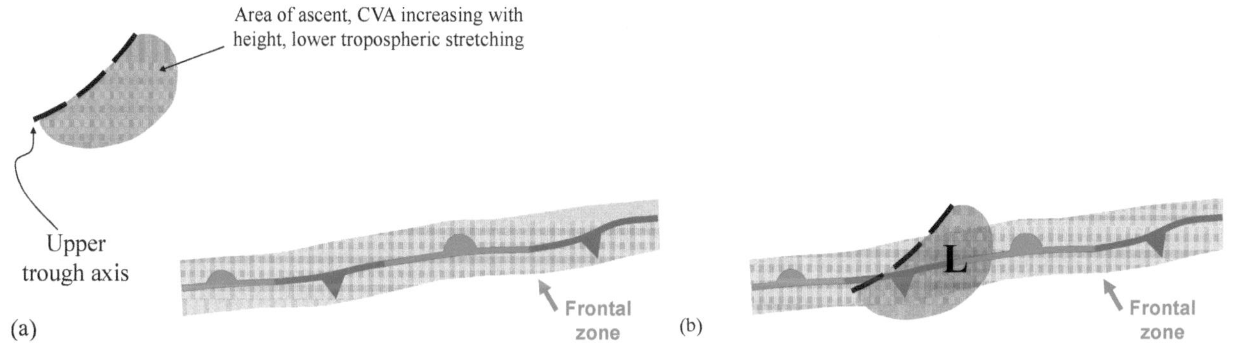


**Figure 5.8.** Idealized schematic of an upper trough (dashed black line denotes trough axis) and associated area of QG forcing for ascent (blue shading) overtaking a preexisting stationary front, associated with a band of enhanced cyclonic relative vorticity (green shading): (a) initial time and (b) later time with an upper trough overtaking surface front, and formation of surface low pressure center (L).

reaches the surface front, and again once it overtakes the frontal zone. We find that the effectiveness of the stretching term within the frontal zone is *double* that outside the zone. This is because the absolute vorticity itself multiplies the stretching. A given amount of stretching (upward vertical motion) is more effective in spinning up vorticity in regions where the vorticity is already large. In fact, this positive feedback mechanism can be described mathematically if we solve the frictionless vorticity equation for a point moving with the surface cyclonic vorticity center, only considering stretching on the right-hand side:

$$\frac{d\zeta_a}{dt} = -\zeta_a(\nabla\cdot\vec{V}); \quad \frac{d\ln\zeta_a}{dt} = -(\nabla\cdot\vec{V}). \tag{5.2}$$

If we take the antilog of each side and integrate to obtain an expression for the vorticity as a function of time, the result is

$$\zeta_a(t) = \zeta_a(0)\exp(-(\nabla\cdot\vec{V})t). \tag{5.3}$$

For a constant value of convergence, (5.3) yields an exponential growth of vorticity with time. In nature, friction limits this growth rate; however, the value of convergence will also tend to increase with the intensity of a system, which partially offsets the spindown. Thus, during some portion of their life cycles, strongly developing systems can exhibit exponential cyclonic vorticity growth. The important lesson here is that regions of preexisting vorticity must be carefully monitored for the rapid spinup of cyclonic systems. Another aspect that makes frontal zones favorable sites for cyclogenesis is that convergence into the frontal trough and accompanying ascent also enhances vortex stretching there.

Conversely, because of cancellation between the planetary vorticity, which is cyclonic, and anticyclonic relative vorticity in high pressure systems, the stretching term is *ineffective* near the center of surface anticyclones (or any region of significant anticyclonic vorticity). This mechanism can explain the observation of cyclones approaching lower-tropospheric ridges that re-form on the other side of the ridge, as the stretching mechanism will not operate efficiently within the ridge. A sequence of sea level pressure analyses in which this process appears to be at work is provided in Fig. 5.9.

As a low pressure system moves northeast from Louisiana at 0000 UTC 16 February 2003, a strong arctic high pressure system is moving eastward into the northeastern United States and eastern Canada (Fig. 5.9a). A ridge of high pressure extends southward from this arctic high center along the eastern slopes of the Appalachian Mountains (Figs. 5.9b,c), a phenomenon known as cold-air damming, which will be discussed at length in chapter 8. This ridge of high pressure, with anticyclonic relative vorticity, reduces the effectiveness of the stretching term, despite ascent and precipitation across this region (not shown). Meanwhile, along the coast, a separate frontal zone remains stationary; this zone of preexisting vorticity is a favored site for cyclogenesis, and by 1200 UTC 17 February, a new surface cyclone has developed along this front to the east of Virginia (Fig. 5.9d).

### 5.3.2. Pressure view

The atmosphere is observed to remain very close to a state of hydrostatic balance; thus, atmospheric pressure is very nearly equal to the hydrostatic weight of the overlying air column. For the surface pressure to fall at the center of a developing low pressure system, there must therefore be net mass divergence in the overlying air column. Because frictional inflow gives rise to mass *convergence* within the planetary boundary layer, from this perspective, we must focus on mechanisms that are associated with divergence aloft. It should be clarified here that *divergence* is not the same as *diffluence*, the latter being associated with the tendency for streamlines to spread out from a given location; depending on along-flow wind speed variations, diffluent flow can exhibit divergence or convergence.

The early study of Bjerknes and Holmboe (1944) was one of the first to link waves in the upper jet stream to patterns of divergence and convergence aloft and to the development of high and low pressure systems beneath. For a hypothetical upper-level wave pattern in which the spacing of upper-level geopotential height contours is everywhere identical along the main jet (Fig. 5.10a), convergence and divergence will arise from the fact that wind speed in the ridge will be faster (supergeostrophic), while flow in the trough will be weaker (subgeostrophic). This is due to the centrifugal force acting outward from the center of rotation in the trough and ridge—in the ridge, the centrifugal acceleration is directed in the same sense as the pressure gradient force, requiring strong flow to produce a sufficient Coriolis force to balance these combined forces. In the trough, the centrifugal and Coriolis are directed in the same sense, allowing weaker flow to remain in gradient balance.

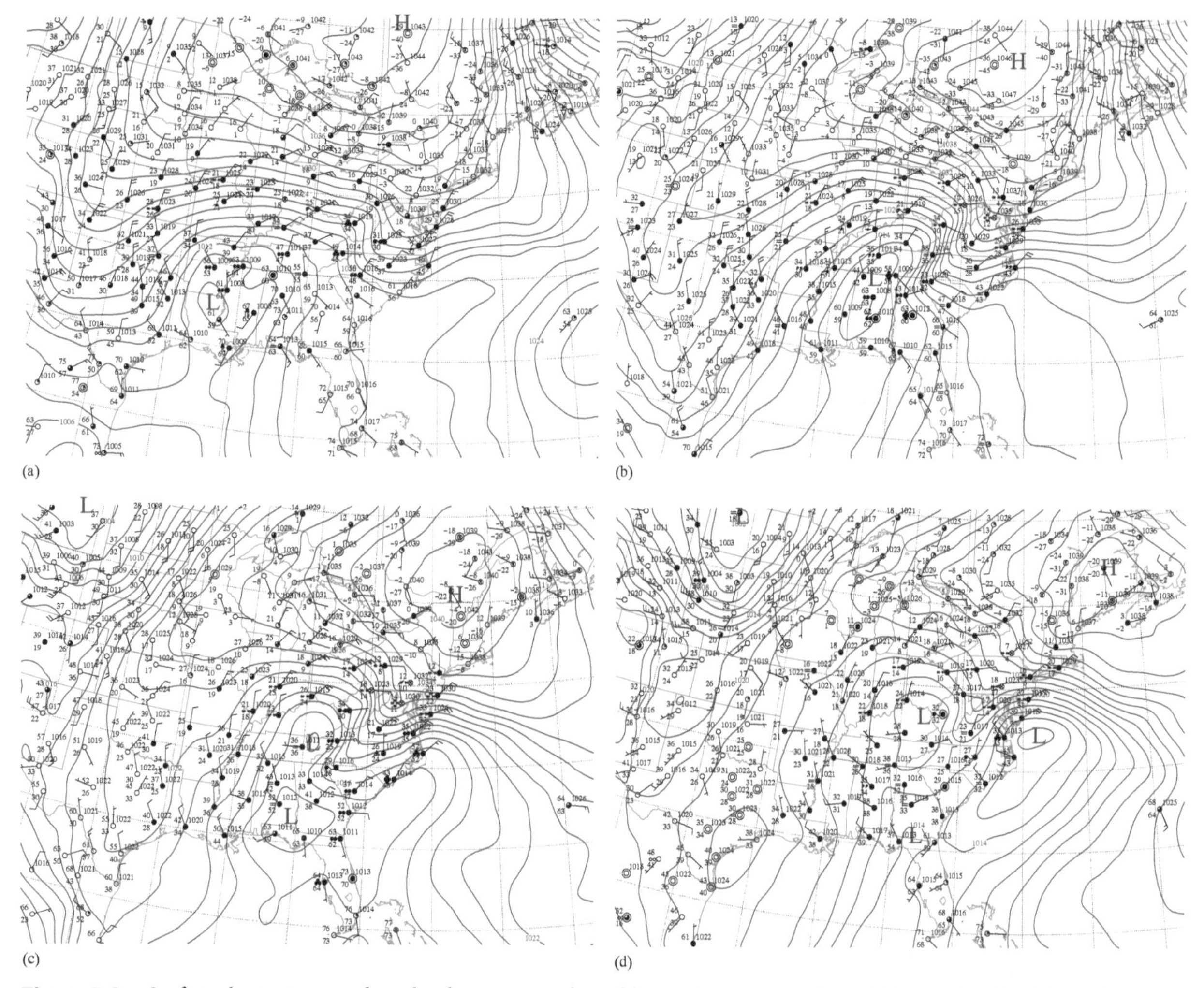

**Figure 5.9.** Surface observations and sea-level pressure analysis (blue contours, interval is 2 hPa) from the North American Mesoscale (NAM) model at (a) 0000 and (b) 1200 UTC 16 Feb and (c) 0000 and (d) 1200 UTC 17 Feb 2003. Dashed lines in (c) correspond to trough axes.

The resulting patterns of divergence and convergence aloft are as shown in Fig. 5.10a, and thus cyclogenesis is favored between an upstream upper-trough axis and the downstream ridge. Similarly, convergence aloft and surface anticyclones are favored downstream of the upper ridge and upstream of a trough; these tendencies are shown for actual upper-air analyses in Figs. 5.10b,c. The tendency for supergeostrophic flow in ridges, subgeostrophic flow in troughs, and the location of surface pressure maxima and minima are consistent with the idealized pattern.

The pressure and vorticity views are exactly consistent with one another, the difference being a focus on lower-tropospheric stretching versus upper-level divergence. As illustrated in Fig. 5.11, each perspective emphasizes different aspects of the same circulation.

### 5.3.3. QG interpretation

The pressure and vorticity perspectives are easily related to the QG framework. Forcing for ascent, as diagnosed by the QG omega equation, implies stretching in the lower troposphere beneath the level of maximum ascent, which is typically found in the middle troposphere. The QG vorticity equation does not account for the exponential vorticity growth from Eq. (5.2) and described in Fig. 5.8, because the multiplier in the simplified QG stretching term is the constant $f_0$. Diagnosis of ascent from the QG omega equation is consistent with divergence aloft through mass

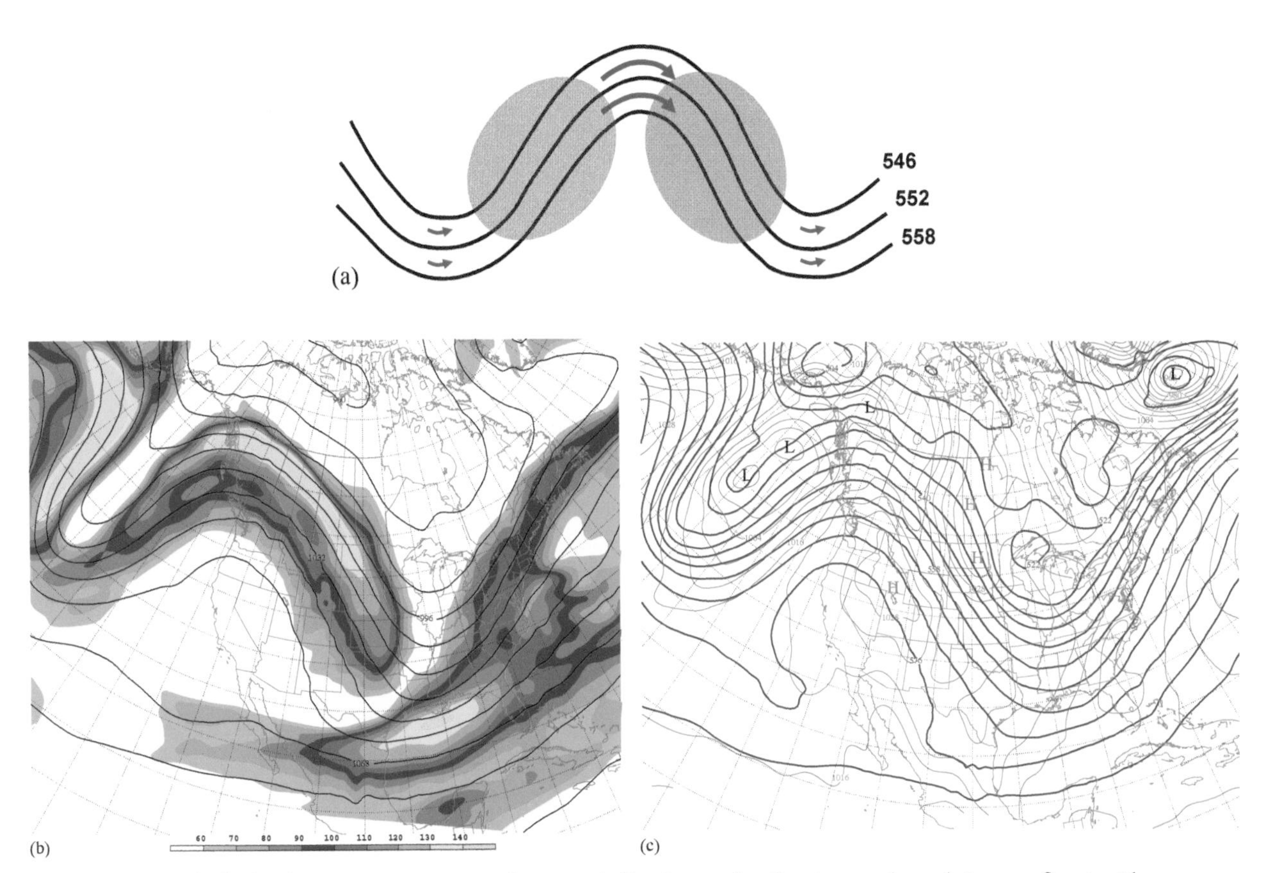


**Figure 5.10.** (a) Idealized upper-wave pattern with arrows indicating weaker flow in troughs and stronger flow in ridges as expected from gradient wind balance; region of pink (green) shading corresponds to a zone of upper-level divergence (convergence); (b) 250-mb height (contours, interval is 12 dam) and isotachs (kt, shaded as in legend) for 0000 UTC 3 Feb 2009; (c) as in (b) but with sea level pressure replacing 250-mb isotachs. The symbols H and L indicate the location of surface high and low pressure systems, respectively.

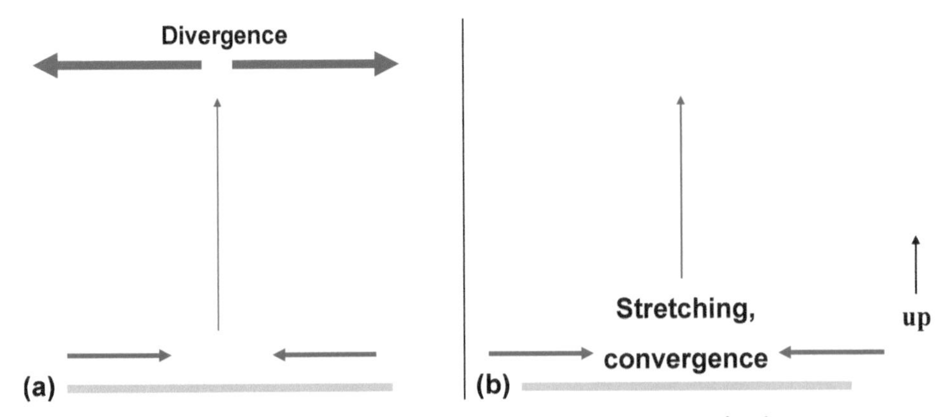


**Figure 5.11.** Cross-sectional comparison of pressure and vorticity views of cyclogenesis: (a) divergence aloft exceeds near-surface convergence for a developing surface low pressure system, and (b) vortex stretching and near-surface convergence result in an increase of cyclonic vorticity in the lower troposphere.

continuity, which is linked to the pressure perspective, so long as the divergence aloft exceeds convergence in the lower levels of the troposphere.

Given that cyclogenesis requires height falls in the lower troposphere (or at the surface), we can utilize the QG height tendency equation to understand this process. As discussed in chapter 2, warm advection increasing with height and cyclonic vorticity advection (CVA) represent forcing mechanisms for height falls, as does diabatic heating increasing with height. Surface low pressure systems often track toward the region of strongest lower-tropospheric warm advection, and isallobaric analyses (contours of constant pressure change) frequently show maximum pressure falls ahead of the surface warm front in a region of strong warm advection. Recall that the height tendency equation is also a statement of PV conservation. This ties in yet another tool for diagnosis of cyclogenesis; the PV perspective will be discussed in section 5.3.6.

### 5.3.4. Upper-trough structure and evolution

The discussion from chapter 2 of the QG height tendency equation demonstrated the importance of vorticity and differential thermal advection to the height tendency. In this context, we discussed "digging" and "lifting" troughs, and the importance of the thermal advection pattern to upper-wave amplification or decay. However, we have not yet discussed another important aspect of the upper waves that is crucial to understanding cyclogenesis: the wavelength dependence of the forcing for ascent in the QG framework. For two upper-wave patterns with equal amplitude (defined here as the maximum north–south deviation of a given geopotential height contour), consider the strength of the vorticity advection in each of these patterns, as shown in Fig. 5.12.

The vorticity advection is clearly stronger in pattern B, despite equal amplitude of the height contours. This is due to both the greater strength of the vorticity extrema and also the closer spacing of the extrema in Fig. 5.11b; both of these aspects serve to increase the vorticity gradient and thus the strength of the vorticity advection. Stronger vorticity advection is consistent with stronger QG forcing for ascent via the differential vorticity advection term in the omega equation. For a perfectly sinusoidal pattern, while the forcing is stronger with shorter wavelength, inverting the Laplacian to solve for omega can largely cancel this effect on the vertical motion itself.

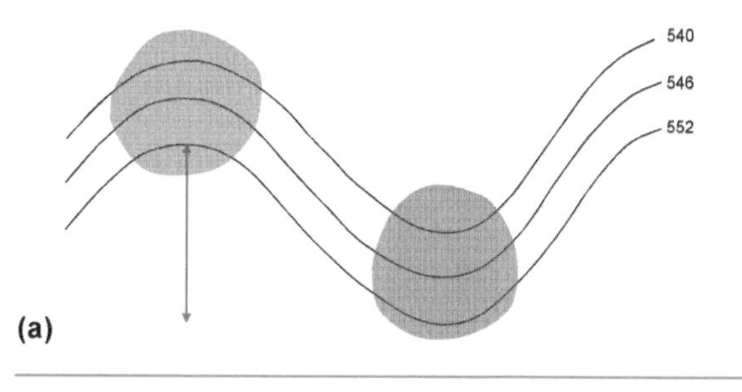


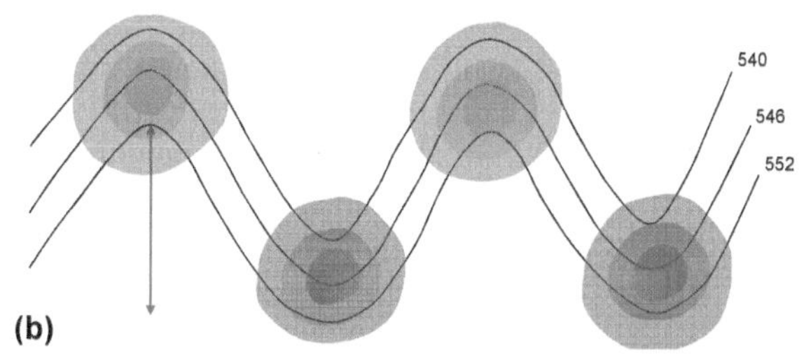


**Figure 5.12.** Idealized upper-wave patterns of equal amplitude (defined as the latitudinal displacement of geopotential height contours). Black solid lines represent geopotential height contours. Shading depicts idealized cyclonic (blue) and anticyclonic (red) relative vorticity centers. (a) Wave pattern one; (b) wave pattern two with shorter wavelength.

### 5.3.5. Sutcliffe-Petterssen "self development"

The preceding discussion demonstrates that forcing for ascent and divergence aloft, which are related to the development of cyclones, depend upon the strength and structure of upper-level disturbances. However, the development of the surface cyclone should not be viewed as a passive response to "upper-level forcing." As is emphasized in both theoretical baroclinic instability and observational cyclone studies, there is an active linkage between lower- and upper-level processes during cyclogenesis. Here, we examine the interaction between a surface cyclone and upper disturbance during the evolution of a cyclone event, following the classic work of Sutcliffe and Forsdyke (1950). Sutcliffe presented a thickness (layer-average virtual temperature) tendency equation in the following form:

$$\frac{\partial [T]}{\partial t} = -[\vec{V} \cdot \nabla T] + [\omega(\Gamma_{ad} - \Gamma_{env})] + \left[\frac{\dot{Q}}{C_p}\right], \qquad (5.4)$$

where the square brackets denote a vertical average over the depth of the layer in question. In (5.4), $\Gamma_{ad}$ is the adiabatic lapse rate of temperature (the dry adiabatic rate for unsaturated air, the moist rate for saturated air), and $\Gamma_{env}$ is the observed environmental lapse rate (such as would be measured by a rawinsonde). The thermal wind equation,

$\vec{V}_T = \vec{V}_{gH} - \vec{V}_{g1000}$, can be rearranged to express the geostrophic wind velocity at any level H, $\vec{V}_{gH}$, as follows:

$$\vec{V}_{gH} = \vec{V}_{g1000} + \vec{V}_T, \tag{5.5}$$

where the layer in question extends from the 1000-mb level to level H. Because the thermal wind $\vec{V}_T$ is parallel to thickness contours, it does not contribute to thermal advection within the layer. Therefore, we write the thickness tendency equation in a form that emphasizes the importance of thermal advection in the lower troposphere as follows:

$$\frac{\partial [T]}{\partial t} \approx -\vec{V}_{g1000} \cdot \nabla [T] + \omega(\Gamma_{ad} - \Gamma_{env}) + \left[\frac{\dot{Q}}{C_p}\right]. \tag{5.6}$$

Equation (5.6) states that the local tendency of layer-averaged virtual temperature, or thickness, is related to the geostrophic thickness advection by the 1000-mb geostrophic wind, a term proportional to the difference between observed and adiabatic lapse rates and the vertical motion, and the net diabatic heating or cooling in the air column. We will discuss each of the right-hand terms in more detail below, within the context of Fig. 5.13.

The typical thermal and geostrophic wind configuration in the vicinity of a developing cyclone is characterized by geostrophic warm thickness advection at the 1000-mb level to the east and northeast of the cyclone center, with cold advection to the west. Considering only the action of the first right-hand term in (5.6), we find a tendency for thickness decrease west of the surface cyclone center and increase to the east.

The thickness tendency results in height changes that are consistent with what one would expect from the QG height-tendency equation, namely, height falls near the center of the upper trough to the west of the surface cyclone and height rises to the east (Fig. 5.13). This pattern of height tendency is aligned with the upper wave so as to amplify it, and the location of the thermal advection pattern often results in a shortening of the wavelength in the upper trough–ridge system, the importance of which was discussed in the previous subsection. The upper-wave amplification also tends to vertically align the upper-wave inflection point with the surface cyclone so as to maximize forcing for ascent, and vortex stretching, in the vicinity of the surface cyclone center.

An amplified upper wave is associated with strengthened QG forcing for ascent, increased divergence aloft over the low center (or, consistently, increased lower-tropospheric vortex stretching), and a subsequent intensification of the surface cyclone. This, in turn, increases the strength of the geostrophic thickness advection, and a positive feedback is established between the lower-tropospheric thermal advection, the upper wave, and the surface cyclone. This feedback was recognized by Sutcliffe and collaborators and is consistent with the linear exponential growth mechanism of baroclinic instability (see chapter 7). It should also be pointed out that the thermal advection additionally drives surface pressure (or lower-tropospheric geopotential height) falls to the north and east of the cyclone center; this serves to propagate the system toward the east and northeast.

This feedback mechanism does not typically operate unopposed; eventually other mechanisms must come into play to limit the intensity of cyclones. One such

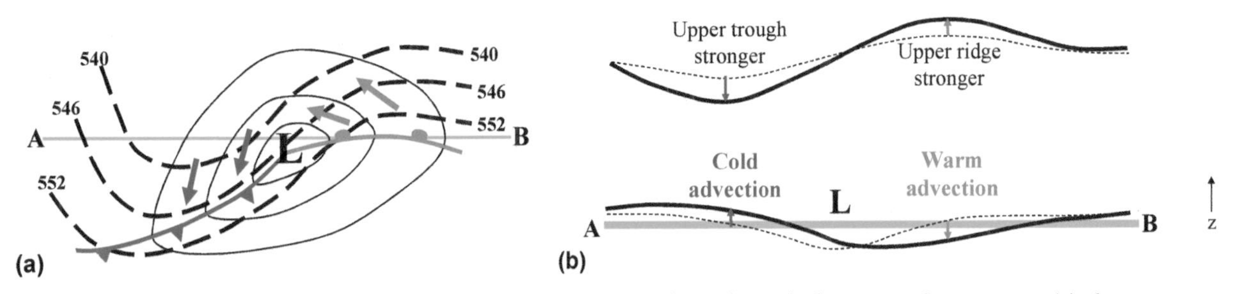


**Figure 5.13.** Schematic depiction of feedback between lower-tropospheric thermal advection and upper wave: (a) plan view of 1000–500-mb thickness (dashed black contours), sea level isobars (gray solid lines), fronts, and selected geostrophic wind vectors [red (blue) indicating warm (cold) thickness advection]. Orange line A–B indicates location of cross section shown in (b); (b) vertical cross section showing before and after positions of the 1000- and 500-mb height surfaces (dashed before, solid after), along with relative locations of warm and cold advection maxima (labeled).

process is represented by the second right-hand term in (5.6), the vertical motion term $\omega(\Gamma_{ad} - \Gamma_{env})$. Provided that the environmental lapse rate is smaller than the adiabatic rate, as is typically the case, this term will usually have the opposite sign as the thermal advection term. For example, in the zone of warm advection east of the low center, $-\vec{V}_{g1000} \cdot \nabla[T] > 0$, and $\omega(\Gamma_{ad} - \Gamma_{env}) < 0$ because for ascent, $\omega < 0$. In the region of cold advection west of the surface low, each term will have the opposite sign. Therefore, the action of the vertical motion term serves to put a brake on the self-development process and is therefore referred to as the "braking term."

Examination of the braking term in (5.6) reveals that this process is strongly sensitive to the static stability. For saturated conditions, the *saturated* adiabatic lapse rate is used for $\Gamma_{ad}$, which, being smaller than the dry adiabatic value, effectively reduces the strength of the brake relative to what it would be in unsaturated conditions. This is important, because it suggests that the positive feedback process described previously in association with the thermal advection is strongly opposed in relatively dry locations. On the other hand, in moist, maritime environments, the positive thermal advection feedback is more effective, given the weaker braking term. Furthermore, if conditions arise in which the environmental lapse rate is large, such as over the warm western oceanic boundary currents during winter, the braking term can be effectively shut off as the environmental lapse rate approaches the moist adiabatic rate. The ability of this term to counteract the self-development feedback plays a significant role in determining the climatological locations of rapidly developing cyclones, as will be shown subsequently.

Use of the moist adiabatic lapse rate in the braking term in (5.6) for saturated conditions implicitly accounts for the impact of latent heat release on the self-development process; however, it should be noted that net heating or cooling due to various mechanisms can exert an additional influence on the thickness tendency, and in some cases these processes can be quite important. Turbulent heat and moisture fluxes in the planetary boundary layer can become large, as discussed further in section 5.3.7. Other latent heating processes can also be important; for instance, the latent heat absorbed by melting is an example of a diabatic process that would not be accounted for by simply using the saturated adiabatic lapse rate in the braking term. However, the dominant feedback and strongest opposing process are represented by the first two right-hand terms in (5.6).

A schematic sequence illustrating this process exemplifies the building of the downstream ridge in the region of warm advection, with the attendant shortening of the trough–ridge wavelength of the upper wave (Fig. 5.14). The building of the ridge to the northeast of the upper trough can also help the upper trough axis tilt

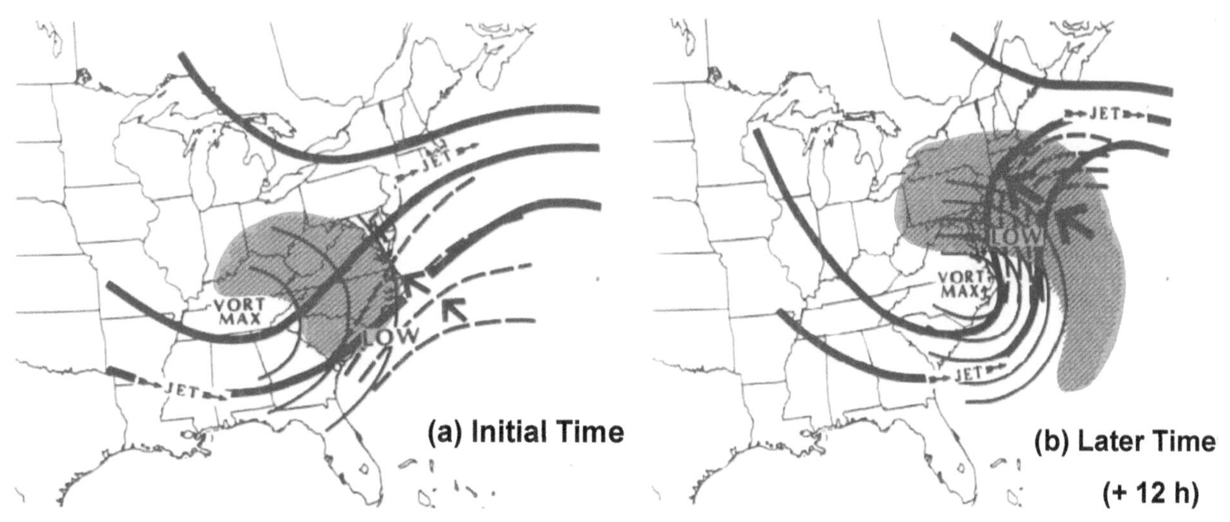


**Figure 5.14.** Schematic representation of Sutcliffe–Petterssen self-development mechanism. Temperature advections, boundary layer heat and moisture fluxes, and latent heat release contribute to the amplification of the upper wave, and also reduce the wavelength between the trough and downstream ridge. This, in turn, increases the QG forcing for ascent and upper-level divergence, further strengthening the cyclone, and thermal advections. (a) Initial time; (b) time approximately 12 h later (from Kocin and Uccellini 1990).

to become negative, aiding the cyclone development in terms of barotropic energy conversion, as discussed in section 2.7.

In addition to the forcing provided by the upper trough–ridge couplet, the presence of *jet streak* entrance and exit regions can also focus regions of divergence aloft (Fig. 5.14). Recall from section 2.3 that the right entrance and left exit regions of straight jet streaks are favored locations for ascent and upper-level divergence. Jet streaks can be amplified by diabatically forced outflow accompanying latent heat release and heavy precipitation associated with the cyclone. The influence of diabatic processes on cyclogenesis is clarified using a conservative variable, such as PV, as a diagnostic tool.

### 5.3.6. PV framework

We will now apply the PV framework outlined in chapter 4 to the problem of cyclogenesis. Recall from this earlier discussion that the *invertibility principle* of PV allows us to partition and associate the balanced wind field with specific portions of the PV distribution. One can recover the pressure or geopotential field from PV inversion, and thus, we can think of a cyclone as a region associated with cyclonic PV anomalies; by inverting these anomalies, the balanced circulation of the cyclone can be recovered to a very large extent.

An additional aspect of the PV framework, discussed in more detail in chapter 7, is the equivalence of boundary potential temperature anomalies as "equivalent" PV anomalies: warm (cold) boundary anomalies assume the role of cyclonic (anticyclonic) PV anomalies, as demonstrated in the idealized PV inversions shown in Fig. 5.15. In this diagram, adapted from Thorpe (1986), a 10-K surface warm potential temperature anomaly is inverted, and the resulting cyclonic wind field extends all the way to the tropopause. This implies that surface disturbances of this strength are capable of producing advections at the tropopause, and the reverse is also true, as evident in Fig. 4.1. The vertical extension of the flow associated with a given PV anomaly is inversely proportional to the static stability (this will be demonstrated in chapter 7); in unstable environments, the circulation associated with a given PV anomaly will extend a greater distance in the vertical relative to what would occur with greater static stability. Thus, the ability of upper and lower PV anomalies to interact is affected by the static stability of the intervening air, as discussed below. Two other factors that determine the vertical extension of circulations associated with PV anomalies are the horizontal scale of the anomaly as well as the disturbance amplitude. Large and strong disturbances produce circulations with a greater vertical extent relative to small and weak ones, as one would expect.

Given that the highly stable stratosphere is the primary source region of cyclonic PV for upper-tropospheric disturbances, observations reveal the presence of preexisting upper-tropospheric cyclonic PV anomalies for almost all extratropical cyclogenesis events. These cyclonic PV maxima can be viewed as "stratospheric extrusions" of cyclonic PV into the upper troposphere. How do these preexisting upper-level PV anomalies form? Vertical motions associated with upper-level frontal systems are one mechanism, and climatological studies demonstrate that upper troughs of this type form preferentially in northwesterly flow, which can be a favored region for upper

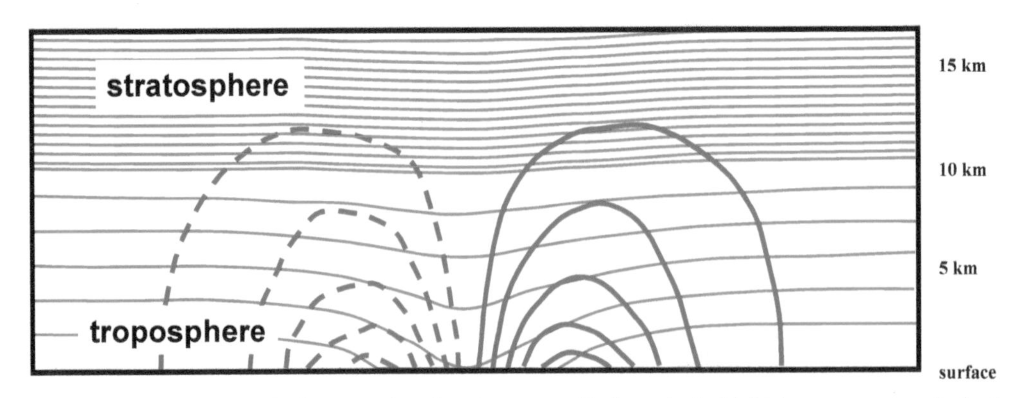


**Figure 5.15.** Isentropes (red contours) and section-normal balanced wind field (green contours, dashed negative) obtained via inversion of a 10-K surface warm potential temperature anomaly (adapted from Thorpe 1986, his Fig. 4).

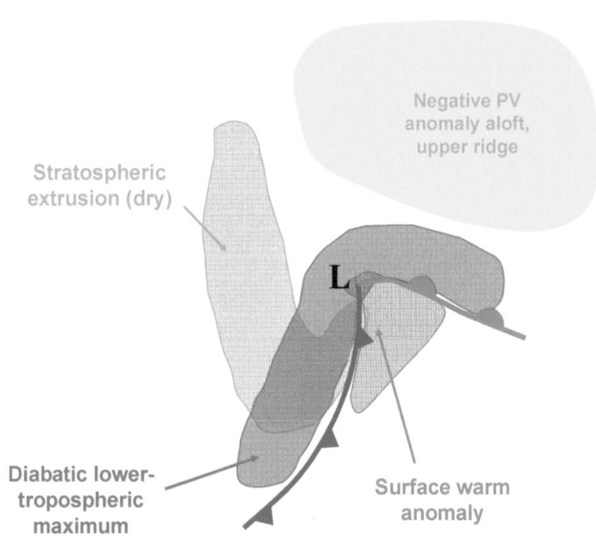


**Fig. 5.16.** Idealized representation of the primary PV anomalies in an extratropical cyclone. Modified from Reed et al. (1992) and Brennan et al. (2008).

frontogenesis. Differential vertical motion associated with upper-level frontal systems can force stratospheric air to descend and distort or even fold the dynamic tropopause in the process.

In chapter 4, we examined the manner in which diabatic processes can lead to the redistribution of PV, generally resulting in the formation of a cyclonic PV maximum in the lower troposphere, and the development of anticyclonic PV aloft, above the level of maximum heating. Typically, there are four important anomalies in a PV view of a midlatitude cyclone: (i) the stratospheric, "dry" upper-level cyclonic PV maximum, (ii) a surface-based warm potential temperature anomaly, (iii) a diabatically produced cyclonic lower-tropospheric PV anomaly, and (iv) a diabatically produced anticyclonic upper PV anomaly. These PV features are represented schematically in Fig. 5.16.

The cyclogenesis process can be viewed as a manifestation of interactions between these PV anomalies and processes that act to cause these anomalies to amplify or superimpose (constructively interfere). There are several processes that can cause PV anomalies to amplify, but here we will focus on the two that are most important for cyclogenesis: *mutual amplification* and *diabatic growth*; these terms were coined by Davis and Emanuel (1991).

In our discussion of Sutcliffe–Petterssen self-development (section 5.3.5), a mutual amplification and positive feedback takes place involving the cyclone in the lower troposphere and the upper-level wave. With the PV framework, this feedback can be quantified using inversion techniques. When inverting a portion of the PV field, it is the strength of a PV *anomaly* that determines the strength of the associated circulation. Thus, self-development is represented by the mutual amplification of the *surface warm anomaly* and the upper-cyclonic PV anomaly, the latter of which is equivalent to a region of *lower potential temperature on the dynamic tropopause* (see section 4.2 and Figs. 4.2 and 4.3).

Consider the cyclonic upper PV maximum, represented as a cold anomaly of potential temperature on the dynamic tropopause shown in Fig. 5.17. This feature is associated with a cyclonic circulation, as depicted by the blue vectors. If the trough amplitude and scale are sufficiently large, and the intervening static stability is not too great, then this cyclonic circulation will extend to the surface. Given the presence of a meridional temperature gradient, as would be found in the vicinity of the jet stream, the branch of southerly flow at the surface associated with the upper PV anomaly creates a surface warm potential temperature anomaly by advection. In the case of a preexisting lower warm anomaly, and provided that there is a favorable horizontal offset between the upper and lower disturbances in the zonal direction, the southerly flow to the east of the upper trough can result in warm advection and *amplification* of the surface warm anomaly.

In a similar fashion, a surface warm anomaly, with its own associated cyclonic circulation denoted by the red vectors, can extend vertically to the tropopause, producing *cold* potential temperature advection in the center of the upper-level cold anomaly (trough), therefore amplifying that cyclonic anomaly (Fig. 5.17). Provided that the tropopause and surface disturbances are embedded in typical meridional potential temperature gradients, and that the vertical extension of their cyclonic circulations is large enough, the disturbances will amplify each other in a process that has been described as the "essence" of baroclinic instability by Hoskins (1990).

If the upper and lower potential temperature (equivalent PV anomalies) are considered in isolation, then note that at the upper level, the *propagation* of the disturbance will be toward the west, against the mean flow. This can be envisioned by considering that the cold potential temperature advection (due to the cyclonic circulation of the anomaly itself) to the west of the upper trough center is

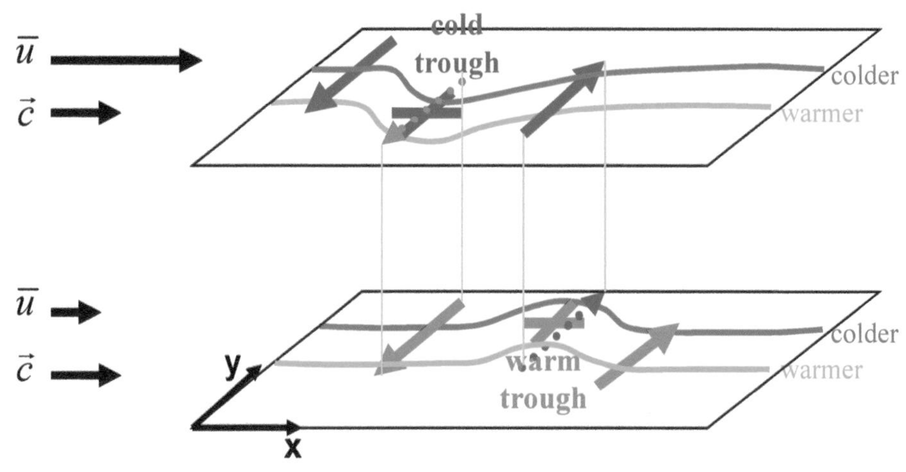


**Figure 5.17.** Idealized schematic illustrating the process of mutual amplification of tropopause-level and surface potential temperature anomalies, equivalent to cyclonic PV anomalies. At upper and lower levels, two isentropes are drawn (green and orange contours). The plus sign at either level indicates the position of the equivalent cyclonic PV anomaly center, and the red and blue arrows represent the associated circulations with the lower and upper PV anomalies, respectively. The vector arrows at left represent the mean flow and wave phase speed at upper and lower levels (adapted from Hoskins 1990).

reinforcing the cyclonic sign of the potential temperature anomaly there, and vice versa to the east. So, at the tropopause, the phase speed of the trough is *slower* than the mean flow, as indicated by the bold vectors on the left side of Fig. 5.17.

Conversely, at the lower boundary, warm advection to the east of the trough draws the cyclonic warm boundary anomaly in that direction, leading to a propagation speed that is *faster* than the mean flow in the lower levels. Because the mean westerly wind at the upper levels in a baroclinic flow is stronger than that near the surface, this counter-propagation effect is critical, because it allows the upper and lower disturbances to move at more similar speeds, which extends the duration of their mutual amplification and constructive cross advections. An additional effect is the change in phase speed in the upper and lower disturbance due to cross advection. Suppose that in Fig. 5.17 the movement of the upper wave was faster than the lower wave. Northerly cold potential temperature advection induced on the upper boundary by the lower wave will act to slow it, while the southerly warm advection on the lower boundary associated with the upper wave will increase its eastward movement. This process, referred to as "phase locking," is also a function of the intervening static stability, along with the wavelength of the upper and lower features.

As discussed in chapter 4, heavy precipitation is typically accompanied by significant diabatic PV redistribution, leading to the development of cyclonic (anticyclonic) PV anomalies below (above) the level of maximum heating. The diabatic lower cyclonic PV feature can play an important role not only in the subsequent evolution of a cyclone, as in the example of the January 2000 East Coast snowstorm discussed in chapter 4, but this anomaly can also contribute strongly to the basic circulation of the cyclone. Piecewise PV inversion studies demonstrate that the diabatic PV feature can contribute up to 50% of the surface cyclonic circulation, and the propagation effects of the diabatic PV anomaly can affect the ability of upper and lower PV anomalies to phase lock (Davis and Emanuel 1991; Davis 1992; Davis et al. 1993; Stoelinga 1996).

We will now present an actual case to demonstrate the utility of the PV perspective and to allow recognition of the specific PV components of a cyclone identified in Fig. 5.16. The often-studied East Coast cyclone that affected the mid-Atlantic region on Presidents' Day in 1979 (Fig. 5.18) exhibited the classic PV signatures, with a hook-shaped cold potential temperature anomaly on the tropopause (Fig. 5.18b), which was associated with a region of higher pressure (lower altitude) on the dynamic tropopause (Fig. 5.18a). Because of the heavy precipitation accompanying the event, a strong diabatically generated cyclonic PV maximum formed in the lower troposphere near the center of the cyclone (Fig. 5.18c). The small, closed

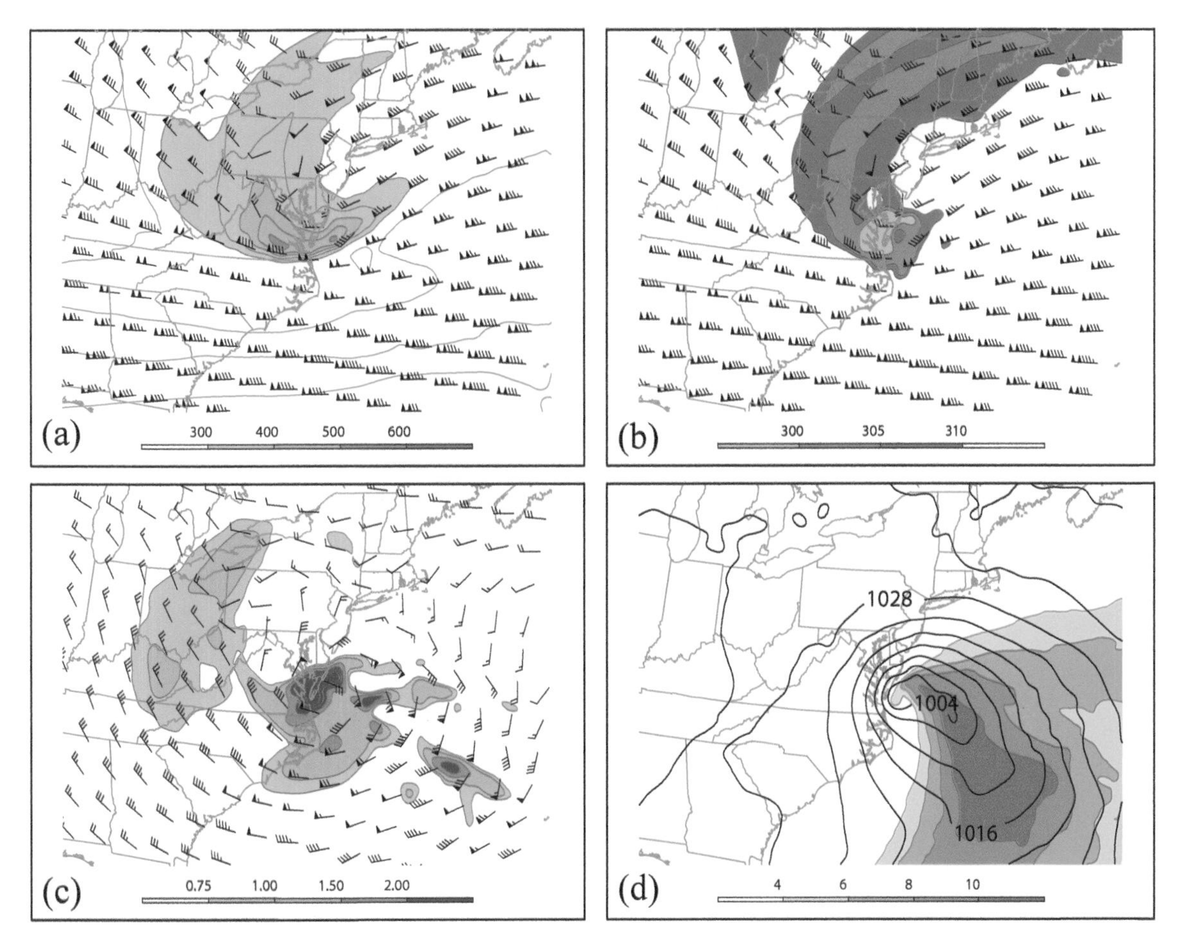


**Figure 5.18.** Summary of PV features for the Presidents' Day cyclone of 1979, from 30-h simulation of Weather Research and Forecasting (WRF) model, valid 1800 UTC 19 Feb 1979: (a) wind and pressure on the dynamic tropopause; (b) wind and potential temperature on the dynamic tropopause; (c) layer-average wind and PV in 800–700-mb layer; and (d) sea level pressure and 950-mb potential temperature anomaly.

sea level pressure minimum is seen to be collocated with a prominent near-surface warm potential temperature anomaly (Fig. 5.18d), which also exhibits a hook shape but with the opposite orientation to that of the tropopause potential temperature minimum. The potential temperature patterns evident in Figs. 5.18b,d are suggestive of phase locking between the upper and lower thermal waves, which are equivalent to PV anomalies. The development of the baroclinic instability theory in chapter 7 will provide a more in-depth treatment of phase locking and cyclogenesis.

**5.3.6.1. Upper-trough genesis.** Given the view of cyclogenesis as the mutual amplification of preexisting upper- and lower-tropospheric disturbances (e.g., Fig. 5.17), it is obvious that a complete treatment of the cyclogenesis problem must account for the origin of these antecedent disturbances. The formation of fronts, which are favored sites for cyclogenesis, will be discussed at length in chapter 6; here we examine the formation of upper-tropospheric disturbances.

Baroclinic instability theory can be held up as an explanation for the development of longwave disturbances in the upper westerlies, but the upper troughs that frequently accompany cyclogenesis are characterized by relatively short wavelengths. The significance of this important problem apparently eluded the atmospheric research community until the pioneering study of Sanders (1988), who asked, "where then do upper-level troughs come from? The question appears hardly to have been addressed in the literature." It is remarkable that this statement could have been made as recently as 1988! Sanders found that upper vorticity maxima at the 500-mb level

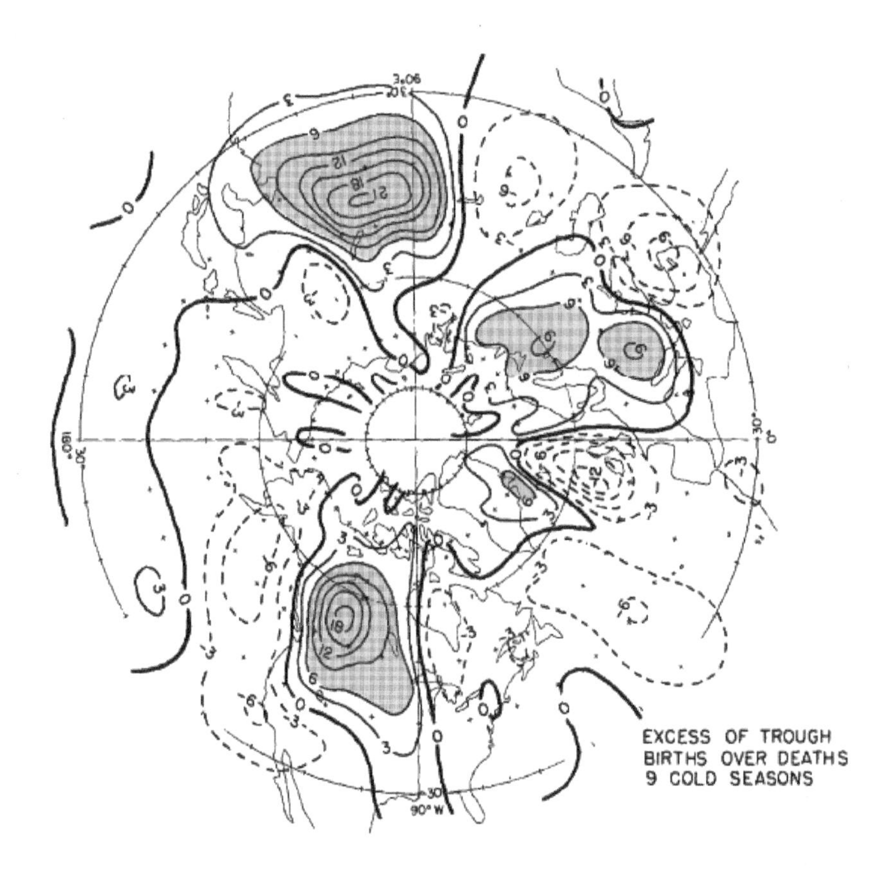


**Figure 5.19.** Contour plot of formation minus termination of mobile upper troughs at the 500-mb level based on data from nine cold seasons. Contours are normalized frequency in 10° latitude–longitude boxes. Blue shading corresponds to normalized frequency greater than six events per 10° box. From Sanders (1988, his Fig. 6c).

formed preferentially downwind of major orographic barriers (Fig. 5.19), and in northwesterly flow. Furthermore, he found that these disturbances attained maximum amplitude at the tropopause level, and that they exhibited little vertical phase tilt in the upper troposphere, suggesting that baroclinic energy conversion may not be their primary energy source.

The author analyzed a number of upper-trough genesis events over North America, and documented examples of upper-level frontal systems that accompanied the generation of elongated upper-tropospheric cyclonic vorticity maxima, with large increases in cyclonic shear vorticity. The ageostrophic, thermally indirect upper frontal circulation, often found in cold advection with northwesterly flow and in jet exit regions, was found to advect the dynamic tropopause downward in some cases. Analysis of the vorticity equation by Lackmann et al. (1997) demonstrated that the tilting term, as well as stretching, was important to the observed increase in vorticity. This result was consistent with studies of the Presidents' Day cyclone of 1979 by Uccellini et al. (1987) and Whitaker et al. (1988), which also highlighted the descent of the tropopause prior to and upstream of the surface cyclogenesis event.

From a larger-scale dynamical perspective, Nielsen-Gammon and LeFevre (1996) found that downstream propagation of Rossby wave energy was critical to a case of upper-trough genesis from 1980, and this may often be the case for the genesis of larger-scale troughs. Schultz and Sanders (2002) examined a large number of cases of trough genesis based on the Sanders (1988) case sample and found a variety of synoptic settings in which trough genesis occurred. While the cases that formed in diffluent flow often exhibited along-flow cold advection, these constituted only about one-quarter of the total events. Thus, there is not a single conceptual picture that can explain the full spectrum of upper-trough genesis events. However, the PV framework, with the view that processes that result in the lowering of the dynamic tropopause can amplify or generate upper disturbances, is a convenient means of monitoring and conceptualizing the upper-trough genesis process.

#### 5.3.6.2. The low-level jet and moisture transport.

Extratropical cyclones effectively transport water vapor poleward, primarily through the warm, moist poleward-directed airstreams that form near the western edge of the cyclone warm sector. During many heavy precipitation events, a deep plume of tropical moisture is observed to extend poleward, often in association with a midlatitude cyclone. Previous studies have referred to this phenomenon using a variety of nomenclature, including the terms *moisture plume*, in the isentropic framework as *warm conveyor belts*, in the eastern North Pacific as *Pineapple Express* events, and more recently as *atmospheric rivers*. Zhu and Newell (1998) demonstrated that a disproportionately large fraction of the total poleward moisture transport takes place within relatively narrow bands, which they termed *atmospheric rivers*.

A positive feedback takes place involving cold-frontal rainbands, moisture transport, and diabatic PV generation. During heavy precipitation events in the Pacific Northwest, latent heat release along a cold front was found to produce a lower-tropospheric cyclonic PV maximum in the vicinity of the cold front. For a case study analyzed from January 1986, piecewise inversion of the PV field demonstrated that the lower cyclonic PV maximum contributed to a low-level jet and maximized moisture transport immediately ahead of the lower PV maximum (Fig. 5.20).

Using QGPV inversion, Lackmann and Gyakum (1999) computed the *piecewise moisture transport* to confirm the dominant role of this PV feature in driving moisture northward in a strong low-level jet. This feedback was identified in a central U.S. event by Lackmann (2002), and a similar configuration of the low-level jet and lower, diabatically produced cyclonic PV anomaly is found in many heavy rainfall events, including the Tennessee flooding event of 1–3 May 2010 (Fig. 5.21).

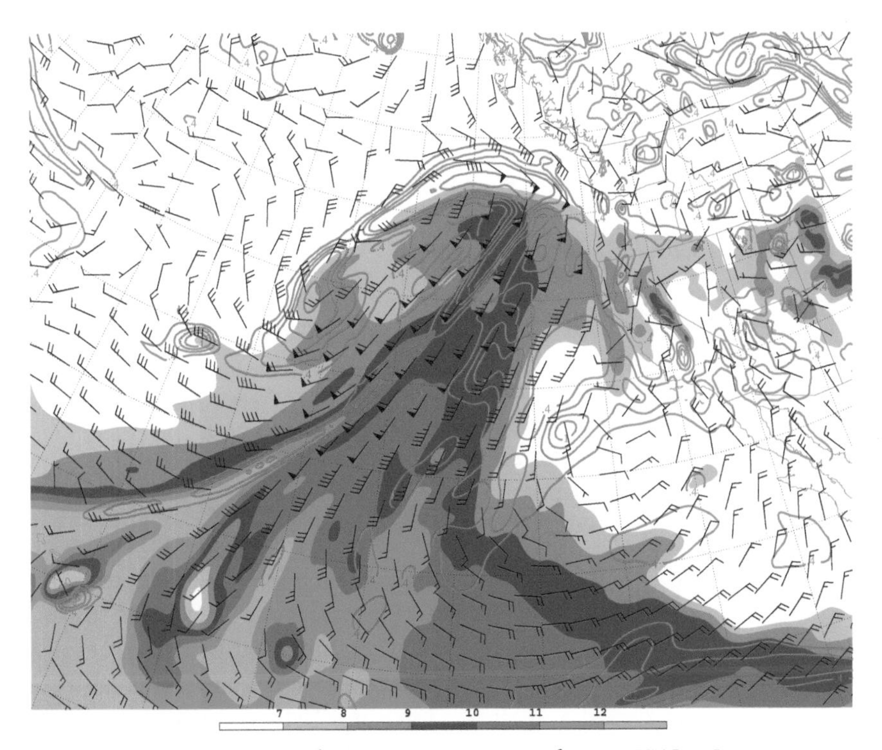


**Figure 5.20.** Lower-tropospheric moisture summary for 0000 UTC 18 Jan 1986, from Climate Forecast System (CFS) reanalysis. Layer-average wind barbs and mixing ratio (shaded above 3 g kg$^{-1}$), and potential vorticity (red contours, interval is 0.2 PVU beginning 0.4 PVU).

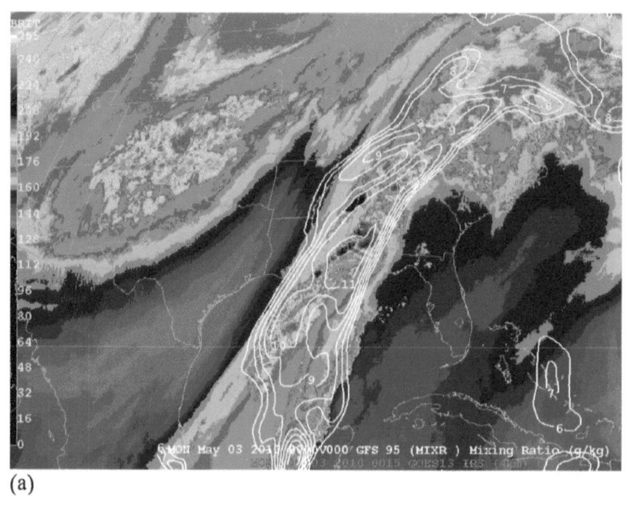

(a)

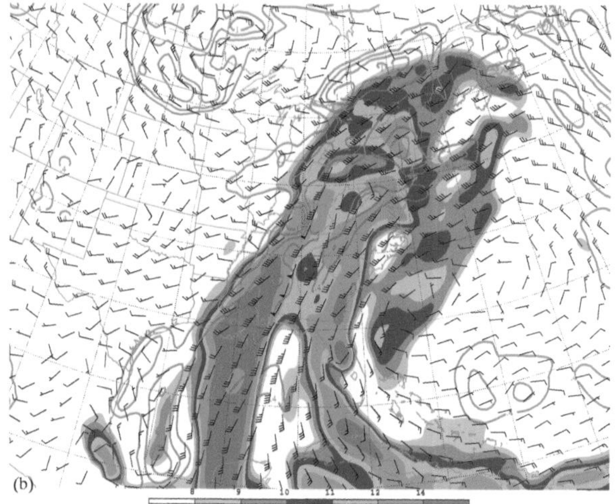

(b)

**Figure 5.21.** Lower-tropospheric water vapor transport during the Tennessee flooding event of 1–3 May 2010: (a) Geostationary Operational Environmental Satellite (*GOES-13*) water vapor imagery for 0015 UTC 3 May 2010, with GFS 700-mb mixing ratio analysis (contour interval is 1 g kg$^{-1}$, contoured above 6 g kg$^{-1}$); (b) 850-mb mixing ratio (shaded greater than 8 g kg$^{-1}$ as in legend), 850-mb wind barbs, 900–700-layer PV (contour interval is 0.2 PVU greater than 0.4 PVU).

Anomalously moist southerly flow became established over the Yucatán Peninsula and Gulf of Mexico in early May 2010. Examination of enhanced water vapor imagery reveals a well-defined tropical moisture plume extending from the deep tropics into the eastern United States (Fig. 5.21a). Note that great care must be exercised in using water vapor imagery to examine moisture transport, because the sampling layer for this imagery is in the *upper* troposphere, whereas the strongest moisture transport takes place in the lower troposphere. However, when superimposed with the 700-mb mixing ratio analysis from the Global Forecast System (GFS) model, it is clear that the plume evident in water vapor imagery in this case corresponds to a deep layer of tropical moisture. Plotting the lower-tropospheric PV along with the 850-mb wind and mixing ratio demonstrates that, as in previous cases, strong moisture transport is taking place immediately to the east of a cyclonic PV maximum, along the axis of a low-level jet (Fig. 5.21b), reminiscent of the Pineapple Express events in the eastern North Pacific.

### 5.3.7. Explosive cyclogenesis

We have now examined many of the dynamical mechanisms that are important to cyclogenesis, including upper trough and jet configuration, energetics, and the interplay between surface cyclones and the upper waves in which they are embedded. How do these mechanisms align during cases of rapid, or "explosive," cyclogenesis? In terms of both prediction and societal impacts, rapidly intensifying storms present important challenges.

Sanders and Gyakum (1980) coined the term *explosive cyclogenesis* and provided a quantitative definition of these systems as those exhibiting a central sea level pressure decrease greater than a specific threshold of 1 Bergeron (b), which is equivalent to a 24-h deepening rate of 24 mb at 60° latitude. More precisely,

$$1\ \text{b} = 24\ \text{mb}\ [\sin\varphi / \sin(60°)]\ 24\ \text{h}^{-1}. \qquad (5.7)$$

Why does *latitude* factor in this definition? Recall that geostrophic wind speed is inversely proportional to latitude [e.g., Eq. (1.13)]; the latitudinal dependence in the definition of explosive cyclogenesis accounts for the enhanced wind speeds that can characterize a storm at lower latitudes for a given pressure-gradient magnitude.

To illustrate, suppose that a change in sea level pressure of 4 mb over a 300-km distance is observed for a weather system located at 60°N. An identical pressure field is located at 30°N. Taking the ratio of the geostrophic wind speed at 30°N to that at 60°N using (1.13) yields a value of 1.73; the geostrophic wind speed is 73% *stronger* at 30°N! Physically, this is reasonable, because the Coriolis force is proportional to the sine of the latitudinal angle, and therefore with $f$ being stronger at 60°N, a weaker wind speed is required for the Coriolis force that

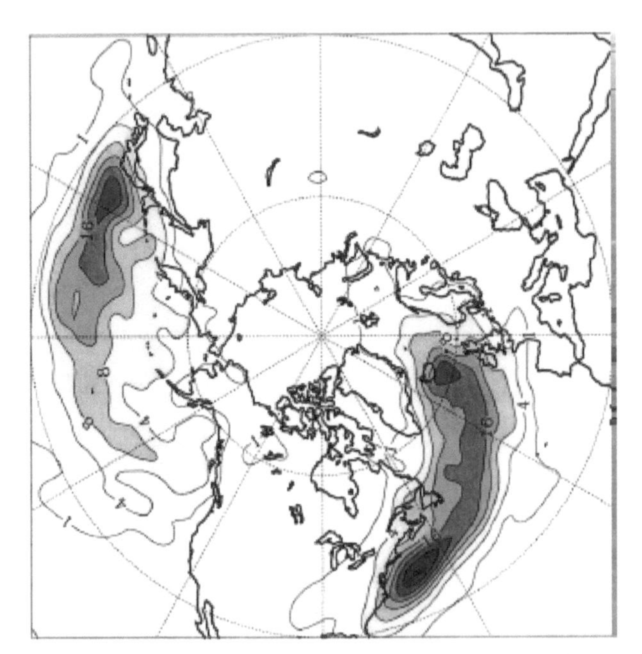

**Figure 5.22.** Northern Hemisphere explosive cyclone density [contour interval is $4\times10^{-5}$ explosive cyclones (°lat$^{-2}$), with $1\times10^{-5}$contour added, and shaded greater than $8\times10^{-5}$] based on NCAR–NCEP reanalysis data from 1979 to 1999 (from Lim and Simmonds 2002, their Fig. 11).

is needed to balance the pressure gradient force geostrophically.

The regions in which explosive cyclogenesis is climatologically favored differ somewhat from those of cyclogenesis in general (Figs. 5.6b, 5.22); the regions of explosive cyclogenesis match the locations of the Northern Hemisphere storm tracks shown in Fig. 5.4. We can explain why these areas are favored for rapid cyclogenesis using concepts from any of the frameworks discussed previously. From the standpoint of *energetics*, these regions are characterized by strong horizontal temperature gradients, meaning that an abundant source of baroclinic potential energy is available for conversion to eddy kinetic energy in developing cyclones. These regions also align with the upper jet stream, so that embedded upper troughs and jet streaks provide a frequent source of mobile upper disturbances. The passage of these upper-level disturbances is associated with QG forcing for ascent and lower-tropospheric vortex stretching.

The coastal regions of North America and Asia are characterized by preexisting vorticity in lower-tropospheric frontal zones often found there, facilitating the effectiveness of vortex stretching in those regions. Another significant factor that makes these regions favorable for explosive cyclogenesis is their maritime location; such environments are characterized by a warm lower boundary due to the western oceanic boundary currents, resulting in reduced static stability, rendering the braking term inefficient and favoring the positive self-development feedback. From the PV perspective, lower static stability increases the vertical extension of cyclonic circulations associated with surface- and tropopause-based disturbances, which facilitates phase locking. The mutual amplification process of cyclogenesis is favored in these regions, and the effectiveness of opposing processes is minimized there. The relative absence of rapid cyclogenesis over land is likely due to reduced diabatic contributions, generally greater static stability, and stronger friction.

Roebber (1984) speculated that distinct dynamical mechanisms were at work during explosive cyclogenesis. Subsequent research quantifying the importance of diabatic and nonlinear process interactions demonstrate that the dynamics are not necessarily different for these rapidly intensifying storms, but that exceptionally favorable conditions can develop when nonlinear feedbacks arise. There are some additional feedback mechanisms at work in marine cyclones that we have not yet discussed.

The importance of *boundary layer processes* (section 1.6), specifically the turbulent transfer of heat and moisture, has been shown to be of primary importance to marine cyclones. These processes can play a critical role in *preconditioning* the atmosphere prior to a cyclone event, and they can also exert an influence during cyclogenesis.

Consider the situation in Fig. 5.23a, in which turbulent warming and moistening of the marine boundary layer is taking place over the warm Gulf Stream current to the east of a developing cyclone. During this preconditioning stage, horizontal gradients of turbulent flux are found in regions of strong sea surface temperature gradient, as well as at the coast. Subsequent cyclone evolution in regions of enhanced temperature gradient results in increased temperature advection. This serves to amplify the upper wave and surface cyclone, which feed back to stronger winds speeds and stronger turbulent fluxes (Figs. 5.23b,c).

On Presidents' Day 1979, a surprise snowstorm blanketed portions of the mid-Atlantic region of the U.S. East Coast with heavy snow. A large volume of research has provided a detailed examination of the processes at work

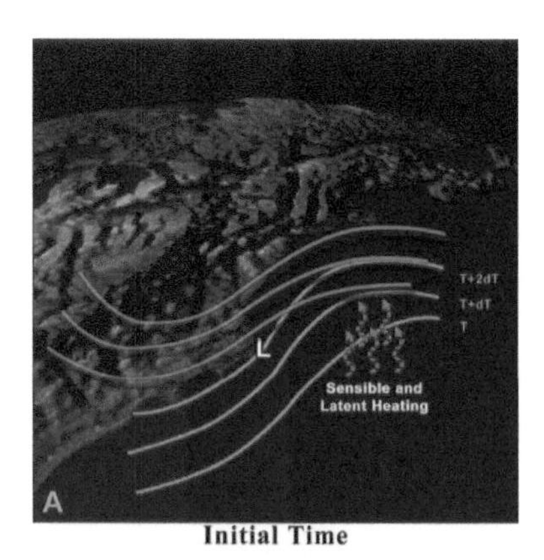


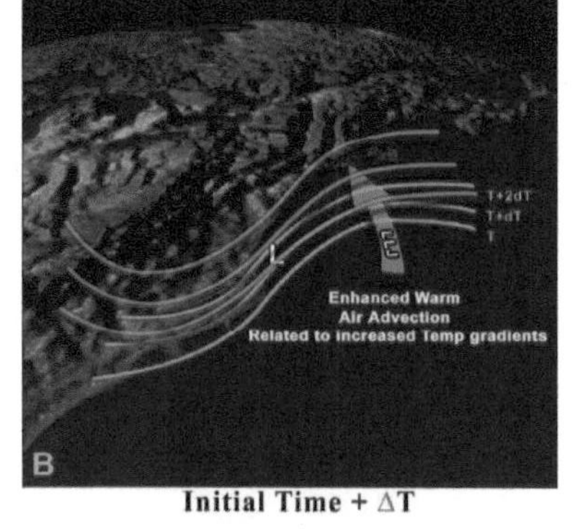


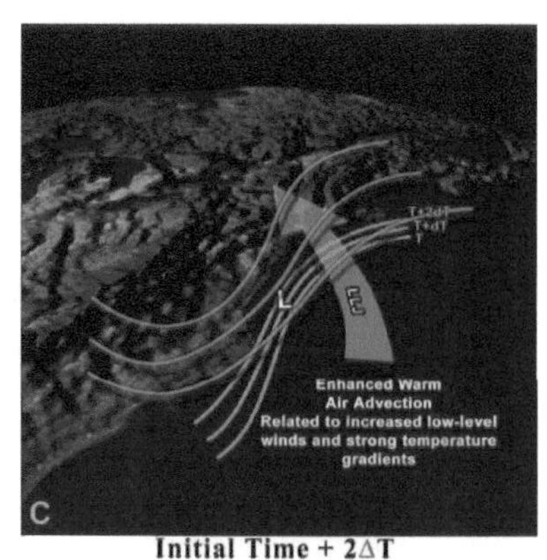


**Figure 5.23.** Idealized depiction of boundary layer feedbacks on the lower-tropospheric thermal gradient and U.S. East Coast cyclogenesis: (a) preconditioning phase; (b) during cyclogenesis, with enhanced thermal advection due to strengthened gradients from boundary layer processes; and (c) positive feedback as strengthened cyclone further increases strength of thermal advection. Blue contours represent upper-tropospheric height, orange contours represent lower-tropospheric isotherms, orange arrow marked "LLJ" indicates the location and strength of the low-level jet (from Kocin and Uccellini 2004, their Fig. 7-8).

during this event, including some advanced numerical modeling studies that isolated nonlinear interactions between the processes during this event. Analyses from this event were presented in Fig. 5.18 to illustrate phase locking in the PV perspective. This case is used below to illustrate the synergy between physical processes during cyclogenesis.

Studies undertaken to elucidate the various mechanisms responsible for this high-impact cyclone event emphasize the role of (i) upper-level processes, as evident from the compact region of lower tropopause in Figs. 5.18a,b and including jet streaks; (ii) a coastal front and cold-air damming, topics that are explored in depth in chapters 6 and 8 but that provide a vorticity-rich and baroclinic lower-tropospheric environment (evident in Fig. 5.18d); and (iii) diabatic and frictional processes, also evident as a potent lower-tropospheric cyclonic PV anomaly in Fig. 5.18c. Numerical simulations were able to capture the development of this storm, allowing revealing sensitivity experiments to be undertaken (Whitaker et al. 1988).

A 24-h control simulation, in which all model physics were included, produced a realistic cyclone minimum central pressure of ~1005 mb at 1200 UTC 19 Feb 1979 (Fig. 5.24d). Model experiments in which both boundary layer turbulent fluxes and latent heat release were

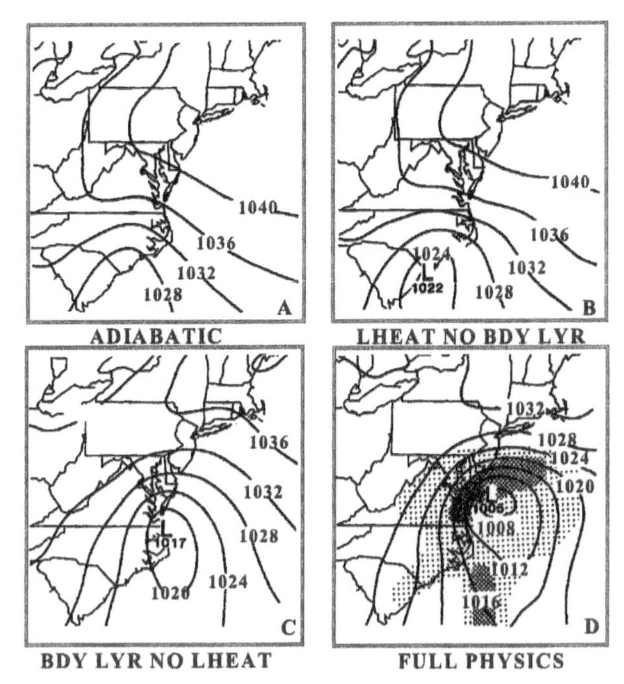


**Figure 5.24.** Sea level pressure at hour 24 from numerical model simulations, valid 1200 UTC 19 Feb 1979: (a) adiabatic simulation; (b) including latent heating, but no boundary layer fluxes; (c) including boundary layer fluxes, but no latent heating; and (d) full-physics simulation (from Kocin and Uccellini 2004, their Fig. 7-10).

omitted (Fig. 5.24a), only boundary layer fluxes were omitted (Fig. 5.24b), and with the exclusion of only latent heating (Fig. 5.24c) reveal the nonlinearity of process interactions taking place during this event. In the adiabatic simulation, which does not exhibit any closed isobars at this time, the cyclone central pressure is ~1027 mb. Comparing this to the adiabatic simulation, when latent heating is included, the central pressure is about 5 mb lower than that in the adiabatic simulation. When only the boundary layer fluxes are included, the central pressure is about 12 mb lower than in the adiabatic run. If the processes were linear, one would expect that with both processes included, the central pressure should be ~17 mb lower, yielding a central pressure of ~1010 mb. However, we see from Fig. 5.24d that a central pressure of ~1005 mb is obtained, a difference of 22 mb. This suggests a synergy between these processes in the model atmosphere, which presumably mimics interactions taking place in the real atmosphere. Difference fields reveal that the pressure was more than 32 mb higher in the adiabatic simulation for this case (not shown).

Based on either QG or PV concepts presented in chapters 2 and 4, we should be able to anticipate the differences between lower- and upper-tropospheric height and wind fields between the adiabatic and full-physics model simulations. Take a minute to think about the impact that latent heating and surface heat and moisture fluxes would have on upper and then lower height fields in the vicinity of this system. How would the wind field differ in association with these changes in the geopotential height pattern? In what location relative to the surface cyclone would upper heights be higher in the full-physics simulation relative to the adiabatic simulation?

As anticipated, the geopotential height in the upper troposphere is considerably higher in the full-physics simulation, especially to the east and north of the cyclone center. The 300-mb wind field exhibits a strong, anticyclonically curved outflow anomaly corresponding to the heating-induced ridge aloft (Figs. 5.25c,d). In the lower troposphere, the 850-mb height field is considerably lower, with a strong, cyclonically curved low-level jet apparent in the wind field associated with this perturbation (Figs. 5.25a,b). These features are exactly what we would expect from a PV-based diagnosis as well: The upper anticyclone corresponds to the development of an anticyclonic diabatic PV anomaly there, and the increased lower cyclonic flow corresponds to the lower diabatic cyclonic PV maximum.

### 5.3.8. Cyclone classifications

From any of the various frameworks introduced to this point, cyclogenesis is seen to involve the growth and interaction of upper- and lower-tropospheric disturbances, with important contributions from diabatic and frictional processes, such as latent heat release and turbulent fluxes of heat and moisture and complex interactions between physical processes. There are several factors relating to the strength and character of the *upper trough/ridge* that are important to the cyclogenesis problem: (i) the amplitude of a trough or upper wave; (ii) the wavelength and horizontal scale; (iii) diffluence or confluence of the upper flow, and jet streak configurations; (iv) trough axis tilt; (v) the strength of the downstream ridge, including diabatic amplification; (vi) the strength of the jet stream/temperature contrast across the jet (baroclinicity); (vii) the extent of mutual interactions between lower and upper disturbances; and (viii) Rossby wave downstream development and ageostrophic geopotential fluxes. The relative importance of these processes and mechanisms, and other aspects of the large-scale environment, results in a rich spectrum of observed cyclone structure and evolution. This has led researchers to categorize cyclones in various ways, seeking to identify meaningful physical differences in cyclone evolution. This subsection summarizes some of the cyclone classification schema proposed over the years.

As discussed subsequently in section 5.4, the early Bergen school work established a strong focus on the polar front as playing a central role in the cyclogenesis process. Perhaps because the upper-air observation network was not established at that time, the role of the upper jet stream was not emphasized in these early studies. Those seeking theoretical explanations for frontal-wave instabilities were met with limited success. However, in the 1930s and 1940s, with the advent of the upper-air network, emphasis shifted to processes aloft. Around this time, Jule Charney (1947) and Eric Eady (1949) advanced convincing theoretical papers, demonstrating that growing instabilities could occur in the presence of very simple baroclinic flows. The disturbances described in these baroclinic instability theories grew spontaneously at all vertical levels, from infinitesimally small perturbations. Perhaps to accommodate theoretical findings as well as explain some observed cyclones that did not seem to have a well-defined upper trough, Petterssen and Smebye (1971) defined *type A* cyclones as those that

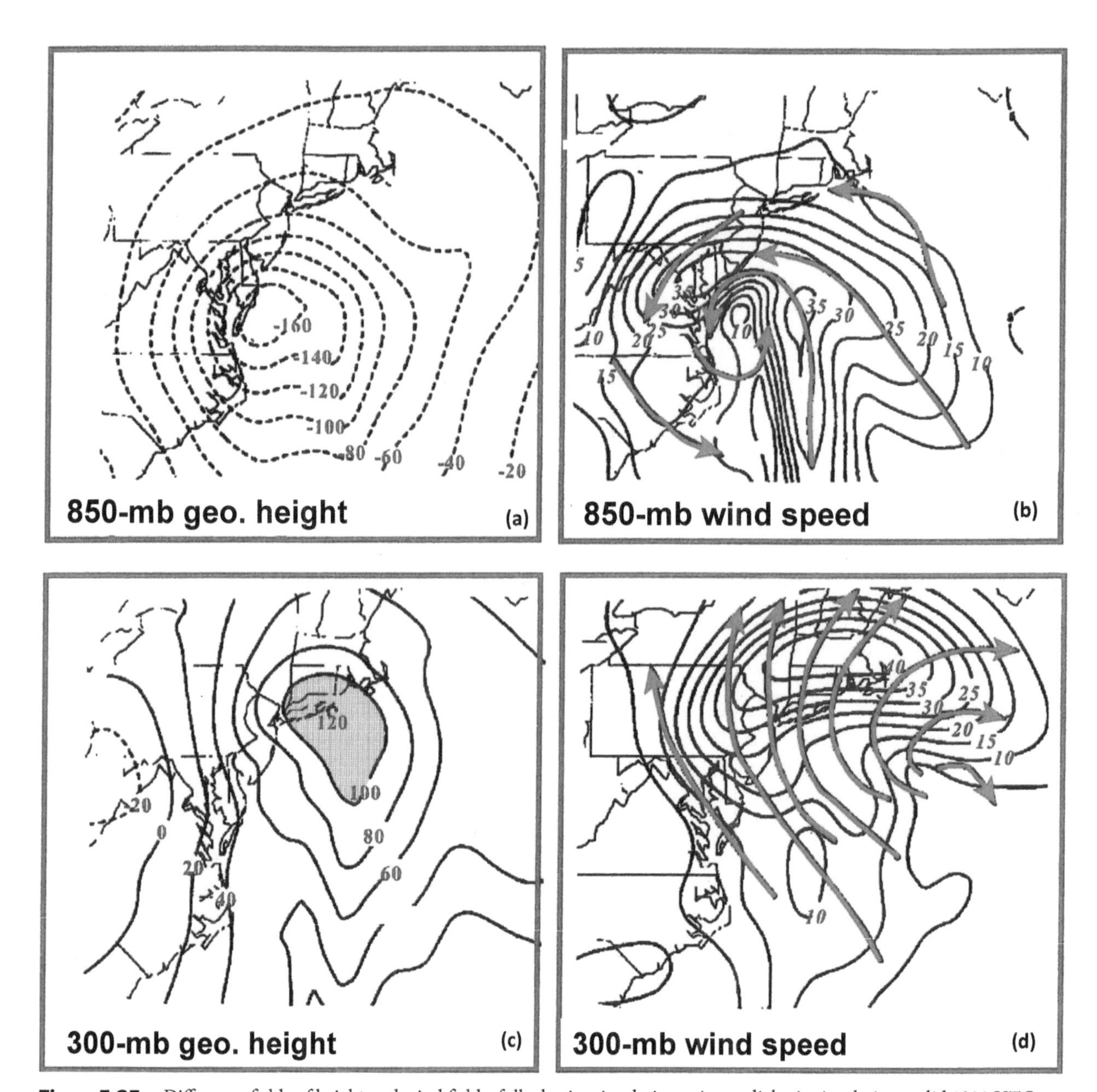


**Figure 5.25.** Difference fields of height and wind fields, full-physics simulation minus adiabatic simulation, valid 1200 UTC 19 Feb 1979: (a) 850-mb geopotential height; (b) 850-mb wind speed and difference streamlines; (c),(d) as in (a),(b), but for 300-mb level. Shaded and hatched region in (c) corresponds to height differences in excess of 100 m (from Kocin and Uccellini 2004).

formed simultaneously at all levels without a clearly defined preexisting upper trough.

Meanwhile, observational studies by Reginald Sutcliffe, Sverre Petterssen, Erik Palmén, and others documented a different type of cyclogenesis process whereby a preexisting, finite-amplitude upper trough overtook a frontal zone, resulting in cyclogenesis, as schematically described in Fig. 5.26. The cyclogenesis sequence with a prominent preexisting upper trough was dubbed a *type B* cyclone by Petterssen and Smebye (1971). Later observational work by Sanders (1986) suggests that most, if not all, true extratropical cyclones are of the Petterssen type B variety.

In a baroclinic environment, there is nearly always some type of upper disturbance, be it in the form of a

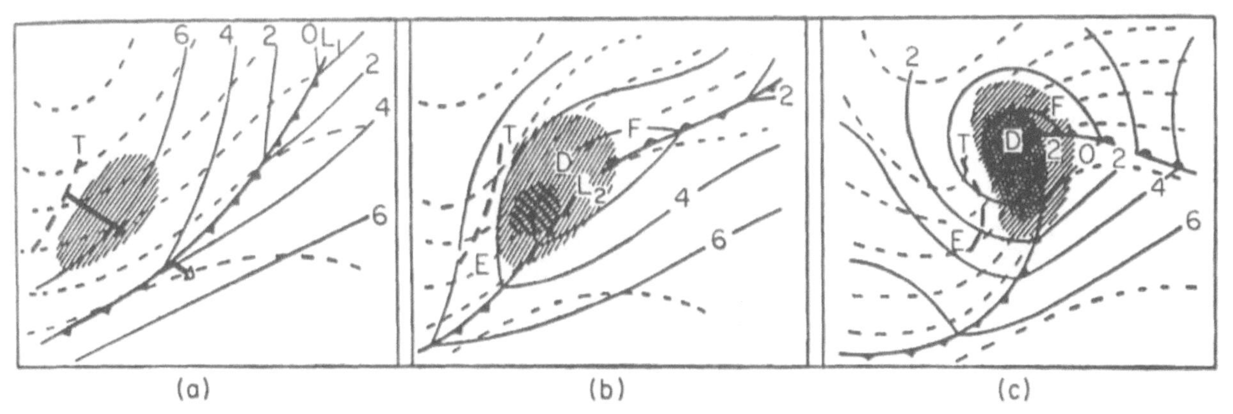


**Figure 5.26.** Idealized depiction of cyclogenesis. Short dashed contours represent 1000–500-mb thickness; solid contours represent sea level pressure. Dash line denoted "T" indicates location of upper trough axis; shaded area ahead of trough indicates region of CVA. (a) Upper trough approaches lower-tropospheric frontal zone; (b) frontal wave forms; and (c) cyclone amplifies. From Petterssen (1956).

jet streak or upper-level trough. Near the surface, such environments are often characterized by preexisting frontal zones. In chapter 7, baroclinic instability theory will be considered in detail, and the theory will be reconciled with observational work within the PV framework. With the advent of the PV view of cyclogenesis, the key elements of type B cyclones—namely, a mobile upper trough and surface thermal wave—are consistent with the Petterssen (1956) description of cyclones (Fig. 5.26) forming as a region of cyclonic vorticity advection overspreads a surface frontal zone. In each view, a preexisting upper disturbance interacts with a preexisting lower disturbance.

Studies of cyclones characterized by strong latent heat release, such as those during the Fronts and Atlantic Storm Track Experiment (FASTEX) over the North Atlantic Ocean, recognized that another class of cyclones existed that did not fit neatly into either the type A or type B classifications of Petterssen and Smebye (1971). Plant et al. (2003) dubbed these systems "type C" and document their key characteristics as (i) an important role for midlevel latent heat release, (ii) weak or nonexistent surface potential temperature anomalies, and (iii) a cancellation of the surface signature of the upper disturbance by the lower-tropospheric diabatic PV anomaly.

There is also a *Miller* type A and type B cyclone classification that is specifically relevant to the eastern United States. The Miller classification describes whether or not the system redevelops on the coast ("jumping" over the coastal plain). An example of a redeveloping storm is provided in Fig. 5.9. Miller type A storms do not exhibit redevelopment, type B storms do. There is more than one redevelopment scenario, as described in Fig. 5.27; cyclones can also redevelop in an along-track sense farther north along the coast.

## 5.4. CYCLONES AND THE RECENT HISTORY OF METEOROLOGY

Undergraduate and graduate curricula in the atmospheric sciences are pressured to keep pace with technological advances and emerging areas of study while maintaining an emphasis on the fundamentals. Faced with a limited number of credit hours to cover a growing body of material, students currently studying the atmospheric sciences often receive limited information about the history of our field. Given the emphasis on extratropical cyclones in the recent evolution of meteorology, this is an appropriate place to introduce some historical material.

### 5.4.1. Prior to 1900

The accumulation of meteorological knowledge has taken place over millennia and in many different regions of the world. Before the development of meteorological instruments, and prior to the development of a meteorological foundation based in mathematics, meteorology could best be described as a *branch of knowledge* rather than as a *science* (Frisinger 1977). *Meteorologica* by Aristotle (350 B.C.) stood for hundreds of years as the primary reference for meteorological knowledge. Despite the significance of this contribution, a period of stagnation followed

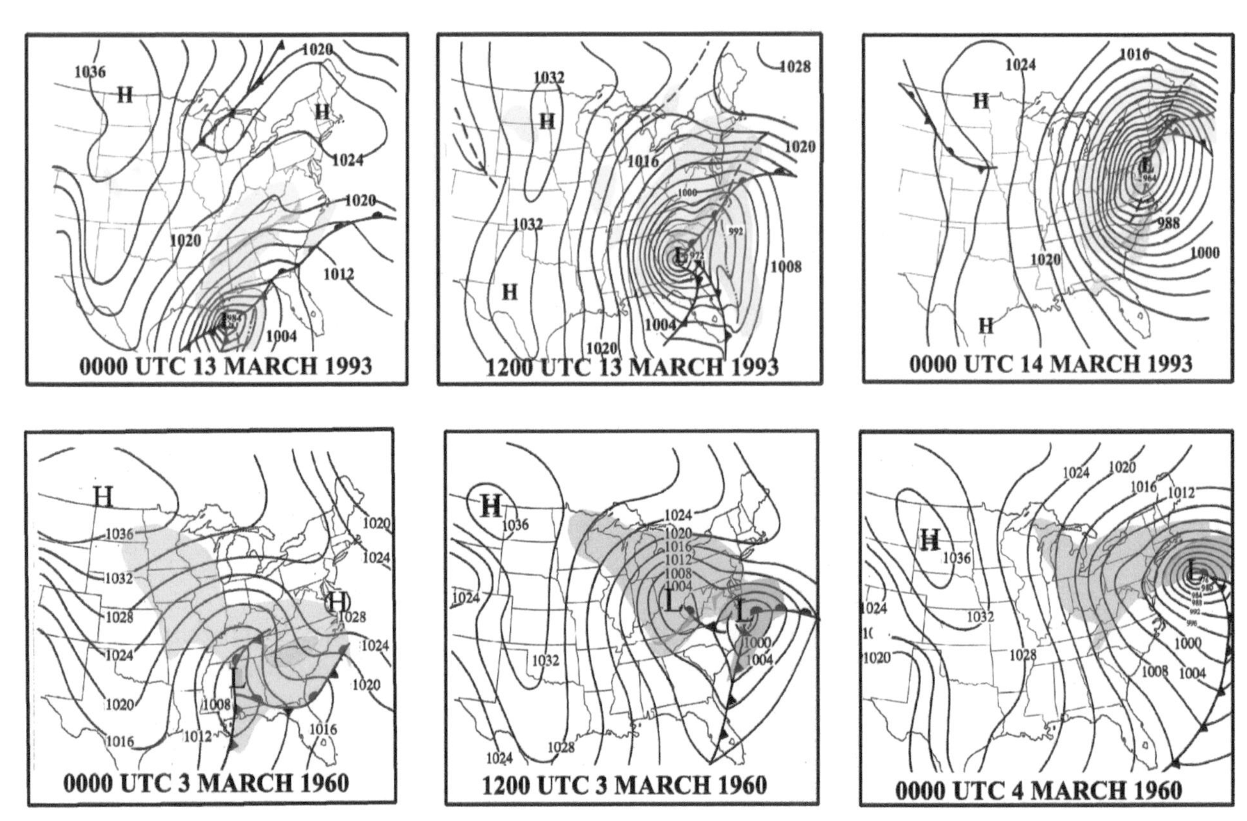


**Figure 5.27.** Sequence of sea level pressure analyses for examples of Miller (top) types A and (bottom) B storms. Top row shows analyses at 12-h intervals beginning at 0000 UTC 13 Mar 1993; bottom row shows analyses at 12-h intervals beginning on 3 Mar 1960 (from Kocin and Uccellini 2004).

in the development of meteorology. During the seventeenth and eighteenth centuries, true scientific progress began in earnest with (i) the development of meteorological instruments, such as the thermometer, barometer, and hygrometer; (ii) the establishment of regular meteorological observations; and (iii) the development and application of mathematics and physics to the atmosphere.

Recognizing that differential solar heating could drive atmospheric circulations, Edmond Halley (1686), George Hadley (1735), and Jean Le Rond d'Alembert (1747) developed theories for the atmospheric general circulation. Issac Newton's *Principia* (1687) contained the elements of calculus and led to the subsequent application of mathematical equations to the atmosphere by Leonhard Euler. By 1755, Euler had applied a set of partial differential equations to the atmosphere, although the set of equations was not closed. The stage was set for a relatively rapid advance in the science of meteorology during the 1800s, and at this time, atmospheric thermodynamics also came to the forefront in completing the set of tools needed for application to atmospheric phenomena (see, e.g., Kutzbach 1979, her Fig. 2 for a summary of progress during this period).

Despite limited technology by today's standards, a great deal of meteorological knowledge was deduced from observations, laboratory experiments, mathematics, and physics. For example, James Espy, following work by John Dalton, computed the dry and moist adiabatic lapse rates and hypothesized that the latent heat of condensation played a crucial role in the development of cyclonic storms (Espy 1841). William Ferrel, in the 1850s, derived the momentum equations for a rotating fluid and correctly deduced the force balance at work in hurricanes. Vladimir Köppen, who is famous for his climate classification work, also advanced a conceptual model for cyclonic storm systems that featured spiraling near-surface flow, with a trough aloft shifted to the west of the surface circulation center. He noted the asymmetric temperature structure accompanying cyclones and used cloud motion observations to ascertain the character of the upper-level flow.

Frank Bigelow, also using wind measurements derived from cloud motion observations, recognized that potential energy was converted to kinetic energy as vertical air motion in storms lowered the atmospheric center of gravity. He further recognized the need for divergence aloft in the development of storms. Additional important early contributions concerning fronts were published by Max Margules, Sir William Napier Shaw, Elias Loomis, Heinrich von Ficker, Felix Exner, and others.

Here we focus on the development of meteorology since around the year 1900. As a graduate student at the University at Albany, I was fortunate to attend an international symposium held in Bergen, Norway, in the summer of 1994 commemorating the 75th anniversary of a paper presenting the Norwegian *frontal-cyclone model.* I was inspired by this symposium; highlights from this event are presented in *The Life Cycles of Extratropical Cyclones,* Shapiro and Grønås (1999). For those seeking additional information, recommended readings include *The History of Meteorology to 1800* by Frisinger (1977), *Appropriating the Weather* by Friedman (1989), *The Thermal Theory of Cyclones* by Kutzbach (1979), and *Weather by the Numbers: A Genesis of Modern Meteorology* by Harper (2008).

### 5.4.2. The situation at the time of the birth of the Bergen School

Imagine a time when there was no Internet, no electronic computers, and not even handheld electronic calculators. There was no radar network, no satellites, and no upper-air sounding network. Communications systems centered on the telegraph and written letters; radio networks did not yet exist. The first powered aircraft had not yet flown, and manned flight was limited to dirigibles that were limited in range and altitude. University degrees were not offered in meteorology or the atmospheric sciences. Weather forecasting was characterized by the application of "rules of thumb," with little consideration of scientific reasoning. Theoretical explanations for atmospheric dynamical phenomena and the availability of relevant conceptual models were lacking. This was the situation in the year 1900, which marks the onset of a series of events that would accelerate the maturation of the fledgling field of meteorology. Table 5.1 provides a timeline of meteorological and non-meteorological advances and events during the period under consideration.

**Table 5.1.** Timeline of events and advances both within and outside the field of meteorology.

| | | |
|---|---|---|
| | 1850 | 1850 British Meteorological Society (later Royal Meteorological Society) founded |
| U.S. Civil War (1861-65); First Trans-Atlantic Telegraph | 1860 | |
| | 1870 | 1870 U.S. Weather Bureau established; 1872 publication of *Monthly Weather Review* by U.S. Weather Bureau |
| 1887 Hertz: first radio transmission | 1880 | |
| | 1890 | 1890 U.S. Weather Bureau becomes civilian |
| First zeppelin airship; Dec. 1903 Wright Bros. 1st flight; First AM radio broadcast | 1900 | 1904: V. Bjerknes outlines numerical weather prediction |
| World War I (1914-1918) | 1910 | 1919 Founding of American Meteorological Society |
| 1927: Lindbergh- First solo trans-Atlantic flight | 1920 | 1922 Richardson publishes results of first NWP effort; 1928: CFL condition for numerical stability published |
| | 1930 | First U.S. radiosonde launch |
| World War II (1939-1945); 1st electronic computers | 1940 | |

As discussed in section 5.4.1, by 1900 there was a solid foundation of knowledge regarding atmospheric thermodynamics, the role of water vapor phase changes, the equations of motion, and wind-pressure relationships. Theories for the energetics of storms as well as the general circulation of the atmosphere had been developed, as had knowledge of the structure and dynamics of midlatitude weather systems. However, because university training and education in meteorology were not available, this knowledge was limited in scope, and the weather forecasters of the day were more likely to use experience and empirical rules rather than the results of scientific publications in their work.

### 5.4.3. The Bergen contributions

Norwegian physicist and mathematician Carl Bjerknes (1825–1903) studied *hydrodynamics* (the study of liquids in motion) until his death in 1903 and introduced his son Vilhelm Bjerknes (1862–1951) to this field of study. Vilhelm Bjerknes, who at one point worked with Heinrich Hertz, sought a mechanical description for electromagnetic phenomena in collaboration with his father, whose work in physics was based on finding hydrodynamic analogs that could explain electromagnetic phenomena. In the 1890s, V. Bjerknes recognized the commonality in these efforts and envisioned a unification of the field of physics via a mechanical explanation for electrical phenomena. V. Bjerknes's work in trying to generalize the hydrodynamic experiments his father had developed led him to realize that by accounting for baroclinicity in fluids, one could understand the development of circulations arising from pressure-density solenoids (Bjerknes 1900). The resulting circulation theorem named for V. Bjerknes is still widely used and can be applied to the development of land–sea breeze circulations, for example. The success of the circulation theorem led Vilhelm to recognize that low-hanging fruit was abundant when one applied rigorous physics and mathematics to atmospheric and oceanic problems. With encouragement from Scandinavian oceanographers, V. Bjerknes continued his foray into the geosciences.

In 1904, V. Bjerknes wrote a remarkable paper presenting a vision for numerical weather prediction (NWP) that included a closed set of governing equations for the atmosphere, along with an outline for a numerical forecast process. The equations had been known previously, but evidently nobody had yet assembled them into a closed system for this type of application. In this paper, V. Bjerknes states the following:

> If it is true, as every scientist believes, that subsequent atmospheric states develop from the preceding ones according to physical law, then it is apparent that the necessary and sufficient conditions for the rational solution of forecasting problems are the following:
>
> 1. A sufficiently accurate knowledge of the state of the atmosphere at the initial time.
> 2. sufficiently accurate knowledge of the laws according to which one state of the atmosphere develops from another.

These comments are followed by a description of what resources would be needed to secure the needed initial conditions and a description of the governing equations, including the equations of motion, the continuity equation, the equation of state, and the two laws of thermodynamics. This system of equations, with the exception of the second law of thermodynamics, still forms the basis of today's powerful computer models for weather and climate prediction.

V. Bjerknes was proactive in promoting the ideas from his 1904 paper. He traveled widely, met with political leaders as well as scientists, and gave numerous lectures to share his vision and to seek support for the implementation of quantitative weather prediction. He was optimistic and believed that a sufficient number of people could manually integrate the governing equations forward in time with the requisite speed to produce viable weather forecasts. There were several reasons why his predictions materialized much later than originally anticipated, such as his lack of awareness of a numerical stability requirement that necessitates many more calculations than he had originally envisioned [the Courant–Friedrichs–Lewy (CFL) condition, discussed in chapter 10]. Nevertheless, his optimism and outgoing approach inspired many who came into contact with him, and this ultimately proved to be critical to the promotion of the atmospheric sciences.

In 1905 V. Bjerknes presented the NWP idea in the United States and was granted support from the Carnegie Institute to allow him to hire a sequence of assistants to pursue atmospheric research. Over a period of many years, a large fraction of these assistants went on to successful careers of their own; they include Halvor Solberg, Tor Bergeron, Carl-Gustaf Rossby, Sverre Petterssen, Erik Palmén, Carl Godske, Johan Sandström, and Harald Sverdrup, to name some of them. Bjerknes and Sandström (1910) published a book entitled *Dynamic Meteorology and Hydrography, Part I. Statics*. This book included units such as the millibar, introduced the concept of geopotential height, and included the hydrostatic equation and the concept of isobaric charts. They used kite data and the concept of thickness to build their upper-air analyses vertically upward from the surface for retrospective case studies.

As a professor in Oslo, V. Bjerknes (1911) then published *Dynamic Meteorology and Hydrography, Part II. Kinematics*. This text presented novel means of representing the wind field using streamlines and isogons (lines of constant wind direction), along with the technique of computing vertical motion from the continuity equation. Analyses featuring confluent flow along what appear to be frontal boundaries were presented, although these features were not yet known as fronts. V. Bjerknes and his collaborators evidently recognized that ascending air motion, clouds, and precipitation were tied to these confluent zones.

In 1913 V. Bjerknes moved to Leipzig, Germany, to serve as director of a new Geophysical Institute. By this time, he had more realistic visions for NWP; however, he pursued the work nevertheless and was prepared to write a third text presenting the associated weather forecasting techniques. However, World War I (1914–18) brought this work to a halt, as several of the students in his research

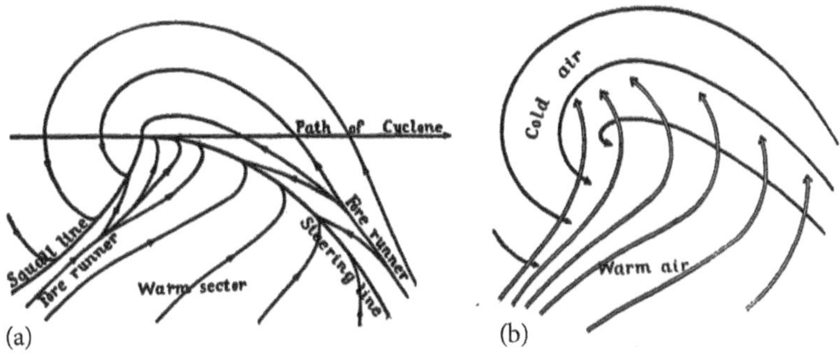


**Figure 5.28.** Schematic diagrams of extratropical cyclones showing (a) streamlines in a moving cyclone and (b) lower-tropospheric airflow. From J. Bjerknes (1919).

group fought and died in the war. At this point V. Bjerknes shifted the focus of his research to the confluence lines that had been identified in their earlier research. German doctoral student Herbert Petzold, who was directing this work, was killed in the war; therefore, responsibility for this project was turned over to Vilhelm's son, Jacob Bjerknes (1897–1975). Because of the war, V. Bjerknes moved his group to Bergen, Norway, and established what is now known as the Bergen School of Meteorology; this group greatly advanced our understanding of fronts and cyclones, and applied these findings to practical weather forecasting.

The war, and the emerging field of aviation, allowed the Bergen group to focus on weather forecasting. Norway was neutral in the war; however, when the United States became involved, food shipments to Norway were suspended, which significantly impacted the Norwegian economy and applied pressure to the agricultural sector. Recognizing the potential role of weather forecasting in support of agricultural and fishing activities, V. Bjerknes approached Norway's minister of defense, as well as the prime minister, to obtain support and approval to implement expanded weather services for Norway. The burgeoning aviation industry demanded more precise and accurate forecast information than had been attempted in the past, and the Bergen group extended their forecasting repertoire to include fog and other aviation hazards, and aggressively pursued short-term forecasting applications derived from their analyses. During this period, rapid advances led V. Bjerknes to conclude that the interplay between practical forecasting and theoretical and observational research was beneficial to the entire meteorological enterprise.

The confluence lines noted in the earlier analyses of Bjerknes' group became the key focus of the activities of the Bergen School. Detailed, yet idealized schematics of the three-dimensional structure of cyclones were developed and published in J. Bjerknes (1919), as shown in Fig. 5.28.

Owing in part to terminology from World War I, the term *polar front* was used to describe the "battle zone" of temperature contrast between polar and tropical air masses. The evolution from a frontal wave to an occluded cyclone was envisioned to take place along the so-called polar front, which was thought to girdle the entire globe, although limited observations precluded its complete representation in early analyses (Fig. 5.29; Bjerknes 1921). The perceived importance of the polar front is underscored in this passage from Bjerknes (1921):

> Along the whole of this front-line, we have the conditions, especially the contrasts, from which atmospheric events originate: the strongest winds, the most violent shifts of wind, and the greatest contrasts in temperature and humidity. Along the whole of the line formation of fog, clouds, and precipitation is going on, fog prevailing where the line is stationary, clouds and precipitation where it is moving.

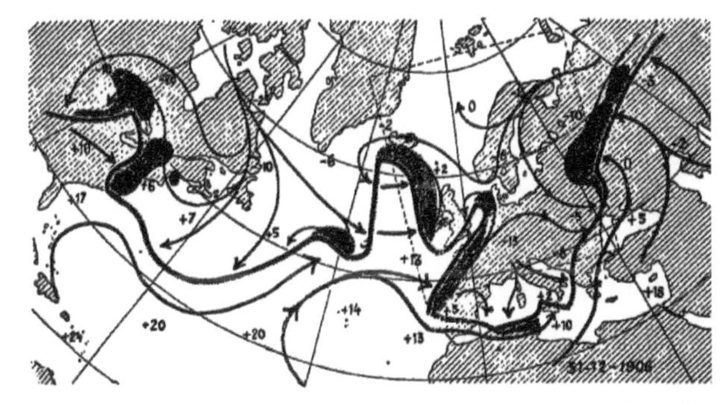

**Figure 5.29.** Surface analysis, including air streams, frontal boundaries, and areas of precipitation, valid 5 Jan 1907 (from Bjerknes 1921).

In addition,

> Cyclone and anticyclone and all meteorological events of the Temperate Zone are in the most intimate way related to the polar front and its motion.

### 5.4.4. Reception of the Bergen ideas

The Bergen group was proactive in advancing its viewpoint to the broader community, and this approach paid dividends in the form of external support and recognition. However, a combination of scientific concerns, resistance to change, and an organizational structure and culture that did not facilitate the transfer of research to operations led to a delay in the incorporation of these ideas in the European and American weather services.

A conference was held in Bergen in 1920, attended by many leading meteorologists of the day; this meeting was a significant step in securing a firm footing for the Bergen ideas in the meteorological community. At the 1920 conference, "the Bergen School asserted dogmatically the superiority of the polar front meteorology over contemporaneous analysis techniques" (Friedman 1989). That many of the foundational ideas behind the Bergen cyclone model were well known in scientific circles *prior* to the development of the Bergen school was noted by some conference attendees, such as Sir Napier Shaw of Britain, and was also acknowledged by V. Bjerknes. Shaw, and Felix Exner, who was the leading Austrian meteorologist of the day, were of the view that upper-level processes were critical to cyclogenesis (Davies 1997), and this tempered their enthusiasm for the Bergen methods. Furthermore, a theoretical explanation for cyclones had not been provided by the Bergen group. Nevertheless, the unique contributions of the Bergen group were to organize the previous ideas into a simple, practical, and cohesive framework that could be used as a basis for weather forecasting, and also the broad and active dissemination of their results.

In addition to Shaw and Exner, Köppen and others had also recognized that upper-level troughs preceded and were located to the west of surface cyclones. The importance of upper-level processes took a back seat in the Bergen view, with a strong emphasis on the polar front. Some of the claims of the Bergen group turned out to be overstated, and some of the skepticism toward the Bergen ideas in other regions was justified. Many of the Bergen publications feature idealized schematics, often excluding actual data, and this may have contributed to skepticism in some scientific circles (Friedman 1989).

The Bergen group brought their ideas to the United States in a series of several articles on the topic in the *Monthly Weather Review* (V. Bjerknes 1919a,b, 1920a,b, 1921, and J. Bjerknes 1919). Accompanying these papers was an "interpretation" provided by A. J. Henry (1922), who was head of the U.S. Weather Bureau at the time. Henry's comment may explain the sluggish reception of the Norwegian ideas by the Bureau. This article, also in the *Monthly Weather Review*, suggested that the latitudinal differences between Norway and the United States limited the relevance of the Bergen concepts to American forecasters. Henry also suggested that a prohibitively large number of new observational and telegraphic sites would be needed to put the Bergen ideas into place (Henry 1922). However, U.S. Weather Bureau meteorologist C. L. Meisinger (1920) recognized the importance of the Bergen papers and published an article highlighting the conformity of a strong U.S. cyclone to the Bergen model. Unfortunately, Meisinger, a forceful advocate for improved upper-air measurements, died in a meteorological ballooning accident, and his research into applying the Bergen methods in the United States was not continued immediately thereafter.

In 1926 an influential, engaging, and energetic member of the Bergen School, C.-G. Rossby, visited the United States; he gained support from key allies such as F. Reichelderfer of the U.S. Weather Bureau. Rossby found that few in the U.S. Weather Bureau had heard of, let alone practiced, the Bergen analysis methods in 1926, despite the earlier publication of several articles on the topic in the *Monthly Weather Review* (1919a,b,c, 1920a,b, 1921, and 1922). When Rossby made an excellent, but unauthorized, forecast at the request of pilot Charles Lindbergh, the Weather Bureau administration took offense and made it clear that Rossby's services were no longer needed. However, partly because of support from the Guggenheim Foundation, Rossby was invited to California to set up a forecasting service to support aviation there. Shortly thereafter, Rossby was invited to lead the development of the first university training program in meteorology in the United States at the Massachusetts Institute of Technology.

### 5.4.5. Numerical weather prediction

Follow-up efforts to the original outline of Bjerknes (1904) for numerical weather prediction were evidently not published by Bjerknes and his collaborators, perhaps because of difficulties involved in this effort, as well as disruptions from war, and a more easily applicable and verifiably useful forecasting technique in their polar-front cyclone model.

However, others did pursue these ideas, with a remarkable effort toward numerical weather forecasting published by L. F. Richardson in 1922. That he undertook this research while working as an ambulance driver during World War I makes his accomplishment even more impressive. Unfortunately, because of noise in his initial data and other complicating factors, the tendency calculations he made starting from observations analyzed by V. Bjerknes from 20 May 1910 were not accurate. He obtained a 6-h pressure tendency of 145 mb! However, modern reanalyses of Richardson's forecast by Lynch (1992, 2006) have shown that filtering the same initial data results in a much more reasonable result. Richardson made important contributions to the field of meteorology, but his pacifist views led him to turn to other fields after learning that his meteorological research may have had unintended military applications.

The idea of quantitative numerical weather prediction was thus set to rest until the advent of electronic computers. Around 1950, when such computers became available for research, several leaders in atmospheric science—such as Charney, John von Neumann, and Norman Phillips—recognized the potential afforded by this invention for numerical weather forecasts. Chapter 10 presents additional historical discussion of NWP.

### 5.4.6. Developments after 1930

With the advent of an upper-air network and, for the first time, detailed upper-air analyses over large areas, the jet stream was discovered, and along with it came an appreciation for the role of upper-level waves and the 3D structure of cyclones during the late 1930s and 1940s. The role of upper-level waves in cyclogenesis (Bjerknes and Holmboe 1944), cyclone self-development (Sutcliffe and Forsdyke 1950), and Rossby waves (Rossby 1945) are some of the landmark discoveries that followed the development of the upper-air network.

In seeking a theoretical basis for cyclogenesis, the earlier emphasis on the polar front by the Bergen school had led to a search for frontal instabilities. Investigations by Margules and Solberg (e.g., Solberg 1936) sought such instabilities but failed to yield convincing results. As the upper-air network became established, discovery of the jet stream and a growing appreciation for the role of upper-level disturbances in cyclogenesis pointed to a different type of instability. In the late 1940s, Eady and Charney demonstrated that growing disturbances could develop spontaneously in a basic baroclinic westerly current that did not include a sharp frontal discontinuity. The theoretically obtained wavelength of maximum instability matched fairly well with that of observed long waves in the atmosphere, and the theory of baroclinic instability took center stage.

These theories led to significant advances in recognizing the conditions necessary for baroclinic instability and in clarifying the distinction between baroclinic instability and baroclinic energy growth. Observational studies of frontal zones and jet streams by Palmén, Newton, Petterssen, Sutcliffe, Richard Reed, and Frederick Sanders advanced understanding of these phenomena, their vertical structure, and the processes leading to their development.

A renewed appreciation for the role of diabatic processes in cyclogenesis came about in the 1980s and 1990s, partly based on numerical modeling studies in association with observational field experiments, such as the Experiment on Rapidly Intensifying Cyclones over the Atlantic (ERICA). The emergence of the PV framework for the interpretation of cyclone development builds on ideas from the earlier baroclinic instability work as well as observational results. Downstream baroclinic development and recognition of the planetary-scale propagation of energy in Rossby wave packets in the midlatitude storm tracks have also come to the forefront in recent decades; these ideas originated in the early works of Rossby and collaborators (see section 2.7).

Development of remote sensing technology, in particular meteorological radar and satellites, the advance of electronic computers, the availability of high-speed electronic communications, and the creation of powerful numerical weather prediction models, have altered the field of atmospheric science over the last 50 years. However, technology alone does not advance scientific understanding. Theory and observation must be synthesized to allow for the realization of the potential afforded by improved technology.

### 5.4.7. Concluding thoughts

How can today's students benefit from the study of meteorological history? Relevant lessons can be gleaned from analysis of past scientific advances in the development of our field. To emphasize one example, consider the interplay between weather forecasting, collection and analysis of meteorological observations, and theoretical development. In the case of the Bergen school, World War I, with the heightened meteorological needs of aviation, shipping, and agriculture, motivated V. Bjerknes and his research group to shift their focus to practical forecasting. The rapid progress that ensued led Bjerknes to conclude that "close contact between theoretical dynamic meteorology and daily practical forecasting was essential for the progress of both" (Friedman 1989).

Today, the rift between research and operations in meteorology is so large that it has been referred to by some as "the valley of death" (NRC 2000). Once again, we recognize the need to integrate theory and application. The modernization of the U.S. National Weather Service included steps to bridge this gap. Efforts are currently underway to optimize the transfer of research to operations, a step known as "R2O" in some circles. However, lessons from the past would suggest a more balanced, two-way interaction; perhaps the acronym, and process, might better be described as "O2R2O."

## REVIEW AND STUDY QUESTIONS

1. Explain why the cyclogenesis process is favored in the vicinity of surface frontal zones.
2. Can the QG vorticity equation be used to explain the favored occurrence of cyclogenesis on frontal zones? Discuss.
3. Suppose that a future climate regime exhibits a drastically weakened equator-to-pole temperature gradient in the lower troposphere, with little change in the gradient aloft. Speculate how this might influence the strength of the jet stream and the climatology of cyclogenesis.
4. Climatologically, there is a minimum frequency of *high pressure systems* in the Great Lakes region in winter, but they are frequently found there in summer. Explain this observation within the context of principles discussed in this chapter.
5. Discuss the mechanism of Sutcliffe–Petterssen self-development, including an account of the processes that oppose it.
6. Explain the differences in the geographical distribution of explosive cyclogenesis as compared to the overall cyclogenesis distribution.
7. Read the early *Monthly Weather Review* publications of V. and J. Bjerknes, L. Meisinger, and A. J. Henry from the 1910s and 1920s. These are available online through the American Meteorological Society (AMS) journals Web site. Specifically, "On the Structure of Moving Cyclones" (Bjerknes 1919), "The Meteorology of the Temperate Zone and the General Atmospheric Circulation" (Bjerknes 1921), and "The Great Cyclone of Mid-February 1919" (Meisinger 1920) are illuminating. For a more in-depth historical review, see Shapiro and Grønås (1999), Friedman (1989), Kutzbach (1979), and Harper (2008).

## PROBLEMS

1. Using Eq. (5.3), determine how long it would take for the vorticity at the center of a midlatitude cyclone to double, given a typical synoptic-scale value of convergence.
2. For the previous problem, evaluate the assumptions that went into the derivation of (5.3), and discuss how these processes might change the answer obtained in that problem.
3. Consider the 300-mb height and isotach analysis (kt) shown below.
   i. For the trough feature indicated with the letter A, is this trough "lifting" or "digging"? _________
   ii. We wish to determine whether this trough will eventually affect the southeastern United States. Based on the information here, indicate the expected location of the trough center or axis for a time 48 h later than that.
   iii. Would you expect the strength of the trough (defined in terms of upper-level height) to be weaker, stronger, or about the same as at this time?

## REFERENCES


Aristotle, 350 BC: *Meteorologica*. Trans. E.W. Webster, available from http://classics.mit.edu/Aristotle/meteorology.html.

Bergeron, T., 1959: Methods in scientific weather analysis and forecasting: An outline in the history of ideas and hints at a program. *The Atmosphere and the Sea in Motion*, B. Bolin, Ed., Rossby Memorial Vol., The Rockefeller Institute Press, 440–474.

Bjerknes, J., 1919: On the structure of moving cyclones. *Mon. Wea. Rev.*, **47**, 95–99.

——, and H. Solberg, 1922: Life cycle of cyclones and the polar front theory of atmospheric circulation. *Geofys. Publ.*, **3**, 3–18.

——, and J. Holmboe, 1944: On the theory of cyclones. *J. Meteor.*, **1**, 1–22.

Bjerknes, V., 1900: The dynamic principle of the circulatory movements of the atmosphere. *Mon. Wea. Rev.*, **28**, 434–443.

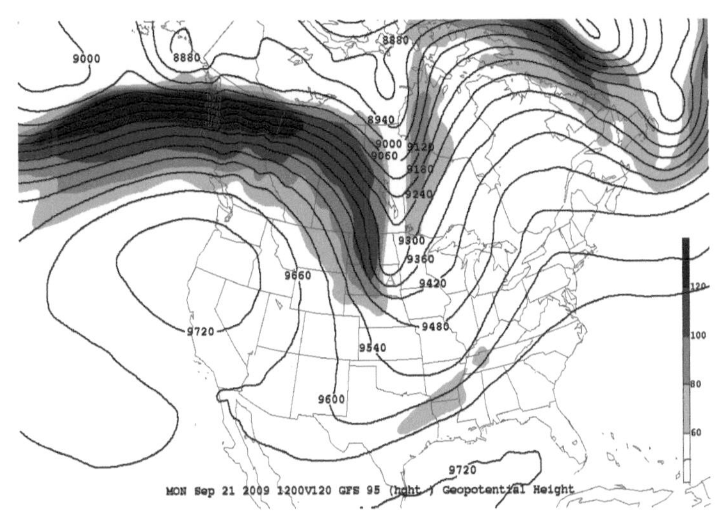


**Figure for Problem 3**

——, 1904: Weather forecast as a problem in mechanics and physics. *Meteor. Z.*, **21,** 1–7.

——, 1919a: Possible improvements in weather forecasting. *Mon. Wea. Rev.,* **47,** 99–100.

——, 1919b: Weather forecasting. *Mon. Wea. Rev.,* **47,** 90–95.

——, 1920a: On the relation between the movements and the temperatures of the upper atmosphere. *Mon. Wea. Rev.,* **48,** 159–159.

——, 1920b: On the temperature of the upper strata of the atmosphere. *Mon. Wea. Rev.,* **48,** 160–160.

——, 1921: The meteorology of the temperate zone and the atmospheric general circulation. *Mon. Wea. Rev.,* **49,** 1–3.

——, and J. W. Sandström, 1910: Dynamic Meteorology and Hydrography: Part I. Statics. Carnegie Institution, Washington D.C., Publication No. 88., 211pp.

——, T. Herselberg, and O. Devik, 1911: *Dynamic Meteorology and: Hydrography. Part II. Kinematics.* Carnegie Institution, Washington D.C., Publication No. 88., 166pp.

Brennan, M. J., G. M. Lackmann, and K. M. Mahoney, 2008: Potential vorticity (PV) thinking in operations: The utility of nonconservation. *Wea. Forecasting,* **23,** 168–182.

Chang, E. K. M., 2009: Diabatic and orographic forcing of northern winter stationary waves and storm tracks. *J. Climate,* **22,** 670–688.

——, and I. Orlanski, 1993: On the dynamics of a storm track. *J. Atmos. Sci.,* **50,** 999–1015.

Charney, J. G., 1947: The dynamics of long waves in a baroclinic westerly current. *J. Meteor.,* **4,** 135–162.

Davies, H. C., 1997: Emergence of the mainstream cyclogenesis theories. *Meteorol. Z.,* **6,** 261–274.

Davis, C. A., and K. A. Emanuel, 1991: Potential vorticity diagnostics of cyclogenesis. *Mon. Wea. Rev.,* **119,** 1929–1953.

——, 1992a: A potential-vorticity diagnosis of the importance of initial structure and condensational heating in observed extratropical cyclogenesis. *Mon. Wea. Rev.,* **120,** 2409–2428.

——, M. T. Stoelinga, and Y.-H. Kuo, 1993: The integrated effect of condensation in numerical simulations of extratropical cyclogenesis. *Mon. Wea. Rev.,* **121,** 2309-2330.

Eady, E. T., 1949: Long waves and cyclone waves. *Tellus,* **1,** 33–52.

Espy, J. P., 1841: *Philosophy of Storms.* Little and Brown, 552 pp.

Fasullo, J. T., and K. E. Trenberth, 2008: The annual cycle of the energy budget. Part II: Meridional structures and poleward transports. *J. Climate,* **21,** 2313–3325.

Friedman, R. M., 1989: *Appropriating the Weather,* Cornell University Press, 251 pp.

Frisinger, H. H., 1977: *The History of Meteorology to 1800.* Science History Publications, 148 pp.

Gill, A. E., 1982: *Atmosphere–Ocean Dynamics.* Academic Press, 662 pp.

Harper, K. C., 2008: *Weather by the Numbers: The Genesis of Modern Meteorology.* MIT Press, 308 pp.

Henry, A. J., 1922: J. Bjerknes and H. Solberg on meteorological conditions for the formation of rain. *Mon. Wea. Rev.,* **50,** 402–404.

Hoskins, B. J., 1990: Theory of extratropical cyclones. *Extratropical Cyclones: The Erik Palmén Memorial Volume,* C. W. Newton and E. O. Holopainen, Eds., American Meteorological Society 129–153.

——, and P. J. Valdes, 1990: On the Existence of Storm-Tracks. *J. Atmos. Sci.,* **47,** 1854–1864.

Kocin, P. J., and L. W. Uccellini, 1990: *Snowstorms along the Northeastern Coast of the United States: 1955 to 1985. Meteor. Monogr.,* No. 44, 280 pp.

——, and ——, 2004: *Northeast Snowstorms.* Vols 1 and 2, *Meteor. Monogr.*, No. 54, American Meteorological Society, 818 pp.

Kutzbach, G., 1979: *The Thermal Theory of Cyclones: A History of Meteorological Thought in the Nineteenth Century.* American Meteorological Society, 255 pp.

Lackmann, G. M., 2002: Potential vorticity redistribution, the low-level jet, and moisture transport in extratropical cyclones. *Mon. Wea. Rev.*, **130,** 59–74.

——, and J. R. Gyakum, 1999: Heavy cold-season precipitation in the northwestern United States: Synoptic climatology and an analysis of the flood of 17–18 January 1986. *Wea. Forecasting,* **14,** 687–700.

——, D. Keyser, and L. F. Bosart, 1997: A characteristic life cycle of upper-tropospheric cyclogenetic precursors during the Experiment on Rapidly Intensifying Cyclones over the Atlantic (ERICA). *Mon. Wea. Rev.*, **125,** 2529–2556.

Lim, E.-P., and I. Simmonds, 2002: Explosive cyclone development in the Southern Hemisphere and a comparison with Northern Hemisphere events. *Mon. Wea. Rev.*, **130,** 2188–2209.

Lynch, P., 1992: Richardson's barotropic forecast: A reappraisal. *Bull. Amer. Meteor. Soc.*, **73,** 35–47.

——, 2006: *The Emergence of Numerical Weather Prediction: Richardson's Dream.* Cambridge University Press, 279 pp.

Meisinger, C. L., 1920: The great cyclone of mid-February 1919. *Mon. Wea. Rev.*, **48,** 582–586.

Nielsen-Gammon, J. W., and R. J. LeFevre, 1996: Piecewise tendency diagnosis of dynamical processes governing the development of an upper-tropospheric mobile trough. *J. Atmos. Sci.*, **53,** 3120–3142.

NRC, 2000: *From Research to Operations in Weather Satellites and Numerical Weather Prediction: Crossing the Valley of Death.* National Academy Press, 80 pp.

Petterssen, S., 1956: *Weather Analysis and Forecasting.* 2nd ed. McGraw-Hill, 428 pp.

——, and S. J. Smebye, 1971: On the development of extratropical cyclones. *Quart. J. Roy. Meteor. Soc.*, **97,** 457–482.

Plant, R. S., C. G. Craig, and S. L. Gray, 2003: On a threefold classification of extratropical cyclogenesis. *Quart. J. Roy. Meteor. Soc.*, **129,** 2989–3012.

Reed, R. J., M. T. Stoelinga, and Y.-H. Kuo, 1992: A model-aided study of the origin and evolution of the anomalously high potential vorticity in the inner region of a rapidly deepening marine cyclone. *Mon. Wea. Rev.*, **120,** 893–913.

Richardson, L. F., 1922: *Weather Prediction by Numerical Process, 2nd ed.*, Cambridge, 236 pp.

Roebber, P. J., 1984: Statistical analysis and updated climatology of explosive cyclones. *Mon. Wea. Rev.*, **112,** 1577–1589.

Rossby, C.-G. 1945: On the propagation of frequencies and energy in certain types of oceanic and atmospheric waves. *J. Meteor.*, **4,** 187–204.

Sanders, F., 1986: Explosive cyclogenesis over the west-central North Atlantic Ocean, 1981–84. Part I: Composite structure and mean behavior. *Mon. Wea. Rev.*, **114,** 1781–1794.

——, 1988: Life history of mobile troughs in the upper westerlies. *Mon. Wea. Rev.*, **116,** 2629–2648.

——, and J. R. Gyakum, 1980: Synoptic-dynamic climatology of the "bomb." *Mon. Wea. Rev.*, **108,** 1589–1606.

Schultz, D. M., and F. Sanders, 2002: Upper-level frontogenesis associated with the birth of mobile troughs in northwesterly flow. *Mon. Wea. Rev.*, **130,** 2593–2610.

Shapiro, M. A., and S. Grønås, 1999: *The Life Cycles of Extratropical Cyclones.* American Meteorological Society, 359 pp.

Sickmöller, M., R. Blender, and K. Fraedrich, 2000: Observed winter cyclone tracks in the Northern Hemisphere in reanalysed ECMWF data. *Quart. J. Roy. Meteor. Soc.*, **126,** 591–620.

Solomon, S., D. Qin, M. Manning, M. Marquis, K. Averyt, M. M. B. Tignor, H. L. Miller Jr., and Z. Chen, Eds., 2007: Climate Change 2007: *The Physical Science Basis.* Working Group I Contribution to the Fourth Assessment Report of the Intergovernmental Panel on Climate Change, Figure 10.8. Cambridge University Press, 996 pp.

Stoelinga, M. T., 1996: A potential vorticity-based study of the role of diabatic heating and friction in a numerically simulated baroclinic cyclone. *Mon. Wea. Rev.*, **124,** 849–874.

Surcliffe, R. C., and A. G. Forsdyke, 1950: The theory and use of upper air thickness patterns in forecasting. *Quart. J. Roy. Meteor. Soc.*, **76,** 189–217.

Thorpe, A. J., 1986: Synoptic disturbances with circular symmetry. *Mon. Wea. Rev.*, **114,** 1384–1389.

Uccellini, L. W., R. A. Petersen, P. J. Kocin, K. F. Brill, and J. J. Tuccillo, 1987: Synergistic interactions between an upper-level jet streak and diabatic processes that influence the development of a low-level jet and a secondary coastal cyclone. *Mon. Wea. Rev.*, **115,** 2227–2261.

Whitaker, J. S., L. W. Uccellini, and K. F. Brill, 1988: A model-based diagnostic study of the rapid development phase of the Presidents' Day cyclone. *Mon. Wea. Rev.*, **116,** 2337–2365.

Winston, J. S., A. Gruber, T. I. Gray Jr., M. S. Vanadore, C. L. Earnest, and L. P. Mannello, 1979: Earth's atmosphere radiation budget analyses derived from NOAA satellite data, June 1974–February 1978, NOAA Rep.

Zhu, Y., and R. E. Newell, 1998: A proposed algorithm for moisture fluxes from atmospheric rivers. *Mon. Wea. Rev.*, **126,** 725–735.

Zishka, K. M., and P. J. Smith, 1980: The climatology of cyclones and anticyclones over North America and surrounding ocean environs for January and July, 1950–1977. *Mon. Wea. Rev.*, **108,** 387–401.

Zolina, O., and S. K. Gulev, 2002: Improving the accuracy of mapping cyclone numbers and frequencies. *Mon. Wea. Rev.*, **130,** 748–759.

## FURTHER READING

Bergeron, T., 1937: On the physics of fronts. *Bull. Amer. Meteor. Soc.,* **18,** 265–275.

Businger, S., and R. J. Reed, 1989: Cyclogenesis in cold air masses. *Wea. Forecasting,* **4,** 133–156.

Davies, H. C., and H. Wernli, 1997: On studying the structure of synoptic systems. *Meteorol. Appl.,* **4,** 365–374.

Evans, M. S., D. Keyser, L. F. Bosart, and G. M. Lackmann, 1994: A satellite-derived classification scheme for rapid maritime cyclogenesis. *Mon. Wea. Rev.,* **122,** 1381–1416.

Lackmann, G. M., D. Keyser, and L. F. Bosart 1999: Energetics of an intensifying jet streak during the Experiment on Rapidly Intensifying Cyclones over the Atlantic (ERICA). *Mon. Wea. Rev.,* **127,** 2777–2795.

Lefevre, R. J., and J. W. Nielsen-Gammon, 1995: An objective climatology of mobile troughs in the Northern Hemisphere. *Tellus,* **47A,** 638–655.

Martin, J. E., and N. Marsili, 2002: Surface cyclolysis in the North Pacific Ocean. Part II: Piecewise potential vorticity analysis of a rapid cyclolysis event. *Mon. Wea. Rev.,* **130,** 1264–1281.

Sutcliffe, R. C., 1947: A contribution to the problem of development. *Quart. J. Roy. Meteor. Soc.,* **73,** 370–383.

Uccellini, L. W., P. J. Kocin, R. A. Petersen, C. H. Wash, and K. F. Brill, 1984: The Presidents' Day cyclone of 18–19 February 1979: Synoptic overview and analysis of the subtropical jet streak influencing the pre-cyclogenetic period. *Mon. Wea. Rev.,* **112,** 31–55.

# CHAPTER 6

# Fronts

*Present surface frontal analyses suffer from the defect that frontal positions are typically not collocated with zones of intense temperature contrast. Further, individuals typically do not agree as to the existence, type, and location of fronts.*

*The author argues that the lack of a surface temperature analysis is mainly responsible for these flaws, and it is proposed that such analysis, preferably of potential temperature in regions of variable terrain elevation, become part of routine procedure.*

—F. Sanders,
"A Proposed Method of Surface Map Analysis" (1999)

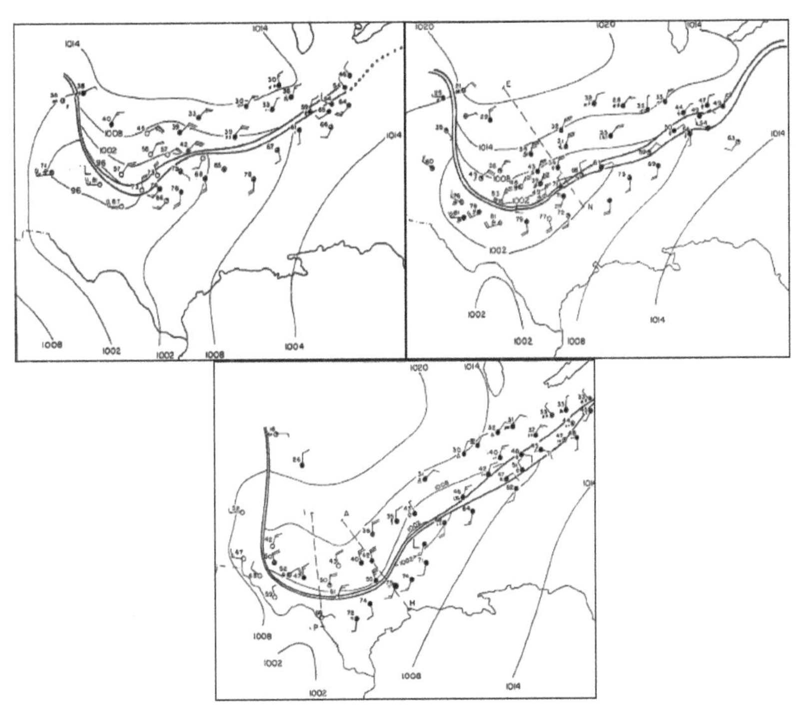


Surface analyses for 17–18 Apr 1953 depicting an intense U.S. Great Plains cold front (from Sanders 1955).

Inspection of geostationary satellite imagery, such as that shown in Fig. 2.1 or at the beginning of chapter 5, frequently reveals long, narrow cloud bands that often correspond to frontal zones. The Bergen school meteorologists, as discussed in section 5.4, recognized the crucial role that fronts play in day-to-day weather forecasting and built their forecasting methods upon analysis of the polar front. In seeking a theoretical explanation for cyclogenesis, they sought frontal instabilities, in keeping with the observation that cyclones formed along preexisting frontal boundaries. With the emergence of baroclinic instability theory in the 1940s, it became evident that cyclones could form spontaneously without fronts, and when they did, they would develop frontal features. Rather than fronts leading to cyclones, it became evident that cyclones could produce fronts.

Whether the systematic lowering and thickening of stratiform clouds accompanying the approach of a warm front, or the formation of convective storms along a cold front, it is clear that fronts can strongly influence weather conditions. To understand and predict frontal weather, we must first understand the mechanisms that lead to frontal formation and structure.

## 6.1. WHAT ARE FRONTS?

### 6.1.1. Definition

Given the prominence of fronts on satellite imagery or surface map analyses, one would think that a universally accepted definition of a front should exist. However, it turns out that there remains considerable disagreement as to what constitutes a front, as well as how to analyze them; Hoskins (1982) concluded that "it is probably a mistake to attempt a rigorous definition of a front," but it is also evident that a working definition is needed.

Fronts can be defined as regions of enhanced horizontal temperature or potential temperature gradient; "enhanced" could be taken to be an order of magnitude larger than the typical synoptic-scale value. Sanders (1999) proposed to classify temperature gradients of 8°C per 220 km as "moderate" and 8°C per 110 km as "strong." However, Sanders also noted, based on examination of many surface potential temperature analyses, that a large fraction of zones of strong temperature contrast are not frontal; he reserved the term *front* to apply only to those baroclinic zones exhibiting a strong cyclonic wind shift at their warm edge.

Perhaps the simplest definition of a front is that it represents an *airmass boundary*. This definition can lead to debates over terminology for boundaries such as dry lines, for example, which are not typically recognized (or analyzed) as fronts but do mark the boundary between warm, moist maritime tropical air and hot, dry continental tropical air. The lack of a surface temperature analysis on many operational analyses was implicated by Sanders (1999) as a major factor leading to difficulty in frontal analysis. Provided that a surface temperature or potential temperature analysis is available, drawing fronts on weather maps may not be necessary; Sanders and Hoffman (2002) argue for "letting the isotherms speak for themselves."

The *Glossary of Meteorology* (Glickman 2000) defines a front as the "**interface** or transition zone between two **air masses** of different **density**" [Glickman's bold]. This requires use of a density variable, such as virtual temperature, virtual potential temperature, or density, to identify a front. However, if one accepts the broader definition of a front as an airmass boundary, then inconsistencies can still arise.

Consider a hypothetical situation in which a dry, clear, continental air mass lies to the west of a warm, moist, cloudy airmass on a late-spring day over the central United States; the boundary between these airmasses is analyzed as a cold front on synoptic frontal analyses. In the early morning, temperatures in the dry airmass are relatively cool because of overnight radiational cooling under clear skies, and a clearly defined cold front exists in the surface isotherm or isentrope field. However, as the sun rises, solar heating causes surface temperatures in the dry, clear, postfrontal air mass to warm more rapidly than in the humid, cloudy prefrontal airmass, and the cross-frontal temperature (and density) gradient weakens; by late afternoon, the frontal temperature contrast no longer exists. A zone of convergence and precipitation may persist in this situation because of frictional convergence into the frontal trough, despite the weakened temperature gradient. In this case, plotting equivalent potential temperature would better identify the frontal zone throughout the day, if we can agree that this feature represents a true front. Most meteorologists would agree that just because the gradient may weaken or vanish during the warmest part of the day, a boundary of substantial meteorological importance would still exist there. The frontal circulation, which is driven by density contrasts, would also exhibit a diurnal pattern in strength; nevertheless, a frontal definition requiring a density contrast would argue for only drawing this front at certain times of day! The preceding argument is hypothetical, and to some extent perhaps a semantic one; however, it does reflect the complications inherent in frontal analysis.

Here, we adopt the less restrictive definition of fronts as airmass boundaries, with acknowledgment of the caveats discussed above and *without* advocating the analysis of all airmass boundaries as fronts on surface analyses. The purpose of this definition is to draw meteorological attention to important baroclinic features, including those that exhibit a diurnal variation in strength, but allowing omission of shallow, semipermanent thermal gradients that are locked in place by topography, such as are found along the California coast during summer. The analysis of a surface temperature field, as advocated by Sanders (1999) and Sanders and Doswell (1995), is strongly encouraged, which lessens the need for a strict frontal definition; by showing the surface isentropes, important features will become apparent in the analysis.

Why do fronts matter? Fronts can strongly impact local weather conditions, and forecasts must account for their movement, type, intensity, and influence on cloud and precipitation. Depending on the static stability of air in the vicinity of a frontal system, fronts can trigger severe, organized convective storms. The presence of enhanced vertical wind shear in frontal zones, as required by thermal wind balance, contributes to the organization and severity of convective storms that form in their vicinity. In some situations, convective storms can develop in succession along frontal boundaries, resulting in flash flooding as sequential cells move over the same areas repeatedly. Both the horizontal wind shift and vertical wind shear accompanying frontal zones have important implications for aviation forecasting, and the timing of fronts can be critical to determining the amount and type of precipitation in a given location.

### 6.1.2. Frontal properties

Given the enhanced horizontal gradients of temperature associated with frontal zones, thermal wind balance indicates that fronts should be zones of strong vertical wind shear. Strong and deep frontal zones would necessarily accompany the midlatitude jet stream, hence the term *polar-front jet*. Drawing on the concept that ageostrophic circulations arise in response to the disruption of thermal wind balance, as discussed in chapter 2, one would expect strong ageostrophic circulations to be associated with frontal zones, as air parcels moving into or out of the frontal zone would experience a marked change in the magnitude of the horizontal virtual temperature gradient, potentially disrupting thermal wind balance. On a basic level, this is one reason why fronts are often associated with bands of cloud and precipitation.

Noting that frontal zones are long and narrow, we can define spatial scales characteristic of frontal zones as synoptic scale (1000 km) in the along-front direction, and one order of magnitude less than this in the cross-front direction (100 km). Hoskins (1982) defines the along-front scale as similar to the Rossby radius of deformation (see chapter 8), which is ~1000 km for synoptic-scale motions. Thus, fronts are in a sense both *mesoscale* and *synoptic-scale* features (Bluestein 1993, section 2.1). This implies that the quasigeostrophic (QG) framework, with characteristic synoptic-scale gradients, will *not be formally valid* in the cross-front direction for frontal systems. Alternate dynamical frameworks are needed to explain the dynamics of frontal systems.

Observations demonstrate that fronts display the following characteristics:

- Enhanced horizontal contrasts of temperature and/or moisture; moisture gradients alone may suffice if we accept the "airmass boundary" definition of fronts;
- A relative minimum of pressure (trough) and maximum of cyclonic vorticity along the front;
- Strong vertical wind shear, and a horizontal wind shift consistent with a pressure trough and cyclonic vorticity;
- Large static stability within the frontal zone;
- Ascending air, clouds, and precipitation near the front (depending on moisture availability and other factors); and
- Greatest intensity near the surface, weakening with height.

Intense frontal zones are not only found at the surface; observational and theoretical studies have documented the formation of strong fronts near the tropopause as well (e.g., Reed and Sanders 1953; Reed 1955; Bosart 1970; Hoskins and Bretherton 1972). The strong vertical wind shear accompanying these systems can result in clear-air turbulence, which is of interest to the aviation industry, among others. Tropopause folding, the formation of upper-tropospheric disturbances, and troposphere–stratosphere mass exchange are also linked to upper-level frontogenesis. The latter mechanism is especially important to the study of stratospheric chemistry and dynamics.

### 6.1.3. An example

To illustrate some of the properties listed above, diagnostic analyses from a typical continental cold front from 17 November 2009 are presented (Fig. 6.1). A low-pressure center is located near southern Illinois and eastern Missouri, with a trough of lower pressure extending southward from the center, suggesting the presence of a cold front; however, it can be dangerous to assume that all troughs are accompanied by fronts; additional data are needed. The same is true for the trough evidently indicating the presence of a warm front extending eastward from the cyclone center (Fig. 6.1a). The potential and equivalent potential temperature fields exhibit very strong gradients, with the packing of isentropes found on the cold side of the frontal boundary (Fig. 6.1b), demonstrating that the troughs in this case do mark frontal locations. The $\theta_e$ field reveals the presence of a zone of enhanced gradient

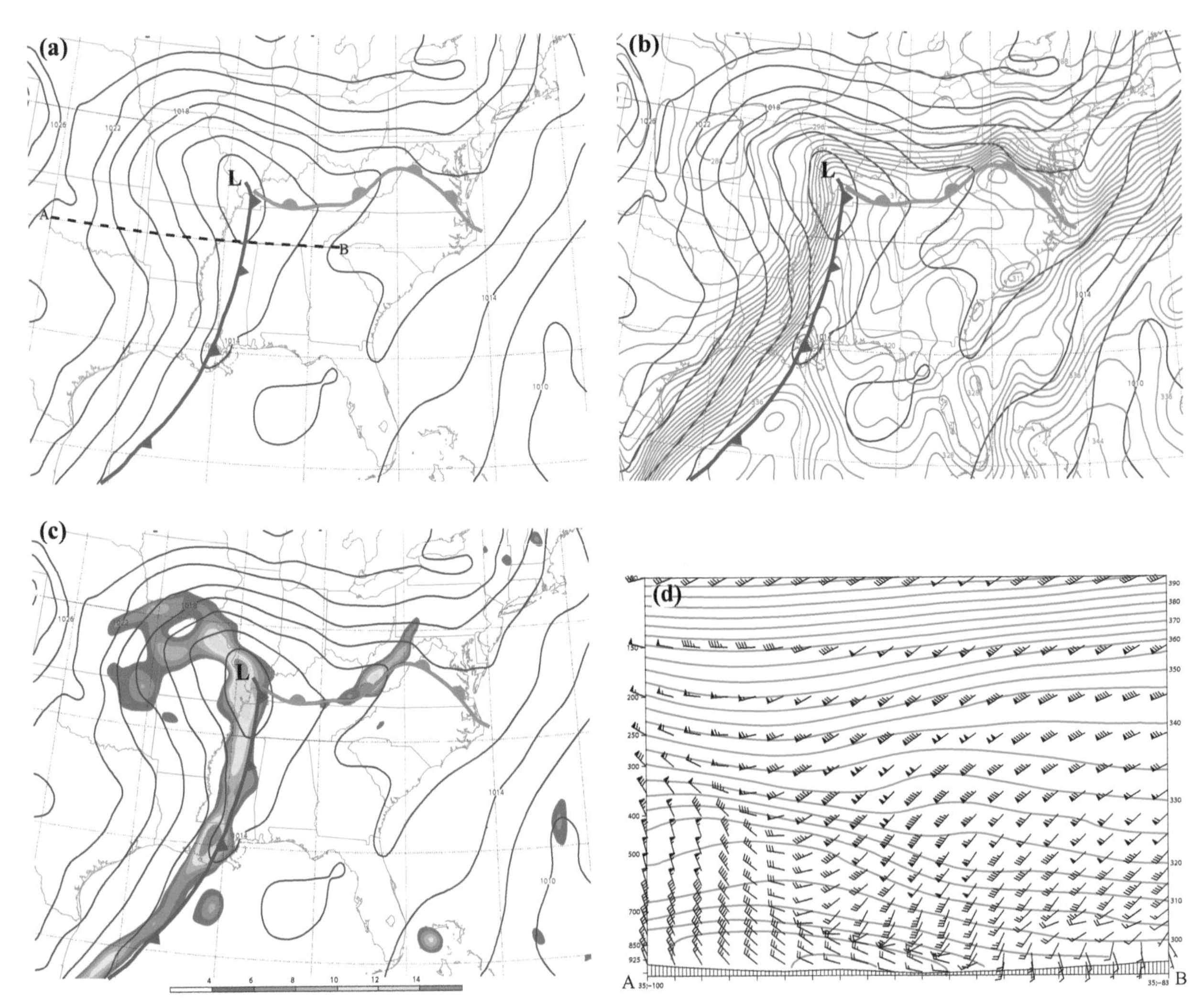


**Figure 6.1.** Diagnostic summary for 0000 UTC 17 Nov 2009, based on North American Mesoscale (NAM) analysis: (a) sea level pressure and subjective frontal analysis, bold line shows orientation of cross section shown in (d); (b) as in (a) with 1000-mb equivalent potential temperature contours (interval is 2 K); (c) as in (a) with 950-mb relative vorticity (shade interval is $2\times10^{-5}\,\mathrm{s}^{-1}$, beginning $4\times10^{-5}\,\mathrm{s}^{-1}$); and (d) cross section of wind and potential temperature (interval 5K, red contours); orientation shown in (a).

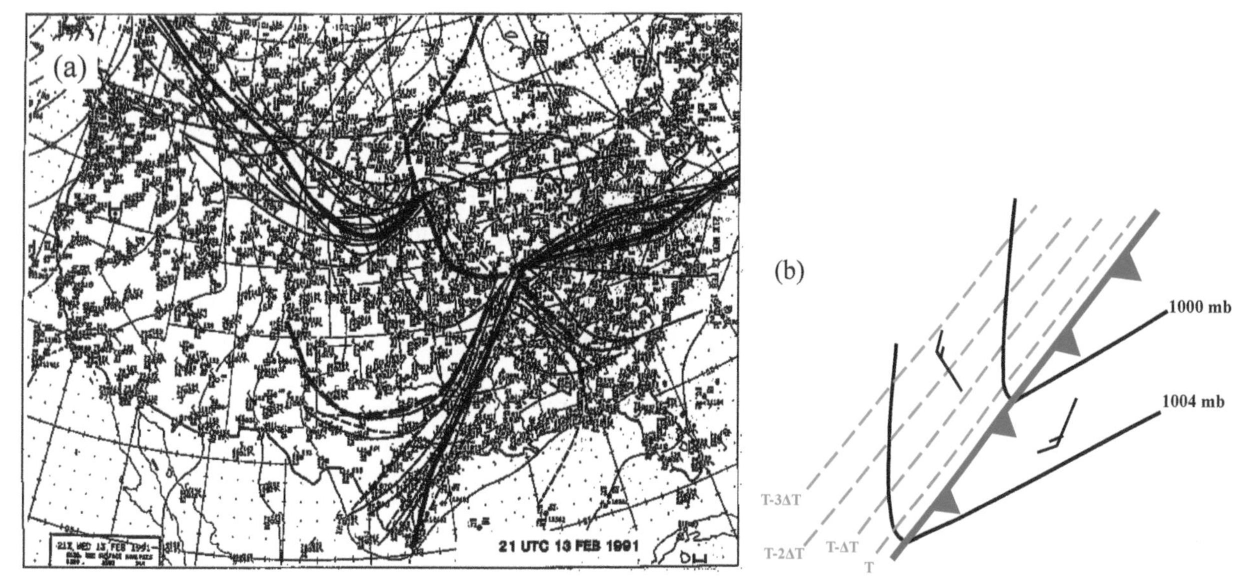


**Figure 6.2.** (a) Superimposed frontal analyses from participants in a workshop on frontal analysis, all using data from 2100 UTC 13 Feb 1991 (from Uccellini et al. 1992). (b) Idealized schematic for isotherms, isobars, and cold-frontal location.

extending to the east of the low center, corresponding to the location of a warm front. The pressure field shown in Fig. 6.1a suggests a maximum of cyclonic vorticity along the frontal zone, which is indeed evident in the analysis (Fig. 6.1c). A vertical cross section, oriented approximately east–west through the frontal zone, demonstrates that the front is a region of strong vertical wind shear as well as large static stability (Fig. 6.1d).

In the cross section, the location of the surface front is evident by a change in the character of the vertical wind profile (Fig. 6.1d). East of the surface front, the winds veer strongly with height; however, this changes abruptly to a backing wind profile to the west of the location of the surface front. The vertical structure of the frontal zone is somewhat ambiguous in this particular cross section, but the zone of most strongly backing winds slopes westward and upward from the location of the surface front. The representation of the front in this relatively coarse (40-km grid spacing) analysis limits the ability to represent fine-scale features of the frontal zone, but it nevertheless gives a consistent picture of typical cold-frontal characteristics. The warm front, a shallow feature, is not characterized by as pronounced a trough in the sea-level pressure field; nevertheless, it exhibits a strong gradient in $\theta_e$.

Several of the National Centers for Environmental Prediction (NCEP), including the Hydrometeorological Prediction Center (HPC), Tropical Prediction Center (TPC), and Ocean Prediction Center (OPC), routinely draw surface analyses, including fronts, as do meteorologists at national centers in other nations. However, despite frontal analysis being a common practice, there is variability in how different analysts treat each situation (Fig. 6.2a). The lack of consensus in frontal placement evident in Fig. 6.2a, which shows the collective analyses of a group of experts, demonstrates that there is no widely agreed upon procedure for frontal analysis. The fact that surface isotherm or isentrope analyses are not routinely conducted in operational surface analyses of the type depicted in Fig. 6.2a has been implicated as contributing to inconsistency between analyzed frontal location and the actual temperature field (e.g., Sanders 1999). Furthermore, actual data do not often exhibit the well-behaved characteristics shown in the idealized Fig. 6.2b. We will return to the topic of frontal analysis in chapter 12.

## 6.2. KINEMATIC FRONTOGENESIS

A traditional measure of frontogenesis was introduced by Petterssen (1936) to explore the kinematic processes leading to changes in the strength of the gradient of a scalar field following a moving air parcel. If we choose potential temperature as our scalar field, the *frontogenesis function* (*F*) is defined as the time rate of change of the magnitude of the horizontal potential temperature gradient following the flow:

$$F \equiv \frac{d|\nabla_p \theta|}{dt}. \tag{6.1}$$

The advantage of defining $F$ with potential temperature, rather than temperature, is that in regions of complex terrain, the potential temperature can help to isolate synoptic features (including fronts) by correcting the temperature to a common pressure level, thereby reducing the portion of the temperature gradient due to elevation. Also, potential temperature is a conservative variable for adiabatic flow.

Here, the different physical mechanisms that can lead to changes in the potential temperature gradient following the flow will be examined on a conceptual level. Accordingly, rather than treat the full frontogenesis equation, we will work with a simplified version; however, the physically relevant processes are accounted for in a clear manner. Here we follow Miller (1948), Reed and Sanders (1953), and Sanders (1955), in that we restrict consideration to the change in temperature gradient in the front-normal direction. This is accomplished by rotating the coordinate system so that the $y'$ axis is perpendicular to the front, and the $x'$ axis is aligned parallel to the frontal zone. This does not necessarily mean that the $y'$ direction is normal to *isentropes*, however. Therefore, $F$ is defined here as $d/dt(-\partial\theta/\partial y')$, with the negative sign accounting for a positive value of $F$, given that $y'$ is positive toward lower values of potential temperature. For convenience, the prime notation will be dropped subsequently. Taking the total derivative of $-\partial\theta/\partial y$ and expanding while using the product rule, the simplified form of the frontogenetical equation is

$$F = \underbrace{\left[\frac{\partial\theta}{\partial x}\left(\frac{\partial u}{\partial y}\right)\right]}_{\text{Shearing}} + \underbrace{\left[\frac{\partial\theta}{\partial y}\left(\frac{\partial v}{\partial y}\right)\right]}_{\text{Confluence}} + \underbrace{\left[\frac{\partial\theta}{\partial p}\left(\frac{\partial \omega}{\partial y}\right)\right]}_{\text{Tilting}} - \underbrace{\left[\frac{\partial}{\partial y}\left(\frac{d\theta}{dt}\right)\right]}_{\text{Diabatic}}. \tag{6.2}$$

It is important to recognize that (6.2) describes frontogenesis in a *Lagrangian* sense; that is, it expresses the change in potential temperature gradient experienced by moving air parcels *following the flow*. The change in potential temperature gradient experienced by moving parcels does not indicate whether the overall front is strengthening or weakening in an absolute sense. A change in the horizontal potential temperature gradient *at the location of the maximum gradient* does indicate whether the overall frontal strength is increasing or decreasing. When air parcels are experiencing a change in horizontal temperature gradient, this holds important implications for thermal wind balance disruption, and as will be shown subsequently, for an ageostrophic response. While the terms on the right side of (6.2) contain information about the kinematic properties of the flow that impact the overall strength of a frontal zone, plotting frontogenesis often exhibits positive values—even for a weakening front. Often meteorological analysis software only accounts for the shearing and confluence term on the right side of (6.2). In this case, positive frontogenesis values at the location of the strongest potential temperature gradient can either indicate a strengthening front, or that the total frontogenesis including diabatic effects has not been calculated. As lower-tropospheric air parcels converge on a frontal zone, they will still experience an increasing gradient relative to that of the background synoptic setting.

In (6.2) the *shearing frontogenesis* describes the change in frontal strength due to *differential (potential) temperature advection* by the *front-parallel* wind component (Fig. 6.3). In evaluating the front-relative wind and temperature derivatives in the shearing term for the cold-frontal region in Fig. 6.3a, we find that both $\partial\theta/\partial x$ and $\partial u/\partial y$ are negative, consistent with a positive contribution and frontogenesis. The strengthened temperature gradient in Fig. 6.3b results from cold advection in the cold air and warm advection in the warm air, both of which act to increase the frontal contrast.

Shearing is not always frontogenetical, as evident from Fig. 6.4. In this case, $\partial\theta/\partial x$ is positive (north of the warm front) and $\partial u/\partial y$ is negative, yielding a negative tendency and frontolytical impact on the warm front. Here, easterly flow to the north of the warm front is associated with warm advection there, warming the cold air relative to the warm air and weakening the front, as evident in Fig. 6.4b.

The *front-normal* wind component can also contribute strongly to $F$, as described by the *confluence term* in (6.2). Given the focus on only the front-normal flow component, this term could also be labeled as *stretching*. A scale analysis of (6.2) indicates that the confluence term scales an order of magnitude larger than the shearing term, due to the smaller cross-front distance scale. A schematic example is provided in Fig. 6.5; as with the shearing term, it is easy to visualize in this example how advection is acting to warm the air on the warm side of the front and cool it on the cold side, strengthening the front. Mathematically,

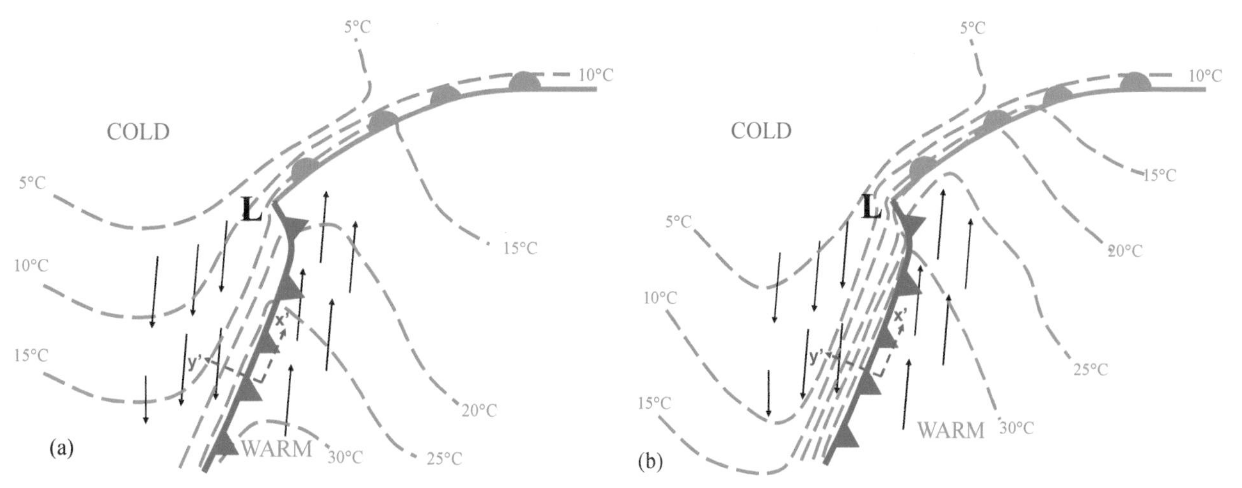


**Figure 6.3.** Idealized horizontal plot of near-surface wind vectors (black arrows), isotherms (red dashed contours), and frontal zones for (a) initial time, and (b) the same front-relative view at a later time (~24 h later). The rotated front-relative coordinate axes are shown.

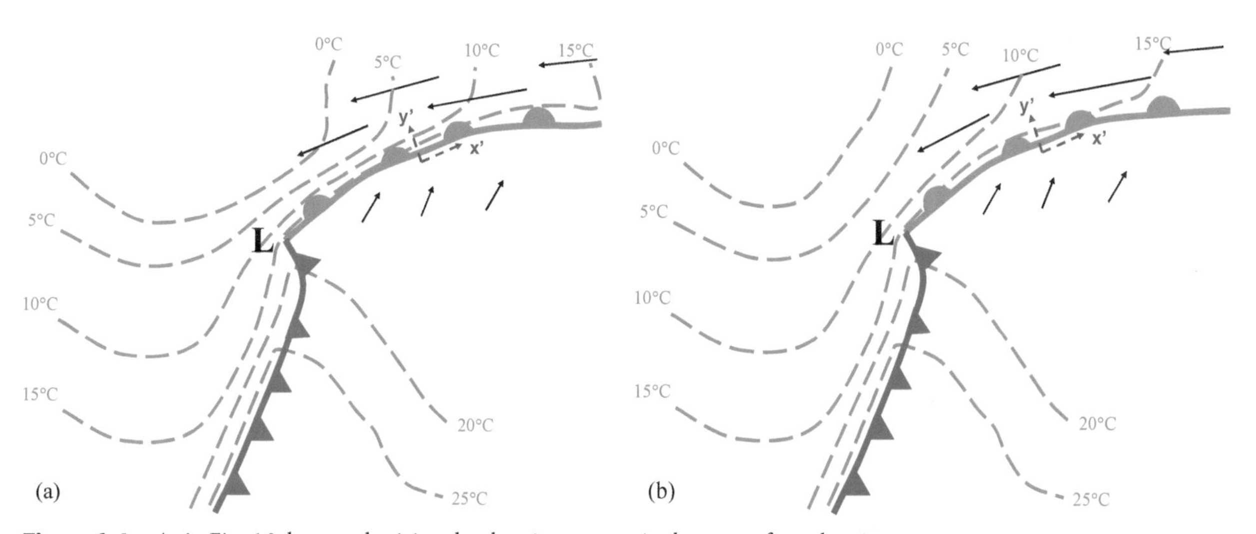


**Figure 6.4.** As in Fig. 6.3, but emphasizing the shearing process in the warm-frontal region.

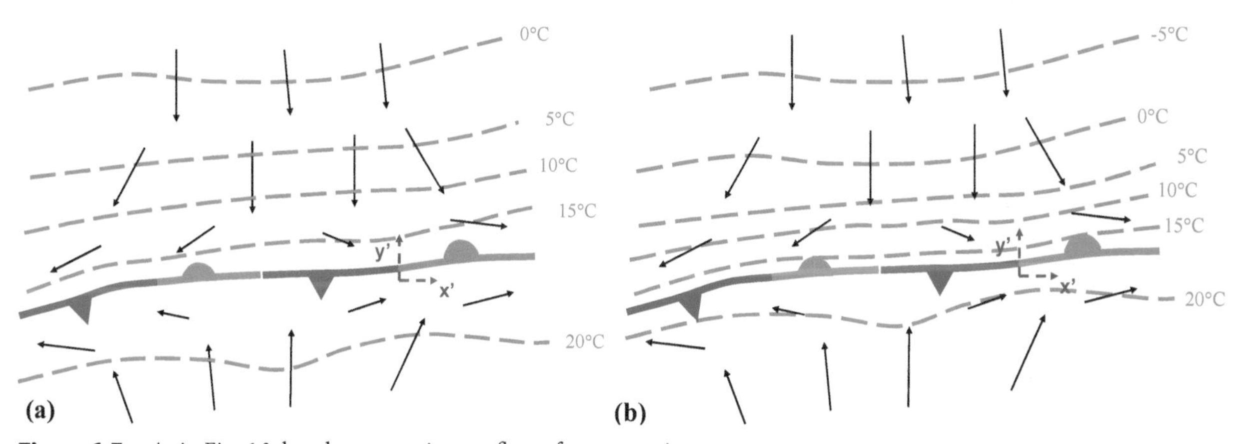


**Figure 6.5.** As in Fig. 6.3, but demonstrating confluent frontogenesis.

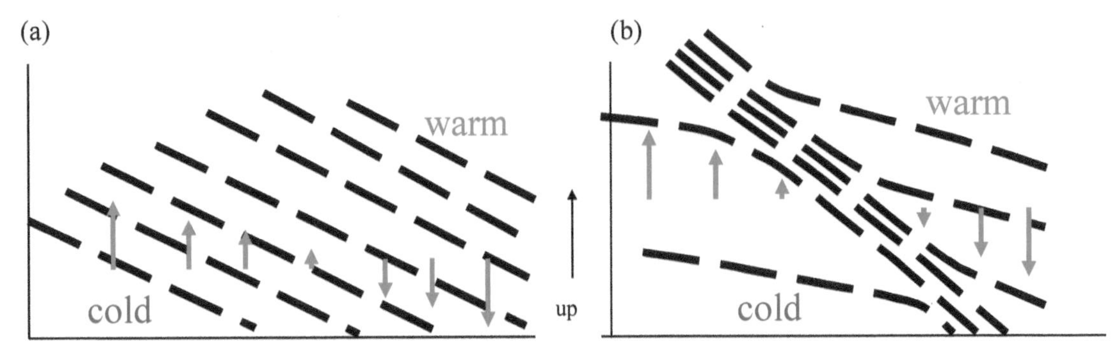


**Figure 6.6.** Idealized cross-sectional diagrams demonstrating frontogenetical tilting. Dashed lines are isentropes and red arrows represent vertical motion for (a) an initial time and (b) a later time (after Carlson 1998).

$\partial \theta / \partial y < 0$ as usual and $\partial v / \partial y < 0$ because of the confluent front-normal flow (changing from a positive $v$ component to negative as one moves in the $+y$ direction across the front). A confluent flow will not always act to strengthen the temperature gradient; the relative angle between the isotherms and the deformation flow is critical, as will be discussed subsequently.

Near the earth's surface, the vertical motion is small, and the effect of tilting [term c in (6.2)] is often relatively weak. An exception to this can occur over sloping terrain, where differential upslope and downslope flow on either side of a front that is oriented parallel to the terrain gradient can exert a significant influence on frontal strength. More often, however, the tilting term contributes to frontogenesis *aloft*, where differential vertical motions are often larger.

Consider the effects of differential vertical motion on the isentrope field shown in Fig. 6.6a. In this case, cold air is rising and warm air is sinking, a *thermally indirect* circulation. This situation is typically observed in the exit region of a jet streak, for example. In this case, differential vertical advection of the isentrope field leads to a cooling of the cold air and warming of the warm air and a strengthening of the frontal zone, as indicated in Fig. 6.6b.

Often, the tilting effect works to oppose changes due to confluence or shear, as a means of maintaining thermal wind balance. When we discuss the *dynamics* of fronts, we will revisit this idea; however, for now, the focus is on kinematic frontogenesis mechanisms.

The remaining term in (6.2) is the *differential diabatic heating* term. There are a number of diabatic processes that must be considered in evaluating this term in practice, including differential solar heating, differential surface heating due to gradients in soil characteristics or surface properties (e.g., coastlines), and differential turbulent heat flux. The example described by Fig. 6.7 involves differential solar heating, which can be related to cloud cover gradients, or gradients in land surface characteristics.

If the heating rate in the warm air, as shown in Fig. 6.7a, exceeds that in the colder air, then the differential diabatic heating process will be frontogenetical. Because the heating becomes smaller in the positive $y$ direction, $-\partial / \partial y \, (d\theta / dt) > 0$ in this example, and a positive value (frontogenesis) results. The situation above can take

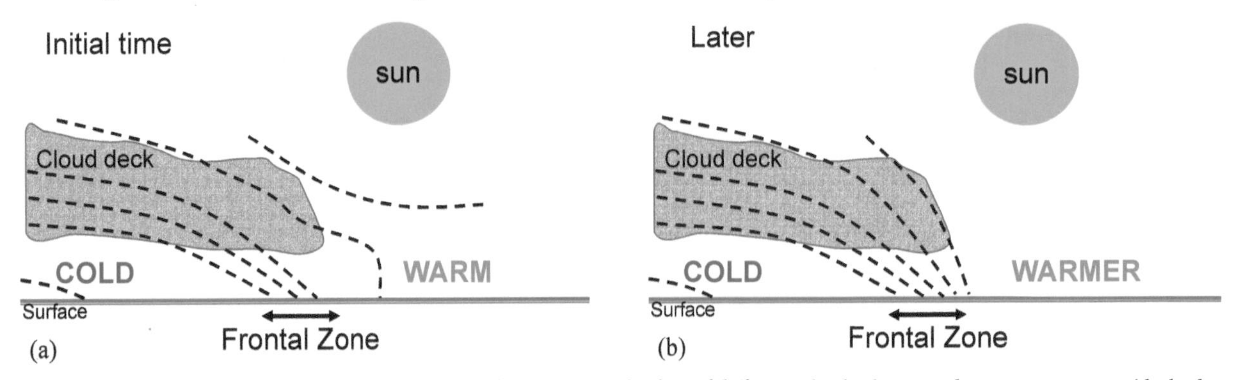


**Figure 6.7.** Idealized cross-sectional depiction of frontogenetical effect of differential solar heating showing isentropes (dashed contours) and cloud cover (gray shading) for (a) an initial time and (b) a later time (after Carlson 1998).

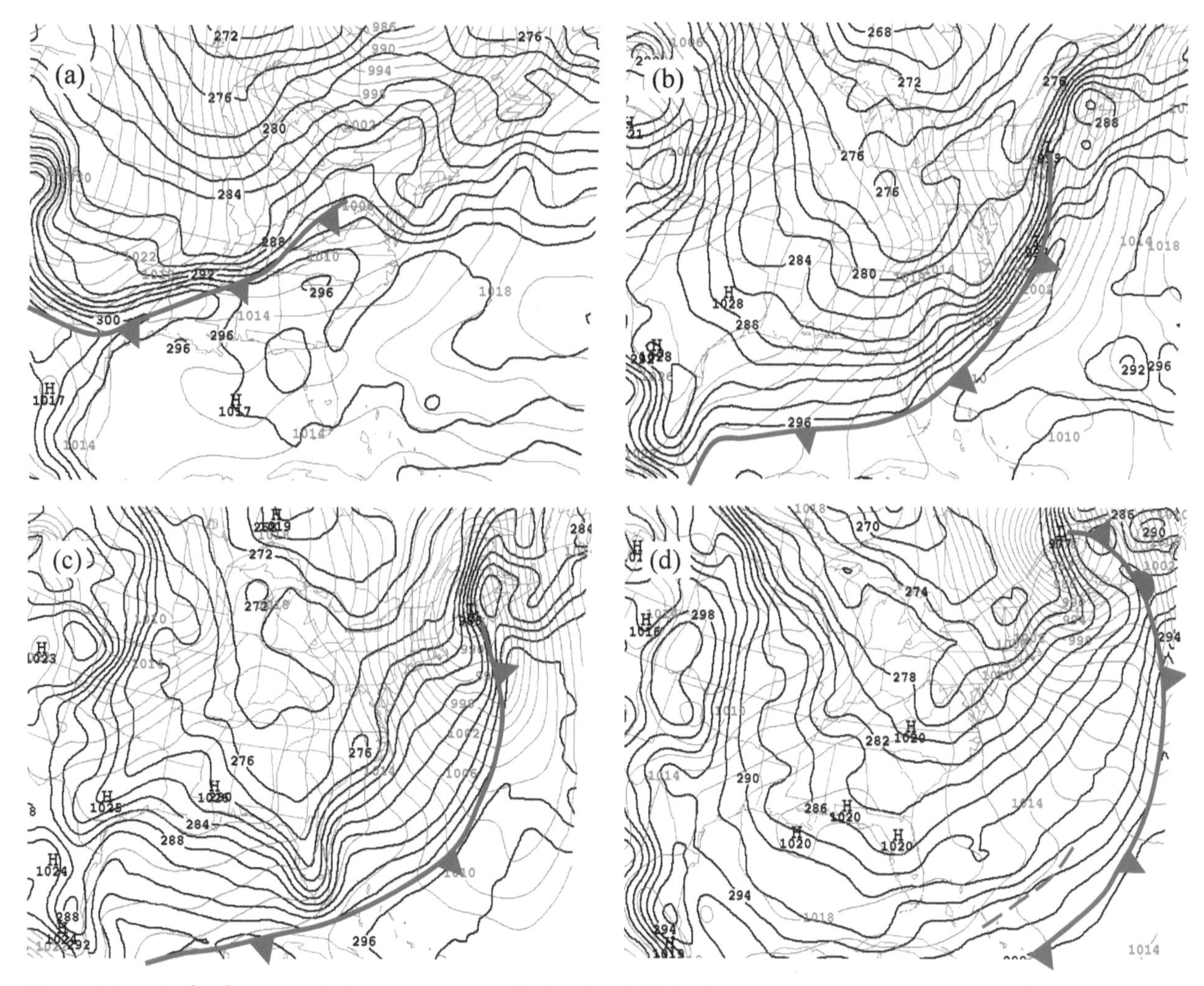


**Figure 6.8.** Sea level pressure (red contours, interval is 2 mb) and 1000-mb equivalent potential temperature (black contours, interval is 2 K) from NAM analyses valid for (a) 0000 UTC 15 Nov; (b) 0000 and (c) 1200 UTC 16 Nov; and (d) 0000 UTC 17 Nov 2007.

place when air is ascending along the sloping isentropes, forming the stable frontal zone. How might the situation depicted in Fig. 6.7 change during an overnight period? Because the moisture content is often greater in the warm air, another situation that can arise features more cloudiness in the warm, moist air relative to the cold air mass. This obviously has the opposite implications for frontal strength to that evident in Fig. 6.7.

Diabatic processes other than radiational heating and cooling are often frontogenetically important. Analyses of a cold front moving off the East Coast of the United States on 15–17 November 2007 exhibit an interesting change in the isentrope distribution after the front crosses the East Coast (Fig. 6.8). A strong gradient of potential temperature extends from Texas to Tennessee at 0000 UTC 15 November (Fig. 6.8a). Once the cold front crosses the Gulf Coast, the packing of isentropes behind the front weakens markedly there, while the gradient is maintained along the front over the eastern United States (Fig. 6.8b). A similar weakening of the potential temperature gradient behind the front occurs as the system crosses the East Coast (Figs. 6.8c,d). This potential temperature evolution is strong evidence for warming of the postfrontal boundary layer over the warm waters of the Gulf of Mexico and Gulf Stream due to an upward turbulent heat flux behind the front. Plots of the frontogenesis function only

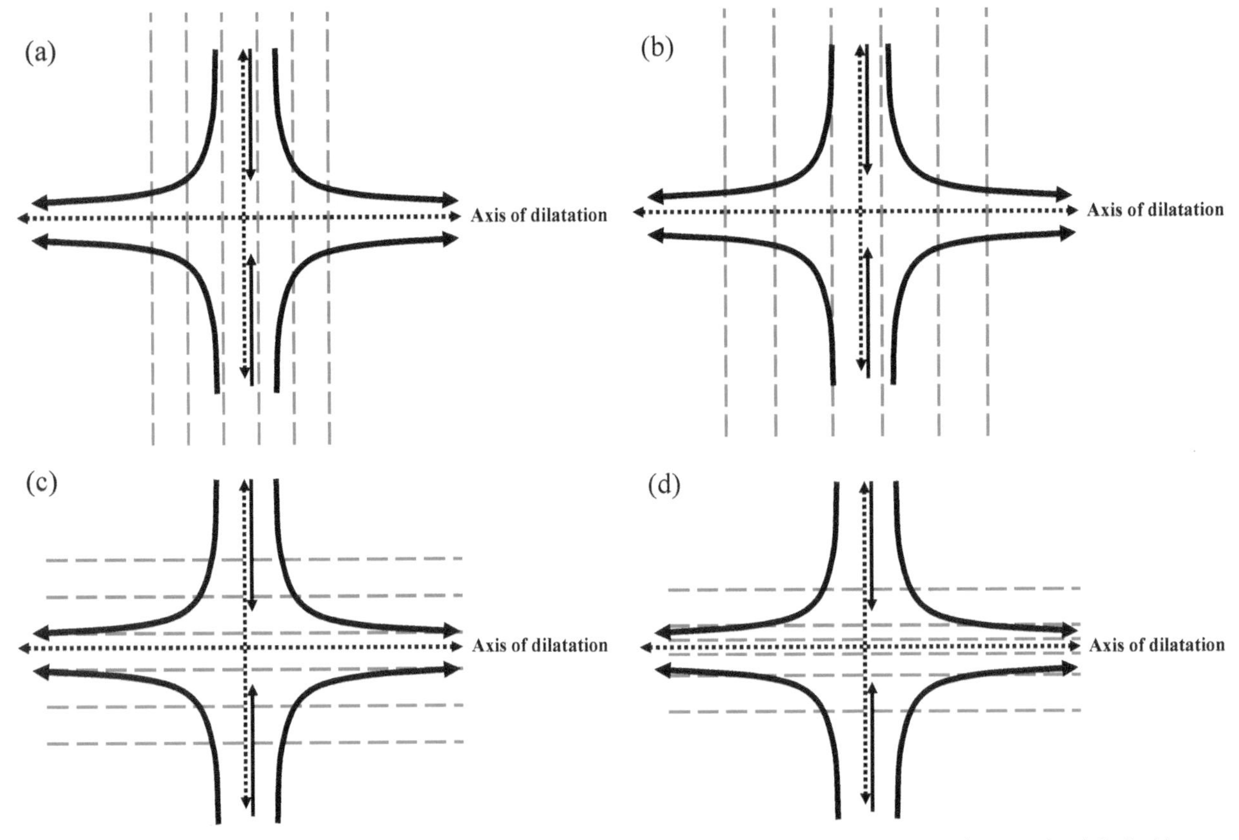


**Figure 6.9.** Idealized schematics of isentrope evolution in a deformation flow. Black lines are streamlines, and red dashed lines are isentropes: (a) initial time, isentropes oriented perpendicular to axis of dilatation; (b) as in (a), but at a later time. (c),(d) As in (a),(b), but with isentropes oriented parallel to the axis of dilatation.

accounting for the shear and confluence terms continue to show positive frontogenesis values even as the front weakens offshore (not shown).

To summarize the kinematics of frontogenesis, there are several distinct physical processes that can act to change the intensity of a front. The mechanisms of shearing and confluence are similar, in that they can both be interpreted as differential thermal advection: in the case of shearing, by the front-parallel wind component and by the front-normal component for confluence. The tilting term, which will be considered again when we discuss upper-level frontogenesis, can act over sloping terrain but is more often significant above the surface. This process can be viewed as differential vertical advection of the potential temperature field. Situations exhibiting a gradient of diabatic heating or cooling can also produce frontogenesis or frontolysis, including radiational, boundary layer, and thermodynamic processes.

The mechanisms in the simplified 1D definition of frontogenesis presented here can be extended to a 2D framework. For a pattern of isentropes in a deformation flow, whether frontogenesis or frontolysis results will depend on the angle of orientation between the isotherms and the *axis of dilatation* of the deformation flow, as indicated in Fig. 6.9. If the isotherms are oriented within an angle of 45° of the axis of dilatation, then the flow is frontogenetical and the isotherms will tend to collect along this axis.

Many meteorological software packages can calculate and plot various measures of the frontogenetical function, often only including the shear and confluence terms. Such graphics can be useful in weather forecasting to determine if sufficient frontogenetical forcing is present to generate a zone of ascent, possibly resulting in clouds and precipitation. This is also an accurate and objective means of tracking frontal systems, even in situations where the basic variables present complex signals. The horizontal

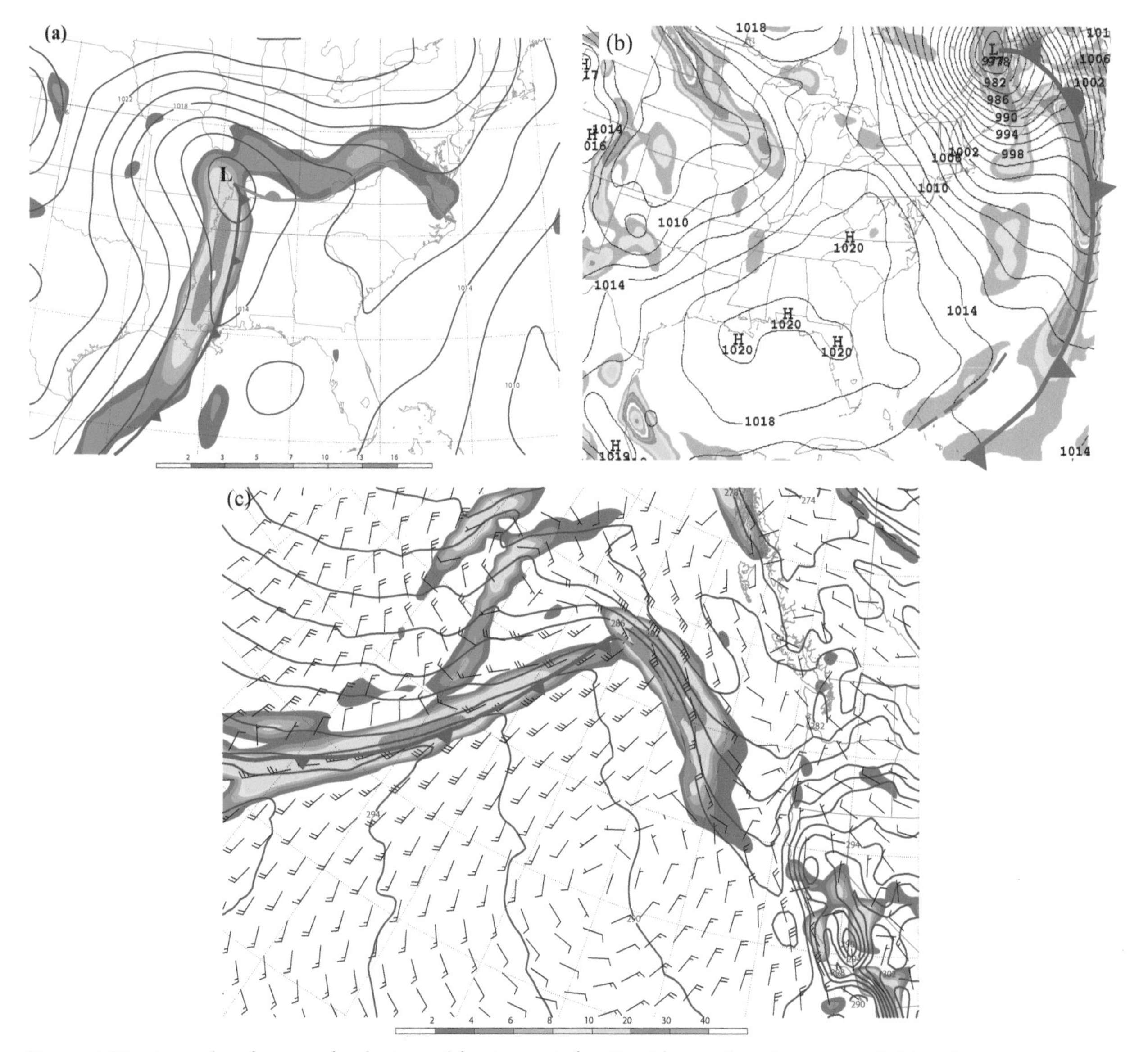


**Figure 6.10.** Examples of near-surface horizontal frontogenesis function (shear and confluence terms): (a) sea level pressure, subjectively analyzed fronts, and frontogenesis (shaded as in legend, beginning at 3 K $(100\ \mathrm{km})^{-1}$ $(3\ \mathrm{h})^{-1}$), corresponding to Fig. 6.1; (b) as in (a) but corresponding to Fig 6.8d, and frontogenesis shade interval is 1 K $(100\ \mathrm{km})^{-1}$ $(3\ \mathrm{h})^{-1}$; (c) 10-m wind (barbs) and 1000-mb potential temperature and frontogenesis (shaded as in legend) from Global Forecast System (GFS) analysis valid 0000 UTC 14 Nov 2008.

frontogenesis field corresponding to the fronts shown in Figs. 6.1 and 6.8d, along with an example from the North Pacific Ocean from November 2008, are shown to provide examples of how this diagnostic relates to the basic pressure and potential temperature fields (Fig. 6.10). In Fig. 6.10c, the pattern of frontogenesis is useful in locating a warm front approaching Washington State, as well as a trailing cold front extending to the southwest. The wind field to the north of the cold front exemplifies a classic pattern of confluent frontogenesis.

## 6.3. DYNAMIC FRONTOGENESIS

In discussing the QG material in chapter 2, it was emphasized that geostrophic advections can disrupt thermal wind balance. Ageostrophic (horizontal and vertical) motions come about in response, which serve to bring the atmosphere back toward thermal wind balance. Because of the enhanced temperature gradients present in frontal zones, we would expect areas where air parcels are entering or exiting a frontal zone to experience a potentially rapid change in temperature gradient and thermal wind balance disruption. In fact, we can relate the frontogenesis expression (6.2) to the **Q** vector. Frontogenesis is associated with a disruption of thermal wind balance and therefore an ageostrophic "frontal circulation" arises. The upward vertical branch of this circulation can be characterized by rising air, clouds, and precipitation. However, because of the subsynoptic cross-front length scale, we cannot expect QG theory to provide an adequate description of frontal dynamics.

In the previous section, our *kinematic* treatment of frontogenesis was useful in describing the mechanisms and flow patterns that can lead to frontogenesis, but this approach failed to account for the *interaction* between changes in the temperature gradient and the ageostrophic frontal circulation itself (in essence, the frontal dynamics). In reality, this interaction between the thermal field and ageostrophic circulation is important. This concept fits within the context of the QG system set up in chapter 2; however, as discussed above, in the cross-front direction, fronts are characterized by a length scale that is an order of magnitude smaller than the typical synoptic length scale. The Rossby number, $U/fL$, which is a measure of geostrophy, will therefore scale as 1 or more in the cross-front direction, making it clear that the QG system, which requires a Rossby number of order 0.1, does not apply in the cross-front direction in frontal zones. This is consistent with the frontal "secondary" circulation serving to influence frontal intensity through ageostrophic advections. Recall that ageostrophic advections are neglected in the QG system, consistent with a small Rossby number. A modified form of QG theory, known as *semigeostrophic* (SG) theory was developed specifically to diagnose the dynamics of frontal zones. To illustrate the basic idea, we will build from the earlier QG equations and modify them for relevance to fronts.

### 6.3.1. Frontal circulations

A convenient framework for the representation of ageostrophic frontal circulations will be presented here, based on the *Boussinesq approximation* with height as the vertical coordinate. With the Boussinesq approximation, density is treated as a constant, except in computing the buoyancy term. The atmosphere is separated into a *basic state*, which can be taken as a long-term time average of midlatitude conditions, and deviations from this state. The basic state will be a function of height only, with perturbations defined with respect to this state as follows:

$$\phi' \equiv \frac{p-p_0(z)}{\rho_0}; \quad \theta' = \theta - \theta_0(z). \tag{6.3}$$

In (6.3), note that $\phi'$ has units of m$^2$ s$^{-2}$, and that the geostrophic wind relations take an analogous form to those in pressure coordinates: $u_g = -1/f\,(\partial\phi'/\partial y)$, $v_g = 1/f(\partial\phi'/\partial x)$. We retain front-relative coordinates as for the kinematic frontogenesis equation, with $u$ ($v$) representing the along (across) front wind component. The frictionless momentum equations, similar to (1.5) and (1.6) but in Boussinesq form, are given by

$$\frac{du}{dt} - fv + \frac{\partial\phi'}{\partial x} = 0 \quad \text{and} \tag{6.4}$$

$$\frac{dv}{dt} + fu + \frac{\partial\phi'}{\partial y} = 0. \tag{6.5}$$

A convenient thermodynamic variable is the buoyancy, which is related to the potential temperature deviation from the basic state:

$$b \equiv \frac{g\theta'}{\theta_0(z)} = \frac{\partial\phi'}{\partial z}. \tag{6.6}$$

Using (6.6), we can write the thermodynamic equation using $b$ as the thermodynamic variable, replacing the perturbation potential temperature:

$$\frac{db}{dt} = -w\frac{g}{\theta_{00}}\frac{d\theta_0}{dz} = -wN_0^2, \tag{6.7}$$

where $N_0^2$ is the basic-state Brunt–Väisälä frequency, and $\theta_{00}$ is the basic-state potential temperature at the surface, a constant. The Boussinesq continuity equation is given by

$$\frac{\partial u}{\partial x} + \frac{\partial v}{\partial y} + \frac{\partial w}{\partial z} = 0. \tag{6.8}$$

If we use a constant value of the Coriolis parameter $f$ in defining the geostrophic wind, then the geostrophic wind is nondivergent, and because the front-parallel wind is assumed to be geostrophic based on the smallness of the corresponding along-front Rossby number, we can further write

$$\frac{\partial u_g}{\partial x} + \frac{\partial v_g}{\partial y} = 0, \text{ and } \frac{\partial v_{ag}}{\partial y} + \frac{\partial w}{\partial z} = 0. \tag{6.9}$$

Based on the determination of the Rossby numbers in the across- and along-front direction, we can approximate the along-front flow as geostrophic; however, we cannot justify this assumption for the cross-front flow component. The cross-front thermal gradient is then in thermal wind balance with the vertical shear of the along-front geostrophic wind component:

$$f\frac{\partial u_g}{\partial z} = -\frac{\partial b}{\partial y}. \tag{6.10}$$

Our motive here is to obtain a diagnostic equation that describes frontogenetical forcing, and the structure of the cross-front ageostrophic circulation arising in response to this forcing. Because the cross-front temperature gradient is close to a state of thermal wind balance, changes in the strength of this gradient will lead to a disruption of balance, giving rise to ageostrophic circulations. The derivation of an equation that describes both the forcing and response in this situation, known as the *Sawyer–Eliassen equation*, is left as an exercise at the end of the chapter. Using the geostrophic wind relation to rewrite (6.4), and expanding the total derivative, yields

$$\frac{\partial u_g}{\partial t} + u_g\frac{\partial u_g}{\partial x} + v_g\frac{\partial u_g}{\partial y} + v_{ag}\frac{\partial u_g}{\partial y} + w\frac{\partial u_g}{\partial z} - fv_{ag} = 0. \tag{6.11}$$

In (6.11), which describes the evolution of the front-parallel ($u$) wind component, some of the familiar QG assumptions are evident, and the small Rossby number for this wind component allows us to approximate the front-parallel wind and advection by their geostrophic values. However, ageostrophic advections are also included, but only for the cross-front and vertical components. This is a critical departure from QG dynamics, in that the frontal circulation itself, which will feature a strong vertical and cross-front ageostrophic flow, is able to modify the temperature and momentum fields via advection. The corresponding form of the thermodynamic equation reflects similar considerations:

$$\frac{\partial b}{\partial t} + u_g\frac{\partial b}{\partial x} + v_g\frac{\partial b}{\partial y} + v_{ag}\frac{\partial b}{\partial y} + w\frac{\partial b}{\partial z} + wN_0^2 = 0. \tag{6.12}$$

It is useful to derive a frontogenesis equation analogous to (6.2) but for this Boussinesq system of equations. We can obtain such an equation by differentiating (6.12) with respect to $y$, using the chain rule on each of the product terms and rearranging:

$$\frac{d}{dt}\left(\frac{\partial b}{\partial y}\right) = \underbrace{-\frac{\partial u_g}{\partial y}\frac{\partial b}{\partial x} - \frac{\partial v_g}{\partial y}\frac{\partial b}{\partial y}}_{Q_2} - \frac{\partial v_{ag}}{\partial y}\frac{\partial b}{\partial y} - \frac{\partial w}{\partial y}\frac{\partial b}{\partial z} - \frac{\partial w}{\partial y}N_0^2, \tag{6.13}$$

where the first two right-hand terms represent the cross-front component of the **Q** vector, which is similar in form to the meridional component of **Q** given by (2.31). This equation can be compared directly to (6.2), and the terms indeed look familiar; the first right-hand term is shearing frontogenesis, except with the geostrophic flow replacing the full along-front wind, the combined second and third terms corresponding to confluence, with the last two terms representing tilting. Now, we differentiate (6.11) with respect to $z$, for reasons which will become apparent if they are not already. The resulting equation describes the time rate of change of the vertical shear of the along-front geostrophic wind component:

$$\frac{d}{dt}\left(f\frac{\partial u_g}{\partial z}\right) = \underbrace{-\frac{\partial u_g}{\partial y}\frac{\partial b}{\partial x} - \frac{\partial v_g}{\partial y}\frac{\partial b}{\partial y}}_{Q_2} + f\frac{\partial v_{ag}}{\partial z}\frac{\partial u_g}{\partial y} - f^2\frac{\partial v_{ag}}{\partial z} + \frac{\partial w}{\partial z}\frac{\partial b}{\partial y}. \tag{6.14}$$

To obtain (6.14), we utilize the nondivergence of the geostrophic wind, and the thermal wind relations to make the right-hand terms consistent with those in (6.13) where possible. It is interesting that $Q_2$ appears on the right side of both (6.13) and (6.14); this shows what was indicated in chapter 2, that geostrophic advections, as described by the **Q** vector, disrupt thermal wind balance; a term in both the shear and thermal gradient tendency equations quantifies this disruption.

Compare the left-hand side of (6.13) and (6.14). These terms represent the time derivatives of the vertical along-front geostrophic shear and the cross-front thermal gradient. These components constitute thermal wind balance (6.10). Given that the observed state of the atmosphere is close to a state of thermal wind balance, and that the smallness of the along-front Rossby number indicates that the along-front flow should be quasigeostrophic, it is reasonable to add (6.13) and (6.14) using (6.10) to eliminate the time derivatives. Essentially, we assume that a frontal circulation arises that will maintain thermal wind balance, despite the forcing described by $Q_2$, which serves to disrupt this balance. Thus, we obtain a diagnostic equation that includes a forcing term, describing how geostrophic advections may disrupt thermal wind balance, and an ageostrophic circulation, which is exactly the response needed to maintain balance. To assist in visualizing ageostrophic frontal circulations, we can introduce an ageostrophic streamfunction, $\psi_{ag}$, that is everywhere parallel to the ageostrophic flow in the $y$–$z$ plane,

$$v_{ag} = -\frac{\partial \psi_{ag}}{\partial z}; \quad w = \frac{\partial \psi_{ag}}{\partial y}. \tag{6.15}$$

Using (6.15) in the resulting sum of (6.13) and (6.14), we arrive at the Sawyer–Eliassen equation

$$\left(N_0^2 + \frac{\partial b}{\partial z}\right)\frac{\partial^2 \psi_{ag}}{\partial y^2} + f\left(f - \frac{\partial u_g}{\partial y}\right)\frac{\partial^2 \psi_{ag}}{\partial z^2} - 2\frac{\partial b}{\partial y}\frac{\partial^2 \psi_{ag}}{\partial z \partial y} = 2Q_2. \tag{6.16}$$

To illustrate a situation of frontogenetical forcing and response, we will first consider an idealized and then an actual case. In Fig. 6.11, the geostrophic flow, shown by red vectors, is acting to concentrate the background thermal gradient. If we evaluate the forcing term in (6.16) for this situation, recalling that $b$ is proportional to the potential temperature perturbation from (6.6) and thus $Q_2 \propto -\partial u_g/\partial y(\partial\theta'/\partial x) - \partial v_g/\partial y(\partial\theta'/\partial y)$, then we find that the first term is zero because $u_g$ is constant in $y$ and that the only contribution is due to $-\partial v_g/\partial y(\partial\theta'/\partial y)$, which is negative in the vicinity of the frontal zone (the leading negative multiplies two negative quantities).

One application of QG theory that can still be used in a qualitative sense in frontal analysis is to note the isentrope-normal component of the **Q** vector ($Q_n$): If $Q_n$ points toward warmer isentropes, this indicates that the geostrophic flow is acting in a frontogenetical sense, as in Fig. 6.10. In situations where $Q_n$ is directed toward colder isentropes, it indicates that the geostrophic flow is acting frontolytically.

It is clear that $Q_2$ is relatively small at the points indicated by dots on either side of the frontal zone, which indicates that **Q** vector convergence and forcing for QG ascent would be found on the warm side of the frontal zone, with descent on the cool side in the zone of **Q** vector divergence. This is what we would expect based on the QG arguments outlined in chapter 2: If the primary geostrophic flow is acting to intensify the thermal gradient, then a circulation that cools the warm air, through adiabatic expansion and ascent, and warms the cool air, through compression and descent, would oppose the effects of the primary

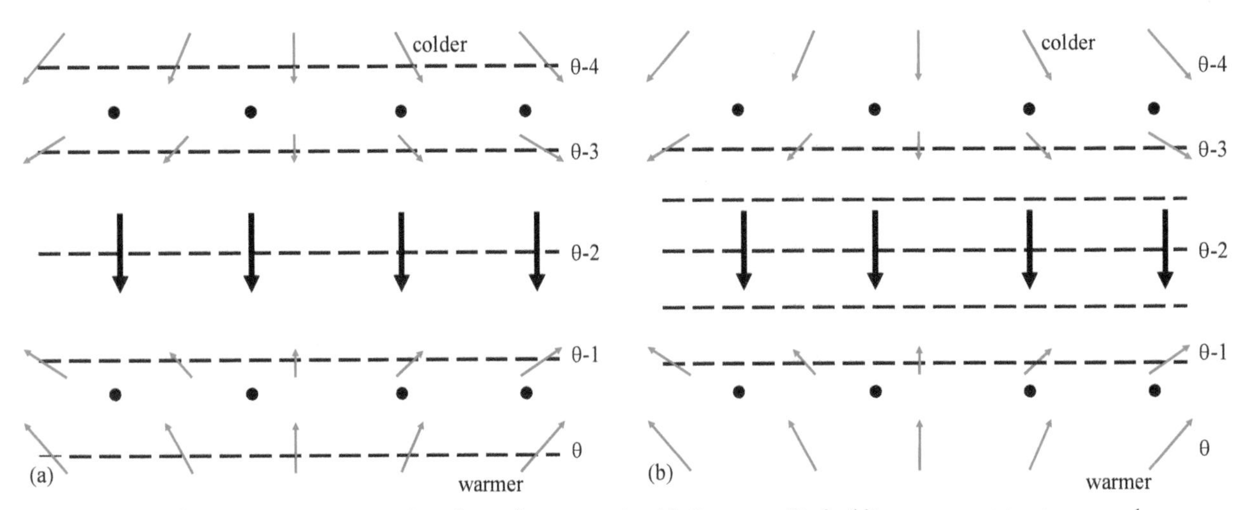


**Figure 6.11.** Schematic representation of confluent frontogenesis with **Q** vectors. Dashed lines represent isentropes, red arrows represent horizontal wind vectors, and thick arrows and dots represent **Q** vectors for (a) initial time and (b) later time. Dotted line A–B in (b) denotes location of cross section shown in Fig. 6.12.

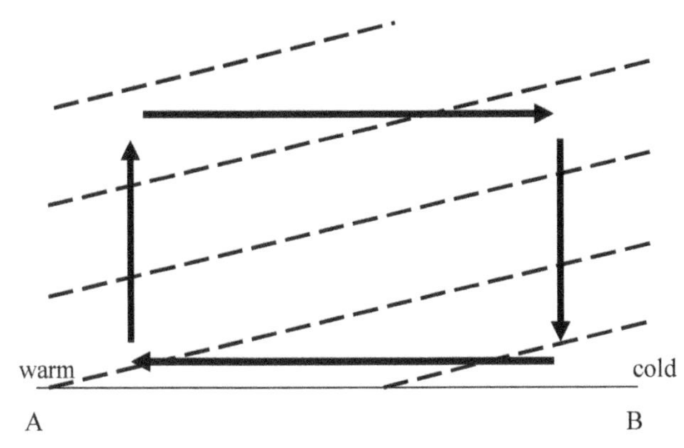


**Figure 6.12.** Schematic cross section of isentropes (dashed sloped lines) and ageostrophic circulation (thick arrows). Section orientation vertical across Fig. 6.11.

circulation and work in the sense needed for thermal wind balance restoration. In this case, the result is a thermally direct circulation, which is both expected and observed for frontogenetical flows. A cross-sectional view of this circulation is provided in Fig. 6.12. Unlike the QG equations, in the current framework, ageostrophic advections *are* accounted for in the cross-front direction. This means that the advective action of the secondary circulation, which is also acting frontogenetically near the surface, is important in the evolution of the front.

As the thermal gradient tightens, because of the frontogenetical forcing (as indicated in Fig. 6.11), the magnitude of $Q_2$ increases, because of a strengthened temperature gradient. This requires an even stronger ageostrophic frontal circulation, and a feedback is established. Why does the secondary circulation not succeed in offsetting the frontogenetical effects of the geostrophic forcing? Near the surface, the vertical motion is weak, meaning that the impact of adiabatic temperature changes is small there. Thus, there are several mechanisms working to intensify the front near the surface, including the primary geostrophic frontogenesis, described by the forcing term $Q_2$, as well as the ageostrophic cross-frontal circulation, which is advecting cold air toward the leading edge of the frontal zone, as evident in the lower branch of the circulation in Fig. 6.12. With no mechanisms to counteract the strong surface frontogenesis, rapid development of very intense fronts may take place. The processes described here also explain the observation that fronts are typically most intense near the surface and weaken with height; this situation arises because of the inability of the ageostrophic circulation to counteract the primary frontogenesis near the surface. Hoskins and Bretherton (1972) provide a theoretical treatment of this process and also demonstrate how a similar mechanism works in the formation of upper-level frontal zones in the vicinity of the tropopause.

The Coriolis torque on the upper and lower branches of the frontal circulation in Fig. 6.12 must also be considered. These act to increase the vertical shear of the geostrophic along-front wind component, and thus these branches of the circulation also serve to bring the flow in the frontal zone closer to thermal wind balance. In this example, the lower (upper) branch of the circulation is directed southward (northward), meaning that Coriolis torque would accelerate easterly (westerly) flow there. This combination acts to increase the vertical shear of the along-front geostrophic wind.

Can we observe this type of circulation using relatively coarse analyses or forecasts from numerical weather prediction (NWP) models? The dynamical equations in all modern NWP models are more complete and accurate than the semigeostrophic Boussinesq equations presented here and, provided sufficient resolution, these models are capable of predicting frontal circulations. In fact, for models to produce credible precipitation and temperature forecasts, fronts must be represented adequately. An oceanic case was selected, using data from a rather coarse-resolution (~95-km grid spacing) GFS analysis, to evaluate the extent to which a frontal circulation can be resolved using standard operational data. A strong deformation zone, with a pronounced potential temperature gradient, was oriented in a southwest-northeast sense across the eastern North Pacific (Fig. 6.13a). Examination of the 10-m wind field indicates confluent frontogenesis, with a classic deformation pattern evident on the cold side of the front and the apparent axis of dilatation aligning with the frontal zone. Shearing frontogenesis is also evident in the warm advection by the front-parallel wind component toward the warm side of the frontal zone. Farther northeast, immediately west of British Columbia, there is evidence of a warm front in the isentrope pattern. Plotting the frontogenesis function (as before, only the confluence and shear terms) confirms the location and intensity of these frontal zones (Fig. 6.13b). A weakness in the eastern end of the cold front, a feature known as a *frontal fracture*, is clearly evident in the frontogenesis field. Consideration of the 850-mb $Q_n$ vectors indicates that, as expected, the vectors are pointing toward the warm side of the front, consistent with frontogenesis in the primary geostrophic flow.

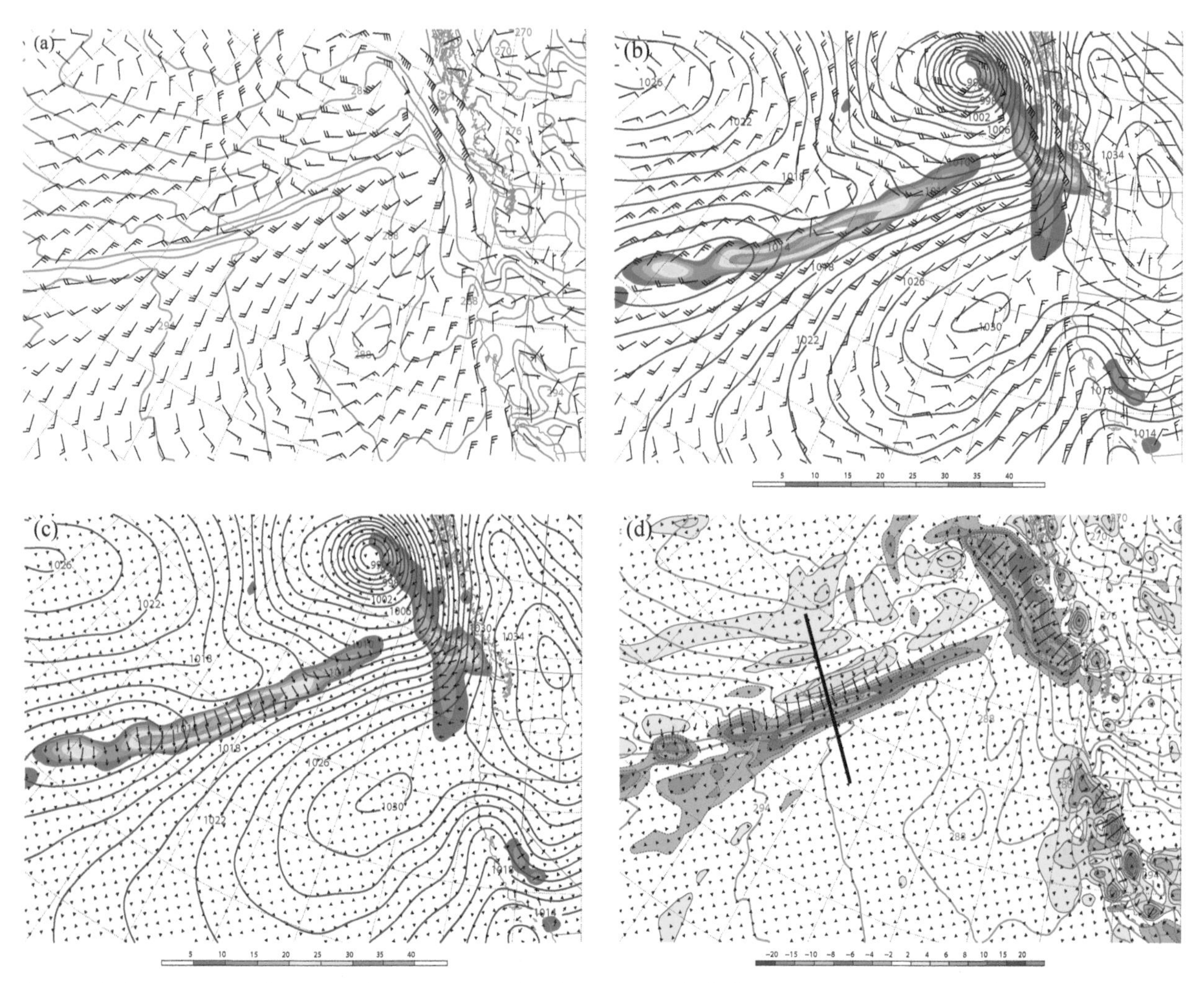

**Figure 6.13.** Diagnosis of North Pacific front with GFS analysis valid 1200 UTC 14 Nov 2008: (a) 10-m wind barbs, 1000-mb isentropes (red, interval is 2 K); (b) 10-*m* wind barbs, sea level pressure, 1000-mb horizontal frontogenesis (shade interval is 5 K (100 km)$^{-1}$ (3 h)$^{-1}$; (c) as in (b) but 850-mb $Q_n$ (black arrows) replaces 10-m wind; and (d) as in (a), but with 850-mb omega (contour interval is $2\times10^{-2}$ Pa s$^{-1}$, dashed for ascent, solid for descent, and shaded as in legend) and 850-mb $Q_n$ replacing 10-m wind.

The ageostrophic response, especially the vertical component, is of interest for forecasting cloud and precipitation patterns. As implied by **Q**-vector convergence, a thermally direct circulation, with rising warm air and sinking cold air, is anticipated. This is also the sense of the circulation depicted in Figs. 6.13d and 6.14. The use of the full omega, rather than the QG omega, implies that the frontogenetical forcing mechanism was dominant in determining the vertical motion in this case.

Given the importance of ageostrophic advections in the cross-front direction, we can compare what would happen in a QG framework to the semigeostrophic framework used here. Suppose that a numerical model based purely on the QG equations was run with a frontogenetical primary flow and compared to the results obtained running a semigeostrophic model. How would the comparison evolve? A cross section taken perpendicular to the front would, based on the frontogenesis of the geostrophic flow, lead to the strengthening of the cross-front temperature gradient, as indicated in Figs. 6.15a,b. The advection of isentropes by the ageostrophic secondary circulation is not accounted for in the QG model, and the front requires a substantial amount of time to intensify. In contrast, the semigeostrophic model, which includes the effects of ageostrophic advection in the cross-front direction, exhibits rapid intensification of the frontal zone, as well as a stronger slope of the frontal zone back toward the cold air, due to the tilting

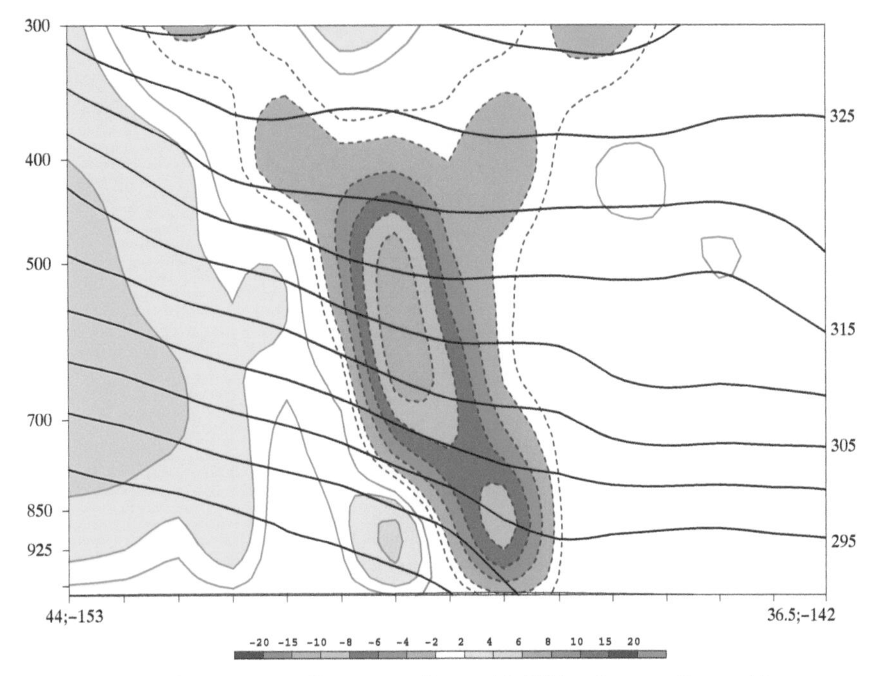


**Figure 6.14.** Cross section of isentropes (interval is 5 K) and omega (interval is $2\times10^{-2}$ Pa s$^{-1}$, dashed blue for ascent, solid red for descent, and shaded as in legend) for time and section indicated in Fig. 6.13d.

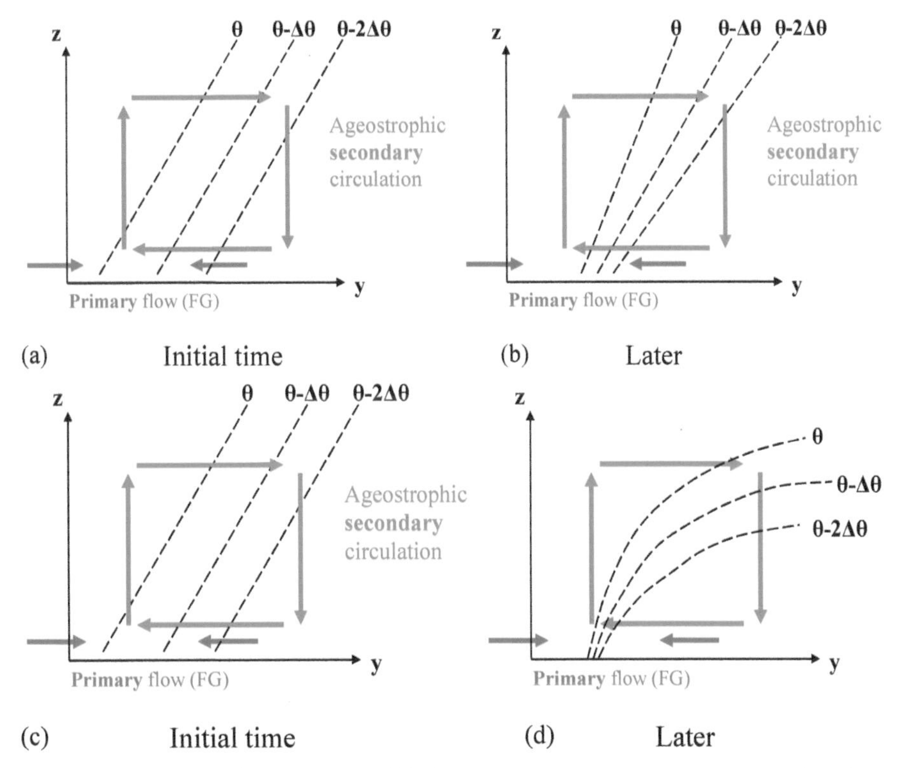


**Figure 6.15.** Idealized comparison of (a),(b) quasigeostrophic frontogenesis with (c),(d) semigeostrophic frontogenesis. Green arrows show primary geostrophic flow, red arrows represent ageostrophic frontal circulation, and dashed black lines are isentropes. Adapted from Bluestein (1986).

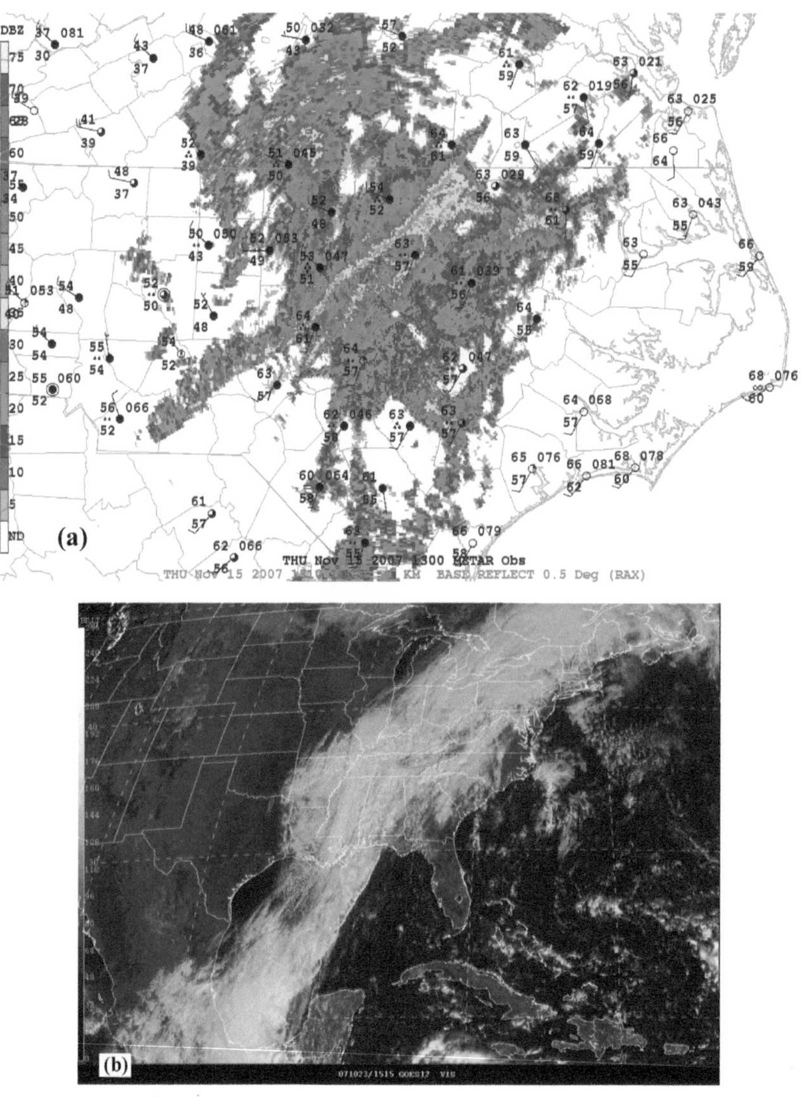


**Figure 6.16.** Observational examples of frontal collapse: (a) radar image depicting narrow cold-frontal rain band over North Carolina around 1300 UTC 15 Nov 2007 and (b) rope cloud over Gulf of Mexico at 1500 UTC 23 Oct 2007.

action of the ageostrophic circulation on the isentropes (Figs. 6.15c,d).

### 6.3.2. Frontal collapse

Based on the previous discussion, we can explain why fronts are able to develop rapidly; why they are often characterized by ascent, clouds, and precipitation; and why they are strongest at the surface. The frontogenetical action of the secondary frontal circulation near the surface leads to the formation of very sharp horizontal discontinuities there, limited only by mixing due to turbulence in the frontal zone. These discontinuities can be associated with a narrow cold-frontal rainband, which may be evident as a narrow band in radar imagery (Fig. 6.16a), or as a "rope cloud" in satellite imagery (Fig. 6.16b). However, observations of dry fronts also reveal such concentrated gradients at mature fronts. The observational studies of Sanders (1955) and Shapiro (1984) present examples of collapsed cold-frontal systems.

## 6.4. TYPES OF FRONTS

Now that we have reviewed the general properties of fronts and have examined the kinematic and dynamical processes responsible for their formation and evolution, it remains to illustrate examples of different types of frontal structure and associated weather. Our goal is to extend beyond a basic discussion of the main frontal types—to focus on some of the key structural differences leading to contrasting weather in their vicinity.

The basic frontal types—cold, warm, stationary, and occluded—are loosely defined based on the sense of local movement and temperature tendency in the frontal zone. Fronts that are moving slowly can be categorized as stationary fronts, and other fronts may require examination of sequential analyses to discern their movement. The mechanism of movement of fronts varies with type, but the frontal dynamics outlined in the preceding section are generally relevant to all types of fronts. Section 6.5 will be dedicated to middle- and upper-tropospheric frontal zones; in this section we will concentrate on fronts that are typically analyzed on surface weather maps but we will give attention to their vertical structure as well.

### 6.4.1. Cold fronts

Cold fronts are characterized by the clear advance of the cold airmass with time, and their stereotypical passage involves brief, violent weather in the form of showers and thunderstorms. However, observations demonstrate that there is tremendous variability in the weather accompanying cold fronts, ranging from dry, cloud-free frontal passages to heavy downpours with severe weather. Stronger fronts are not necessarily accompanied by more or heavier precipitation than weaker ones. Cold-frontal weather is sensitive to the pattern of airflow in the vicinity of the front, among other factors. The storm- (or front-) relative isentropic flow framework presented in chapter 3 lends itself to aiding the understanding of how airflow in a frontal system and the associated weather are related.

One important distinction between different types of cold fronts is the front-relative orientation of the *warm conveyor belt* (the main flow of warm, moist air that is typically located in the warm sector of a cyclone, ahead of the cold front; also discussed in section 5.3.6). The two basic cold-frontal structures are the *katafront* and the *anafront*; these terms are attributed to Bergeron (1937) and were employed by Sansom (1951) and several subsequent investigators. A katafront structure favors precipitation ahead of the surface front, whereas anafronts feature precipitation at and toward the *cold side* of the surface frontal boundary, with rearward-sloping ascent (Fig. 6.17). These

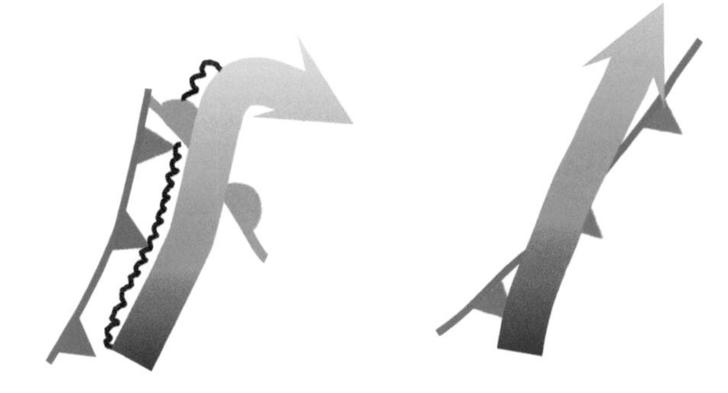


**Figure 6.17.** Idealized schematics of front-relative orientation of the warm conveyor belt accompanying different types of cold front: (a) katafront and (b) anafront. Broad shaded arrow indicates warm conveyor belt (adapted from Browning 1990).

differences in front-relative precipitation have obvious implications for weather forecasting.

#### 6.4.1.1. Katafront structure.

Cold fronts with a katafront structure are characterized by a lack of ascent in the warm conveyor belt along the surface cold front. Aloft, a push of dryer, cooler air often moves out ahead of the surface cold front. The leading edge of this low-$\theta_e$ or $\theta_w$ air is known as a *cold front aloft* (CFA), and the leading edge of dry and/or cold advection aloft may not be collocated with the surface frontal zone (Fig. 6.18). As a result of this structure, the actual surface cold front may pass with only light precipitation, or none at all, because of the limited vertical extent of the cloud layer, as indicated in Figs. 6.18b and 6.19. If the cloud-top temperatures in this shallow moist zone are not sufficiently cold to give rise to mixed-phase clouds, then light showers and drizzle may be found in this region.

Because of the presence of dry air above moist in a katafront, the vertical thermodynamic structure behind the cold-front aloft may be convectively unstable (characterized by $\theta_e$ decreasing with height). If a lifting mechanism is present, as is often found ahead of an upper-level trough centered west of the surface cold front, or the frontal circulation associated with the surface cold front, then convection can develop in this region in some circumstances.

#### 6.4.1.2. Anafront structure.

When the front-relative flow in the warm conveyor belt is oriented with a component directed toward the cold side of the frontal

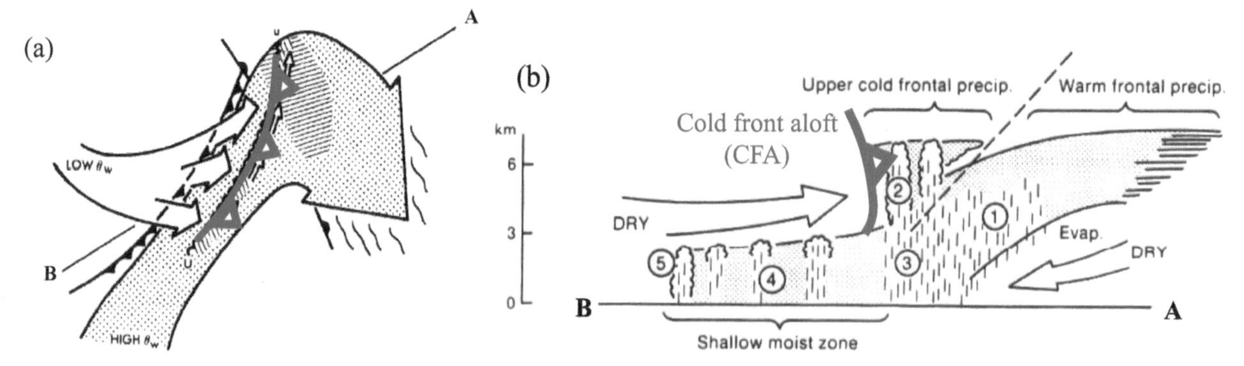


**Figure 6.18.** Schematic of a katafront or split-front structure with a cold-front aloft. (a) Plan view showing orientation of warm conveyor belt as large, shaded arrow, with surface fronts and indication of cool, dry low-$\theta_e$ air aloft given by hollow arrows. Line A–B indicates orientation of vertical cross section shown in (b). In (b), the numbers correspond to (1) warm-frontal precipitation, (2) convective cells ahead of the cold front aloft, (3) precipitation from upper cold front falling into lower warm-advection region, (4) shallow moist zone, and (5) light precipitation with surface cold front (from Browning and Monk 1982, their Fig. 5; Browning 1990, his Fig. 8.10).

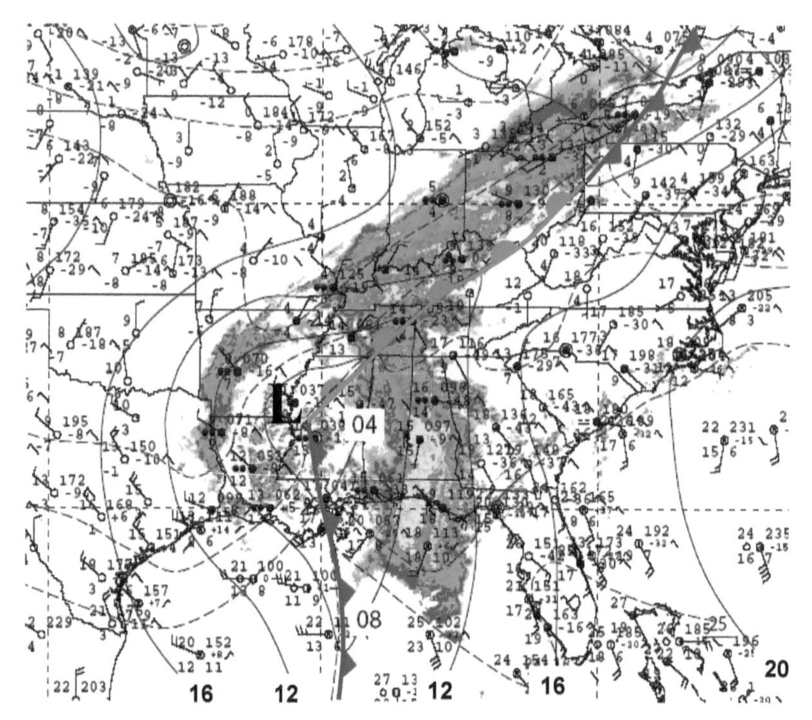

**Figure 6.19.** Example of katafront, including manual surface analysis for 1800 UTC 31 Dec 2002 with composite radar reflectivity mosaic (shaded) superimposed (from Mahoney and Lackmann 2007).

boundary, as indicated in Figs. 6.17b and 6.20, a zone of rearward-sloping ascent with lift over the cold air will develop. Because the precipitation is falling into the cold air, there is an increased risk for wintry precipitation during the cold season. In general, the prediction of weather conditions relative to the time of frontal passage is clearly sensitive to anafront–katafront structure. Anafronts are often accompanied by a southerly low-level jet (LLJ, see section 5.3.6) marking the warm conveyor belt, which typically attains maximum intensity immediately ahead of the associated cold front. A *narrow cold-frontal rainband* (NCFR) may be found in the immediate vicinity of

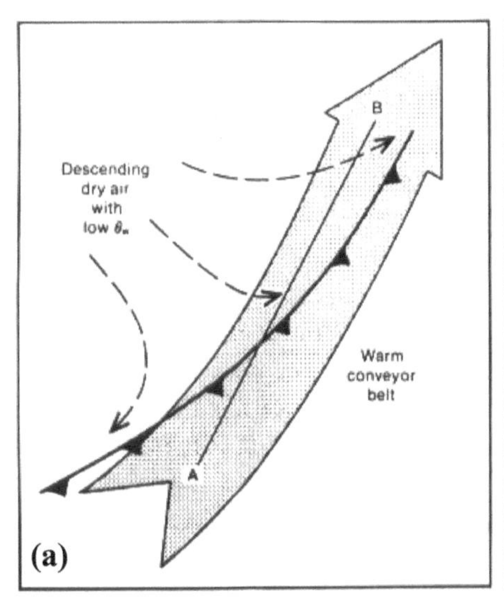


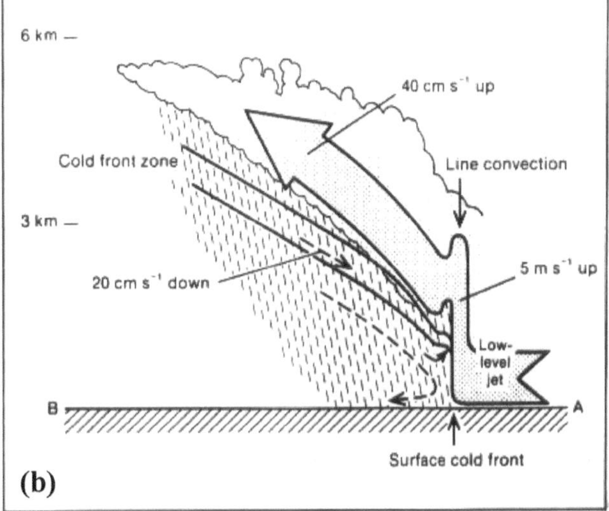


**Figure 6.20.** Schematic of anafront structure. (a) Plan view showing orientation of warm conveyor belt as large, shaded arrow, with surface cold front and indication of cool, dry low-$\theta_e$ air aloft given by dashed arrows. Line A–B indicates orientation of vertical cross section shown in (b) (from Browning 1990, his Fig. 8.8).

the frontal zone, with a swath of lighter stratiform precipitation located farther toward the cold side of the front.

Detailed observational studies form the basis of the idealized schematic shown in Fig. 6.21, which depicts a cross-sectional view of airflow and hydrometeors through an anafront. Regions of moderate precipitation can be found well behind the leading NCFR, as indicated here. An observational example of an anafront was shown in the satellite image Fig. 6.16b, where the zone of ascent and cloud are located exclusively to the west of the rope cloud, which marks the location of the cold front.

An early observational study by Sansom (1951) for the British Isles found generally heavier precipitation accompanying anafronts relative to katafronts, and also that the

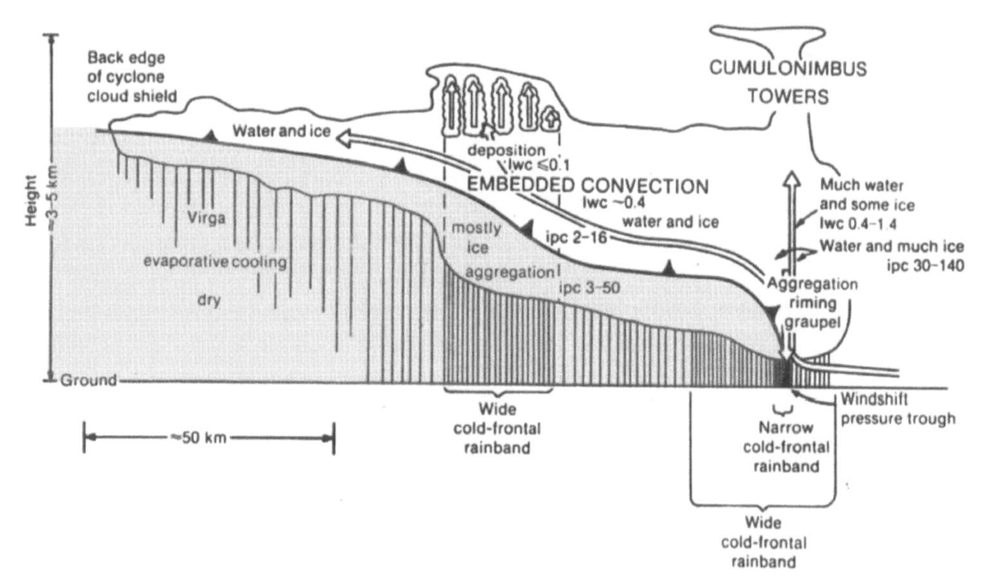


**Figure 6.21.** Idealized cross section of clouds and precipitation in an anafront. Vertical lines represent precipitation, and arrows represent airflow relative to front. Ice particle concentration (IPC) in pounds per liter, and liquid water content (LWC) is given in g m$^{-3}$ at select locations. Figure taken from Browning (1990), originally from Matejka et al. (1980).

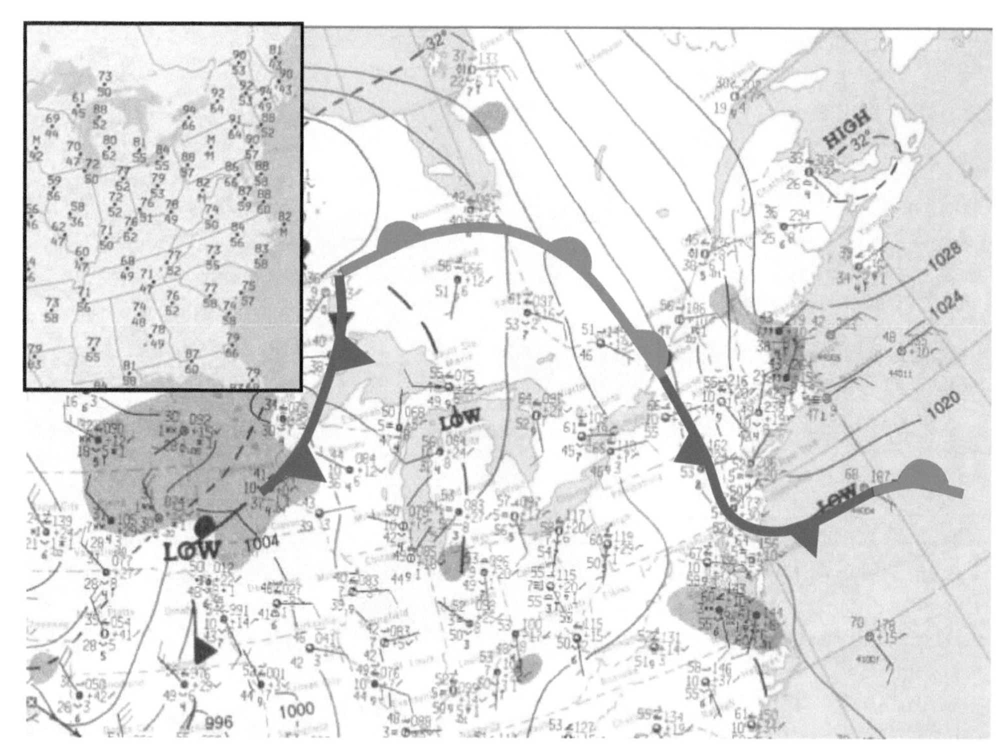


**Figure 6.22.** Surface analysis for New England backdoor cold front of 29 Apr 1990; analysis valid 1200 UTC. Surface observations in standard station model format; frontal analyses superimposed to emphasize eastern United States. Inset shows maximum and minimum temperatures (°F) for 28 Apr 1990. Maps obtained online via historic weather map series (at http://docs.lib.noaa.gov/rescue/dwm/data_rescue_daily_weather_maps.html).

two types mark the stages in the evolution of a front—the katafront structure is more common later in the frontal life cycle. Winds aloft were found to exhibit a smaller cross-front angle for anafronts relative to katafronts. In the author's experience, anafront structures are more often found to accompany lifting troughs, with katafront structures more common with digging troughs.

**6.4.1.3. Arctic fronts, "back door" cold fronts.** Not all cold fronts can be categorized as having either katafront or anafront structures, and in fact, some cold fronts may exhibit both of these structures along different portions of the frontal zone. A point of emphasis here is that there exist a wide variety of frontal structures and accompanying weather conditions, and it is not sufficient to simply look at a cold front analyzed on a surface map and make conclusions about what type of weather to expect during its passage.

In addition to the aforementioned structures, other cold-frontal structures that are sometimes observed include very shallow, intense *arctic fronts*. These features may not be accompanied by precipitation, but they are often characterized by very intense temperature gradients and a rapid drop in temperature during their passage. Owing to the very large density contrast associated with such fronts, they tend to be quite shallow, and their movement may be controlled more by local density contrasts than by larger-scale processes. There may be no semblance of a warm conveyor belt airstream anywhere in the vicinity of an arctic front, and thus they are not easily categorized as either a katafront or anafront.

In other situations, cold fronts may approach from unusual directions, such as the New England backdoor cold front shown in Fig. 6.22. In this situation, an airmass that had originated over the cool waters of the Gulf of Maine flowed southwestward, bringing to a sudden end a period of unseasonably warm New England weather. The maximum temperature of 90°F from 28 April 1990 in Boston must have seemed like a distant memory to those who attended the Boston Red Sox home baseball game against the Oakland Athletics on 29 April, experiencing temperatures in the 40s with drizzle and a brisk northeasterly wind!

### 6.4.2. Warm fronts

By definition, if the boundary of warmer air is advancing steadily, the front can be classified as a warm front; however, the mechanism of frontal movement for a warm front can differ from that of a cold front. The process of warm-frontal movement is more dependent on turbulent mixing than for cold fronts, resulting in a relatively gradual advance relative to cold-frontal cases. As the warm front approaches, the depth of the cold air may become very shallow, meaning that in zones of complex terrain, the hills and mountains may break into the warm air long before valleys and other low-lying areas. In some cases, the flow on the cold side of a warm front will begin to retreat in the direction of frontal motion (indicating that the process of advection is contributing to frontal movement); however, in many other cases, the winds immediately north of the warm-frontal boundary are light, and mixing is the primary mechanism of frontal movement.

Much research remains to be done regarding the mechanisms and prediction of warm-frontal advance, including studying the interaction between turbulent mixing and frontal movement. However, as discussed at the end of chapter 1, we know that given sufficient vertical wind shear and small static stability, turbulent mixing may develop. Diurnal heating may contribute to a more rapid frontal advance, because of enhanced mixing and reduced stability during the afternoon hours. Because the frontal zone itself is characterized by large static stability, the mixing and warm-frontal movement process can be quite slow.

Some warm fronts take on a very shallow slope, meaning that hydrostatically, we should not expect a pronounced frontal pressure trough relative to cold fronts. This circumstance can make warm fronts difficult to analyze, as the wind shift and pressure trough are two useful signatures in frontal analysis. In some cases, the most prominent trough in the sea level pressure field is found in the vicinity of the strongest warm advection at the 850-mb level.

A warm-frontal example from 24 Jan 2010 is presented in Fig. 6.23 to illustrate some of the warm-frontal properties discussed above. Warm air advancing northward from the Gulf of Mexico encountered a remnant cold air mass, leading to the formation of a warm front (Fig. 6.23a). This frontal boundary was aligned with the coast over the Carolinas, reminiscent of a coastal front, to be discussed in section 6.4.4. At the 850-mb level, the zone of strongest warm advection was located well to the north of the surface warm front over the Carolinas (Fig. 6.23b). A sounding taken at Peachtree City, Georgia, located to the north of the surface warm front, indicates a shallow layer of cold air near the surface, with a strong frontal inversion evident in the 925–850-mb layer. A strongly veering wind profile is observed over the layer corresponding to the frontal inversion, indicating that the strongest lift is confined to a shallow layer near the surface. This is also consistent with the relative-humidity profile in the sounding, as saturated conditions are found from the surface to near the 700-mb level, with dry air above. The shallow, above-freezing cloud layer is consistent with the lack of precipitation observed to the north of this front, as the absence of a mixed-phase cloud reduces the chances of precipitation development. Farther north, where deeper ascent was taking place, rain was observed.

As with cold fronts, warm fronts can sometimes display unorthodox orientations. In the example displayed in Fig. 6.24, relatively warm maritime polar air is advecting west and south across Quebec and the Hudson Bay region, replacing colder continental polar air in its path. It is important to bear in mind the consistency between frontal location and movement and earlier concepts regarding temperature advection. Overlays of height and/or pressure plotted at different vertical levels demonstrate the sense of temperature advection, as discussed at length in chapters 1 and 2. Figure 6.24b presents the sea level pressure along with the 500-mb height, allowing diagnosis of the sense of the lower-tropospheric thermal advection.

The region of strong veering in the vicinity of the warm front confirms the expected sign of the thermal advection. Thermal advection can provide critical information for use in frontal analysis, particularly for unconventional cases such as that presented in Fig. 6.24.

### 6.4.3. Occluded fronts

**6.4.3.1. The occlusion process.** In our discussion of cyclones in chapter 5, we noted how cyclogenesis is favored along preexisting frontal boundaries, in part because of the enhanced effectiveness of vortex stretching in the presence of a vorticity-rich lower frontal zone. However,

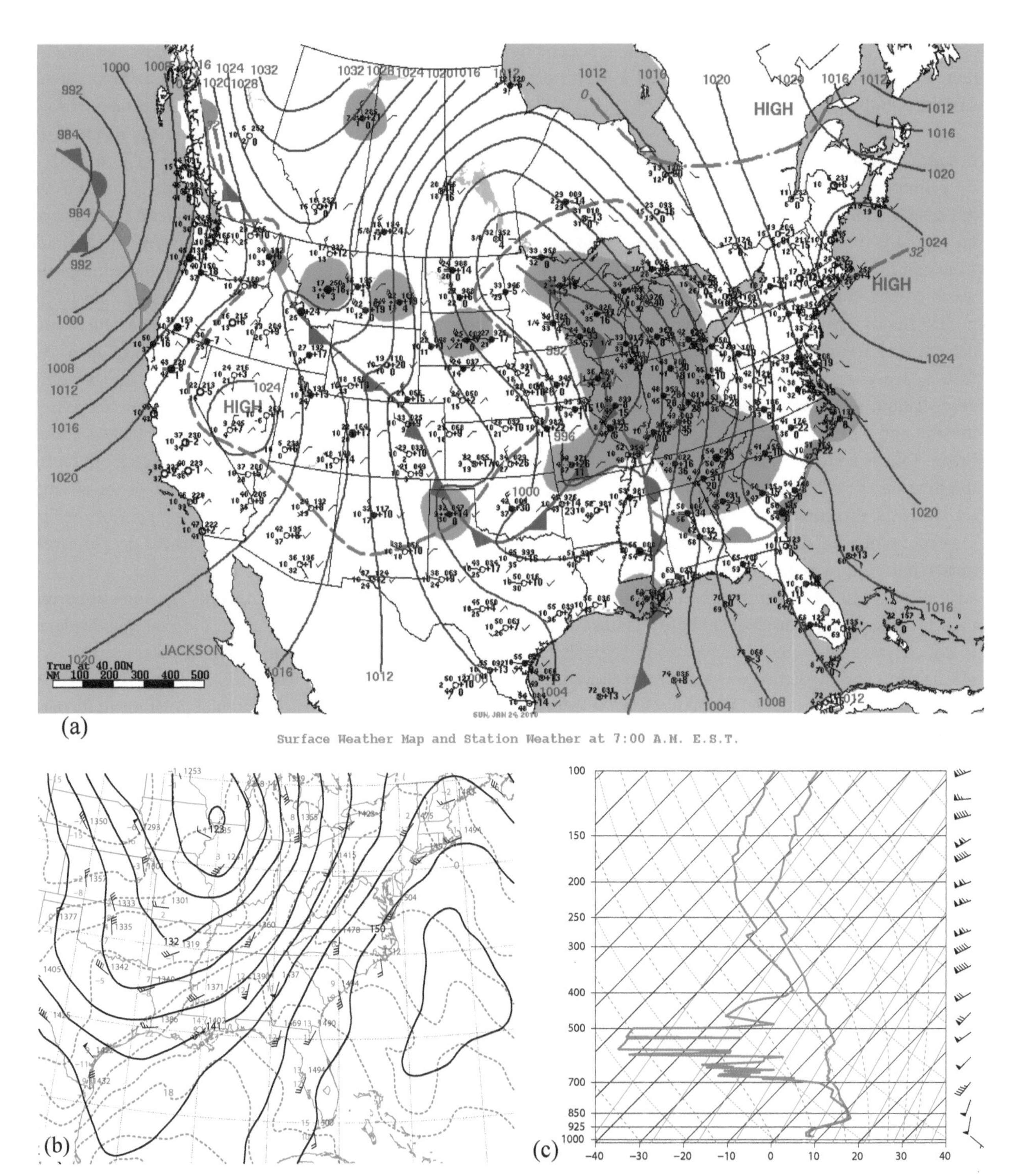


**Figure 6.23.** Diagnostics of a warm front from 1200 UTC 24 Jan 2010 over the southeastern United States: (a) HPC surface analysis showing isobars (contour interval is 4 mb), fronts (standard symbols), precipitation (green shading), and surface observations (standard station model plot); (b) 850-mb height (contour interval 3 dam), isotherms (red dashed contours, interval 3°C), and rawinsonde observations (upper-air station model format); and (c) skew $T$–log$p$ diagram for Peachtree City, GA (Falcon Field Airport; FFC).

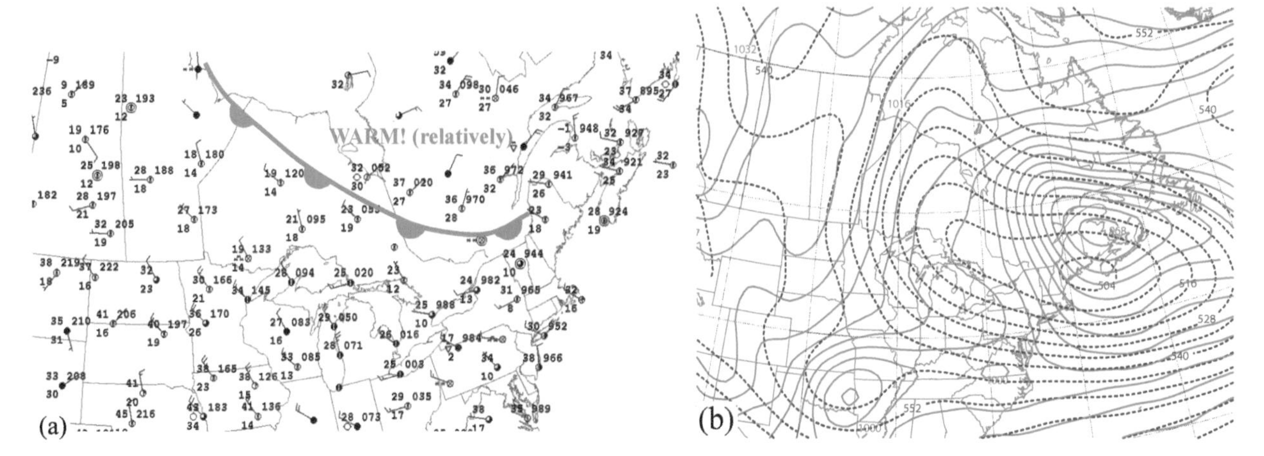


**Figure 6.24.** Example of a backdoor warm front from 1800 UTC 19 Mar 1981 (a) surface observations and subjective frontal analysis and (b) sea level pressure (red contours, interval 2 mb) and 500-mb height analysis (blue dashed contours, interval 6 dam) from 1800 UTC Mar 1981.

we did not discuss the typical frontal evolution accompanying a cyclone life cycle, and we will do so now. The Bergen school meteorologists, as explained in section 5.4, are generally recognized as being the first to assemble a coherent picture of the evolution of fronts over the life cyclone of an extratropical cyclone. The *Norwegian cyclone model*, a schematic of which is presented in Fig. 6.25, describes the amplification of ån initially small frontal wave into a full-fledged cyclone, and ultimately, as the cold front overtakes the warm front from the west, a dissipating occluded cyclone. During the course of this *occlusion process*, it was envisioned that the warm sector air is lifted above the surface, leaving a zone of weakened thermal contrast at the surface, marking the occluded front. A more recent synthesis of extensive field experiment data and high-resolution numerical simulations has resulted in the modification of the earlier Bergen cyclone life cycle, presented in the *Shapiro–Keyser* frontal-cyclone conceptual model (Fig. 6.26). Here are the key points from a comparison of these two schematic models:

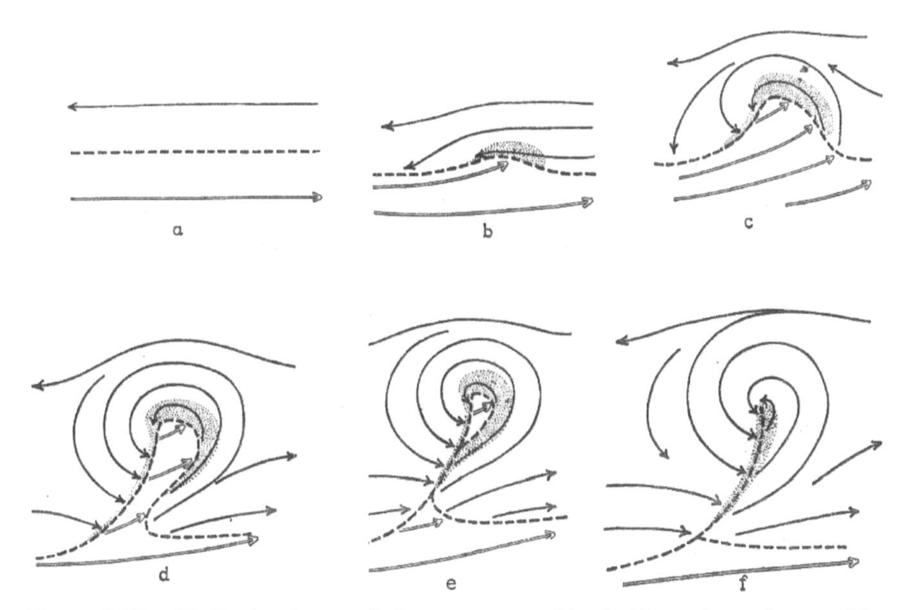


**Figure 6.25.** Idealized cyclone evolution as represented by the Norwegian cyclone model, shows progression with time. Solid arrows depict surface streamlines; dashed lines indicate locations of frontal boundaries (from Bjerknes and Solberg 1922).

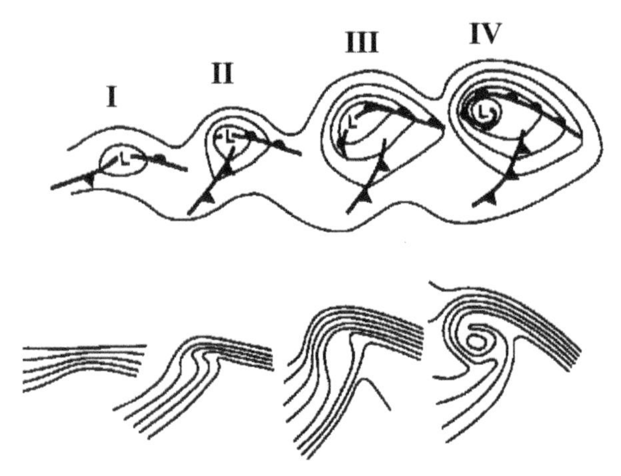


**Figure 6.26.** Life cycle of a maritime cyclone: I) incipient, II) frontal fracture, III) bent-back warm front, and IV) warm-air seclusion stage. (top) Depiction of sea level pressure, areas of stratiform cloud (shaded), and fronts; (bottom) display of isotherms and streamlines (from Schultz et al., 1998, adapted from Shapiro and Keyser, 1990).

- in both models, initial cyclone development takes place along a stationary front;
- as the frontal wave amplifies, cold air retreats to the east and advances to the west;
- when the occluded front develops, the cyclone is near peak intensity; and
- local topographic features can cause departures from the idealized evolutions.

Some fundamental aspects of the occlusion process continue to be debated. The original Norwegian cyclone model explained occlusion as a faster-moving cold front overtaking a warm front. However, this view fell out of favor, with several prominent studies stating that the catch-up process did not occur; for example, the Wallace and Hobbs (1977) textbook states:

> ...there are few, if any, well-documented examples of cold fronts overtaking warm fronts to form occlusions. Rather, it appears that most occluded fronts are essentially new fronts which form as surface lows separate themselves from the junctions of their respective warm and cold fronts and deepen progressively further back into the cold air.

In an effort to settle the issue, Schultz and Mass (1993) undertook a careful analysis of the occlusion process and found that frontal catch-up *was* observed, at least in some cases (Fig. 6.27).

The cold-frontal catch-up evident in Fig. 6.27 is likely influenced in this case by the presence of the Appalachian Mountains, which are retarding the movement of the warm front to their east, a phenomenon known as cold-air damming (CAD, to be discussed in depth in chapter 8). The presence of the Scandinavian Mountains in Norway may also have contributed to the southerly location of the warm front evident in the schematic 6.25d, similar to what was observed in the Schultz and Mass (1993) study for the Appalachian Mountains. Several other recent studies have also documented the catch-up mechanism in observations and model simulations of the occlusion process (e.g., Market and Moore 1998; Martin 1998).

Recent work by Schultz and Vaughan (2011) summarizes recent findings relating to the occlusion process.

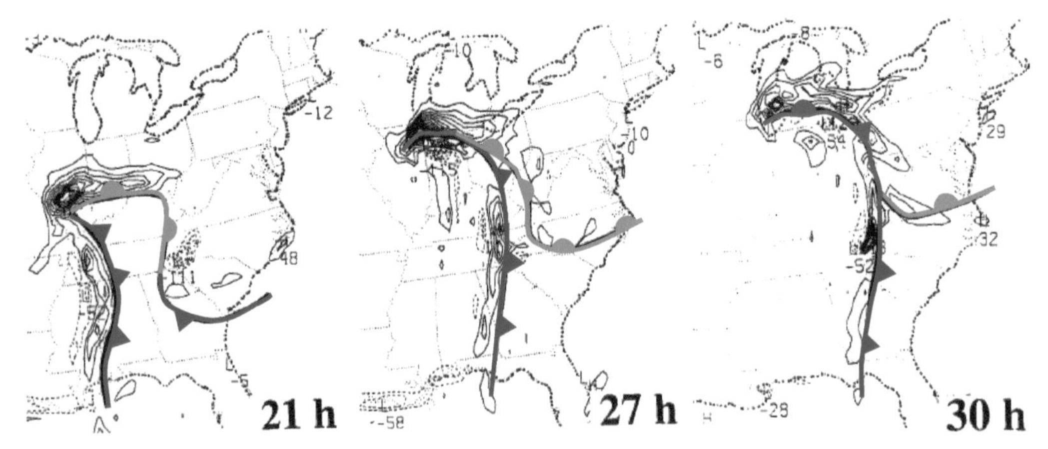


**Figure 6.27.** Surface frontogenesis computed using output from a numerical model simulation of an occluding cyclone at 21, 27, and 30 h into the simulation. Contour interval is $0.2 \times 10^{-8}\ \mathrm{K\,m^{-1}\,s^{-1}}$, or approximately $2°\mathrm{C}(100\,\mathrm{km})^{-1}\,(3\mathrm{h})^{-1}$. Subjective frontal locations are also shown [from Schultz and Mass (1993), their Fig. 14].

Several misconceptions regarding this process are evident in the literature, including textbooks. Although it is true that cold fronts move faster than warm fronts, there is more to the occlusion process than the frontal catch-up mechanism. Studies of isentrope evolution in idealized barotropic flows (in which potential temperature is treated as a passive tracer) exhibit filamentation of the thermal field that strongly resembles the occlusion process, even when the warm and cold fronts are moving at the same speed. Another apparent misconception is that cyclones are at peak intensity at the time of occlusion. Several studies have documented cases where cyclones continued to deepen for 24 h or more after the occlusion process began (e.g., Sanders 1986; Reed and Albright 1997; Martin and Marsili 2002).

**6.4.3.2. Warm- and cold-type occlusions.** In discussing the occlusion process depicted in Fig. 6.25, Bjerknes and Solberg (1922) present two types of occluded frontal structure, which depends on the relative temperature of the pre- and postfrontal air masses. If the air behind an occluded front is colder than the air ahead of the boundary, then a "cold type" occlusion structure will result (Fig. 6.28b). Conversely, warmer postfrontal air leads to a warm-type occlusion structure (Fig. 6.28a).

Schultz and Mass (1993) scoured the literature to find examples of warm- and cold-type occlusions but could find no clear examples of cold-type occlusion structures, at least

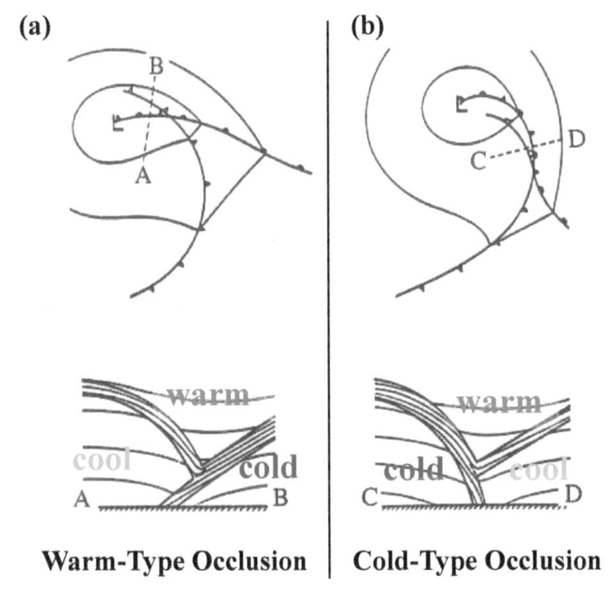


**Figure 6.28.** Cold- and warm-type occlusion structures (from Schultz and Mass 1993).

as presented in idealized schematics. A more recent investigation into this question by Stoelinga et al. (2002) indicates that it is the static stability, rather than the temperature contrast, that appears to determine the occluded structure. The tendency for greater static stability ahead of warm fronts (which are characterized by shallow boundary layers and cloud cover) relative to behind cold fronts (which are characterized by deeper boundary layers and often clearing skies) could explain the preponderance of warm-type occlusions in observations (Stoelinga et al., 2002). However, additional research is required to resolve the question of the existence of cold-type occlusions.

**6.4.3.3. Instant occlusion.** A different type of frontal structure that bears the name *occlusion* is the nonclassical frontal evolution referred to as *instant occlusion*. During these events, an occluded-like frontal structure develops through a mechanism that differs from that described in the previous subsection. Rather, as an upper-air disturbance located well to the cold-air side of an existing surface front approaches a preexisting frontal zone, the cloud band associated with the upper disturbance merges with that from the initial front to form an occluded structure (Fig. 6.29d). Satellite imagery is useful in the identification of these events, and examples can be found easily through routine examination of imagery, especially in maritime regions. The schematic classification scheme presented in Fig. 6.29 is based largely on examination of satellite imagery of marine cyclogenesis.

### 6.4.4. Coastal fronts

A common feature along the southeastern U.S. coast, the Texas coast, and in coastal New England is a frontal zone that forms preferentially along the coast. These *coastal fronts* form in place along the zone of strong temperature contrast between land and water and can play an important role in a variety of weather phenomena. The mechanisms that favor the development of coastal fronts are in part related to the gradient in surface properties found at the coastline. For instance, coastal fronts, with preexisting vorticity, can play a major role in the track and formation of coastal cyclones. The importance of such fronts was emphasized in studies of the Presidents' Day blizzard of 1979 by Bosart (1981), and this work helped to motivate subsequent field programs, such as the Genesis of Atlantic Lows Experiment (GALE) in 1986. Based on these field programs and other observational and modeling studies,

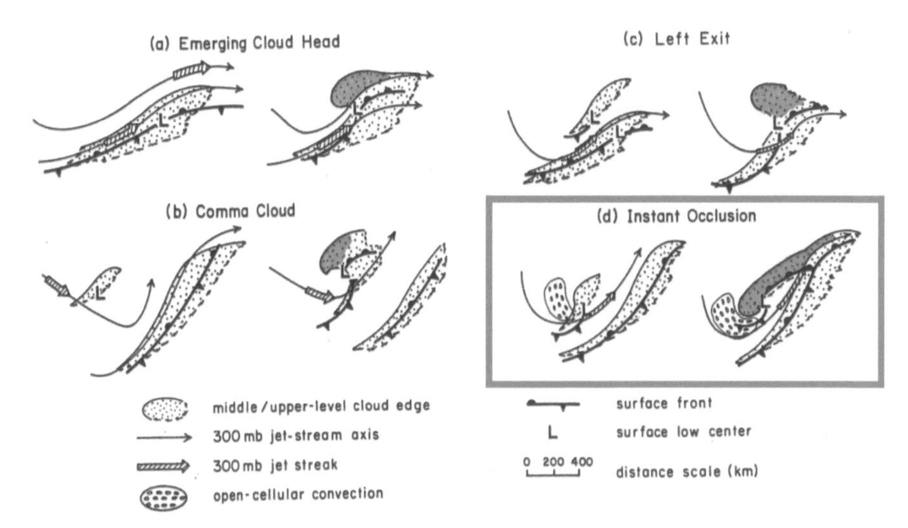


**Figure 6.29.** Satellite-based schematic representation of four types of cyclogenesis presented by Evans et al. (1994): (a) emerging cloud head, (b) comma cloud, (c) left-exit, and (d) instant occlusion.

the characteristics of coastal fronts have been shown to include the following:

- the presence of a shallow front separating cold continental air from warm maritime air;
- most common occurrence during the cold season (November–March) and along concave coastlines (e.g., New England, Carolinas, Texas);
- quasistationary or warm-frontal structure and movement, often migrating inland with time;
- heaviest precipitation on cold side of front due to enhanced isentropic lift there, especially with an extratropical, or a tropical cyclone, located to the south;
- coastal fronts often form the eastern boundary of a cold-air damming event (discussed in depth in chapter 8);
- an "inverted trough" in the sea level pressure pattern can accompany the coastal front;
- coastal fronts may serve as the site of primary or secondary cyclogenesis because of the presence of a lower-tropospheric zone of enhanced preexisting vorticity; and
- coastal fronts can be accompanied by convection, even severe weather in some cases.

Consider a north–south-oriented coastline as shown in Fig. 6.30 during a period of northeasterly geostrophic flow, such as would take place with an anticyclone located to the north.

Considering the frontogenesis mechanisms discussed in connection with equation (6.2), the situation described by Fig. 6.30 would clearly exhibit strong confluent frontogenesis. What is unique is that the confluence is enhanced by a mesoscale land-breeze circulation, which owes its existence in part to differential diabatic heating in the coastal zone. Especially for coastlines characterized by warm boundary currents, such as the Gulf Stream in the southeastern United States, there is often a positive temperature tendency due to diabatic processes over the warm water. These would include upwelling infrared radiation from

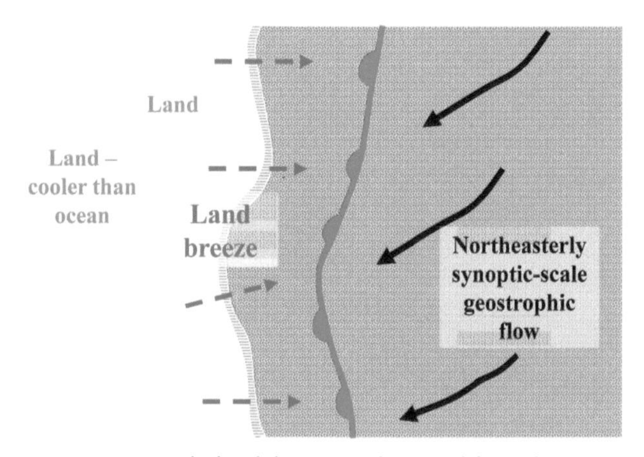


**Figure 6.30.** Idealized depiction of a coastal front along a north–south-oriented coastline. Green dashed arrows indicate the near-surface flow associated with a land-breeze circulation; black solid arrows represent a synoptic-scale northeasterly flow.

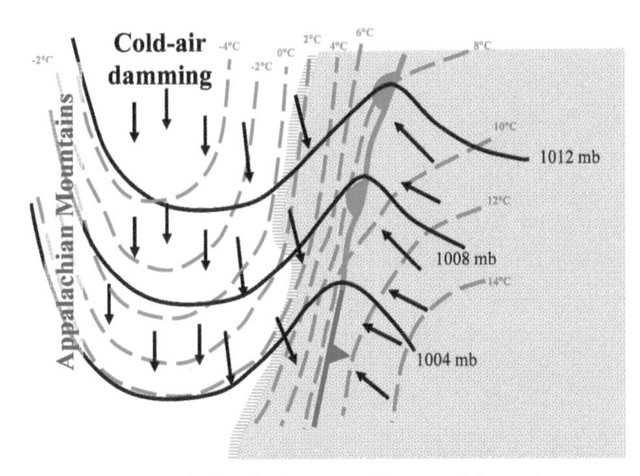


**Figure 6.31.** Idealized schematic of a coastal front along the U.S. southeastern coast, with CAD to the west of the front. Black arrows represent near-surface winds, red dashed lines are isotherms, and black solid lines are sea level isobars.

the warm water surface, along with the turbulent transfer of latent and sensible heat to the overlying atmosphere. If the situation corresponds to an overnight period, then radiational cooling over the land at night would further enhance the cross-front heating gradient. Based on these arguments, one might expect a diurnal signal in the intensity and movement of coastal fronts. Indeed, climatological studies have shown that many coastal fronts are in fact diurnal phenomena (Nielsen 1989; Appel et al. 2005).

Another factor that can contribute to coastal frontogenesis along the U.S. East Coast is the presence of the Appalachian Mountains to the west. This mountain barrier can be associated with CAD (discussed in chapter 8), during which cold air pushes southward to the east of the mountains. This can further enhance baroclinicity in the coastal front, as shown in Fig. 6.31.

The presence of ageostrophic mountain-parallel cold advection associated with CAD is consistent with shearing frontogenesis. The cold dome to the west of the coastal front in this example can contribute to frontogenesis in yet another indirect way—the presence of sloping isentropes within the cold dome, and southeasterly or southerly flow above it, can give rise to clouds and precipitation, which can cool the cold side of the front by evaporation and via sheltering the cold air from solar heating.

Changes in the cross-isobar angle of boundary layer winds at coastlines (due to differential frictional roughness) and also the shape of a coastline can work to favor frontogenesis there as well, but these factors may play a secondary role compared to those processes discussed above.

The concept of frontogenesis in the primary geostrophic flow with an ageostrophic response requires some modification in the case of coastal fronts. The land breeze pictured in Fig. 6.30 is typically a mesoscale ageostrophic flow, as is cold advection near the mountain barrier during cold-air damming. However, these frontogenetical effects can still be interpreted as primary forcing, and an ageostrophic response in the form of a frontal circulation may still result. In fact, the thermally direct ageostrophic circulation that would arise in response to coastal frontogenesis would work to reinforce the land-breeze circulation.

Forecasting challenges associated with coastal fronts are in part related to their movement. In the southeastern United States, when a coastal front moves inland, rapidly changing weather conditions frequently accompany its passage. Studies of the large-scale synoptic patterns that favor onshore migration of the coastal front, versus patterns associated with fronts that remain offshore, demonstrate the importance of strong synoptic-scale onshore geostrophic flow for cases that feature inland movement (Fig. 6.32). For coastal-front events in which the poleward surface anticyclone shifts eastward, an onshore geostrophic wind component develops, favoring the inland movement of the coastal front in the southeastern United States.

## 6.5. MIDDLE- AND UPPER-LEVEL FRONTS

Mature frontal zones are often most intense near the surface and weaken with height; however, as discussed previously, intense upper-level fronts can form as well (e.g., Fig. 6.33). Large-scale patterns characterized by frontogenetic confluence and shear are often associated with upper-level fronts; however, unlike the dynamics of frontal zones near the surface, tilting can also play a major role in contributing to the strength of upper fronts. Initially, there was some debate in the literature concerning the question of the secondary circulation accompanying upper fronts. Some studies had documented thermally direct frontal circulations, while others had observed the strongest subsidence on the warm side of the front in a frontogenetic, thermally indirect circulation (e.g., Reed and Sanders 1953; Bosart 1970). The controversy was partly resolved using numerical simulations; with along-flow cold advection, Keyser et al. (1986) demonstrated

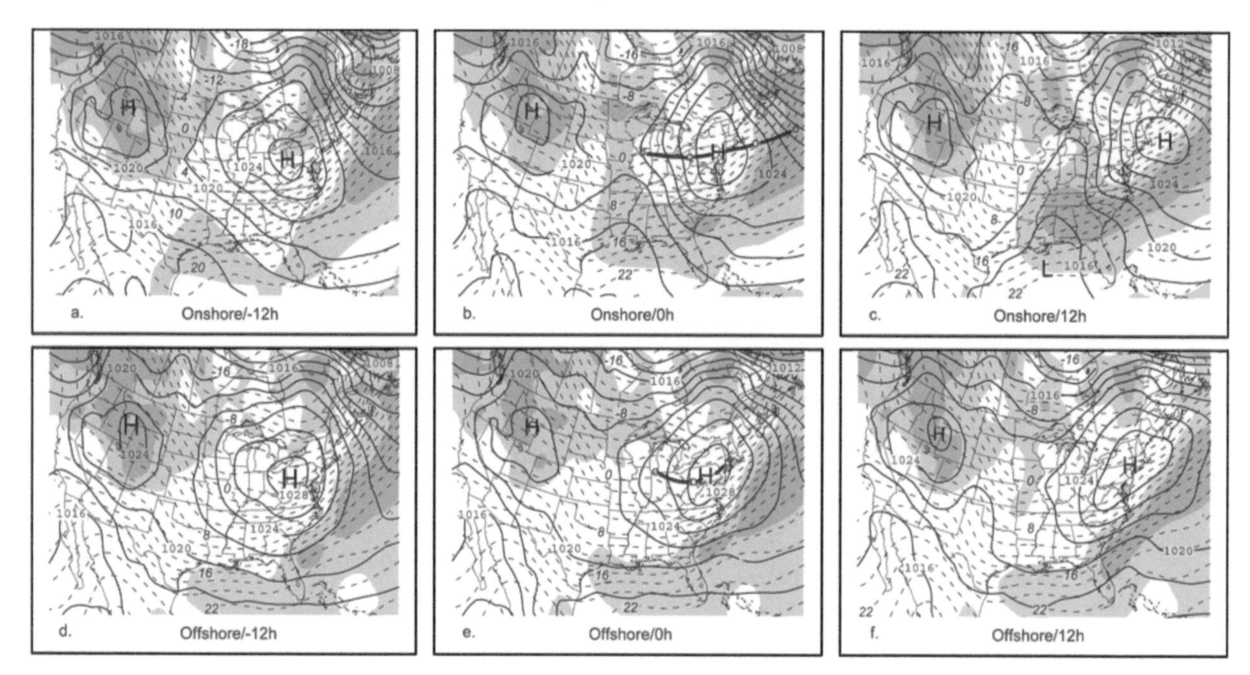


**Figure 6.32.** Composite analysis of (a)–(c) onshore versus (d)–(f) offshore coastal front movement in 12-h intervals. Sea level pressure (solid black contours), 1000-hPa isotherms (dashed lines, labeled in °C), and 925-mb relative humidity (shade intervals are 80% and 90%) [from Appel et al. (2005), their Fig. 12].

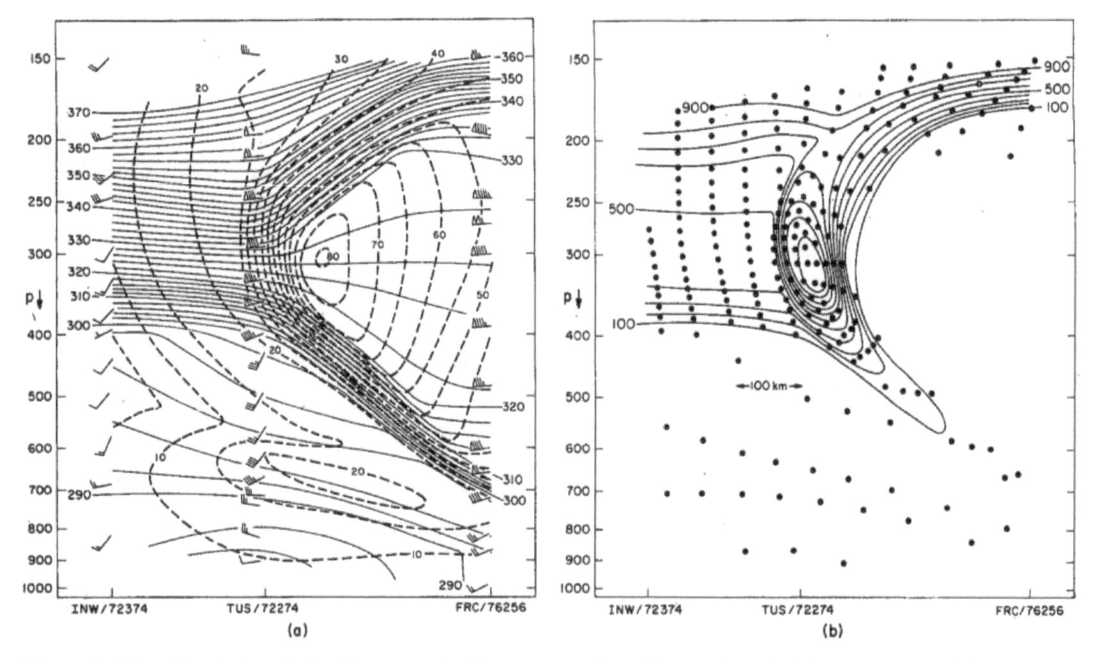


**Figure 6.33.** Classical analysis of an upper jet/front system from Shapiro (1981): (a) isentropes (solid contours), isotachs (m s$^{-1}$, dashed contours), and wind observations (barbs) are shown. Section based on rawinsonde observations supplemented with National Center for Atmospheric Research (NCAR) Sabreliner aircraft data. (b) PV (solid contours).

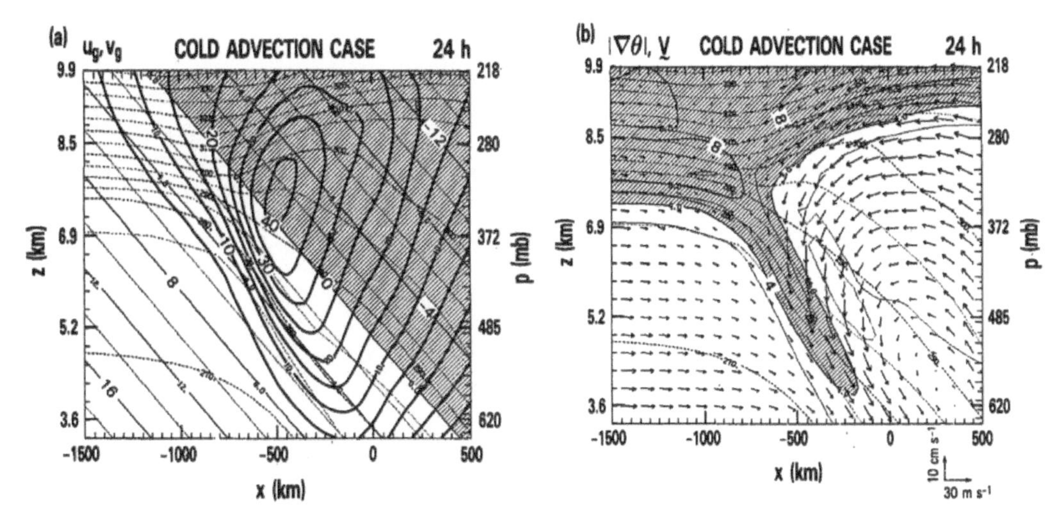


**Figure 6.34.** Cross sections from a 2D numerical model of upper-level frontogenesis forced by horizontal shear. (a) Cross-front geostrophic wind (thin solid contours, shaded negative) and along-front geostrophic wind (thick solid contours); (b) magnitude of potential temperature gradient (shaded greater than $4\times10^{-5}\,\mathrm{K\,m^{-1}}$), and vector arrows representing the total flow in the plane of the cross section. Isentropes are shown every 5 K as dotted contours (from Keyser et al. 1986).

that the frontal circulation is shifted toward the warm side of the front, resulting in frontogenetical tilting (Fig. 6.34).

Ordinarily, the ageostrophic response to frontogenetical forcing by the primary geostrophic flow serves to weaken the thermal contrast. Near the surface, the ageostrophic circulation cannot function in this way because of limited vertical air motion. Similarly, near the tropopause, the large static stability of the stratosphere can also act to preclude a compensating effect (e.g., Hoskins and Bretherton 1972). For the case of along-flow cold advection, the ageostrophic circulation is again rendered inefficient in compensating for the primary frontogenesis, and the frontal circulation can shift in a way that allows tilting to act frontogenetically.

The importance of upper-level fronts is related to the role of these systems in the exchange of air between the troposphere and stratosphere, in their linkage to upper jet streaks, and the role that upper frontal circulations can play in pulling stratospheric air with large potential vorticity (PV) toward the surface, resulting in stretching and vorticity production in upper-level troughs (section 5.3.6). The strong vertical wind shear associated with intense upper-frontal zones can be associated with clear-air turbulence and can present a hazard to aircraft.

As with fronts in general, the dynamics of upper fronts involve an interaction between the primary forcing mechanism (e.g., confluence or shear) and the secondary ageostrophic circulation that arises in response to the forcing. In some situations, confluence or shearing leads to the initiation of an upper front. When conditions are right, such as with *cold advection along the flow*, the subsiding branch of the resulting secondary circulation is shifted toward the warm side of the jet, with ascent (or weaker descent) on the cold side, as shown in Fig. 6.34.

As with surface fronts, the ageostrophic circulation itself can play a significant role in upper-frontal evolution. Based on the earlier discussion of surface frontogenesis, should we expect that a *QG model* could handle the development of an *upper* front? Given the absence of ageostrophic advections in the QG system, tilting would not be represented in such a model, and therefore an important mechanism for upper frontogenesis would not be included. Because upper fronts are characterized by strong adiabatic vertical motions, it is useful to describe their evolution in terms of a quantity that is conserved during such motions and that links the dynamics to the thermodynamics of the atmosphere, namely, the PV. In chapter 2, we reviewed the derivation of the QG version of this quantity, finding that quasigeostrophic potential

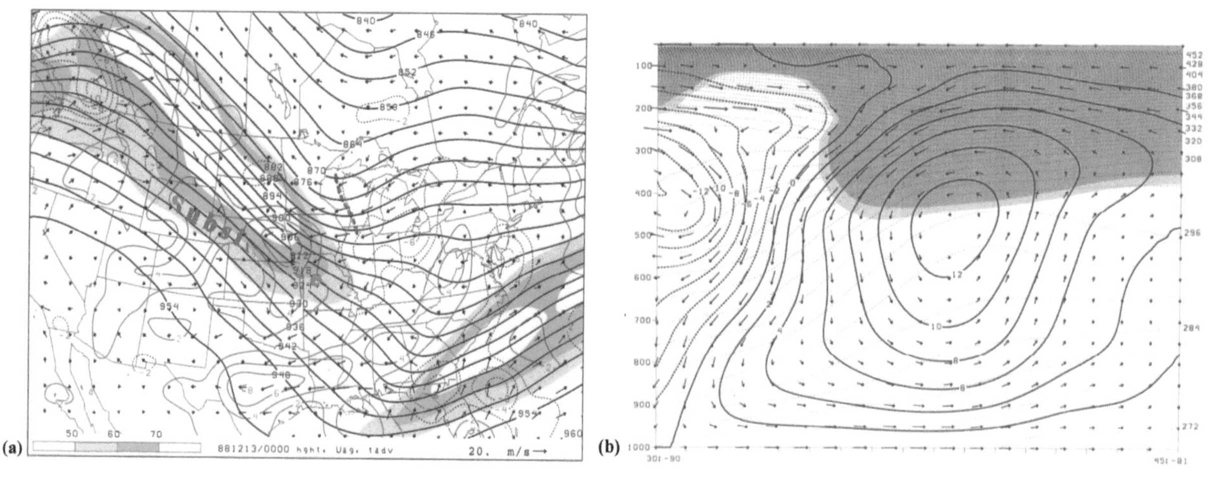

**Figure 6.35.** Diagnosis of an upper-level front at 0000 UTC 13 Dec 1988: (a) 300-mb isotachs (m s$^{-1}$, shaded as in legend at bottom of panel), geopotential height (solid contours), vertical motion (thin solid contours: red for descent, blue for ascent), and ageostrophic flow vectors (solid arrows). (b) Ageostrophic streamfunction (solid contours), ageostrophic velocity vectors in plane of section (black arrows), PV (shaded beginning at 1.5 PVU and for 2 PVU) and isentropes (thin dotted lines) (adapted from Lackmann et al. 1997).

vorticity (QGPV) conservation was obtained directly from the QG height tendency equation. Here, we will use the Ertel form of PV as defined by equations (4.1) or (4.2).

The Ertel PV is given by the product of the absolute vorticity on an isentropic surface and the static stability (see chapter 4). Given that the tropopause is the lower boundary of a zone of increased static stability corresponding to the stratosphere, we can utilize the PV to define the *dynamic tropopause*, as discussed in section 4.2. As shown, in the Northern Hemisphere, large values of PV correspond to cyclonic flow, upper troughs, and locations where the tropopause is locally lower. In some situations, the dynamic tropopause can actually fold upon itself, and these situations are almost invariably associated with upper-level frontal zones. The ageostrophic circulation associated with an upper front can contribute to tropopause folding, as seen in the case study analysis shown in Figs. 6.35a,b. In this case, an upper trough is forming in the left exit region of a strong upper-level jet streak (Fig. 6.35a); the thermally indirect circulation in the jet exit, characterized by cool air rising and warm air sinking (Fig. 6.35a) demonstrates that tilting is acting in a frontogenetical sense in this region. The ageostrophic streamfunction, plotted for a cross section through the jet exit, reveals tropopause folding, tilting, and the downward advection of high PV air from the stratosphere (Fig. 6.35b). This thermally indirect circulation is acting frontogenetically to strengthen the upper-level frontal zone in this case.

It is useful to integrate the different concepts reviewed previously into the conceptual picture of the upper-level jet-front system. The gradient in dynamic tropopause height is consistent with the reversal of the cross-jet temperature gradient, which marks the altitude of the jet-stream core via thermal wind arguments. The PV distribution is also consistent with the strongest flow being located there, as a strong horizontal gradient of PV exists across the frontal zone (Fig. 6.36). The change in vertical slope of isentropes across the jet core corresponds to this discontinuity or fold in the dynamic tropopause. The change in static stability across the upper jet is only partially responsible for the large horizontal PV gradient there—the change in vorticity also contributes to the PV gradient. The large stability found on the cold side of the jet also coincides with a region of cyclonic shear vorticity, with the opposite situation on the warm side of the jet.

Given a cross-sectional plot of isentropes, the wind field (isotachs) can thus be sketched. The potential vorticity (and location of the dynamic tropopause) can then be estimated as well, with consistency between the fields guaranteed by their dynamical relations. Provided any one of these three quantities (the isentropes, isotachs, or PV), the other two fields can be estimated in a qualitatively consistent fashion.

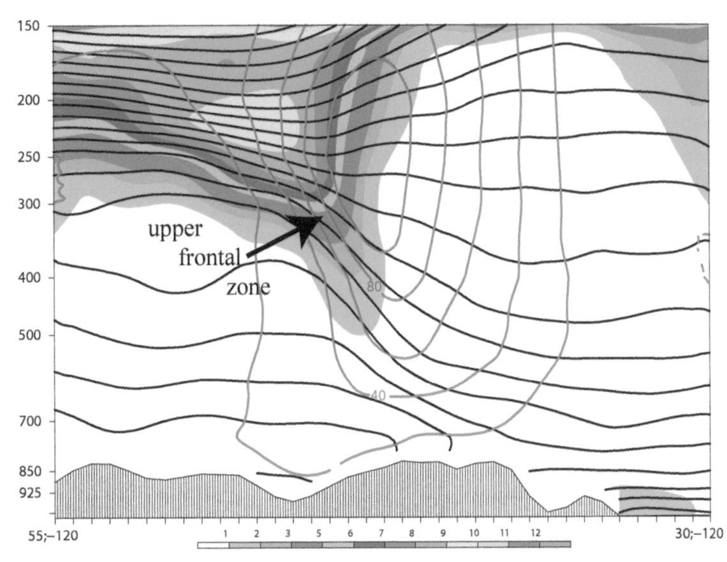


**Figure 6.36.** North–south cross section through an upper-level jet-front system, based on a short-term NAM forecast valid 18 UTC 25 Aug 2004. PV (shaded as in legend at bottom of panel), isentropes (black solid contours every 5K), and isotachs (red contours every 20 kt).

## REVIEW AND STUDY QUESTIONS

1. Perform a scale analysis of the terms in the frontogenesis equation (6.2) for a typical midlatitude synoptic-scale setting.
2. For terms A–D in the frontogenesis equation (6.2), sketch a simple situation in which each term would be *negative* (frontolytical).
3. Why is QG theory not well suited to completely explain the dynamics of frontal zones? Can QG concepts be used to understand any aspects of frontogenesis?
4. Explain why fronts are often characterized by clouds and precipitation. In other words, why is ascent frequently observed in the vicinity of fronts? Explain using the concept of primary and secondary circulations.
5. How do the dynamics of upper-level frontal zones compare to surface fronts? Discuss.

## PROBLEMS

1. Using the definition of the Lagrangian derivative, the chain rule from calculus, and the definition of $F$ shown below derive the frontogenesis equation (6.2),

$F$ is $\equiv$

$$\frac{d}{dt}\left(-\frac{\partial\theta}{\partial y'}\right) \quad \frac{d}{dt}=\frac{\partial}{\partial t}+u\frac{\partial}{\partial x'}+v\frac{\partial}{\partial y'}+\omega\frac{\partial}{\partial p} \quad \text{and} \tag{1}$$

$$F=\underbrace{\left[\frac{\partial\theta}{\partial x'}\left(\frac{\partial u}{\partial y'}\right)\right]}_{term\,A}+\underbrace{\left[\frac{\partial\theta}{\partial y'}\left(\frac{\partial v}{\partial y'}\right)\right]}_{term\,B}+\underbrace{\left[\frac{\partial\theta}{\partial p}\left(\frac{\partial\omega}{\partial y'}\right)\right]}_{term\,C}-\underbrace{\left[\frac{\partial}{\partial y'}\left(\frac{d\theta}{dt}\right)\right]}_{term\,D}. \tag{2}$$

2. Derive the Sawyer–Eliassen equation (6.16) from the preceding equations.

## REFERENCES

Appel, K. W., A. J. Riordan, and T. A. Holley, 2005: An objective climatology of Carolina coastal fronts. *Wea. Forecasting*, **20**, 439–455.

Bergeron, T., 1937: On the physics of fronts. *Bull. Amer. Meteor. Soc.*, **18,** 265–275.

Bjerknes, J., and H. Solberg, 1922: Life cycle of cyclones and the polar front theory of atmospheric circulation. *Geofys. Publ.*, **3,** 3–18.

Bluestein, H. B., 1993: *Observations and Theory of Weather Systems.* Vol. 2, *Synoptic-Dynamic Meteorology in Midlatitudes*, Oxford University Press, 608 pp.

Bosart, L. F., 1970: Mid-tropospheric frontogenesis. *Quart. J. Roy. Meteor. Soc.*, **96,** 442–471.

——, 1981: The Presidents' Day snowstorm of 18–19 February 1979: A subsynoptic-scale event. *Mon. Wea. Rev.*, **109**, 1542–1566.

Browning, K. A., 1990: Organization of clouds and precipitation in extratropical cyclones. *Extratropical Cyclones: The Erik Palmén Memorial Volume*, C. W. Newton and E. O. Holopainen, Eds., American Meteorological Society, 63–80.

——, and G. A. Monk, 1982: A simple model for the synoptic analysis of cold fronts. *Quart. J. Roy. Meteor. Soc.*, **108,** 435–452.

Carlson, T. N., 1998: *Mid-Latitude Weather Systems.* American Meteorological Society, 507 pp.

Evans, M. S., D. Keyser, L. F. Bosart, and G. M. Lackmann, 1994: A satellite-derived classification scheme for rapid maritime cyclogenesis. *Mon. Wea. Rev.*, **122,** 1381–1416.

Glickman, T., Ed., 2000: *Glossary of Meteorology.* 2nd ed. American Meteorological Society, 855 pp.

Hoskins, B. J., 1982: The mathematical theory of frontogenesis. *Ann. Rev. Fluid Mech.*, **14,** 131–151.

——, and F. P. Bretherton, 1972: Atmospheric frontogenesis models: Mathematical formulation and solution. *J. Atmos. Sci.*, **29,** 11–37.

Keyser, D., M. J. Pecnick, and M. A. Shapiro, 1986: Diagnosis of the role of vertical deformation in a two-dimensional primitive equation model of upper-level frontogenesis. *J. Atmos. Sci.*, **43,** 839–850.

Lackmann, G. M., D. Keyser, and L. F. Bosart, 1997: A characteristic evolution of upper-tropospheric cyclogenetic precursors during the Experiment on Rapidly Intensifying Cyclones over the Atlantic (ERICA). *Mon. Wea. Rev.*, **125,** 2529–2556.

Mahoney, K. M., and G. M. Lackmann, 2007: The effect of upstream convection on downstream precipitation. *Wea. Forecasting*, **22,** 255–277.

Market, P.S., and J. T. Moore, 1998: Mesoscale evolution of a continental occluded cyclone. *Mon. Wea. Rev.*, **126,** 1793–1811.

Martin, J. E., 1998: The structure and evolution of a continental winter cyclone. Part I: Frontal structure and the occlusion process. *Mon. Wea. Rev.*, **126,** 303–328.

——, and N. Marsili, 2002: Surface cyclolysis in the North Pacific Ocean. Part II: Piecewise potential vorticity diagnosis of a rapid cyclolysis event. *Mon. Wea. Rev.*, **130,** 1264–1281.

Matejka, T. J., R. A. Houze Jr., and P. V. Hobbs, 1980: Microphysics and dynamics of clouds associated with mesoscale rainbands in extratropical cyclones. *Quart. J. Roy. Meteor. Soc.*, **106,** 29–56.

Miller, J. E., 1948: On the concept of frontogenesis. *J. Meteor.*, **5,** 169–171.

Nielsen, J. W., 1989: The formation of New England coastal fronts. *Mon. Wea. Rev.*, **117,** 1380–1401.

Petterssen, S., 1936: Contribution to the theory of frontogenesis. *Geofys. Publ.*, **11,** 1–27.

Reed, R. J., 1955: A study of a characteristic type of upper-level front. *J. Meteor.*, **12,** 226–237.

Reed, R. J., and F. Sanders, 1953: An investigation of the development of a mid-tropospheric frontal zone and its associated vorticity field. *J. Meteor.*, **10,** 338–349.

——, and M. D. Albright, 1997: Frontal structure in the interior of an intense mature ocean cyclone. *Wea. Forecasting*, **12,** 866–876.

Sanders, F., 1955: An investigation of the structure and dynamics of an intense surface frontal zone. *J. Meteor.*, **12,** 542–552.

——, 1986 Explosive cyclogenesis in the West-Central North Atlantic Ocean, 1981–84. Part I: Composite structure and mean behavior, *Mon. Wea. Rev.*, 114, 1781–1794.

——, 1999: A proposed method of surface map analysis. *Mon. Wea. Rev.*, **127,** 945–955.

——, and C. A. Doswell III, 1995: A case for detailed surface analysis. *Bull. Amer. Meteor. Soc.*, **75,** 505–521.

——, and E. G. Hoffman, 2002: A climatology of surface baroclinic zones. *Wea. Forecasting*, **17,** 774–782.

Sansom, H. W., 1951: A study of cold fronts over the British Isles. *Quart. J. Roy. Meteor. Soc.*, **77,** 96–120.

Schultz, D. M., and C. F. Mass, 1993: The occlusion process in a midlatitude cyclone over land. *Mon. Wea. Rev.*, **121,** 918–940.

——, and G. Vaughn, 2011: Occluded fronts and the occlusion process: A fresh look at conventional wisdom. *Bull. Amer. Meteor. Soc.*, **92,** in press.

Shapiro, M. A., 1981: Frontogenesis and geostrophically forced secondary circulations in the vicinity of jet stream-frontal zone systems. *J. Atmos. Sci.*, **38,** 954–973.

——, 1984: Meteorological tower measurements of a surface cold front. *Mon. Wea. Rev.*, **112,** 1634–1639.

——, and D. Keyser, 1990: Fronts, jet streams and the tropopause. *Extratropical Cyclones, The Erik Palmén Memorial Volume,* C. W. Newton and E. O. Holopainen, Eds., American Meteorological Society, 167–191.

Stoelinga, M. T., J. D. Locatelli, and P. V. Hobbs, 2002: Warm occlusions, cold occlusions, and forward-tilting cold fronts. *Bull. Amer. Meteor. Soc.*, **83,** 709–721.

Uccellini, L. W., S. F. Corfidi, N. W. Junker, P. J. Kocin, and D. A. Olson, 1992: Report on the Surface Analysis Workshop held at the National Meteorological Center 25–28 March 1991. *Bull. Amer. Meteor. Soc.*, **73,** 459–472.

Wallace, J. M., and P. V. Hobbs, 1977: *Atmospheric Science: An Introductory Survey.* Academic Press, 467 pp.

## FURTHER READING

Carlson, T. N., 1980: Airflow through midlatitude cyclones and the comma cloud pattern. *Mon. Wea. Rev.,* **108,** 1498–1509.

Doswell, C. A., III, 1984: A kinematic analysis of frontogenesis associated with a nondivergent vortex. *J. Atmos. Sci.,* **41,** 1242–1248.

Keyser, D., and M. A. Shapiro, 1986: A review of the structure and dynamics of upper-level frontal zones. *Mon. Wea. Rev.,* **114,** 452–499.

Newton, C. W., 1954: Frontogenesis and frontolysis as a three-dimensional process. *J. Meteor.,* **11,** 449–461.

Posselt, D. J., and J. E. Martin, 2004: The effect of latent heat release on the evolution of a warm occluded thermal structure. *Mon. Wea. Rev.,* **132,** 578–599.

Schultz, D. M., and F. Sanders, 2002: Upper-level frontogenesis associated with the birth of mobile troughs in northwesterly flow. *Mon. Wea. Rev.,* **130,** 2593–2610.

——, D. Keyser, and L. F. Bosart, 1998: The effect of large-scale flow on low-level frontal structure and evolution in midlatitude cyclones. *Mon. Wea. Rev.,* **126,** 1767–1791.

# CHAPTER 7

# Baroclinic Instability

*. . . the supposed instability of the zonal current, has been the subject of a large number of theoretical studies in the last twenty-five years. The instability of "infinitely small" disturbances superimposed upon the westerly current has been the usual starting point in these investigations. A study of these attempts to solve the cyclone problem, however, gives the impression that there still does not exist any theoretical solution fully applicable to the cyclone problem. Therefore the question arises whether it is, in principle, permissible to start from infinitely small perturbations in discussing the cyclone problem. If such an infinitely small perturbation could cause the development of strong cyclones, it would indicate that the atmosphere is extremely unstable. How such an instability associated with storing of useful potential energy could develop in an atmosphere where rather strong perturbations of all kinds are always present seems difficult to understand.*

—E. Palmén,
*Compendium of Meteorology (1951)*

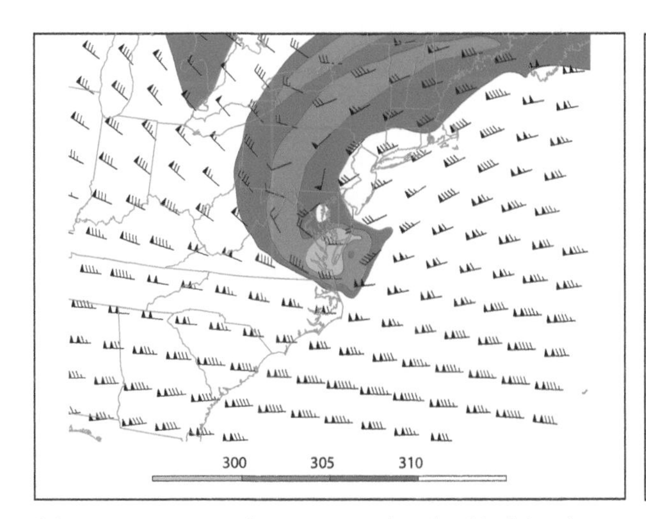


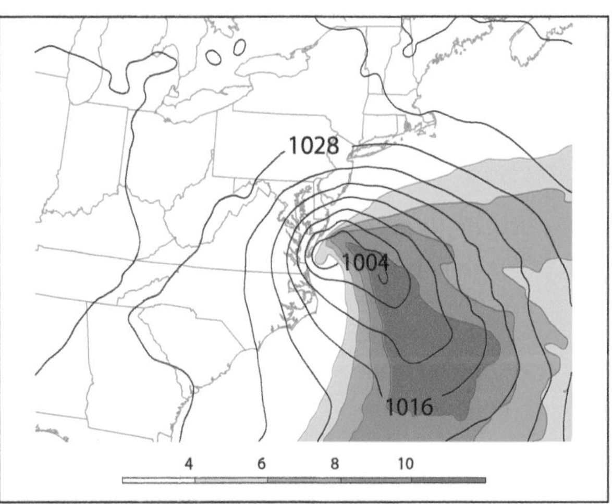


(left) Tropopause potential temperature and wind and (right) surface potential temperature anomaly and sea level pressure derived from a WRF simulation valid 1200 UTC 19 Feb 1979.

Based on the material presented in chapters 2–5, it is clear that upper-level wave patterns exert a powerful influence on day-to-day weather in the midlatitudes. As discussed at the beginning of chapter 5, the transport of heat, moisture, and momentum by transient wave disturbances in the atmosphere is a critical element of the global circulation. The storm tracks have a large influence on the climatology of midlatitude precipitation, which in turn is strongly tied to agriculture and natural ecosystems. The question of how storm tracks may be altered in future climate change is also significant for society. To understand and address these important problems, a theoretical understanding of the origin of atmospheric waves on the synoptic scale is helpful.

The atmosphere is characterized by many different types of instability, including upright and slantwise convective instability, small-scale Kelvin–Helmholtz instabilities that arise in the presence of strong vertical wind shear, and larger-scale barotropic and *baroclinic instability*. But what exactly does *instability* mean? A basic-state environment upon which small-scale perturbations would spontaneously amplify can be considered unstable with respect to the growth of those disturbances. If the energy source for the growth is drawn from the kinetic energy of the background flow, the instability can be labeled as barotropic. Baroclinic instability involves the growth of perturbations that derive energy from the basic-state potential energy in a baroclinic environment.

As the field of synoptic–dynamic meteorology has matured over the past 25 years, a healthy blend of theoretical, observational, and numerical modeling activities has led to rapid advances in the understanding of atmospheric instabilities, cyclogenesis, storm-track dynamics, and atmospheric predictability. The emergence of potential vorticity (PV) as an analysis tool has helped to bridge the gap between theory and observation. Despite this progress, some important problems in synoptic–dynamic meteorology remain unsolved. For example, observational studies have conclusively demonstrated that cyclogenesis involves the interaction and amplification of *preexisting* upper- and lower-tropospheric disturbances, but few studies have sought to identify the genesis mechanism for the mobile upper-level precursor disturbances, which often take the form of cyclonic vorticity maxima coincident with depressions of the dynamic tropopause.

Chapter 5 presented an analysis of the cyclogenesis problem based primarily on observational and numerical modeling studies. An important omission to this point has been discussion of theoretical approaches to the problem of how cyclones and upper-level waves form in the atmosphere. In this chapter, we present a classical theoretical approach to the baroclinic instability problem, followed by a PV-based diagnosis to establish the extent to which the theory corresponds to observation. The quote by E. Palmén at the beginning of this chapter demonstrates the difficulty in linking studies of the growth of infinitesimal perturbations in idealized flows to observed cyclones and waves in the real atmosphere. A goal of this chapter is to demonstrate that perturbation theory holds some relevance to the real-world cyclone problem, but only if connections are made by using compatible diagnostic tools and if some of the theoretical assumptions are relaxed.

Suppose that the earth was devoid of topography and covered by a globally uniform surface and that we could somehow establish zonal jets in thermal wind balance in both hemispheres. Would these jets remain zonal, or would wave-like perturbations develop in response to instability? If so, what would determine the size, structure, and intensity of these perturbations? What conditions are required for the instability to operate, and what is its mechanism? If a theory can be developed that answers these questions, how well will it match observations?

Consider the two 500-mb height and vorticity analyses presented in Fig. 7.1. A long-wave trough is present over eastern North America in Fig. 7.1a, with relatively smooth cyclonic flow and only weak short-wave troughs embedded in this large trough. In contrast, the height field in Fig. 7.1b is also characterized by a long-wave trough, but also a set of superimposed short-wave troughs and ridges is evident over the Gulf of Alaska, with several distinct cyclonic vorticity maxima embedded in the flow. These differences clearly have implications for the expected pattern of cyclone activity, because the short-wave troughs provide concentrated areas of quasigeostrophic (QG) forcing for ascent and lower-tropospheric vortex stretching. What is responsible for the difference in the extent of short-wave activity in these two situations? How do the long-wave and short-wave disturbances originate?

Although many theoretical problems are relevant to synoptic-scale phenomena, we elect here to explore the Eady (1949) baroclinic instability problem. This selection is motivated by the elegance and simplicity of Eady's analysis, the abundant insight derived from it, and its

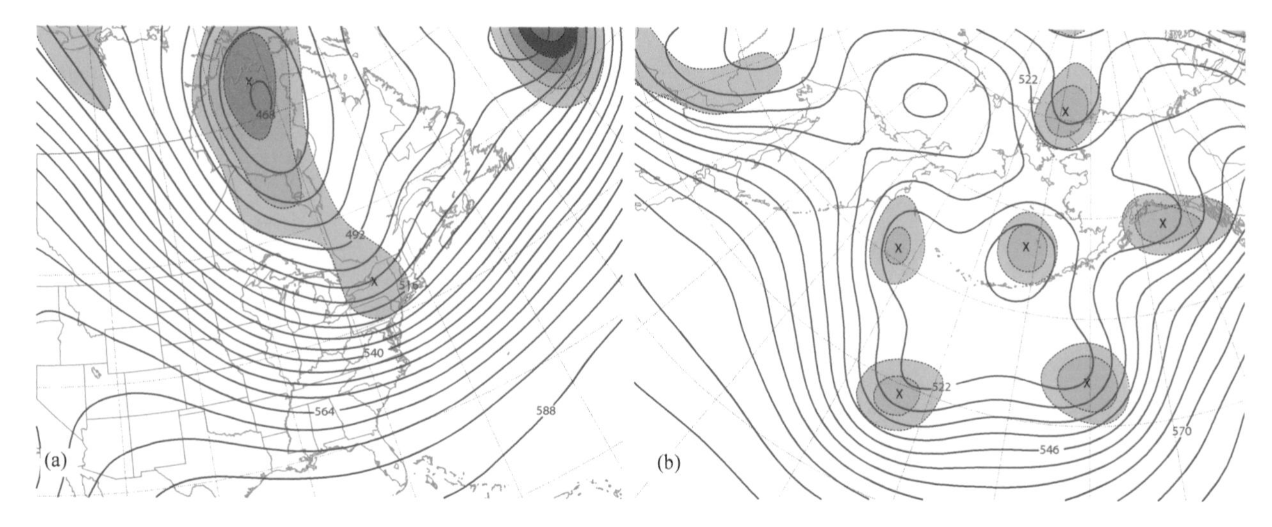


**Figure 7.1.** 500-mb height (solid blue contours every 6 dam) and absolute vorticity (dashed contours, shaded above $16\times10^{-5}$ s$^{-1}$), based on National Centers for Environmental Prediction (NCEP) reanalyses for (a) 1200 UTC 12 Feb 2003 and (b) 1200 UTC 8 Dec 1988.

relevance to the cyclone problem discussed previously in chapters 4 and 5.

Prior to baroclinic instability studies by Charney (1947) and Eady (1949), the Bergen school meteorologists emphasized the *polar front* in their analysis and prediction of extratropical cyclones (see section 5.4). The need for a theoretical foundation for the cyclone model was appreciated, and a natural starting point for their theoretical work on the cyclone problem involved the polar front. Thus, the Bergen group sought *instabilities of the frontal surface* as the mechanism for cyclogenesis (e.g., Solberg 1936). Despite considerable effort, these studies did not yield a convincing theoretical explanation for cyclone-scale disturbances. During the 1930s and 1940s, in conjunction with the establishment of the upper-air network, there developed a growing appreciation for the role of upper-level processes and the jet stream in cyclogenesis. As will subsequently be shown, theoretical studies by Eady (1949) and Charney (1947) were based on simplified atmospheric basic states that were baroclinic but *did not rely on the presence of preexisting frontal zones.*

The success of these early baroclinic instability studies shifted attention away from frontal-type instabilities but observations demonstrate that cyclones often originate as "frontal waves," and in this sense the study of instabilities of broad zonal currents does not constitute a complete treatment of the cyclone problem. However, some aspects of the theory are clearly reflected in observations. In addition, more recent studies of smaller-scale frontal instabilities (e.g., Orlanski 1968; Joly and Thorpe 1990; Schär and Davies 1990; Parker 1998) have demonstrated the presence of additional short-wave modes. However, the fronts are now known to form as the result of frontogenetical flow within the larger-scale disturbances and not the reverse.

Following Eady (1949), the purpose of the following analysis is to investigate the types of wave-like disturbances that can grow on a very simple westerly current. The theoretical design is not meant to precisely mimic the real atmosphere; rather, our goal is to design the simplest possible system that retains the dynamical essence of the baroclinic instability mechanism. If we can understand the conditions under which instabilities arise in a simplified setting, insight into how the real atmosphere works may be gleaned, although additional analysis will be required to translate the results directly to observation. This translation will be summarized in the latter portions of the chapter using PV diagnosis as a means to facilitate the link between theory and observation.

## 7.1. APPROXIMATIONS AND GOVERNING EQUATIONS

To maximize simplification, we will study a quasigeostrophic, hydrostatic, and adiabatic flow. Using height as the vertical coordinate, the primitive equations are

$$\frac{d\vec{U}}{dt} = -f\hat{k}\times\vec{U} - \alpha\nabla p - g\hat{k} + \vec{F}, \qquad (7.1)$$

$$\frac{d\rho}{dt}=-\rho(\nabla\cdot\vec{U}), \quad (7.2)$$

$$\frac{d\theta}{dt}=\frac{\theta}{T}\frac{Q_d}{C_p}, \quad (7.3)$$

$$p=\rho RT, \quad \text{and} \quad (7.4)$$

$$\theta=T\left(\frac{p_{00}}{p}\right)^{R/C_p}, \quad (7.5)$$

where all notation is consistent with that presented in chapter 1, except that a subscript $d$ has been added to $Q_d$ to distinguish it from the basic-state potential vorticity and Q vector utilized later. Also, the basic-state pressure profile is given by $p_0=p_0(z)$, and the constant $p_{00}=p_0|_{z=0}=1000$ mb. A simple, balanced, baroclinic basic-state flow will be specified, with the goal to identify analytical solutions to (7.1)–(7.5) that will determine if, how, and what kind of disturbances can grow in such a flow. The basic state must represent a valid solution to these governing equations, as will be demonstrated in one of the homework questions. The generic basic-state relations are as follows:

$$u=U(y,z);\ v,w=0, \quad (7.6)$$

$$U=-\frac{\alpha}{f}\frac{\partial P}{\partial y}, \quad (7.7)$$

$$\theta=\Theta(y,z), \quad (7.8)$$

$$\frac{\partial P}{\partial z}=-\rho g, \quad (7.9)$$

$$Q_d=0, \quad \text{and} \quad (7.10)$$

$$p,\rho,\alpha=f(y,z). \quad (7.11)$$

Additional simplification is achieved with the *Boussinesq* approximation, (see section 6.3.1) in which the density is treated as constant, except in the buoyancy term in the vertical momentum equation; the buoyancy will subsequently be utilized as a thermodynamic variable in the governing equations. The Boussinesq approximation is most relevant to the study of shallow circulation systems. A similar but less restrictive assumption is the anelastic approximation, in which the *basic-state* density is allowed to vary in the vertical. For either the Boussinesq or anelastic approximations, the following will be utilized:

$$\rho_{00}=const=\rho_0|_{z=0},$$
$$P_{00}=const=P_0|_{z=0}, \quad \text{and}$$
$$p'=p-P_0(z). \quad (7.12)$$

We next separate the momentum equation into horizontal and vertical components:

$$\frac{d\vec{V}}{dt}=-f\hat{k}\times\vec{V}-\alpha\nabla p+\vec{F}_h \quad \text{and} \quad (7.13)$$

$$\frac{dw}{dt}=-\alpha\frac{\partial p}{\partial z}-g+\vec{F}_z. \quad (7.14)$$

The basic-state fields are in hydrostatic balance, and $p=P_0(z)+p'(x,y,z,t)$; therefore,

$$\frac{dw}{dt}=-\frac{1}{\rho}\frac{\partial}{\partial z}(p'+P_0)-g+\vec{F}_z \quad \text{and} \quad (7.15)$$

$$\frac{dw}{dt}=-\frac{1}{\rho}\frac{\partial p'}{\partial z}+g\frac{(\rho_0-\rho)}{\rho}+\vec{F}_z. \quad (7.16)$$

The intervening steps needed to arrive at (7.16) are left as a homework problem. Proceeding now with the Boussinesq approximation, we set $\rho=\rho_{00}=const$ everywhere, except in the second right-hand term in (7.16). Note that the buoyancy term can be expressed in terms of potential temperature,

$$g\frac{(\rho_0-\rho)}{\rho}=g\frac{(\theta-\theta_0)}{\theta_{00}}\equiv b, \quad (7.17)$$

where the buoyancy $b$ is clearly related to the perturbation potential temperature and will be utilized as a thermodynamic variable in our subsequent analysis. At this point we will also assume frictionless motions; utilizing the thermodynamic variable $b$, the governing equation set is as follows:

$$\frac{d\vec{V}}{dt}=-f\hat{k}\times\vec{V}-\frac{1}{\rho_{00}}\nabla_h p'; \quad (7.18)$$

$$\frac{dw}{dt}=-\frac{1}{\rho_{00}}\frac{\partial p'}{\partial z}+g\frac{(\rho_0-\rho)}{\rho}; \quad (7.19)$$

$$\nabla_h \cdot \rho_{00}\vec{V} + \frac{\partial \rho_{00} w}{\partial z} = 0, \text{ or } \nabla_h \cdot \vec{V} + \frac{\partial w}{\partial z} = 0; \quad (7.20)$$

$$\frac{db}{dt} + w\frac{g}{\theta_{00}}\frac{d\theta_0}{dz} = 0; \quad \text{and} \quad (7.21)$$

$$\pi = \left(\frac{p}{P_{00}}\right)^{R/C_p}. \quad (7.22)$$

Some other useful relations include $p/P_{00} = \pi^{C_p/R}$; $P_{00}/p = \pi^{-C_p/R}$; $C_p = C_v + R$. The square of the basic-state Brunt–Väisälä frequency is defined

$$N_0^2 \equiv \frac{g}{\theta_{00}}\frac{d\theta_0}{dz}. \quad (7.23)$$

Now, finally invoking the hydrostatic assumption and utilizing the above definitions, the frictionless, adiabatic, Boussinesq, governing equations can be written as

$$\frac{d\vec{V}}{dt} = -f\hat{k}\times\vec{V} - \frac{1}{\rho_{00}}\nabla_h p'; \quad (7.24)$$

$$\frac{1}{\rho_{00}}\frac{\partial p'}{\partial z} = g\frac{(\rho_0 - \rho)}{\rho_{00}}; \quad (7.25)$$

$$\nabla \cdot \vec{U} = 0; \quad (7.26)$$

$$\frac{db}{dt} + wN_0^2 = 0; \quad \text{and} \quad (7.27)$$

$$P_0 = \rho_0 R T_0; \quad p = \rho R T. \quad (7.28)$$

### 7.1.1. The anelastic equations and PV relations

The original Eady paper was based on the Boussinesq equations, but given the restrictive nature of this assumption, we will develop the anelastic equations to show the path toward treatments that are valid for deeper motions as well. For average atmospheric conditions, the density *scale height* (*e*-folding height) in the atmosphere is roughly 8 km, and therefore the Boussinesq equations would be valid for systems of a depth much smaller than the scale height (over which basic-state density variations are small). In the hydrostatic anelastic equations, the basic-state density profile is allowed to vary in the vertical, and we replace the density by $\rho_0(z)$, except in the buoyancy term,

$$\frac{d\vec{V}}{dt} = -f\hat{k}\times\vec{V} - \nabla_h\frac{p'}{\rho_0}, \quad (7.29)$$

$$\frac{\partial}{\partial z}\frac{p'}{\rho_0} = g\left(\frac{\theta - \theta_0}{\theta_0}\right), \quad (7.30)$$

$$\nabla_h \cdot \rho_0\vec{V} + \frac{\partial \rho_0 w}{\partial z} = 0, \quad (7.31)$$

$$\frac{db}{dt} + w\frac{g}{\theta_{00}}\frac{d\theta_0}{dz} = 0, \quad (7.32)$$

$$\phi \equiv \frac{p'}{\rho_0}, \quad \text{and} \quad (7.33)$$

$$b \equiv g\left(\frac{\theta - \theta_0}{\theta_0}\right). \quad (7.34)$$

In the anelastic system, the continuity equation shows that the *momentum* is nondivergent [(7.31)]; in the Boussinesq system, the *velocity* exhibited this property [(7.26)]. The density-normalized pressure perturbation $\phi$ has the same units and plays a similar role to the geopotential in the isobaric system as a geostrophic streamfunction and is introduced for notational convenience. The anelastic definition of buoyancy differs only slightly from that in the Boussinesq system.

At this point, we have not yet applied the quasigeostrophic assumptions, which are identical to those utilized in chapter 2,

$$\beta \equiv \left.\frac{\partial f}{\partial y}\right|_{y=0}; \quad f = f_0 + \beta y; \quad (7.35)$$

$$\frac{d}{dt_g} = \frac{\partial}{\partial t} + \vec{V}_g \cdot \nabla_h; \quad \text{and} \quad (7.36)$$

$$\vec{V}_g = -\frac{1}{f_0}\hat{k}\times\nabla_h\phi. \quad (7.37)$$

The QG, anelastic, frictionless, hydrostatic, adiabatic equation set is then

$$\frac{d\vec{V}}{dt_g} = -f\hat{k}\times\vec{V} - \nabla_h\phi = -f_0\hat{k}\times\vec{V}_{ag} - \beta y\hat{k}\times\vec{V}_g, \quad (7.38)$$

$$\frac{\partial \phi}{\partial z} = b, \quad (7.39)$$

$$\nabla \cdot \rho_0\vec{U} = 0, \quad \text{and} \quad (7.40)$$

$$\frac{db}{dt_g} = -N^2 w. \quad (7.41)$$

Expanding derivatives and vector relations, we have

$$\frac{du_g}{dt_g} = f_0 v - f_0 v_g + \beta y v_g, \tag{7.42}$$

$$\frac{dv_g}{dt_g} = -f_0 u + f_0 u_g - \beta y u_g, \tag{7.43}$$

$$\frac{\partial u_{ag}}{\partial x} + \frac{\partial v_{ag}}{\partial y} + \frac{1}{\rho_0}\frac{\partial}{\partial z}(\rho_0 w) = 0, \quad \text{and} \tag{7.44}$$

$$\frac{\partial b}{\partial t} + u_g \frac{\partial b}{\partial x} + v_g \frac{\partial b}{\partial y} = -N^2 w. \tag{7.45}$$

Following a similar approach to that in chapter 2, we wish to simplify the equations into a more useful and understandable form. First, we develop the vorticity equation and then the potential vorticity relation for this system of equations. Taking the curl of (7.38) or taking $\partial/\partial x$ (7.43) – $\partial/\partial y$ (7.42) yields the vorticity equation

$$\left(\frac{\partial}{\partial t} + \vec{V}_g \cdot \nabla_h\right)(\zeta_g + f) = \frac{f_0}{\rho_0}\frac{\partial}{\partial z}(\rho_0 w). \tag{7.46}$$

Multiplying (7.45) by $\rho_0/N^2$ and then taking $(f_0/\rho_0)\partial/\partial z$ of the result, a modified form of the thermodynamic equation results,

$$\left(\frac{\partial}{\partial t} + \vec{V}_g \cdot \nabla_h\right)\left(\frac{f_0}{\rho_0}\frac{\partial}{\partial z}\left(\frac{\rho_0 b}{N^2}\right)\right) = -\frac{f_0}{\rho_0}\frac{\partial}{\partial z}(\rho_0 w). \tag{7.47}$$

Adding (7.46) and (7.47), to eliminate $w$, yields the potential vorticity relation. This procedure is analogous to that employed in forming the QG height-tendency equation in chapter 2; as was shown then, that equation is an expression of QG PV conservation, as is the following:

$$\left(\frac{\partial}{\partial t} + \vec{V}_g \cdot \nabla_h\right)\left(\zeta_g + f + \frac{f_0}{\rho_0}\frac{\partial}{\partial z}\left(\frac{\rho_0 b}{N^2}\right)\right) = 0, \tag{7.48}$$

where the second term in brackets is the anelastic, QG form of the PV for height coordinates. For adiabatic, frictionless, geostrophic flow,

$$\frac{dq}{dt_g} = 0, \tag{7.49}$$

where

$$q = \zeta_g + f + \frac{f_0}{\rho_0}\frac{\partial}{\partial z}\left(\frac{\rho_0 b}{N^2}\right). \tag{7.50}$$

By expressing the geostrophic relative vorticity and buoyancy in terms of $\phi$, a linear operator relates the QG PV to the geostrophic streamfunction, and this operator can be inverted in a piecewise fashion to attribute specific portions of the streamfunction to specific PV anomalies,

$$q = \frac{1}{f_0}\nabla^2\phi + f + \frac{f_0}{\rho_0}\frac{\partial}{\partial z}\left(\frac{\rho_0}{N^2}\frac{\partial\phi}{\partial z}\right). \tag{7.51}$$

To clarify the QG PV inversion operation,

$$q - f = \ell(\phi) \quad \text{and} \tag{7.52}$$

$$\phi = \ell^{-1}(q - f), \tag{7.53}$$

where

$$\ell = \frac{1}{f_0}\nabla_h^2 + \frac{f_0}{\rho_0}\frac{\partial}{\partial z}\left(\frac{\rho_0}{N^2}\frac{\partial}{\partial z}\right). \tag{7.54}$$

For a given specification of $q - f$ and boundary condition information, (7.53) can be solved for $\phi$, and this can be done in a piecewise fashion. Because the operator is linear, the PV field can be divided into an arbitrary number of pieces and inverted, with the sum of the resulting fields equaling the total $\phi$ field. This diagnostic operation has been usefully applied in many settings to isolate the influence of specific PV anomalies or to quantify the interactions of different PV centers (Hakim et al. 1996).

In a procedure similar to that described above, a prognostic PV inversion can be accomplished by working with the PV tendency from (7.49), as shown by Hakim et al. (1996):

$$\frac{\partial q}{\partial t} = -\vec{V}_g \cdot \nabla_h q, \tag{7.55}$$

$$\frac{\partial q}{\partial t} = \frac{1}{f_0}\nabla^2\frac{\partial\phi}{\partial t} + f + \frac{f_0}{\rho_0}\frac{\partial}{\partial z}\left(\frac{\rho_0}{N^2}\frac{\partial}{\partial z}\frac{\partial\phi}{\partial t}\right), \quad \text{and} \tag{7.56}$$

$$\frac{\partial q}{\partial t} = \ell\left(\frac{\partial\phi}{\partial t}\right) = -\vec{V}_g \cdot \nabla_h q. \tag{7.57}$$

Given a field of QG PV advection, we can then invert the same linear operator to compute the local pressure or height tendency. This is also exactly consistent with our development in chapter 2 of the QG height-tendency equation. Because the PV advection field is often relatively noisy and may be more difficult to divide into physically meaningful partitions, the use of prognostic PV inversion has remained limited relative to the diagnostic approach.

The QG PV for the Boussinesq equations is obtained in a fashion very similar to that for the anelastic equations, and it is expressed as

$$q = \frac{1}{f_0}\nabla_h^2\phi + f + \frac{f_0}{N_0^2}\left(\frac{\partial^2\phi}{\partial z^2}\right). \quad (7.58)$$

The constant density in the Boussinesq system allows a simplified stability term in (7.58) relative to the anelastic expression in (7.51).

### 7.1.2. The Eady baroclinic instability problem

Returning to our goal of analyzing the growth of instabilities on a simple baroclinic flow, we utilize the Boussinesq equations on an *f* plane to maximize simplicity. By adopting this strategy, note that Rossby wave propagation would not take place as it does in the real atmosphere. Furthermore, the gradient of planetary vorticity is also a gradient of potential vorticity, and its neglect here has implications for our stability analysis, as will be shown later.

The adiabatic, frictionless, hydrostatic, *f*-plane, QG, Boussinesq governing equations are

$$\frac{\partial u_g}{\partial t} + u_g\frac{\partial u_g}{\partial x} + v_g\frac{\partial u_g}{\partial y} = f_0 v_{ag}, \quad (7.59)$$

$$\frac{\partial v_g}{\partial t} + u_g\frac{\partial v_g}{\partial x} + v_g\frac{\partial v_g}{\partial y} = -f_0 u_{ag}, \quad (7.60)$$

$$\frac{\partial u_{ag}}{\partial x} + \frac{\partial v_{ag}}{\partial y} + \frac{\partial w}{\partial z} = 0, \quad (7.61)$$

$$\frac{\partial b}{\partial t} + u_g\frac{\partial b}{\partial x} + v_g\frac{\partial b}{\partial y} = -N_0^2 w, \text{ and} \quad (7.62)$$

$$\frac{\partial \phi}{\partial z} = b. \quad (7.63)$$

Our domain is bounded on top by a rigid lid at $z = H$, possesses a flat lower boundary at $z = 0$, extends indefinitely in the meridional direction, and has *periodic lateral boundary conditions* in the zonal direction at $x = 0, L$

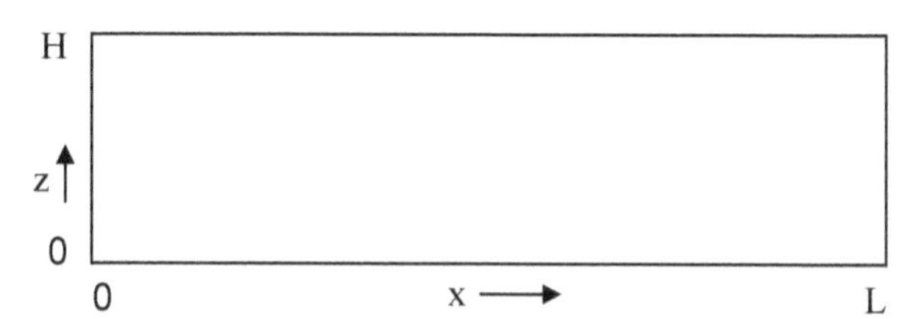


**Figure 7.2.** Zonally oriented vertical cross section of Eady domain.

(Fig. 7.2). The periodic boundary conditions require solutions that match at $x = 0, L$; disturbances passing through the downstream boundary at $x = L$ will reappear at $x = 0$. This domain design will result in further simplifications to the equations through the omission of meridional variation in the disturbance fields.

Next, we specify a basic-state flow, in hydrostatic and geostrophic balance, upon which we seek growing disturbances. A most basic baroclinic flow consists of linear vertical shear,

$$u_g = U_g(z) = U\frac{z}{H}; \quad (7.64)$$

in (7.64), $U$, a constant parameter, represents the maximum speed of the westerly current, found at $z = H$, and the basic-state flow is not a function of the $x$ or $y$ directions. The thermal wind relation then specifies that

$$\frac{\partial U_g}{\partial z} = \frac{U}{H} = const = -\frac{1}{f_0}\frac{\partial B}{\partial y}, \quad (7.65)$$

where $B$, the basic-state thermodynamic variable (buoyancy), is obtained by integrating (7.65) with respect to $y$ to yield,

$$B(y) = -\frac{f_0 U}{H}y. \quad (7.66)$$

The buoyancy relation applied to the basic state is

$$\frac{\partial \Phi}{\partial z} = B, \quad (7.67)$$

allowing the basic-state geostrophic streamfunction to be obtained by integrating (7.67) with respect to $z$ and using (7.66),

$$\Phi(y,z) = -f_0\frac{U}{H}yz. \quad (7.68)$$

The full expressions for the zonal wind, buoyancy, and pressure include disturbances as well, whose structure will be established subsequently,

$$u = U_g + u'_g + u'_{ag};\ \ v = v'_g + v'_{ag};\ \ w = w'(x,z,t) \quad (7.69)$$

and

$$b = B(y) + b'(x,z,t);\ \ \phi = \Phi(y,z) + \phi'(x,z,t). \quad (7.70)$$

Because of our choice of a domain that is unbounded in the meridional direction, we are seeking solutions in the $x$–$z$ plane; the disturbances will have no $y$ dependence. Therefore, $u'_g = 0$ in (7.69); because the zonal geostrophic flow is not a function of $x$ or $y$, (7.59) then requires that $v'_{ag} = 0$. Taking these modifications into account, our final set of Eady governing equations is

$$\frac{\partial v'_g}{\partial t} + U_g \frac{\partial v'_g}{\partial x} = -f_0 u'_{ag}, \quad (7.71)$$

$$\frac{\partial b'}{\partial t} + U_g \frac{\partial b'}{\partial x} = -v'_g \frac{\partial B}{\partial y} - N_0^2 w', \quad (7.72)$$

$$\frac{\partial u'_{ag}}{\partial x} + \frac{\partial w'}{\partial z} = 0, \text{ and} \quad (7.73)$$

$$q = f_0 + \frac{1}{f_0}\nabla_h^2 \phi' + \frac{f_0}{N_0^2}\left(\frac{\partial^2 \phi'}{\partial z^2}\right). \quad (7.74)$$

The zonal momentum equation does not appear for the reasons discussed above. The buoyancy (thermodynamic) equation demonstrates that changes in the disturbance temperature field can arise from advection of the basic-state temperature field as well as via vertical air motions. The potential vorticity field can be viewed as a constant basic-state value $f_0$, along with contributions from vorticity in the meridional perturbation geostrophic flow and stability anomalies.

At first glance, (7.71)–(7.73) do not form a closed set of equations because we have four unknowns but only three equations; however, the perturbation geostrophic streamfunction $\phi'$ is used to replace the perturbation meridional geostrophic velocity via the geostrophic wind relation $[v'_g = 1/f_0(\partial\phi'/\partial x)]$ and perturbation buoyancy via $[b' = \partial\phi'/\partial z]$. Expressed in this way, the system is closed.

### 7.1.3. Energetics for the Eady atmosphere

One of our primary objectives is the identification of disturbance structures that are able to extract energy from the basic-state flow. This motivates the development of a set of energy equations for use in our subsequent analysis. Similar to the QG energy equation discussed in section 2.1.7, we define the disturbance kinetic energy (per unit mass) as

$$K_e \equiv \frac{v'^2_g}{2}. \quad (7.75)$$

The energy equation is formed by multiplying the momentum equation (7.71) by $v'_g$, using the geostrophic wind relation, and rearranging to obtain

$$\frac{\partial}{\partial t}\left(\frac{v'^2_g}{2}\right) + U_g \frac{\partial}{\partial x}\left(\frac{v'^2_g}{2}\right) = -u'_{ag}\frac{\partial \phi'}{\partial x}. \quad (7.76)$$

Through application of the product rule, (7.76) can be cast into a form that resembles that of (2.43),

$$\left(\frac{\partial}{\partial t} + U_g \frac{\partial}{\partial x}\right)\left(\frac{v'^2_g}{2}\right) = -\frac{\partial}{\partial x}(\phi' u'_{ag}) - \frac{\partial}{\partial z} w'\phi' + w'b'. \quad (7.77)$$

The interpretation of this equation is similar to that of (2.43); the eddy kinetic energy tendency following the geostrophic flow is related to the horizontal and vertical ageostrophic geopotential flux convergences and baroclinic conversion. Equation (2.43) also included a barotropic energy conversion term, which does not appear in (7.77). Recall that barotropic conversion requires a horizontal gradient of the mean (basic state) flow; the basic-state wind field here is horizontally uniform as indicated by (7.64). Thus, for the problem being developed here, there is no conversion from basic-state kinetic energy into disturbance kinetic energy, and baroclinic conversion provides the critical energy source.

In developing the potential energy equation, recall that $b' \propto \theta' - \theta_0$, which represents a potential temperature perturbation. Thus, $b'^2$ is analogous to potential energy. A more formal definition of the eddy potential energy for the Eady system is

$$P_e \equiv \frac{b'^2}{2N_0^2}, \quad (7.78)$$

which has units of $m^2\,s^{-2}$, or energy per unit mass. The derivation of a potential energy equation involves multiplication of the thermodynamic equation (7.72) by the eddy potential energy and rearranging,

$$\left(\frac{\partial}{\partial t}+U_g\frac{\partial}{\partial x}\right)\left(\frac{b'^2}{2N_0^2}\right)=-w'b'-\frac{v_g'b'}{N_0^2}\frac{\partial B}{\partial y}. \tag{7.79}$$

The appearance of the baroclinic term as in (7.77), but with opposite sign, is consistent with the expectation that eddy kinetic energy is drawn from the eddy potential energy. When warm air rises and cold air sinks, eddy potential energy is lost and eddy kinetic energy is gained. But what is the source of the eddy potential energy? The rightmost term in (7.79) includes the meridional eddy heat flux $v_g'b'$, multiplying the basic-state meridional thermal gradient. Equation (7.72) indicates that warm (cold) anomalies are produced with perturbation southerly (northerly) flow because of advection of the basic-state meridional temperature gradient. The sign of $\partial B/\partial y$ is $<0$ in keeping with westerly shear in the basic-state flow (westerly flow increasing with height). Therefore, when there is a positive correlation between $v_g'$ and $b'$ (i.e., there is a poleward eddy heat flux), the eddy potential energy is increasing.

A complete eddy energy equation is formed by summing (7.77) and (7.79),

$$\left(\frac{\partial}{\partial t}+U_g\frac{\partial}{\partial x}\right)\left(\frac{v_g'^2}{2}+\frac{b'^2}{2N_0^2}\right)$$
$$=-\frac{v_g'b'}{N_0^2}\frac{\partial B}{\partial y}-\frac{\partial}{\partial x}(\phi'u_{ag}')-\frac{\partial}{\partial z}w'\phi'. \tag{7.80}$$

Despite the use of an equation of this form for an analysis of local energetics, it will prove instructive for the problem at hand to integrate this equation over the entire Eady domain: that is, over the $x$–$z$ domain presented in Fig. 7.2. Integrating from $0<x<L$ and $0<z<H$, we find that several terms in the integrated version of (7.80) are zero,

$$\int_0^L\frac{\partial}{\partial x}(\phi'u_{ag}')dx=0 \quad \text{and}$$

$$\int_0^L U_g\frac{\partial}{\partial x}\left(\left[\left(\frac{v_g'^2}{2}+\frac{b'^2}{2N_0^2}\right)\right]\right)dx=0, \tag{7.81}$$

because of the periodic boundary conditions in $x$; the integral involves subtracting the same value at $x=0$, $L$. Similarly,

$$\int_0^H\frac{\partial}{\partial z}(\phi'w')dz=0, \tag{7.82}$$

because $w'=0$ at both $z=0$ and $z=H$ by our assumption of a rigid upper and lower boundaries. The remaining equation identifies the ultimate source of eddy energy,

$$\frac{\partial}{\partial t}\int_0^L\int_0^H\left(\frac{v_g'^2}{2}+\frac{b'^2}{2N_0^2}\right)dz\,dx=-\frac{1}{N_0^2}\frac{\partial B}{\partial y}\int_0^L\int_0^H v_g'b'\,dz\,dx. \tag{7.83}$$

This total eddy energy equation equates the domain-average change in eddy energy to the domain-averaged meridional heat flux. To assemble a complete picture of the energy exchanges within the Eady atmosphere under consideration, Fig. 7.3 provides a schematic representation of the energy flow based on the energy equation analysis.

Because of the lack of time dependence in our basic state (all basic-state variables are constant with time), there is essentially an infinite reservoir of potential energy available for conversion to eddy potential energy.

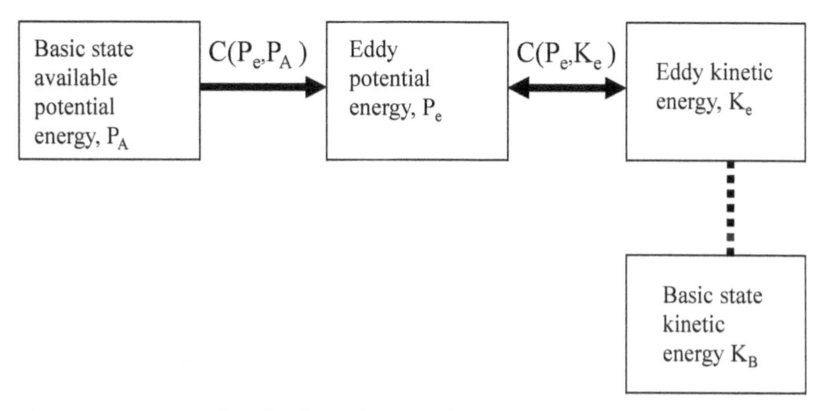


**Figure 7.3.** Energy flow for the Eady atmosphere.

There is no provision for direct transformation of basic-state energy into eddy kinetic energy in (7.77), so the diagram only indicates a one-way flow of energy from the basic-state available potential energy $P_A$ to eddy potential energy $P_e$. Because a baroclinic conversion term appears in both (7.77) and (7.79), a two-way exchange of energy takes place between $P_e$ and $K_e$ in Fig. 7.3. Because of the horizontal uniformity of the specified basic-state westerly flow, $U_g = U_g(z)$, there is no possibility of barotropic energy conversion from the basic-state kinetic energy $K_B$ to $K_e$. The conversion term from $P_A$ to $P_e$ is given by

$$C(P_A, P_e) = -\int_0^L \int_0^H \frac{v_g' b'}{N_0^2} \frac{\partial B}{\partial y} dz\, dx. \qquad (7.84)$$

and the conversion between $P_e$ and $K_e$ is

$$C(P_e, K_e) = -\int_0^L \int_0^H w' b'\, dz\, dx. \qquad (7.85)$$

These conversion terms are related to the meridional and vertical eddy heat fluxes, respectively. Before formally solving the baroclinic instability problem, it is still instructive to further explore the types of disturbance structures that are needed to draw energy from the basic state. Using $v_g' = 1/f_0 (\partial \phi' / \partial x)$ and $b' = \partial \phi' / \partial z$, we can express $C(P_A, P_e)$ in terms of $\phi'$ to examine the structures needed for a given sign of energy conversion,

$$C(P_A, P_e) = -\frac{1}{f_0 N_0^2} \frac{\partial B}{\partial y} \int_0^L \int_0^H \frac{\partial \phi'}{\partial x} \frac{\partial \phi'}{\partial z} dz\, dx. \qquad (7.86)$$

Given that $\partial B / \partial y < 0$, it is evident from (7.86) that the conversion of basic-state potential energy to eddy potential energy requires that the product $\partial \phi' / \partial x (\partial \phi' / \partial z) > 0$. As a homework problem, the reader is asked to demonstrate that this requires trough and ridge axes to slope westward with height. This finding is also consistent with observations of midlatitude cyclones; these systems typically grow as long as the upper trough lies to the west of the surface cyclone center. Although the energy flow in the Eady atmosphere developed here differs from that in the real atmosphere in some respects, the structure and mechanism for baroclinic energy conversion appears to be realistic.

#### 7.1.4. Diagnosis of ageostrophic motions in the Eady atmosphere

The role of ageostrophic secondary circulations was an important theme in our earlier analysis using the QG system of equations. In chapter 2, a derivation of the QG omega equation provided insight into the forcing for vertical motion; in chapter 6, the Sawyer–Eliassen equation (6.16) was developed for use in the diagnosis of ageostrophic frontal circulations. Here, these developments will be mimicked to supply an analysis tool for examination of the Eady solutions that will be obtained in section 7.2.

The thermal wind relations applied to the geostrophic perturbation and basic-state flow are expressed as

$$f_0 \frac{\partial v_g'}{\partial z} = \frac{\partial b'}{\partial x}; \quad f_0 \frac{\partial U_g}{\partial z} = -\frac{\partial B}{\partial y}. \qquad (7.87)$$

Taking $f_0(\partial / \partial z)$ of the momentum equation (7.71) and subtracting $(\partial / \partial x)$ of the thermodynamic equation (7.72), the thermal wind relations (7.87) can be used to cancel terms, leaving a form of the Sawyer–Eliassen equation appropriate for the current equation set,

$$2 \frac{\partial B}{\partial y} \frac{\partial v_g'}{\partial x} - f_0^2 \frac{\partial u_{ag}'}{\partial z} + N_0^2 \frac{\partial w'}{\partial x} = 0. \qquad (7.88)$$

Next, we define an ageostrophic streamfunction,

$$u_{ag}' = \frac{\partial \psi}{\partial z}; \quad w' = -\frac{\partial \psi}{\partial x}. \qquad (7.89)$$

The relation between the ageostrophic flow and streamfunction contours is shown for a hypothetical circulation in Fig. 7.4, with clockwise (counterclockwise) circulation about streamfunction minima (maxima) in the $x$–$z$ plane.

Substituting (7.89) in (7.88), a form of the Sawyer–Eliassen equation results,

$$-f_0^2 \frac{\partial^2 \psi}{\partial z^2} - N_0^2 \frac{\partial^2 \psi}{\partial x^2} = -2 \frac{\partial B}{\partial y} \frac{\partial v_g'}{\partial x} = 2Q_x. \qquad (7.90)$$

The left side of (7.90) describes the ageostrophic motion, and the right side represents the forcing for ageostrophic motions. Although it may not be immediately apparent, the right side of (7.90) is twice the zonal component of the Q vector, as discussed in sections 2.1.3 and 6.3.1 and as found for Eq. (6.16). The forcing for ageostrophic motions is proportional to the Q vector itself rather than the divergence of this vector field. The forcing for ageostrophic circulations is illustrated in Fig. 7.4b, which, given that $\partial B / \partial y < 0$, demonstrates that regions of $\partial v_g' / \partial x > 0$ are associated with ageostrophic streamfunction maxima. The implied vertical motion centers, maximized in regions of the strongest horizontal streamfunction gradient,

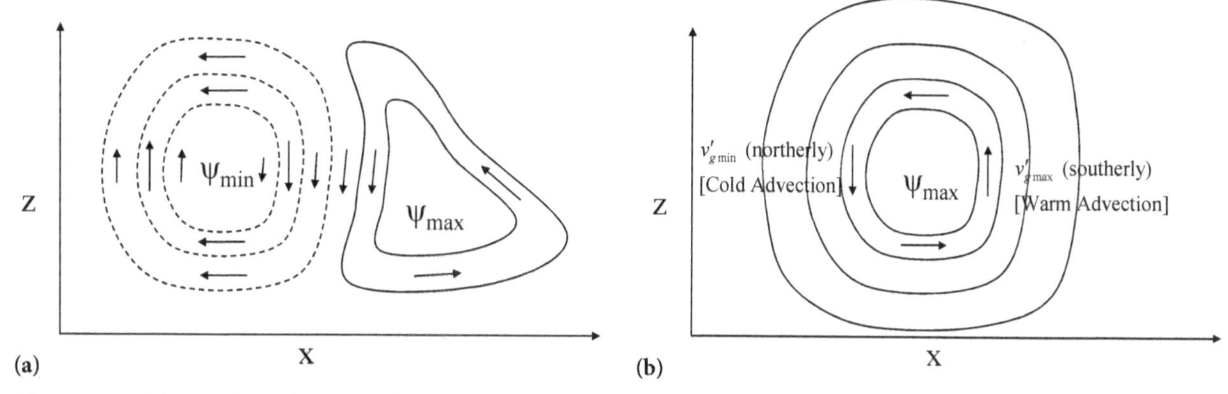


**Figure 7.4.** (a) Hypothetical ageostrophic streamfunction (contours) and vector arrows indicating sense of ageostrophic circulation in the $x$–$z$ plane; (b) illustration of ageostrophic flow branches relative to aspects of the meridional geostrophic flow.

are also consistent with the traditional QG omega equation thermal advection terms (Fig. 7.4b).

If we wish to diagnose the vertical motion field, we can use the continuity equation (7.73) to replace $\partial u'_{ag}/\partial z$ in (7.88), differentiate with respect to $x$, utilize the basic-state thermal wind relation, and rearrange to obtain a "$w$ equation" that is analogous to the QG omega equation for the current system of equations,

$$\left(-\frac{f_0^2}{N_0^2}\frac{\partial^2}{\partial z^2}-\frac{\partial^2}{\partial x^2}\right)w'=\frac{2f_g}{N_0^2}\left[-\frac{\partial U_g}{\partial z}\frac{\partial}{\partial x}\left(\frac{\partial v'_g}{\partial x}\right)\right]. \quad (7.91)$$

The interpretation of (7.91) is exactly consistent with the earlier QG omega equation. Having taken a horizontal derivative, the right side is now related to $-2\nabla\cdot Q$. Another interpretation of the right side of (7.91) is the vorticity advection by the thermal wind, which is exactly the term in square brackets; this is thus the "Trenberth" form of the QG $w$ equation.

## 7.2. SOLUTION TO THE EADY PROBLEM

The traditional starting point for stability analysis is to assume solutions of *normal-mode* form. Normal modes are independent wave solutions that maintain fixed structure with time and operate independently of one another. If a set of unstable normal modes exists in a system, we expect that the one exhibiting the largest growth rate would dominate with time, akin to "natural selection." As will be discussed in section 7.5, however, nonmodal solutions may grow faster than the fastest growing normal mode. Sinusoidal, wave-like solutions of the form

$$\phi'=\mathrm{Re}\left[\hat{\phi}(z)e^{ik(x-ct)}\right] \quad (7.92)$$

are assumed, where the vertical structure of the solution is given by $\hat{\phi}(z)$ and the wave-like portion, giving structure in $x$ and time, is the exponential component. In (7.92), $k=2\pi/\lambda$ is the wavenumber.

Some minimal review of complex variables may be useful here; any number can be represented as a point on the *complex plane*, where the abscissa represents the real component of the number and the ordinate represents the imaginary part, with the number $i=\sqrt{-1}$ being the basis for the imaginary component. In (7.92), $c$ is the *phase speed* of the wave, also a complex number,

$$c=c_r+ic_i. \quad (7.93)$$

A convenient equation for translating between exponential and sinusoidal expressions is Euler's formula,

$$e^{ix}=\cos(x)+i\sin(x). \quad (7.94)$$

Substituting (7.93) in (7.92), the growing or decaying part of the solution can be isolated,

$$\phi'(t)=e^{kc_it}. \quad (7.95)$$

We see that amplifying solutions *depend on the existence of a nonzero imaginary component* of the complex phase speed.

The governing equation for the interior of the Eady domain is the PV relation, which represents a combination of the vorticity and thermodynamic equations. For the current set of equations, the Boussinesq PV, Eq. (7.58), is modified with the $f$-plane approximation,

$$q=f_0+\frac{1}{f_0}\nabla_h^2\phi+\frac{f_0}{N_0^2}\left(\frac{\partial^2\phi}{\partial z^2}\right). \quad (7.96)$$

The full expression for $\phi$ is given by

$$\begin{aligned}\phi &= \Phi(y,z)+\phi'(x,z,t)\\ &=-f_0\frac{U}{H}yz+\phi',\end{aligned} \qquad (7.97)$$

which we can substitute in (7.96) to yield

$$q=f_0+\frac{1}{f_0}\nabla_h^2\phi'+\frac{f_0}{N_0^2}\left(\frac{\partial^2\phi'}{\partial z^2}\right). \qquad (7.98)$$

The linearity of the basic-state $\Phi$ field in $y$ and $z$ means that only the constant $f_0$ remains of the basic-state PV; the two rightmost terms in (7.98) are the PV associated with the disturbance. In other words, $q=Q+q'(x,z,t)$, where $Q=f_0$ is the basic-state PV.

The initial disturbance amplitude is infinitesimally small, because we are investigating the instability of the basic-state current to disturbances that would grow spontaneously in the absence of precursor disturbances. Thus, all of the disturbance quantities, including $\phi'$, are initially zero, and therefore $q'=0$. Given that the basic-state PV is constant, *this requires that the disturbance PV remain zero throughout the interior domain* $(0<z<H, 0<x<L)$ *for all time as well,*

$$q'=\frac{1}{f_0}\frac{\partial^2\phi'}{\partial x^2}+\frac{f_0}{N_0^2}\left(\frac{\partial^2\phi'}{\partial z^2}\right)=0. \qquad (7.99)$$

Substitution of the wave solution (7.92) into (7.99) yields a second-order ordinary differential equation for the vertical structure function,

$$\frac{d^2\hat{\phi}}{dz^2}-\frac{k^2N_0^2}{f_0^2}\hat{\phi}=0. \qquad (7.100)$$

It is left as an end-of-chapter exercise to verify that the following is a solution for the vertical structure,

$$\hat{\phi}=A(k)\sinh\left(\frac{N_0kz}{f_0}\right)+B(k)\cosh\left(\frac{N_0kz}{f_0}\right). \qquad (7.101)$$

Using boundary conditions, we can next solve for the coefficients $A$ and $B$.

Given that the interior PV is zero and cannot change with time, we can ask what kind of disturbances are expected, and where must these disturbances be located? If disturbances are to develop in the interior, effective PV sources must reside on the upper and lower boundaries. The assumption of rigid upper and lower boundaries can be applied to the thermodynamic equation (7.72), resulting in the elimination of the term involving vertical velocity:

$$\frac{\partial b'}{\partial t}+U_g\frac{\partial b'}{\partial x}=-v_g'\frac{\partial B}{\partial y}\quad \text{at } z=0,H. \qquad (7.102)$$

Utilizing the basic-state expressions, as well as the definitions of buoyancy and geostrophic wind, the equation is expressed in terms of $\phi$:

$$\frac{\partial^2\phi'}{\partial z\partial t}+U\frac{z}{H}\frac{\partial^2\phi'}{\partial z\partial x}-\frac{U}{H}\frac{\partial\phi'}{\partial x}=0\quad \text{at } z=0,H. \qquad (7.103)$$

Substituting (7.95) into (7.103) and factoring out common terms yields

$$-c\frac{d\hat{\phi}}{dz}+U\frac{z}{H}\frac{d\hat{\phi}}{dz}-\frac{U}{H}\hat{\phi}=0\quad \text{at } z=0,H, \qquad (7.104)$$

which we can then apply first at $z=0$, then at $z=H$,

$$-c\frac{d\hat{\phi}}{dz}-\frac{U}{H}\hat{\phi}=0\quad \text{at } z=0 \text{ and} \qquad (7.105)$$

$$(U-c)\frac{d\hat{\phi}}{dz}-\frac{U}{H}\hat{\phi}=0\quad \text{at } z=H. \qquad (7.106)$$

Next, the interior solution (7.101) is substituted into (7.105) and (7.106) to provide algebraic equations that can be solved for the coefficients $A$ and $B$. Substituting first for the lower boundary relation, we have

$$A\left(-c\frac{N_0k}{f_0}\right)+B\left(-\frac{U}{H}\right)=0\quad \text{at } z=0. \qquad (7.107)$$

Similarly, substituting in (7.106) for $z=H$ and rearranging to isolate $A$ and $B$,

$$\begin{aligned}&A\left[(U-c)\frac{N_0k}{f_0}\cosh\left[\frac{N_0kH}{f_0}\right]-\frac{U}{H}\sinh\left[\frac{N_0kH}{f_0}\right]\right]\\ &+B\left[(U-c)\frac{N_0k}{f_0}\sinh\left[\frac{N_0kH}{f_0}\right]-\frac{U}{H}\cosh\left[\frac{N_0kH}{f_0}\right]\right]\\ &=0\quad \text{at } z=H.\end{aligned} \qquad (7.108)$$

To solve (7.107) and (7.108) for $A$ and $B$, we make use of the fact that the determinant of coefficients is zero for all nontrivial solutions for $A$ and $B$. The resulting equation relates the phase speed $c$ to the wavenumber $k$, a dispersion relation. To simplify the expression above, we can define the terms multiplying $A$ and $B$ in (7.107) and (7.108) as $a$, $b$, $d$, and $e$,

$$a=-c\frac{N_0 k}{f_0};\quad b=-\frac{U}{H};$$

$$d=\left[(U-c)\frac{N_0 k}{f_0}\cosh\left[\frac{N_0 kH}{f_0}\right]-\frac{U}{H}\sinh\left[\frac{N_0 kH}{f_0}\right]\right]$$

$$e=\left[(U-c)\frac{N_0 k}{f_0}\sinh\left[\frac{N_0 kH}{f_0}\right]-\frac{U}{H}\cosh\left[\frac{N_0 kH}{f_0}\right]\right]. \quad (7.109)$$

Using this simplified notation,

$$a\,A(k)+b\,B(k)=0,$$
$$d\,A(k)+e\,B(k)=0, \quad \text{and}$$
$$ae-bd=0. \quad (7.110)$$

Substituting (7.109) in (7.110) and rearranging yields

$$c^2-Uc+\left(\frac{Uf_0}{HN_0 k}\right)^2\left[\frac{N_0 kH}{f_0}\coth\left(\frac{N_0 kH}{f_0}\right)-1\right]=0; \quad (7.111)$$

this equation can be solved for $c$ using the standard quadratic formula, which, as a reminder, for an equation of the form $ax^2+bx+d=0$ gives the solutions $x=(-b\pm\sqrt{b^2-4ad})/2a$ [note that $a$, $b$, and $d$ here are not the same as those defined in (7.109)]. From (7.111), we then have

$$c=\frac{U}{2}\pm\frac{U}{2}\left[1-4\left(\frac{f_0}{N_0 kH}\right)^2\left\{\frac{N_0 kH}{f_0}\coth\left(\frac{N_0 kH}{f_0}\right)-1\right\}\right]^{1/2}, \quad (7.112)$$

which is the dispersion relation. To simplify the interpretation of this result, we introduce a nondimensional wavenumber $s$ and the Rossby radius of deformation $L_R$,

$$s=\frac{N_0 kH}{2f_0};\quad L_R=\frac{N_0 H}{f_0};\quad s=k\frac{L_R}{2}, \quad (7.113)$$

and substitute into (7.112),

$$c=\frac{U}{2}\pm\frac{U}{2s}[s^2-2s\coth(2s)+1]^{\frac{1}{2}}. \quad (7.114)$$

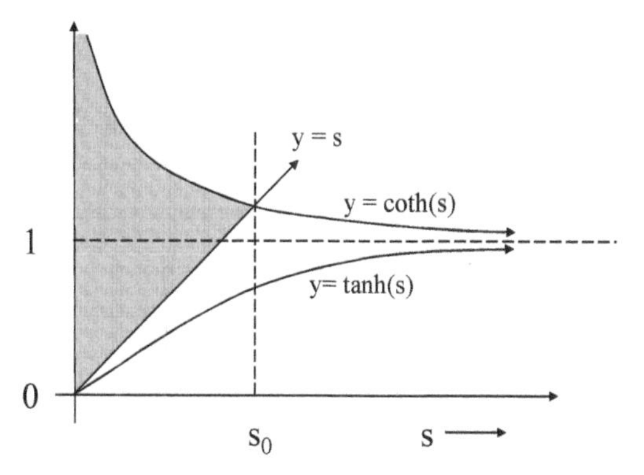


**Figure 7.5.** Graphical representation of conditions for occurrence of a complex phase speed for the Eady baroclinic instability problem. The shaded region of the graph corresponds to the source of imaginary phase speed from (7.115).

Using the identity that $2\coth(2s)=\tanh(s)+\coth(s)$, this can be rewritten in a form that lends itself more readily to graphical representation,

$$c=\frac{U}{2}\pm\frac{U}{2s}[(s-\coth s)(s-\tanh s)]^{\frac{1}{2}}. \quad (7.115)$$

Recall from (7.95) that the growing or decaying portion of the solution depends entirely on the existence of a nonzero imaginary component of the phase speed; therefore, growing solutions can only exist if the discriminant in (7.115) is less than zero. It is clear from Fig. 7.5 that the second term under the radical is always positive, and therefore the imaginary component of $c$ must originate for only values of $s$ that fall below a critical value $s_0$, below which $\coth s > s$; this region is shaded in Fig. 7.5. Because $s$ represents a wavenumber, wavelength increases to the left on the abscissa in Fig. 7.5, meaning that the value $s_0$ can be interpreted as a *short-wave cutoff* to the instability. Substituting typical midlatitude values into this term demonstrates that the short-wave cutoff corresponds to a wavelength of roughly 2,500 km. The physical interpretation of this cutoff for the Eady atmosphere will be presented in subsequent sections.

In addition to the short-wave cutoff, it is also evident from Fig. 7.5 that there exists a wavenumber characterized by maximum growth, because, as the wavenumber becomes very small, $s-\tanh s\rightarrow 0$, leading to a reduction in growth rate at large wavelength.

The dispersion relation (7.115) can be written

$$c=c_r+ic_i=\frac{U}{2}\pm i\frac{U}{2s}\left[(\coth s-s)(s-\tanh s)\right]^{\frac{1}{2}}. \quad (7.116)$$

In our original wave solution,

$$\phi' = \mathrm{Re}\left[\hat{\phi}(z)e^{ik(x-ct)}\right] = \mathrm{Re}\left[\hat{\phi}(z)e^{ik(x-[c_r+ic_i]t)}\right]$$
$$= \mathrm{Re}\left[\hat{\phi}(z)e^{ik\left(x-\frac{U}{2}t\right)}e^{kc_i t}\right], \tag{7.117}$$

there are three parts to the solution evident in the rightmost expression in (7.117). The vertical structure $\hat{\phi}(z)$ has the form of hyperbolic functions as evident in (7.101); the middle term, using Euler's formula (7.94), is the wave part of the solution. The time-dependent growth rate $e^{kc_i t}$ will now be examined in greater detail. Given (7.116), this term is actually $e^{\pm kc_i t}$; however, because we are starting with infinitesimally small initial amplitude, the decaying solutions are of no interest. Recall that $s = k(L_R/2)$; $k = 2s/L_R$, so that we can write the Eady growth rate, which is the coefficient multiplying time in the exponent, as

$$kc_i = \left(\frac{U}{L_R}\right)[(\coth s - s)(s - \tanh s)]^{\frac{1}{2}}$$
$$= \frac{U f_0}{H N_0}[(\coth s - s)(s - \tanh s)]^{\frac{1}{2}}. \tag{7.118}$$

The growth rate is scaled by the basic-state shear $(U/H)$, is inversely proportional to the basic-state static stability $N_0 = (g/\theta_{00})d\theta_0/dz$, and has latitudinal dependence. All of these factors are consistent with observations and with the material discussed in chapter 5; as we will see subsequently, they are easily interpreted in the PV framework. To determine the wavenumber of maximum growth rate, we can take $\partial c_i/\partial s$ and set the result equal to zero to solve for the maximum,

$$kc_{i\max} = 0.31\frac{f_0}{N_0}\frac{U}{H}, \tag{7.119}$$

which corresponds to $s_{\max} = 0.8031$. Plugging in synoptic-scale values ($L_R \sim 1000$ km) yields a wavelength of maximum growth of ~4,000 km, which corresponds to wavenumber 6 or 7 at 45°N latitude.

Before providing physical explanations for the growth and cutoff wavelengths, it remains to examine the vertical structure of the disturbances. We have utilized the boundary conditions to solve for the coefficients $A$ and $B$ in (7.101), which show that

$$B = 1, \ A = -\frac{f_0}{N_0 k}\frac{U}{cH}. \tag{7.120}$$

We can also define the Rossby depth,

$$H_R \equiv \frac{f_0}{N_0 k}, \tag{7.121}$$

and utilize this to write

$$\hat{\phi}(z) = -\frac{H_R}{H}\frac{U}{c}\sinh\left(\frac{z}{H_R}\right) + \cosh\left(\frac{z}{H_R}\right). \tag{7.122}$$

If we express the hyperbolic functions in (7.122) as exponentials using (7.124), it is evident that the Rossby depth $H_R$ describes the e-folding vertical length scale for the disturbance. Intuitively, this scale is inversely proportional to the basic-state static stability and wavenumber and is proportional to latitude. Recall that, with zero interior PV, the amplitude of the disturbance amplitude in the interior will depend on the extent to which the boundary PV anomalies extend in the vertical.

With an expression for the vertical structure of the disturbance, we have what is needed to compute the disturbance streamfunction, geostrophic and ageostrophic wind, and potential temperature fields using $v_g' = 1/f_0/(\partial\phi'/\partial x)$ and $b' = \partial\phi'/\partial z$. Starting with the perturbation geopotential field, we isolate the contribution from the top and bottom boundaries individually (Figs. 7.6a,c); these "edge waves" are presented in section 7.3. For the parameters used here to plot the most unstable wave, the disturbance on either boundary extends through the interior and reaches the opposing boundary with some amplitude remaining. The total geopotential field can then be interpreted as the sum of the vertical extensions of these two boundary disturbances.

There is a phase shift between the upper and lower disturbances such that the trough and ridge axes slope westward with height. This structure is consistent with our expectation based on the energetics of the disturbances discussed previously in section 7.1.3. In fact, the disturbance is leaning against the basic-state shear, which is also consistent with our expectations based on the QG energy analysis in chapter 2.

The perturbation geostrophic wind matches what is dictated by the streamfunction pattern, with southerly wind maximized to the east (west) of trough (ridge) axes in the vertical plane (cf. Figs 7.6b, 7.7). Because there are no interior PV anomalies and friction is neglected, the solution presented here cannot produce isolated interior wind maxima.

From our earlier discussion of energy conversions in the Eady atmosphere, we expect from (7.83) that there will be a northward eddy heat flux, consistent with extraction

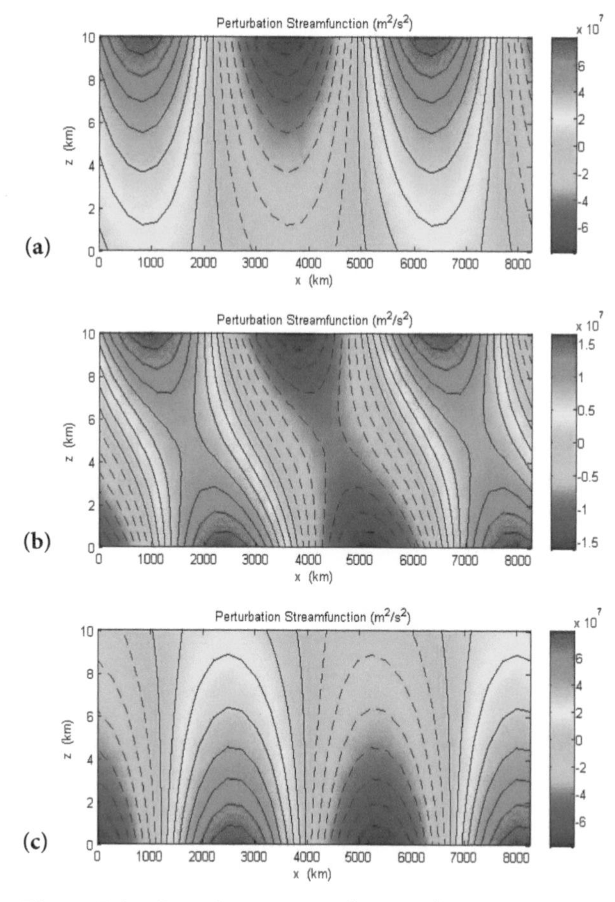


**Figure 7.6.** Perturbation streamfunction for most unstable Eady mode, contoured and shaded: (a) top disturbance only; (b) full solution; and (c) lower disturbance only.

of basic-state potential energy. Examination of the perturbation temperature field along with the meridional disturbance wind confirms that there is indeed a northward eddy heat flux in the solution; areas of southerly (northerly) perturbation flow are characterized by positive (negative) perturbation temperatures (cf. Figs 7.7a,b). The structure of the disturbance temperature leans in the opposite sense to that of the geopotential, consistent with warm lower troughs and cold upper troughs.

Our analysis of the energetics also demonstrated that the disturbance kinetic energy developed from an *upward* heat flux, again consistent with baroclinic energy growth. We therefore expect ascent (descent) in regions of warm (cold) temperature perturbation for several reasons. First, the temperature anomalies must arise from advection of the basic-state temperature gradient, meaning that regions of southerly flow should be relatively warm. The QG $w$ equation would also lead us to expect ascent in regions of maximum warm advection. Comparison of Figs. 7.7 and 7.8 demonstrates that regions characterized by southerly flow, consistent with warm advection and warm perturbation potential temperature, exhibit ascent and vice versa for the cold northerly flow.

At the upper and lower boundaries, there is a 21° phase shift between the thermal and streamfunction fields, which is critical for the existence of a poleward heat flux. This shift is also evident in plan plots. If we compare a streamline and an isentrope on the top surface (Fig. 7.9a) with the same on the bottom surface (Fig. 7.9d), we see that troughs at the upper boundary are cold, whereas troughs at the lower boundary are warm. This has implications for interpreting the results via the PV framework (see section 7.4). Recall from chapters 4 and 5, particularly Fig. 5.17, that the cyclogenesis process involves mutual amplification of upper and lower disturbances that are spatially offset from one another in a manner that allows "phase locking" of the disturbances. The same interpretation is evident from

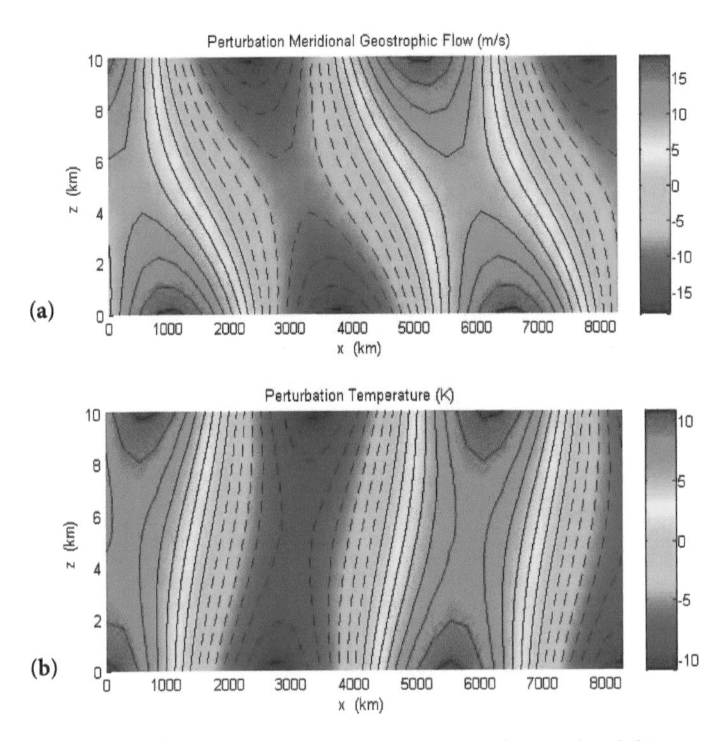


**Figure 7.7.** Perturbation (a) meridional geostrophic wind and (b) temperature components, shaded and contoured with negative values dashed.

Fig. 7.6; the disturbance from one boundary is associated with a circulation that extends to the opposite boundary, and the horizontal offset allows mutual amplification of the upper and lower disturbances.

For the upper and lower disturbances to phase lock, several conditions must be met. These conditions can be interpreted with the Rossby depth, $H_R$ given by (7.121), which controls the vertical extent ($e$-folding distance) of the boundary-based disturbance amplitude. For an environment characterized by large static stability, the Rossby depth would be small, and the disturbances may not be able to phase lock. In unstable environments, the boundary disturbances would more readily reach the opposite boundary.

The normalized wavenumber $s$ is related to the Rossby depth,

$$s = \frac{N_0 k H}{f_0} = \frac{H}{H_R}, \tag{7.123}$$

and we see that disturbances of small wavelength extend a smaller vertical distance into the interior of the domain. Thus, the physical interpretation of the cause for

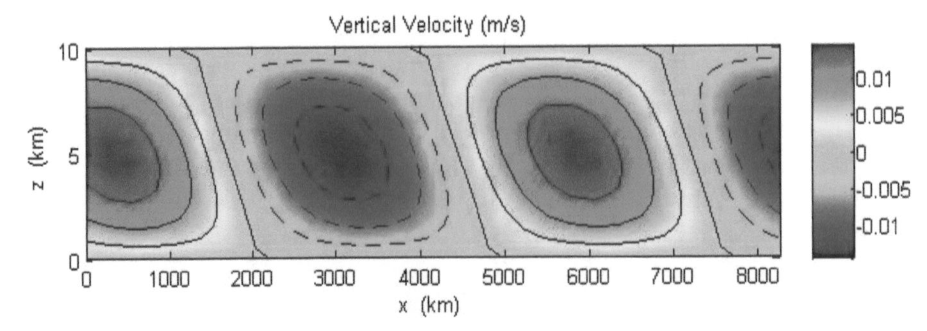


**Figure 7.8.** Vertical velocity for most unstable Eady mode; warm colors and solid contours correspond to ascent, and cold colors and dashed contours correspond to descent.

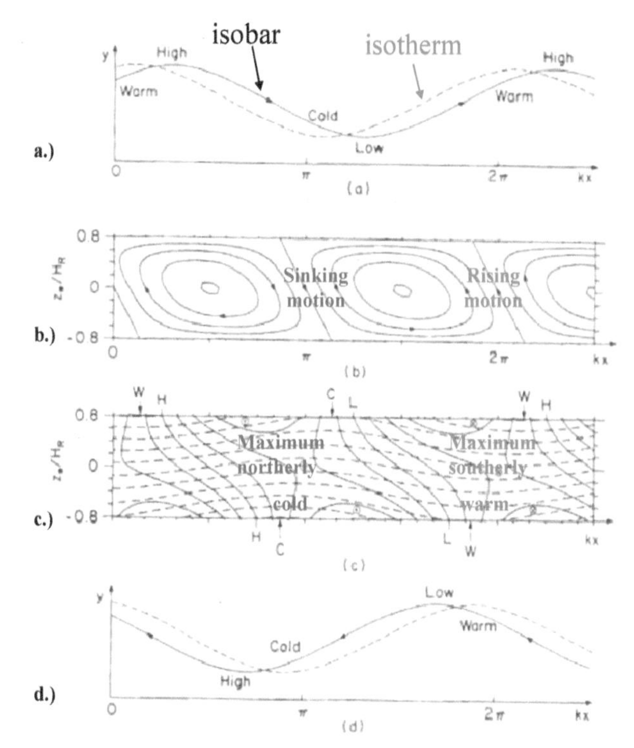


**Figure 7.9.** Structure of most unstable Eady wave: plan view of isobar and stream line at (a) the top of the domain ($z = H$) and (d) the lower boundary ($z = 0$), and cross section in $x$ and $z$ of (b) ageostrophic streamfunction, and (c) meridional wind (solid contours) and potential temperature (dashed contours; from Gill 1982).

the short-wave cutoff is apparent: When the vertical extent of the boundary circulations falls below some critical fraction of the depth of the troposphere, the disturbances no longer interact and mutually amplify. Because of the stability dependence of the Rossby depth, we expect that the short-wave cutoff would vary proportionally to environmental stability. Either large environmental static stability (meaning that the fraction $H/H_R$ is large) or a large wavenumber (small wavelength) could result in an inability of the disturbances to phase lock.

In addition to the vertical extent of the boundary circulations, the wave number also determines the phase speed of the disturbances through the dispersion relation. The advective action of the vertically sheared basic-state flow tends to work against phase locking by advecting the upper and lower waves apart. The unstable upper and lower Eady waves propagate in opposite directions with respect to the basic-state flow. If the propagation is sufficiently strong to counter the action of the basic-state advection, the upper and lower disturbances may travel at similar speeds, facilitating phase locking and growth (see Fig. 5.17). The wavelength of maximum growth rate can be interpreted as the wave that exhibits the most favorable phase-locking speed at the upper and lower boundary and also sufficient vertical extension of the boundary disturbances through to the opposite boundary to allow mutual amplification.

It was mentioned in reference to Fig. 7.5 that there also exists a long-wave cutoff. This can also be interpreted from the phase-locking mechanism: as the wavelength becomes large, the upper and lower disturbances would move apart because of a propagation effect that overcomes the advective tendencies of the basic-state flow, leading to *westward* movement of the upper disturbance with respect to the lower, preventing phase locking. For in-depth discussion of these aspects of the Eady model, see Lindzen (1994) and Liu and Mu (2001).

Given that we have assumed an $f$ plane here, how can we interpret the upper and lower propagating waves as Rossby waves, which were said to result from advection of planetary vorticity? The case where a gradient of planetary vorticity allows Rossby waves to propagate is a specific case of a more general condition that Rossby waves need a *PV gradient* upon which to propagate. There is no basic-state PV gradient in the interior of the Eady domain, but recall that there is equivalence between boundary thermal anomalies and PV anomalies, as discussed in chapters 2 and 5. Therefore, the basic-state thermal gradient on the upper and lower boundaries plays the role of a PV gradient, and the upper and lower thermal waves propagate on this gradient. The structure of an Eady edge wave is provided in Fig. 7.10.

## 7.3. EADY EDGE WAVES

From the preceding interpretation, the importance of the tropopause to the existence of baroclinic instability is clear: the mutual amplification of upper and lower waves requires interaction between disturbances tied to the upper and lower boundaries. We can then ask, if we took away the upper boundary, would Eady-type baroclinic instability still be possible? This could represent a situation in which the tropopause was relatively high in altitude, such as in the tropics, or for disturbances of sufficiently short wavelength that their associated Rossby depth is small relative to the depth of the tropopause. This problem may

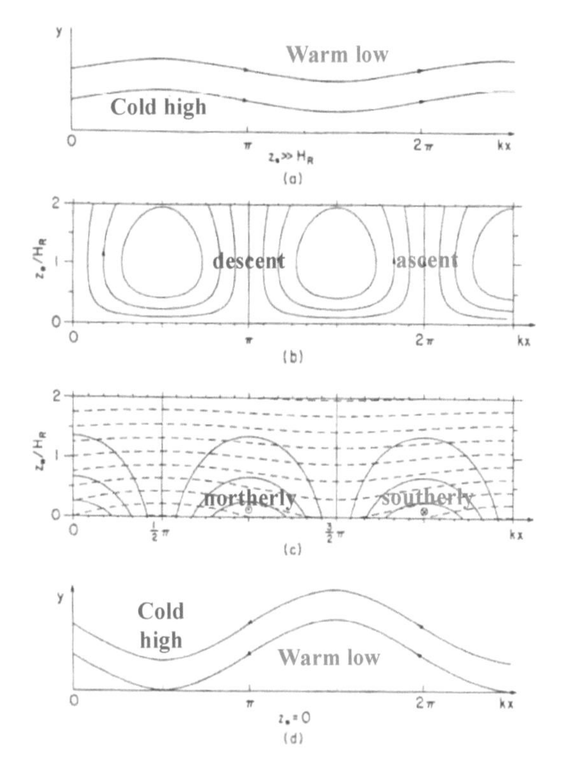


**Figure 7.10.** As in Fig. 7.9, but for Eady edge wave. Adapted from Gill (1982).

also be relevant to other planets, such as Mars, in which there is not enough atmospheric oxygen to allow the development of ozone and a stratospheric layer.

We can approach the problem exactly as before, except that the upper boundary condition will be changed from that of a rigid lid to a condition that guarantees that the disturbance energy will remain bounded as $z \rightarrow \infty$. The governing equation in the interior remains (7.92); assuming the same wavelike disturbance as before, we again obtain (7.100), for which the solution is again (7.101). The lower boundary condition (7.105) remains unchanged as well. If we this time rearrange (7.101) using the relations

$$\sinh x = \frac{1}{2}(e^x - e^{-x}); \quad \cosh x = \frac{1}{2}(e^x + e^{-x}), \qquad (7.124)$$

we can express the solution as

$$\hat{\phi}(z) = A_2 e^{\frac{N_0 k}{f_0} z} + B_2 e^{-\frac{N_0 k}{f_0} z}. \qquad (7.125)$$

The subscripts on the coefficients $A$ and $B$ indicate that these are not the same as those in (7.101) but represent a combination of those earlier coefficients. Given the upper boundary condition here, $\phi' \rightarrow 0$ as $z \rightarrow \infty$, this requires that $A_2 = 0$. Therefore,

$$\hat{\phi}(z) = B_2 e^{-\frac{N_0 k}{f_0} z} = B_2 e^{-\frac{z}{H_R}}, \qquad (7.126)$$

and we again see that the Rossby depth is the $e$-folding vertical length scale for the vertical extent of the disturbance. Letting $B_2 = \Phi_0$, the amplitude of the disturbance geopotential perturbation at the surface, the full disturbance solution is

$$\phi' = \Phi_0 e^{-\frac{z}{H_r}} e^{ik(x-ct)}. \qquad (7.127)$$

Applying the lower boundary condition,

$$\frac{\partial^2 \phi'}{\partial z \partial t} - \frac{U}{H}\frac{\partial \phi'}{\partial x} = 0 \quad \text{at } z = 0, \qquad (7.128)$$

and substituting (7.127), we obtain the dispersion relation,

$$c = \frac{f_0}{N_0 k}\frac{U}{H} = H_R \frac{U}{H}. \qquad (7.129)$$

Unlike the full Eady problem that included an upper boundary, here $c$ has no imaginary component and the solution features only propagating waves that cannot amplify. The dispersion relation (7.129) reveals that the phase speed increases with wavelength, and with the basic-state shear, which is consistent with a faster motion from advection of the disturbance by the basic-state flow. The phase speed also increases with Rossby depth, which can be interpreted as a greater vertical extent of the disturbance, and an effectively higher "steering level" for the system. The interpretation of the parameter $H$ here is different than the case with an upper boundary; now, this represents a vertical scaling distance over which the basic-state shear is evaluated.

## 7.4. NECESSARY CONDITIONS FOR INSTABILITY

The preceding analysis reveals that even an extremely simple baroclinic westerly current is unstable to synoptic disturbances with realistic structure, scale, and energy conversion processes. Provided that the Rossby depth is sufficiently large and that the depth of the troposphere is not too large, growing Eady waves can result. When these conditions are not met, the *short-wave cutoff* describes the situation in which mutual amplification of the upper and lower waves will no longer take place. However, studies by Green (1960), Charney and Stern (1962), Pedlosky (1964), Miles (1964), and Bretherton (1966) extended the simplified Eady model into more general statements of when instability is permitted. By relaxing the $f$-plane assumption, Green (1960) showed that growing modes are present at nearly all wavelengths. By examining the behavior of the system near the vertical level where the wave phase speed $c$ equals the basic-state flow $U$, additional growing modes are possible when $\beta \neq 0$. This *critical layer instability* was cast into a PV context by Bretherton (1966) and will be discussed in detail below. Green (1960) also pointed out that a *necessary condition* for instability was that the parcel trajectories must, on average, follow a slope between the horizontal and the angle of the basic-state isentropes, but at this point we will focus on understanding why relaxing the $f$-plane assumption eliminates the short-wave instability cutoff.

If $\psi = \phi / f_0$ and the $f$-plane assumption is relaxed to allow $\beta \neq 0$, the anelastic PV expression becomes

$$q = f_0 + \beta y + \nabla_h^2 \psi + \frac{f_0}{\rho(z)} \frac{\partial}{\partial z}\left( \frac{f_0^2 \rho}{N^2} \frac{\partial \psi}{\partial z} \right). \qquad (7.130)$$

Consider a domain that includes lateral boundary conditions such that the meridional geostrophic wind component goes to zero at two different latitudes ($\partial \psi / \partial x = 0$ *at* $y = y_1, y_2$) and that includes upper and lower boundaries on which $\partial \psi / \partial z = 0$ at $z = 0, H$. The zonal average of a given quantity in such a domain is not a function of $x$. In such a domain, *the mean, domain-averaged meridional eddy PV flux is zero,*

$$\int_0^H \int_{y_1}^{y_2} \overline{q v_g'}\, dy\, dz = 0. \qquad (7.131)$$

The upper and lower boundary conditions required for (7.131) are restrictive, equivalent to assuming uniform temperature on these boundaries, which is clearly not realistic for many problems of interest. However, the point of this development is to demonstrate that an additional instability mechanism can arise in the interior of the domain.

Consider the development of PV perturbations as the result of meridional flow. For $\beta \neq 0$, a basic-state meridional PV gradient exists. Accordingly, as an air parcel is displaced from its original latitude, PV conservation requires that a PV anomaly will develop. Denoting the meridional distance of departure from the original latitude as $\eta = y - y_0$ and noting that $d\eta / dt = dy/dt = v_g'$, we find that the average meridional eddy PV flux is

$$\overline{q' v_g'} = -\frac{dQ}{dy} \overline{\eta v_g'} = -\frac{dQ}{dy} \frac{d}{dt} \frac{\overline{\eta^2}}{2}. \qquad (7.132)$$

Next, integrating the eddy PV flux over the domain as in (7.131) using (7.132) yields

$$\int_0^H \int_{y_1}^{y_2} \frac{dQ}{dy} \frac{d}{dt} \frac{\overline{\eta^2}}{2} dy\, dz = 0. \qquad (7.133)$$

Turning now to a growing baroclinically unstable disturbance, we expect that, on average, as the disturbance amplifies, the average absolute meridional departure of air parcels from their latitude of origin ($\overline{\eta^2}/2$) would increase with time. In fact, Bretherton (1966) shows that

$$\frac{d}{dt} \frac{\overline{\eta^2}}{2} \propto f(c_i) e^{2kc_i t}, \qquad (7.134)$$

which clarifies that growth of the average displacement with time is consistent with the requirement for an imaginary component to the phase speed, as discussed earlier with (7.95). Examination of (7.134) demonstrates that to satisfy (7.133) requires that *the sign of $dQ/dy$ reverse somewhere within the domain* to allow for cancellation of the eddy PV flux somewhere within the domain. That the basic-state PV gradient reversal is a *necessary condition for instability* has many practical applications and pertains to barotropic as well as baroclinic instability.

However before relating this back to the original Eady problem, consider what happens at a *critical level* in the flow, where $U = c$. At such a level, air parcels ahead of a trough axis will maintain the same system-relative location and thus move continuously poleward (and vice versa behind the trough axis) as shown in Fig. 7.11. Thus, *if a basic-state PV gradient exists at the critical level, then a growing PV perturbation must develop there.* This is true even for neutral (nonamplifying) disturbances. In this case, to satisfy (7.133), there must be a growing disturbance elsewhere in the domain to balance the critical-level PV flux, requiring that $c_i \neq 0$, and instability of the flow.

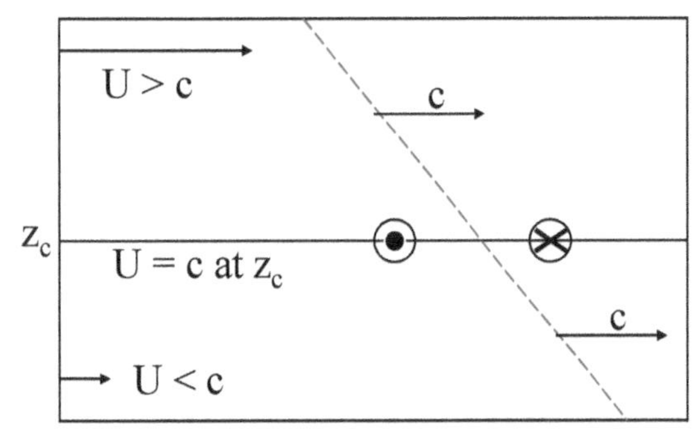


**Figure 7.11.** Idealized $x$–$z$ cross section showing a critical-level $z_c$; the green dashed line represents a hypothetical trough axis.

This type of critical-layer instability must exist whenever $dQ/dy \neq 0$ at $z = z_c$.

The $f$-plane assumption eliminated the meridional PV gradient at the critical level, precluding this type of instability in the original Eady (1949) problem. However, by including a meridional PV gradient in the form of $\beta$ at the critical level (and throughout the domain), the short-wave cutoff is eliminated because a slowly growing disturbance is now required to offset the meridional PV flux at the critical level. There can be neutral modes even with $\beta$, but these modes must have a node at the critical level such that there is no meridional flow there. Aside from these special cases, there are no neutral modes with $\beta \neq 0$ at $z = z_c$, and there are growing modes with a critical layer only if the PV gradient reverses within the domain.

In the original *f-plane* Eady problem, instability was identified for certain wavelengths, despite the absence of a basic-state PV gradient or critical-level instability. The PV in the interior of the Eady domain was zero. How can this be reconciled with the necessary condition stated above? It would seem at first glance that the necessary condition of a basic-state PV gradient reversal is not met. However, this condition must take into account what happens at the *boundaries* of the domain. Bretherton (1966) demonstrates that boundary potential temperature gradients can be mathematically expressed as PV gradients that exist in a very thin layer at the boundary. The potential temperature gradient on both the upper and lower boundary is directed equatorward in the Eady problem; however, because warm perturbations are cyclonic on the lower boundary, whereas cold perturbations are cyclonic on the tropopause, this condition is consistent with a reversal of the PV gradient. Mathematically, at the lower boundary, in a very thin sheet centered about $z = z_1$, we can write

$$q(z) = \left[ \frac{f_0^2}{N^2} \frac{\partial \psi}{\partial z} \right]_{z1-}^{z1+} \delta(z - z_1); \tag{7.135}$$

for the upper boundary,

$$q(z) = - \frac{f_0^2}{N^2} \frac{\partial \psi}{\partial z} \bigg|_{h-} \delta(z - h). \tag{7.136}$$

The change in sign between (7.135) and (7.136) is consistent with opposite-signed PV anomalies representing cyclonic disturbances on the upper and lower boundary and is consistent with the necessary condition of a PV-gradient reversal for meridional potential temperature gradients at the upper and lower boundaries.

For those not comfortable with the mathematical equivalence of boundary potential temperature gradients as PV gradients in thin sheets, an alternate statement for the necessary condition (e.g., Charney and Stern 1962; Pedlosky 1964; Gill 1982) is that *the following expressions must not have the same sign throughout the domain and on the boundaries:*

$$\frac{\partial Q}{\partial y}\bigg|_{\text{interior}}; \quad \frac{\partial \Theta}{\partial y}\bigg|_{\text{lower}}; \quad -\frac{\partial \Theta}{\partial y}\bigg|_{\text{upper}}. \tag{7.137}$$

## 7.5. THEORY AND OBSERVATION: NONMODAL INSTABILITY

As highlighted in the quote from Palmén at the beginning of this chapter, there is some question concerning the relevance of normal-mode baroclinic instability of a simple westerly current to observed cases of atmospheric cyclogenesis. The observational study of Sanders (1986) demonstrates the ubiquity of preexisting upper disturbances in nearly all cases of extratropical cyclogenesis. Thus, the structural properties and interactions between finite-amplitude initial disturbances may be more relevant to observed cyclogenesis than the stability of a hypothetical basic-state current.

Studying cyclogenesis that results from the interaction and growth of finite-amplitude initial disturbances is an *initial-value problem* (e.g., Farrell 1984, 1985, 1989). A complete spectrum of initial disturbances can be described with this approach, but the difficulty lies in how to isolate their structure. In the normal-mode approach adopted in the Eady problem, the structure (of hyperbolic sine and cosine functions) fell out naturally, and there was no need to arbitrarily prescribe a structure other than to assume a wavelike disturbance. In the Eady approach, the fastest growing wave comes to dominate via a natural selection process; thus, by studying the most unstable mode, the problem is simplified. However, as Farrell has argued, it is not at all clear that the initial disturbances are of normal-mode form in the real atmosphere; furthermore, Farrell demonstrates that nonmodal structures often exhibit more rapid growth than the most unstable normal mode. To address the challenge of which of the nonmodal structures would grow most rapidly, Farrell (1989) identified disturbances possessing a structure that would maximize energy growth. Although he found disturbances that grow more rapidly than the most unstable normal mode, some of the structures do not appear realistic in comparison to observations. However, the key point of this analysis is that, in the real world, nonmodal disturbances exist and are able to convert energy more efficiently than a normal mode would.

The beauty of the Eady problem lies in its simplicity. Baroclinic instability is envisioned as a mutual amplification of upper and lower thermal waves, based at the surface and tropopause. Other baroclinic instability treatments, such as Charney (1947) and Green (1960), describe more complicated processes involving interior PV anomalies, such as would develop near a critical level because of advection of a basic-state PV gradient. In Charney's model, there is no upper rigid boundary as Eady assumed, but $\beta$ is included. Robinson (1989) evaluated the PV structures of the Eady, Charney, and Green baroclinic instability problems and concluded that the relevance of the Eady model depends on the interactions between the surface and tropopause waves dominating the effect of advectively generated PV in the interior. The latter was found to characterize the Charney and Green instability problems. The presence of a tropopause in the Charney model was found by Robinson to be insufficient to change the character of the modes; also, the tropospheric PV gradient in the Charney model is larger than that observed in the atmosphere.

How can we synthesize the elements of the theoretical work into a picture that is both consistent with observations and conceptually simple? If we treat the origin of upper and lower disturbances as a separate problem but study the interactions of the preexisting disturbances, we arrive at a simple and useful picture of cyclogenesis. The PV framework advanced by Bretherton (1966), Hoskins et al. (1985), and Robinson (1989) clarifies the necessary conditions for baroclinic instability. The potential vorticity framework is readily applicable to observed data as well; Davis and Emanuel (1991) used observational analyses to investigate the structures of PV growth during cyclogenesis in the real atmosphere. One motive for their work was to quantify the importance of the "Eady view" (surface and tropopause disturbance amplification) versus the "Charney view," in which interior PV anomalies would develop due to advection of a basic-state PV gradient.

The results of the Davis and Emanuel study show that the structural evolution of a cyclone from 5 February 1988 is complex and changes with time. The importance of diabatic PV redistribution in the interior in association with precipitation and latent heat release is important, and this element is not considered in the early theoretical baroclinic instability studies, which assumed adiabatic, frictionless conditions. Nevertheless, despite these complications, Davis and Emanuel used PV inversion to identify the contributions of the different PV anomalies to the total flow field, and they used these to diagnose the presence of mutual amplification between upper and lower disturbances. The surface potential temperature anomaly clearly displayed the character of a propagating surface wave (Fig. 7.12a), and this circulation was sufficiently strong to reach the tropopause and amplify the upper-tropopause cold potential temperature anomaly there (equivalent to a cyclonic PV anomaly; Fig. 7.12b). The case used in this study featured a strong upper-level wave, and the actions of the cyclonic winds associated with this upper disturbance are clearly seen to amplify the surface warm anomaly (Fig. 7.12c).

The PV-based diagnosis of the February 1988 cyclone event revealed that the surface thermal anomaly and a lower-tropospheric cyclonic PV anomaly contributed most strongly to the surface cyclonic circulation and that the lower interior diabatic PV anomaly was important initially in the amplification of the surface thermal wave. Later, as the upper-level PV anomaly approached the lower disturbance, the associated cyclonic flow became important to the amplification of the lower wave. The upper wave amplified because of the combined advective effects of the lower thermal anomaly and the interior diabatic PV anomaly. Thus, during a portion of the cyclone life cycle,

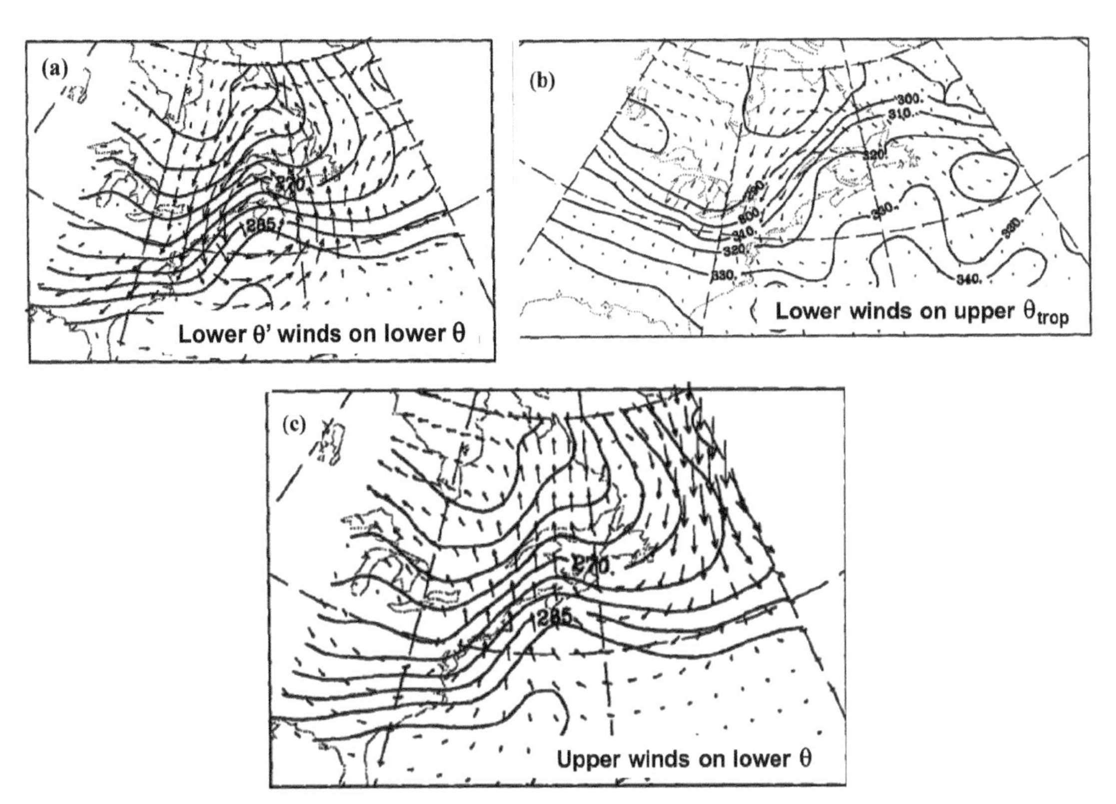


**Figure 7.12.** Results of PV inversion from observational data: (a) wind field recovered from inversion of lower potential temperature and lower-tropospheric PV anomalies (vectors) superimposed on lower potential temperature contours; (b) inverted winds from lower boundary potential temperature and lower-tropospheric PV superimposed on potential temperature at dynamic tropopause; and (c) inverted winds from upper-level PV anomalies superimposed on lower potential temperature (from Davis and Emanuel 1991).

the PV inversion results revealed an "Eady like" mutual amplification of upper and lower disturbances, but with an important added contribution of the interior diabatic PV anomaly. Thus, Davis and Emanuel conclude that, in the case they analyzed, the atmosphere appeared more consistent with the Eady model than that of Charney.

Several additional cases were analyzed with the same general PV approach, many with the benefit of numerical simulations. These studies reveal a high degree of case-to-case variability and suggest different roles for the lower diabatic PV anomaly. In the cases studied by Davis and Emanuel (1991), Davis (1992), and Davis et al. (1993), the diabatic PV anomaly contributed significantly to the lower circulation; however, in a case of continental cyclogenesis studied by Davis (1992), the action of the lower cyclonic PV anomaly led to the eastward acceleration of the surface thermal wave, leading it to outrun the upper disturbance, perhaps ultimately leading to a detrimental influence on the cyclone. Another study of diabatic PV anomalies by Stoelinga (1996) demonstrated a 70% contribution to the cyclonic circulation from this anomaly.

An important finding of Davis and Emanuel (1991) is that the dynamical character of the growing disturbance exhibits distinct phases, and is thus fundamentally nonmodal. However, the normal-mode approach makes no attempt to examine the full life-cycle of cyclones or baroclinic waves. In recent years, a rich variety of baroclinic instability studies have examined the full nonlinear behavior of growing baroclinic waves and related these evolutions to observed systems (e.g., Simmons and Hoskins 1980; Thorncroft et al. 1993; Rotunno et al. 1994).

A summary of some of the most important insights from the Eady baroclinic instability analysis is provided below:

- We have determined that even a very simple baroclinic flow is unstable to certain perturbations.

- The structure of the most unstable mode naturally falls out of the equations and is characterized by mutually amplifying, counter-propagating waves that reside on the upper and lower boundaries.
- The Rossby depth, which is dependent on the static stability and wavenumber, describes the extent to which disturbances on a given boundary are able to "reach" the opposite boundary, potentially leading to mutual amplification.
- A necessary condition for instability is that the basic-state PV gradient reverse sign within the fluid. Equivalently, the following functions must not all have the same sign: $\left.\frac{\partial Q}{\partial y}\right|_{\text{interior}}$ ; $\left.\frac{\partial \Theta}{\partial y}\right|_{\text{lower}}$ ; $-\left.\frac{\partial \Theta}{\partial y}\right|_{\text{upper}}$ .

A weakness of the Eady analysis is that, in the real atmosphere, cyclones are not observed to develop simultaneously at all levels from infinitesimally small perturbations. Furthermore, the fastest growing modal wave disturbances may not represent the fastest growing disturbances from the complete spectrum of possible disturbance structures. Finally, baroclinic energy growth is not necessarily an indication of baroclinic instability. Rotunno and Fantini (1989) showed that a set of neutral modes exhibited baroclinic energy growth simply through superposition.

Analysis of observational data does provide support for an Eady-like growth mechanism during at least some portion of the life cycle of many midlatitude cyclones. PV-based graphics from the Presidents' Day cyclone of 1979, provided at the beginning of this chapter, demonstrate a suggestive phase-locking hook shape of the upper and lower thermal anomalies, also consistent with an Eady-type mechanism.

To reconcile the question of frontal instability with larger-scale baroclinic instability of a broad westerly current, we can consider the observed long-wave disturbances in the atmosphere as being related to baroclinic instability. The frontogenetical properties of the resulting large-scale disturbances give rise to intense frontal zones, which can in turn be associated with smaller-scale frontal-wave-type instabilities. Cyclones initially forming as frontal waves may amplify and exhibit upscale growth.

## REVIEW AND STUDY QUESTIONS

1. How did the *f-plane assumption* affect the Eady baroclinic instability problem?
2. In what ways are the results of the Eady problem realistic, and in what ways are they unrealistic?
3. In early Earth history, prior to the development of significant amounts of oxygen, there was no ozone layer and thus no stratosphere. Based on the Eady analysis, discuss how the cyclogenesis/baroclinic instability process may differ in such an atmosphere.
4. Using a PV-based analysis, to what extent did Davis and Emanuel (1991) find observed cases of cyclogenesis to be "Eady like"? What differences did they find?
5. What are the advantages and disadvantages of the initial-value problem approach to baroclinic instability?

## PROBLEMS

1. Show that the basic-state atmosphere described by (7.6)–(7.11) represents a solution to the primitive equations (7.1)–(7.5).
2. Derive (7.16) from the preceding equations.
3. Demonstrate that the requirement for a poleward heat flux for growth of eddy potential energy as seen in (7.86) is only found for disturbances whose axes slope westward with increasing height.
4. Verify that (7.101) is a solution to (7.100).
5. The Eady problem as presented here contained a large number of simplifying assumptions, including neglect of friction. If we wish to extend our Eady edge-wave analysis to include the effects of a frictionally driven planetary boundary layer (PBL), we can redefine the lower boundary in the Eady domain to be the top of the PBL, at height $h$. Keep all other aspects of the problem the same, but utilize a different lower boundary condition. A simple but realistic parameterization of Ekman friction in the planetary boundary layer in a manner that is consistent with the QG system we are using is

$$w' = \frac{C_d h}{f_0}\zeta_g', \quad \text{at } z = 0 = \text{top of PBL}, \tag{1}$$

where $C_d$ is a linear friction coefficient, $\zeta_g'$ is the disturbance geostrophic relative vorticity, and $h$ is the PBL depth (assumed constant). The interior PV is

$$q' = \frac{1}{f_0}\frac{\partial^2 \phi'}{\partial x^2} + \frac{f_0}{N_0^2}\frac{\partial^2 \phi'}{\partial z^2} = 0. \tag{2}$$

As before, assume a solution of the form $\phi' = \mathrm{Re}\{\hat{\phi}\, e^{ik(x-ct)}\}$; the thermodynamic equation is

$$\frac{\partial b'}{\partial t} + U_g \frac{\partial b'}{\partial x} + v_g' \frac{\partial B}{\partial y} + w' N_0^2 = 0. \tag{3}$$

The Eady basic-state environment remains $U_g = U\frac{z}{H}$. The upper boundary condition remains that the disturbance amplitude must vanish as $z$ goes to infinity to obtain

$$\hat{\phi}(z) = \Phi_0 e^{\frac{-N_0 kz}{f_0}}.$$

i. Utilize (1) as a lower boundary condition in (3) to obtain an expression for the *phase speed.* Note that $z = 0$ on the lower boundary, even though we are not assuming that this is the physical ground surface.
ii. Use the results of (i) to develop an expression for the perturbation geopotential. Express the $x$-dependent part of the solution as a sinusoidal function.
iii. How does this solution compare to that in the original (inviscid) edge-wave problem? Discuss the mathematical form of the solution. What physical (environmental) conditions maximize the frictional influence?
iv. Given $C_d = 1.5 \times 10^{-5}\ \mathrm{s}^{-1}$, $h = 1.0$ km, $N_0 = 2 \times 10^{-2}\ \mathrm{s}^{-1}$, and $f_0 = 10^{-4}\ \mathrm{s}^{-1}$, compute the $e$-folding spindown time for a disturbance of 4,000-km wavelength. In other words, how long would it take for a disturbance of arbitrary initial amplitude to spin down to $1/e$ of that value?

6. Does a critical level exist for the Eady edge-wave problem? Justify your answer. If so, does this mean that Eady edge waves are unstable?
7. Starting from the thermodynamic energy equation

$$\frac{\partial b'}{\partial t} + U_g \frac{\partial b'}{\partial x} + v'_g \frac{\partial B}{\partial y} = -w' N_0^2, \qquad (4)$$

using the Eady edge wave solutions derived earlier in the chapter [repeated as Eqs. (6) and (7) below], the given relation $U_g = U\frac{Z}{H}$, and the corresponding basic-state thermal wind equation, show that

$$w' = -\frac{k}{N_0^2}\frac{U}{H}\frac{z}{H_R}\phi_0 e^{\frac{-z}{H_R}} \sin\left[k\left(x - \frac{H_R}{H}Ut\right)\right]. \qquad (5)$$

8. Show that the net meridional heat flux vanishes when averaged over one wavelength of an Eady edge wave. In other words, show that $\int_0^{\frac{2\pi}{k}} v'_g b' dx = 0.$

Show that the conversion from eddy potential energy to eddy kinetic energy for Eady edge waves is also zero when averaged over one wavelength.

Eady edge wave solutions:

$$v'_g = -\frac{k}{f_0}\phi_0 e^{\frac{-z}{H_R}} \sin\left[k\left(x - \frac{H_R}{H}Ut\right)\right] \qquad (6)$$

$$b' = -\frac{\phi_0}{H_R} e^{\frac{-z}{H_R}} \cos\left[k\left(x - \frac{H_R}{H}Ut\right)\right] \qquad (7)$$

## REFERENCES


Bretherton, F. P., 1966: Critical layer instability in baroclinic flows. *Quart. J. Roy. Meteor. Soc.*, **92,** 325–334.

Charney, J. G., 1947: The dynamics of long waves in a baroclinic westerly current. *J. Meteor.*, **4,** 135–162.

——, and M. E. Stern, 1962: On the instability of internal baroclinic jets in a rotating atmosphere. *J. Atmos. Sci.*, **19,** 159–172.

Davis, C. A., 1992: A potential-vorticity diagnosis of the importance of initial structure and condensational heating in observed extratropical cyclogenesis. *Mon. Wea. Rev.*, **120,** 2,409–2,428.

——, and K. A. Emanuel, 1991: Potential vorticity diagnostics of cyclogenesis. *Mon. Wea. Rev.*, **119,** 1,929–1,953.

——, M. T. Stoelinga, and Y.-H. Kuo, 1993: The integrated effect of condensation in numerical simulations of extratropical cyclogenesis. *Mon. Wea. Rev.*, **121,** 2,309–2,330.

Eady, E. T., 1949: Long waves and cyclone waves. *Tellus,* **1,** 33–52.

Farrell, B., 1984: Modal and nonmodal baroclinic waves. *J. Atmos. Sci.*, **41,** 668–673.

——, 1985: Transient growth of damped baroclinic waves. *J. Atmos. Sci.*, **42,** 2,718–2,727.

——, 1989: Optimal excitation of baroclinic waves. *J. Atmos. Sci.*, **46,** 1,193–1,206.

Gill, A. E., 1982: *Atmosphere-Ocean Dynamics.* Academic Press, 662 pp.

Green, J. S. A., 1960: A problem in baroclinic instability. *Quart. J. Roy. Meteor. Soc.*, **86,** 237–251.

Hakim, G. J., D. Keyser, and L. F. Bosart, 1996: The Ohio Valley wave-merger cyclogenesis event of 25–26 January 1978. Part II: Diagnosis using quasigeostrophic potential vorticity inversion. *Mon. Wea. Rev.*, **124,** 2176–2205.

Hoskins, B. J., M. E. McIntyre, and A. W. Robertson, 1985: On the use and significance of isentropic potential vorticity maps. *Quart. J. Roy. Meteor. Soc.*, **111,** 877–946.

Joly, A., and A. J. Thorpe, 1990: Frontal instability generated by tropospheric potential vorticity anomalies. *Quart. J. Roy. Meteor. Soc.*, **116,** 515–560.

Lindzen, R. S., 1994: The Eady problem for a basic state with zero PV gradient but $\beta \neq 0$. *J. Atmos. Sci.*, **51,** 3,221–3,226.

Liu, Y., and M. Mu, 2001: Nonlinear stability of the generalized Eady model. *J. Atmos. Sci.*, **58,** 821–827.

Miles, J. W., 1964: Baroclinic instability of the zonal wind. *Rev. Geophys.*, **2,** 155–176.

Orlanski, I., 1968: Instability of frontal waves. *J. Atmos. Sci.*, **25,** 178–200.

Palmén, E., 1951: The aerology of extratropical disturbances. *Compendium of Meteorology,* American Meteorological Society, 599–620.

Parker, D. J., 1998: Secondary frontal waves in the North Atlantic region: A dynamical perspective of current ideas. *Quart. J. Roy. Meteor. Soc.*, **124,** 829–856.

Pedlosky, J., 1964: The stability of currents in the atmosphere and the ocean. Part I. *J. Atmos. Sci.*, **21,** 201–219.

Robinson, W. A., 1989: On the structure of potential vorticity in baroclinic instability. *Tellus,* **41A,** 275–284.

Rotunno, R., and M. Fantini, 1989: Petterssen's "type B" cyclogenesis in terms of discrete, neutral Eady modes. *J. Atmos. Sci.*, **46,** 3,599–3,604.

——, W. C. Skamarock, and C. Snyder, 1994: An analysis of frontogenesis in numerical simulations of baroclinic waves. *J. Atmos. Sci.*, **51,** 3,373–3,398.

Sanders, F., 1986: Explosive cyclogenesis in the west-central North Atlantic Ocean, 1981–84. Part I: Composite structure and mean behavior. *Mon. Wea. Rev.*, **114,** 1,781–1,794.

Schär, C., and H. C. Davies, 1990: An instability of mature cold fronts. *J. Atmos. Sci.*, **47,** 929–950.

Simmons, A. J., and B. J. Hoskins, 1980: Barotropic influences on the growth and decay of nonlinear baroclinic waves. *J. Atmos. Sci.*, **37,** 1,679–1,684.

Solberg, H., 1936: Le mouvement d'inertie de l'atmosphere stable et son role dans le theorie des cyclones. *Union Geodesique et Geophysique Internationale Vlieme Assemblee,* Vol. 2, Edinburgh, United Kingdom, 66–82.

Stoelinga, M. T., 1996: A potential vorticity-based study of the role of diabatic heating and friction in a numerically simulated baroclinic cyclone. *Mon. Wea. Rev.*, **124,** 849–874.

Thorncroft, C. D., B. J. Hoskins, and M. E. McIntyre, 1993: Two paradigms of baroclinic-wave life-cycle behaviour. *Quart. J. Roy. Meteor. Soc.*, **119,** 17–55.

## CHAPTER 8

# Cold-Air Damming

*...the belt of high pressure, evidently centered far north in Canada, gave an abundant supply of cold air, sweeping down from the deep snowfields of the north far into the United States toward the general area of low pressure along the southern coast. The return flow of warm moist air, riding over the wedge-like encroaching north wind, was cooled and forced to precipitate rain, sleet, or snow over most of the eastern United States throughout the period.*

—C. L. Meisinger, "The Precipitation of Sleet and the Formation of Glaze in the Eastern United States January 20 to 25, 1920, with Remarks on Forecasting" (1920)

Apex, North Carolina, 5 December 2002

There exist a plethora of synoptic-scale and mesoscale weather phenomena that could be described in this text. The selections made are based on my experiences, reflecting the different geographical regions in which I've lived or visited, and also my "upbringing"—advisors, colleagues, and teachers who have emphasized certain topics. Growing up in Seattle, Washington, I was fascinated by regional weather phenomena such as the Puget Sound convergence zone, the infrequent but disruptive intrusion of arctic air into western Washington, and the "onshore push" of marine air into Puget Sound following summertime heat waves. Living in the northeastern United States and Montreal, Quebec, I experienced heavy snowfall, lake-effect snow, and severe weather. During my time in North Carolina, I've witnessed flooding rains accompanying landfalling tropical cyclones, cold-air damming (CAD), freezing rain events, and heavy snowfall. My past participation in field experiments has given me the opportunity to experience and study high-latitude gap-wind events and arctic meteorology. Although the emphasis of this text is on synoptic-scale weather systems, several of the phenomena mentioned above are related to the influence of terrain on the flow. The purpose of this chapter is to provide an in-depth account of the processes and mechanisms accompanying a terrain-influenced phenomenon known as *cold-air damming.*

In the presence of a significant mountain barrier, an along-barrier component of the pressure gradient, a stably stratified lower atmosphere, and sufficient influence of the earth's rotation, CAD may occur. This phenomenon has been documented in many geographical locations around the world, and the occurrence of CAD has long been recognized as an important orographic influence on regional weather and climate. The term "cold-air damming" is an incomplete description of the dynamics of these topographically driven weather events, which may more generally be categorized as a geostrophic adjustment process for orographically disturbed flow.

Cold-air damming events are characterized by a surface-based layer of shallow cold air, a structure that can be conducive to freezing precipitation during winter. In fact, the climatology of freezing rain and sleet indicate maxima over the Carolinas and Virginia in the southeastern United States, a region prone to CAD (Fig. 8.1). Even before this phenomenon was fully understood and dubbed as CAD, meteorologists were aware of the influence that shallow cold air east of the mountains could

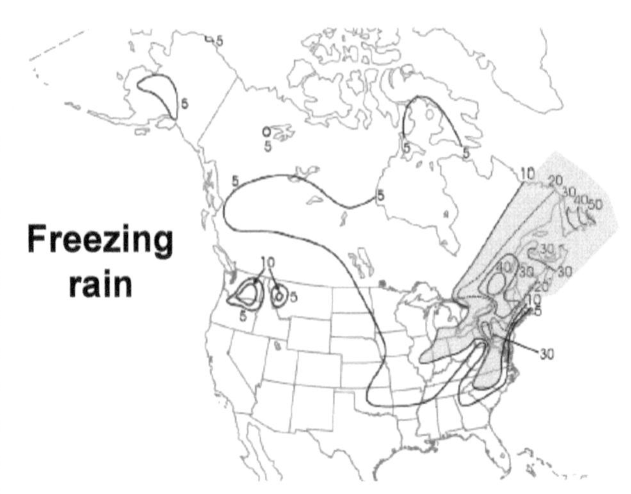


**Figure 8.1.** Median annual hours of freezing rain from 1976 to 1990. Contours drawn for 5, 10, 20, 30, and 40 h (from Cortinas et al. 2004).

exert on winter weather. A surface analysis from 8 p.m. 20 January 1920 shows the southward extension of colder air east of the Appalachian Mountains, in conjunction with the northward bowing of isotherms immediately west of the mountains (Fig. 8.2).

The significance of CAD is not limited to winter weather; virtually all important sensible weather parameters can be impacted by CAD, including several that are critically important to the aviation industry, such as visibility, cloud ceilings, precipitation type, and wind speed and direction. However, not all CAD events are associated with significant sensible weather impacts. How can one predict in advance whether a given CAD event will be associated with strong or weak impacts?

Here, our goals are to (i) investigate the *physical processes* leading to cold-air damming, including the theoretical parameters related to the structure and intensity of CAD; (ii) demonstrate the signatures of cold-air damming and related phenomena; (iii) relate CAD to sensible weather conditions; and (iv) convey an appreciation for the difficulties in forecasting CAD processes and weather impacts.

## 8.1. PHYSICAL MECHANISMS

### 8.1.1. CAD as a geostrophic adjustment process

Consider an isolated, midlatitude, north–south oriented mountain barrier in the presence of a southward-directed pressure gradient force as indicated in Fig. 8.3a. Assume that the easterly flow is in geostrophic balance at some distance upwind of the barrier and that the air is stably stratified.

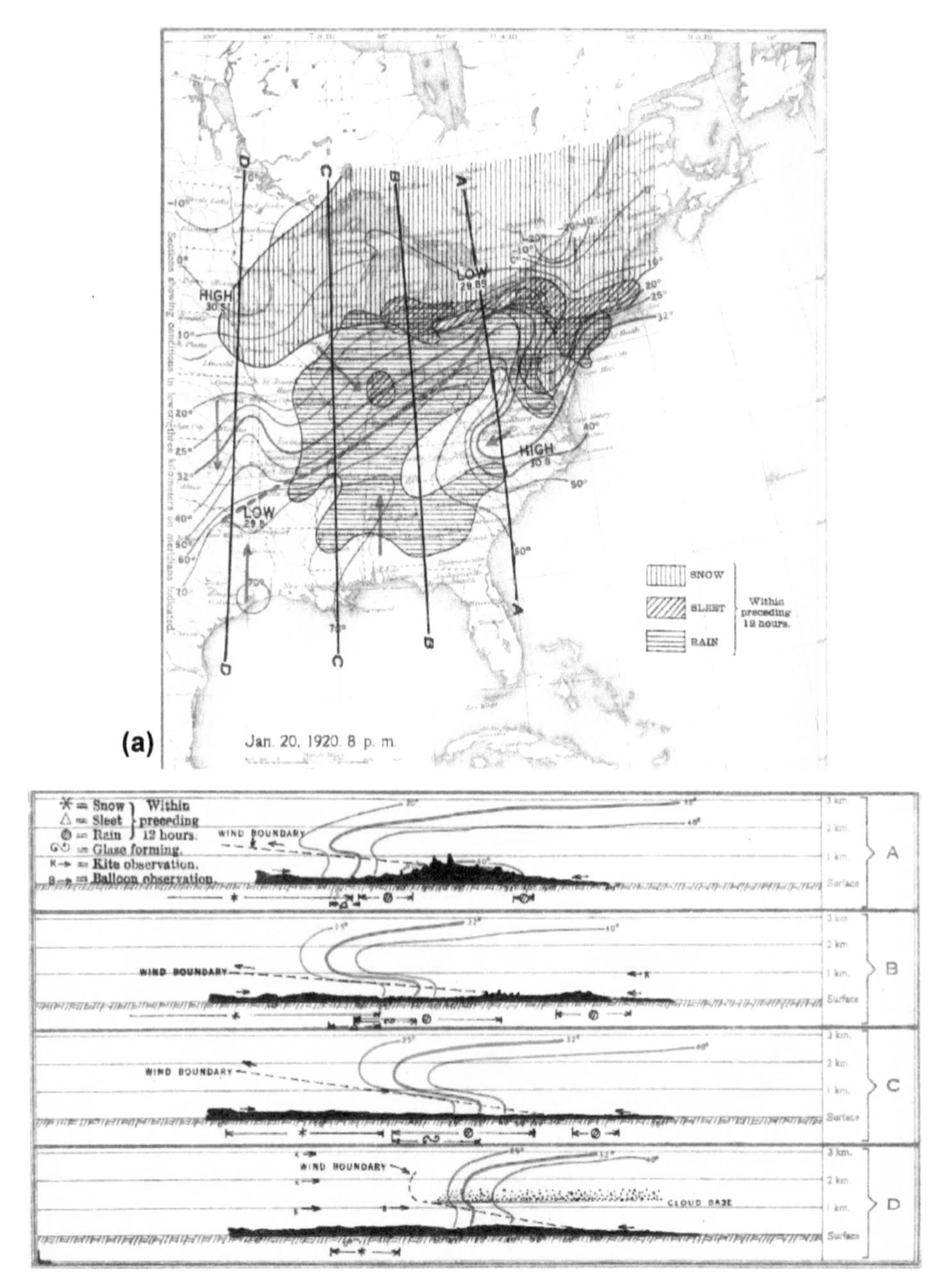


**Figure 8.2.** Surface temperature and precipitation analysis for 8 p.m. eastern time 20 Jan 1920. Inset at lower right shows cross section of isotherms corresponding to the four cross-sectional lines indicated on plan map (from Meisinger 1920).

What will happen as the air approaches the barrier from the east? If the stability of the air approaching the barrier is sufficiently small, if the barrier is not too high, or if the incoming velocity component toward the barrier is large, then the air could pass up and over the barrier. In this case, adiabatic cooling on the upslope side of the mountain range and compressional warming in the lee would give rise to a weak high pressure (low pressure) perturbation upwind (downwind) of the barrier; these features can be referred to as the *upwind high* and *lee trough*.

The situation described above, with air flowing up and over the mountain barrier, is not representative of a typical CAD event. As the name implies, CAD is

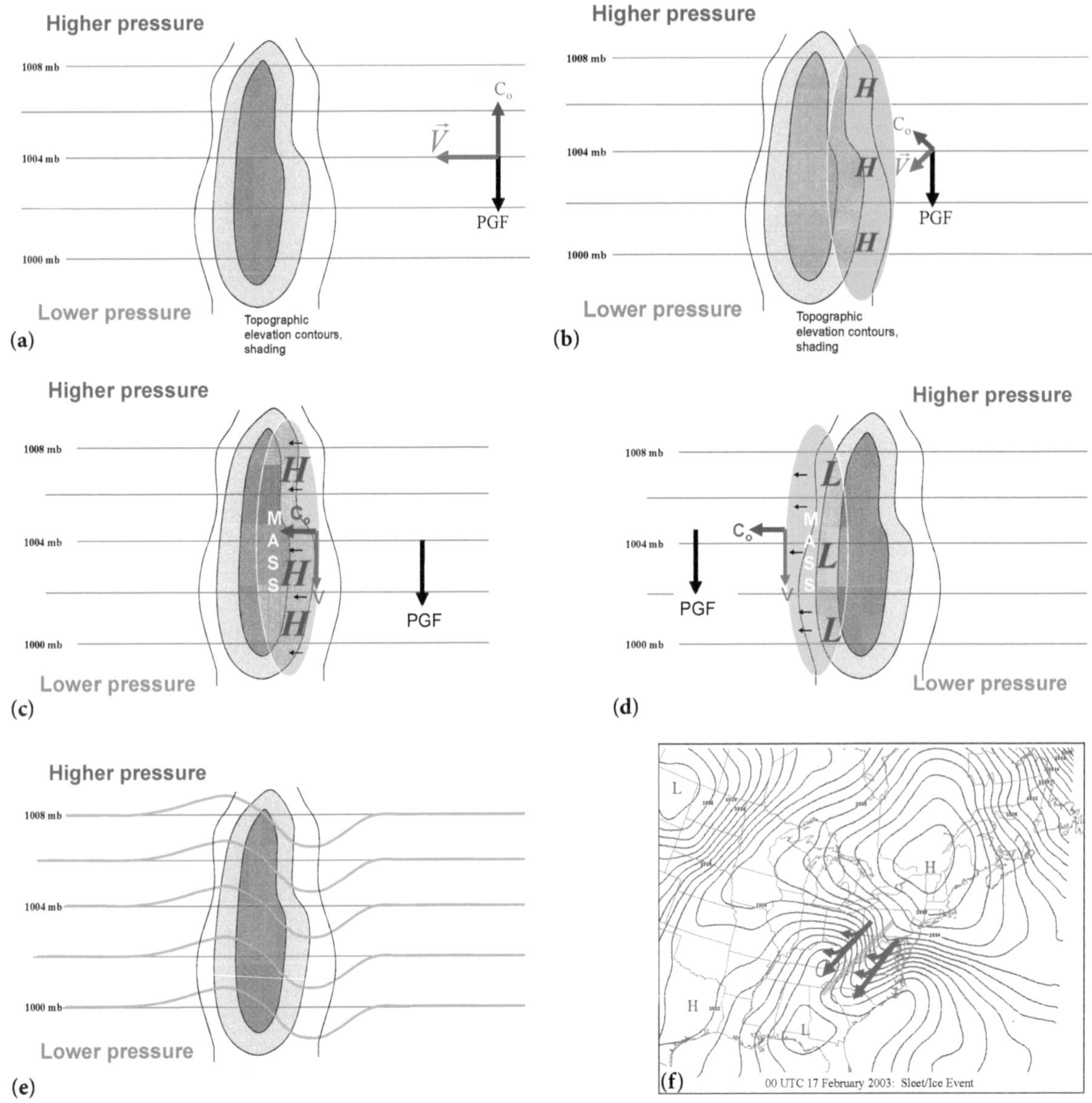


**Figure 8.3.** Schematic evolution of pressure and force balance during onset of CAD for a north–south-oriented mountain range: (a) initial undisturbed configuration, with orographic barrier (gray shading and contours representing terrain elevation), east–west oriented sea level isobars, and force balance shown for a selected air parcel east of the barrier; (b) as in (a), but when parcel has approached barrier and slowed; (c),(d) the effect of Coriolis torque on the near-barrier mass field; (e) adjusted pressure configuration; and (f) sea-level pressure analysis for CAD event from 0000 UTC 17 Feb 2003. Abbreviations for Coriolis force (Co) and pressure-gradient force (PGF) appear in (a)–(d).

characterized by a strong blocking (damming) of the flow, such as would happen if the atmosphere was more stable, the barrier was higher, or the barrier-normal geostrophic flow was weaker than in the previously described situation. Whether the flow is blocked or it flows up and over the barrier is related to the ratio of the inertia of the approaching air to the energy needed to surmount the barrier. This dimensionless ratio is related to the *Froude number*[1] of the flow,

$$Fr = \frac{U}{NH}, \tag{8.1}$$

[1]The *square* of the Froude number, while still dimensionless, would have units of energy per unit mass divided by energy per unit mass.

where $U$ is the barrier-normal wind speed; $H$ is the height of the barrier; and $N$, the Brunt- Väisälä frequency, is a measure of static stability,

$$N^2 = \frac{g}{\theta}\frac{d\theta}{dz}. \quad (8.2)$$

A small Froude number implies blocking of the flow by the mountain barrier, with a diminishing fraction of the flow blocked as the Froude number approaches a value of 1. There are some complications in evaluating the Froude number from actual data. However, in many CAD situations, a dome of cold air in the lower troposphere is separated from the air aloft by an inversion layer. Thus, in practice, we can compute $Fr$ by taking H to be the average height of the mountains and evaluating $N^2 \approx g/\theta_0(\Delta\theta/\Delta z)$, where $\Delta\theta$ is the change in potential temperature across the inversion, $\Delta z$ is the depth over which $\Delta\theta$ is evaluated, and $\theta_0$ is the average potential temperature beneath the inversion. For Appalachian cold-air damming events in the southeastern United States, the value of $U$ can be estimated by the barrier-normal geostrophic wind or by averaging the barrier-normal wind component in the lowest 1 km of the atmosphere using upstream rawinsonde data prior to the development of CAD.

If we assume that the Froude number is sufficiently small to ensure that a significant fraction of the incoming flow is blocked by the mountain barrier, we can ask what impact the mountains would have on the pressure and wind fields. We should not expect geostrophic balance to be maintained as air approaches the mountain barrier as in Fig. 8.3a, because, as the air is blocked on approach to the barrier, the wind speed decreases, weakening the northward-directed Coriolis force as depicted in Fig. 8.3b. As the then-unbalanced pressure-gradient force acts on the decelerated flow, a northerly cross-isobar wind component develops, and eventually friction in this accelerated barrier-parallel flow will balance the southward-directed component of the pressure gradient force.

In this hypothetical example, orographic disruption of geostrophic force balance takes place on *both* sides of the barrier, and the Coriolis force acts on the resulting along-barrier ageostrophic flows. To the east of the barrier in Fig. 8.3c, the Coriolis torque on the ageostrophic northerly flow "banks" mass against the barrier; this mass accumulation leads to an increase in surface pressure along the eastern slopes because of the increased mass in the overlying column. On the western side of the barrier, the opposite situation exists. Here, Coriolis deflection acts to draw mass *away* from the barrier, resulting in a lowering of sea level pressure and troughing there, as depicted in Fig. 8.3d. The final adjusted pressure configuration may appear as in Fig. 8.3e, with a trough–ridge couplet straddling the mountain barrier. Observed pressure distributions during CAD events often resemble that shown in the hypothetical example, as illustrated for the Appalachian Mountains by the analysis presented in Fig. 8.3f.

If we consider the forces acting in the cross-barrier direction, the development of a ridge banked against the eastern slopes of the terrain leads to a component of the pressure-gradient force directed away from the barrier, toward the east in this example. As the adjustment process continues, a balance comes about between the westward-directed Coriolis and the eastward-directed pressure gradient. Hence, the *cross-barrier* forces become closer to geostrophic balance, and *we can view the distorted pressure field in the vicinity of the mountains as a result of geostrophic adjustment* (Figs. 8.3e,f, 8.4).

Observations also show that, during most Appalachian CAD events, the amplitude of the ridge to the east of the mountains exceeds that of the trough to the west. There are several reasons for this. In situations with stable high pressure to the north, colder air is almost invariably found there as well. Thus, the northerly, barrier-parallel ageostrophic flow within the ridge is associated with cold advection; the southward advection of cold, dense air hydrostatically enhances the ridge, but northerly flow and cold advection west of the barrier tends to weaken the trough.

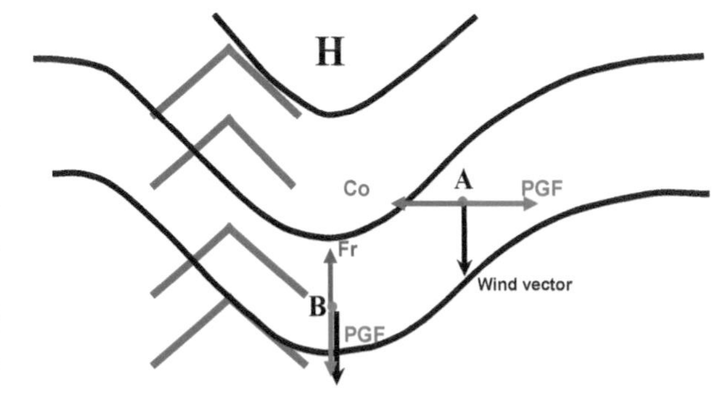


**Figure 8.4.** Schematic illustration of the separate along- and across-barrier force balances at the mature stage of a CAD event. Black solid lines represent sea-level isobars, black vector arrows represent wind velocity at points A and B. At point A, barrier-normal forces are indicated; at point B, barrier-parallel forces are indicated.

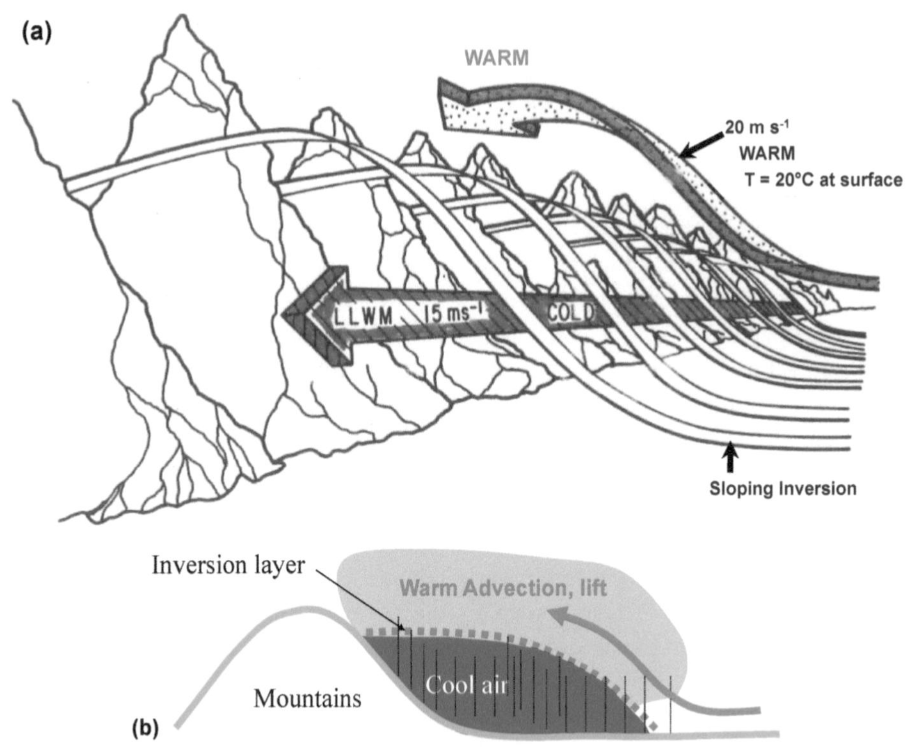


**Figure 8.5.** (a) Idealized, three-dimensional representation of CAD, including low-level wind maximum (LLWM; blue arrow) and depiction of warm airstream rising over the lower cold airflow (red arrow; from Bell and Bosart 1988). (b) Idealized cross section of CAD showing ascent and associated formation of clouds and precipitation above a shallow cold dome.

As discussed earlier, some contribution to the pressure pattern from adiabatic expansion and compression is often present, and this is difficult to separate from that due to geostrophic adjustment. A schematic of the final adjusted state of Appalachian CAD, from the classic Bell and Bosart (1988) paper, is presented in Fig. 8.5a. The characteristic northerly jet within the cold dome is separated from ascending, moist flow above by an inversion layer. The significance of the vertical gradient in temperature advection will be discussed below.

### 8.1.2. Thermal advection, diabatic processes, and CAD

It is tempting to think that warm advection above the CAD cold dome, as depicted in the idealized Fig. 8.5, would lead to a rapid demise of the CAD and displacement or erosion of the cold air. However, this is typically not the case. Warm advection above the capping inversion, with cold advection beneath, serves to strengthen the inversion and increase stability; this can prolong CAD by preventing mixing out of the dome of cold air beneath the inversion. In addition to the stabilizing influence of this differential thermal advection pattern, the warm advection is consistent with forcing for ascent as the flow glides along the upward-sloping isentropes within and above the inversion layer (Fig. 8.5b). The resulting cloud cover can protect the cold dome from solar heating, which can be especially important to maintaining the cold dome during the spring, summer, and fall. If the air within the cold dome is unsaturated, precipitation falling from the warm-advection layer may result in evaporational cooling beneath the inversion, further cooling and stabilizing the lower atmosphere as might occur for the situation depicted schematically in Fig. 8.5b. If dry air is advecting southward beneath the inversion, precipitation generated by the ascent above the cold dome can lead to continued evaporational cooling in the lower atmosphere.

Some CAD events *depend* on the occurrence of diabatic processes of the type discussed above, whereas other events are strongly driven by synoptic-scale processes

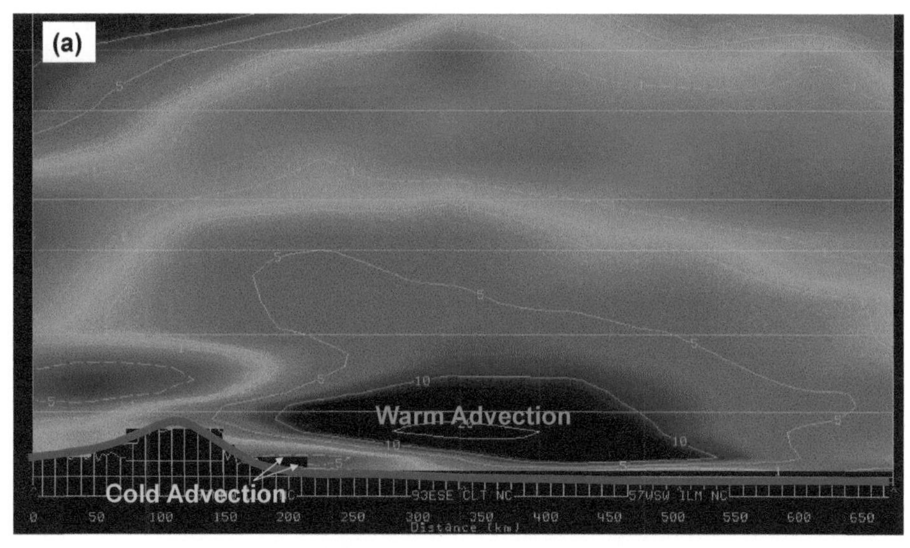


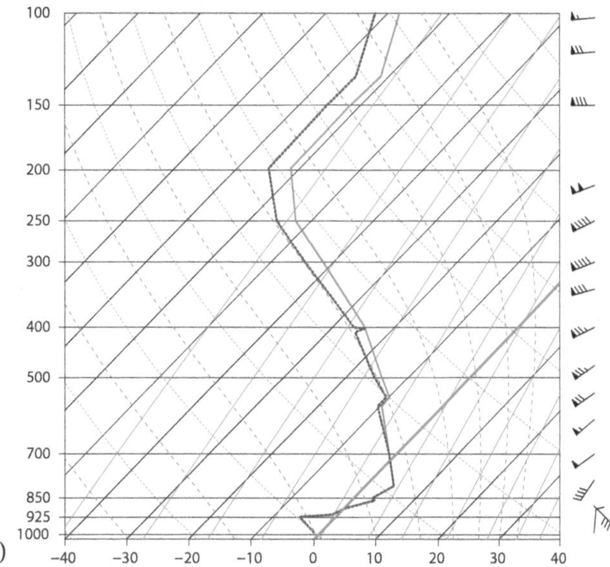


**Figure 8.6.** Data from CAD event of 4 and 5 Dec 2002. (a) Cross section of 12-h NAM temperature advection forecast valid 1200 UTC 5 Dec 2002: red shading represents warm advection, blue shading represents cold advection. Courtesy National Weather Service (NWS), Raleigh, North Carolina. (b) Greensboro, North Carolina (GSO), rawinsonde profile for 0600 UTC 5 Dec 2002, in skew T-log*p*. Zero degree isotherm highlighted red.

and the occurrence of diabatic processes is less crucial. The strength of synoptic forcing can be measured by the strength and location of the "parent high" that is typically located to the north during CAD events in the Northern Hemisphere. The variable contribution of diabatic versus synoptic forcing for CAD events has led forecasters and researchers in the southeastern United States to categorize CAD events on this basis, as discussed in section 8.2.2.

A CAD event that was accompanied by freezing rain in parts of the southeastern United States took place in early December 2002 (Fig. 8.6). Cross-sectional views of the North American Mesoscale (NAM) model 12-h temperature-advection forecast, valid 1200 UTC 5 December 2002, reveals a shallow layer of cold advection within the cold dome immediately east of the Appalachian Mountains, with a strong, deep zone of warm advection above this layer (Fig. 8.6a). This cross section

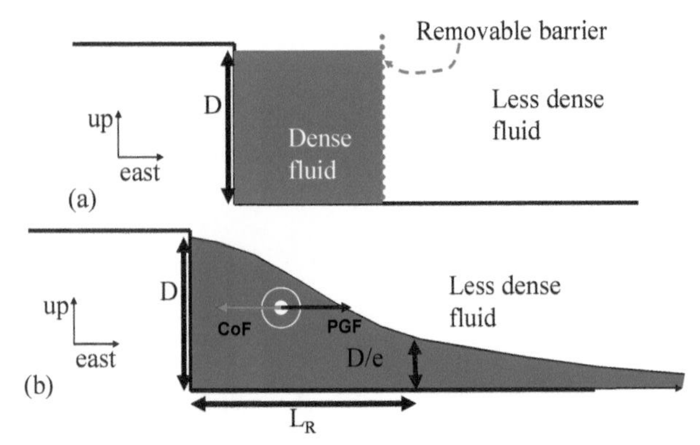


**Figure 8.7.** Idealized thought experiment for geostrophic adjustment of a two-fluid rotating system, with a rigid barrier of depth $D$ to the west. (a) Dense fluid is held against the barrier by removable partition, with less dense fluid to east; (b) after the barrier is removed, interface between dense and lighter fluid spreads laterally and an equilibrium adjusted state is reached. At distance $L_R$, the dense fluid has a depth of $D/e$.

corresponds to a time when the CAD event was approaching maximum strength and freezing rain was falling within portions of the cold dome. An observed sounding from GSO, taken at 0600 UTC, is shown in Fig. 8.6b. Close examination of the wind profile indicates strongly veering flow in the inversion layer, consistent with the model forecast of warm advection. A shallow subfreezing layer is located beneath a deep layer of above-freezing air, setting the stage for freezing rain.

In viewing CAD as a geostrophic adjustment process, the *Rossby radius of deformation* is a useful theoretical parameter that is related to the *horizontal structure* of CAD events. Consider the thought experiment described in Fig. 8.7a, in which two fluids of differing density are oriented side by side in a rotating tank (with the axis of rotation on the right side of the diagram), separated by a removable barrier. What will happen if the barrier separating the fluids is instantly removed? The situation in Fig. 8.7a includes some available potential energy, because the center of mass of the system is higher than it would be if the dense fluid occupied the lowest portion of the domain. If the barrier is removed, some of this potential energy will be converted to kinetic energy, as the more dense fluid flows eastward beneath the lighter fluid, which would rise and move westward.

If the system was not rotating, and assuming limited mixing of the fluids, the final interface between the dense and light fluids would be horizontal, with the maximum amount of kinetic energy gained at the expense of potential. However, in a rotating fluid, the eastward-flowing dense fluid will be acted upon by the Coriolis force, in this example giving rise to a northerly flow in the dense fluid. As the eastward-directed pressure gradient force, which is attributable to the hydrostatic pressure difference along the sloping interface of the dense fluid, comes into balance with the Coriolis force acting on this northerly flow, geostrophic balance comes about for forces in the cross-barrier direction (Fig. 8.7b).

The distance at which the depth of the dense fluid is $1/e$ of the original depth $D$, the "$e$ folding" scaling distance for geostrophic adjustment, is known as the *Rossby radius of deformation* ($L_R$). In other words, if we denote the depth of the dense fluid by $\eta$, a function of the $x$ direction, the equation describing this depth is $\eta(x) = D\exp(-x/L_R)$. The Rossby radius depends on the density difference between the fluids, the strength of rotation in the fluid, and the height of the barrier,

$$L_R = \frac{\sqrt{g'H}}{f}, \tag{8.3}$$

where

$$g' = \frac{g\Delta\theta}{\theta_0} \tag{8.4}$$

is the "reduced gravity" and other symbols have their previous meanings. The quantity $g'$ is related to the static stability and can be computed from routine sounding data in the vicinity of the inversion layer.

The functional dependences given by (8.3) make intuitive sense; with strong rotation, the Coriolis force pushes the fluid more strongly against the barrier, resulting in a narrow tongue of dense fluid (small $L_R$). Low-latitude damming events, all else being equal, would have a broader, shallower cold dome than those at polar latitudes. Concerning the stability dependence, if a very large density difference characterized the two fluids (e.g., air and water), the Rossby radius would be very large as the denser fluid would spread strongly under gravity.

### 8.1.3. Numerical model experiments

Another means of isolating the influence of terrain in CAD events is to utilize numerical models. Although such models are in many ways inexact replicas of nature, they provide an extremely powerful means of testing hypotheses and isolating the importance of specific processes.

Examine the highly distorted sea level pressure field over the Carolinas and mid-Atlantic region from the analysis of 0000 UTC 17 February 2003, displayed in Fig. 8.3f. You may recall this example from chapter 3 in our discussion of cyclogenesis (e.g., Fig. 3.9); a more complete evolution of sea level pressure during this event is shown in Fig. 8.8. Note how the shape of the large, cold anticyclone changes as it builds eastward in the vicinity of the Appalachian Mountains. How might the atmosphere have evolved during this event if the Appalachian Mountains were *not there*, given the same large-scale synoptic pattern evolution?

Here, we utilize the fifth-generation Pennsylvania State University (PSU)–National Center for Atmospheric Research (NCAR) Mesoscale Model (MM5), run with 30-km grid spacing and 31 vertical levels. The model runs are initialized with analyses obtained from the National Centers for Environmental Prediction (NCEP) at the time of Fig. 8.8a (0000 UTC 15 February 2003), prior to the onset of CAD. The *control run* is a simulation designed to reproduce what happened in nature. In addition, *experimental simulations* were run in which the orography of the model was altered. In the *no-mountain* experiment, the Appalachian Mountains (and all other terrain) were flattened in the model. By initializing the model prior to the beginning of the CAD event, we can compare the synoptic evolution between the two simulations to quantify the distortion of pressure, temperature, and wind due to orography. The topography for the control simulation is shown in Fig. 8.9a with that for the no-mountain simulation in Fig. 8.9b.

The control simulation provides a qualitatively realistic depiction of the sea level pressure evolution during the event; the 48-h simulation, valid 0000 UTC 17 February (Fig. 8.10a) can be compared to the analyzed sea level pressure field shown in Fig. 8.8e. The narrow ridge along and to the east of the Appalachian Mountains, along with the two troughs flanking this ridge on either side, are reproduced in approximately the observed locations, although there is a difference in the map projection between Figs. 8.8 and 8.10, accounting for the apparent rotation of these features.

As expected, after 48 h, the simulation without terrain has evolved in a very different manner, with little evidence of pronounced ridging and troughing in the vicinity of the Appalachian Mountains (Fig. 8.10b). A *difference field* of sea level pressure (Fig. 8.10c) demonstrates that the amplitude of the ridge east of the terrain is at this time in excess of 7 mb in amplitude, with the trough to the west of the mountains more than 5 mb in amplitude. This pattern is consistent with our expectation that the ridge would be somewhat stronger than the trough because of the southward advection of cold, dense air in the ridge, which contributes to the increased pressure there.

To eliminate any differences that might be due to the adjustment of pressure to sea level, the difference in potential temperature at the 950-mb level, located above ground in most of the damming region, is shown (Fig. 8.11). We would expect that, in the control simulation, temperatures would be much colder to the east of the mountains because of enhanced along-barrier ageostrophic cold advection within the cold dome; Fig. 8.11 demonstrates a difference of greater than 15°C between the simulations! There is a *positive* difference of greater than 12°C on the west side of the mountains, meaning that the control simulation was significantly warmer than the no-terrain simulation there, also as expected. This difference is likely due in part to the elimination of downslope compression in the easterly flowing air in the no-mountain simulation. An additional effect, which would require additional study to fully quantify, is that, in the control simulation, precipitation and lift above the CAD cold dome release latent heat, which is then realized in the descending flow to the west of the Appalachians. In the no-mountain simulation, with reduced lift and precipitation over and to the east of the Appalachians, the westward-flowing air does not experience this chinook-type flow, contributing to cooler temperatures west of the barrier in that simulation.

To further illustrate the difference in sensible weather conditions that exists between these simulations, model forecast soundings taken in the location of Columbia, South Carolina, for the two simulations are compared (Fig. 8.11b). Although conditions aloft are nearly identical between the runs, the differences in surface conditions are striking: a north wind and temperatures less than 5°C characterize surface conditions in the control simulation, versus southwesterly winds and temperatures approaching 20°C in the no-mountain simulation (Fig. 8.11b). Differences are even larger at locations to the west of Columbia, in upstate South Carolina, and the Charlotte, North Carolina, area.

The distorted sea level pressure pattern evident during this event (Figs. 8.8, 8.10) features a trough–ridge couplet

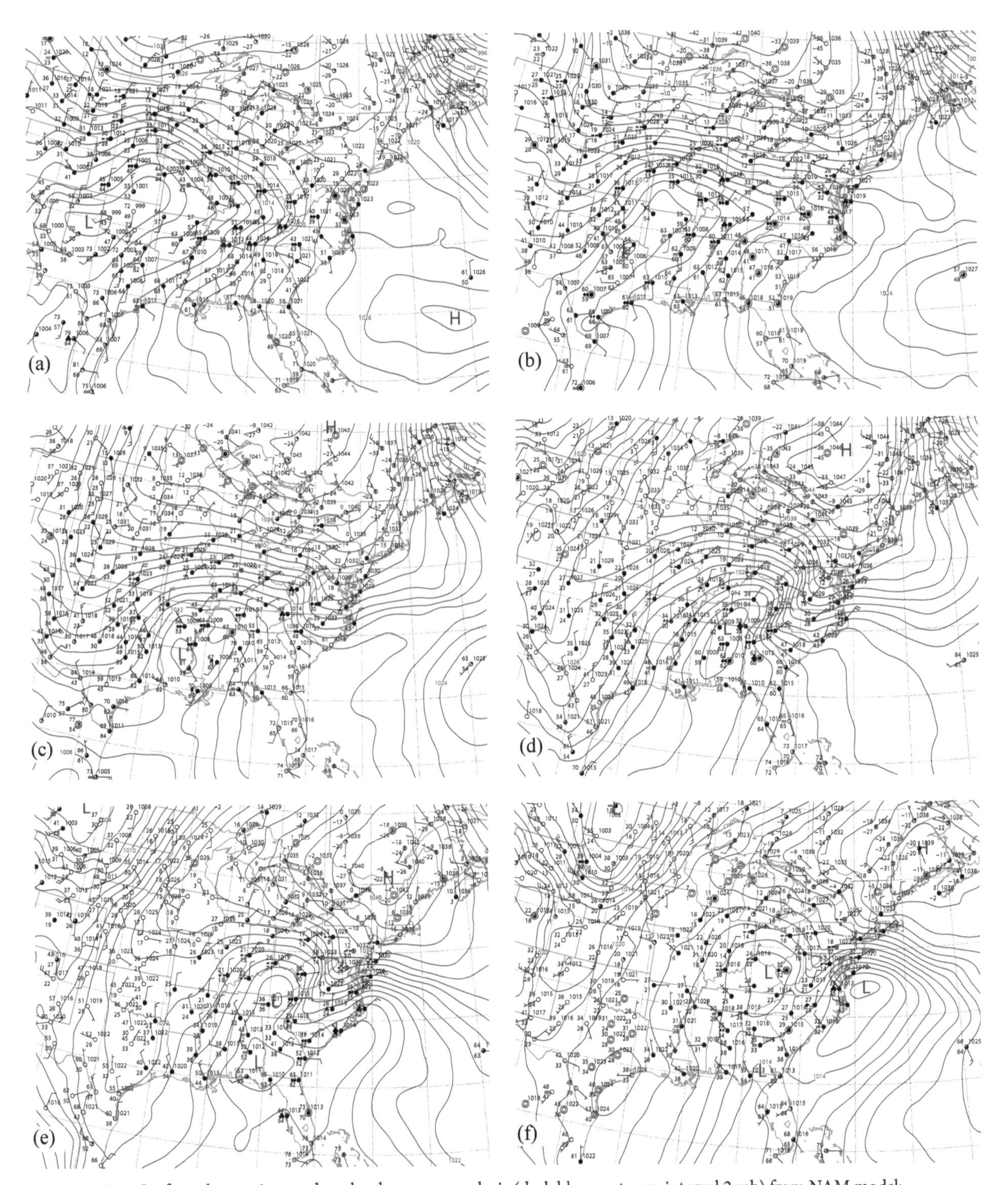


**Figure 8.8.** Surface observations and sea level pressure analysis (dark blue contours, interval 2 mb) from NAM model: (a) 0000 UTC 15 Feb; (b) 1200 UTC 15 Feb; (c) 0000 UTC 16 Feb; (d) 1200 UTC 16 Feb; (e) 0000 UTC 17 Feb; and (f) 1200 UTC 17 Feb 2003.

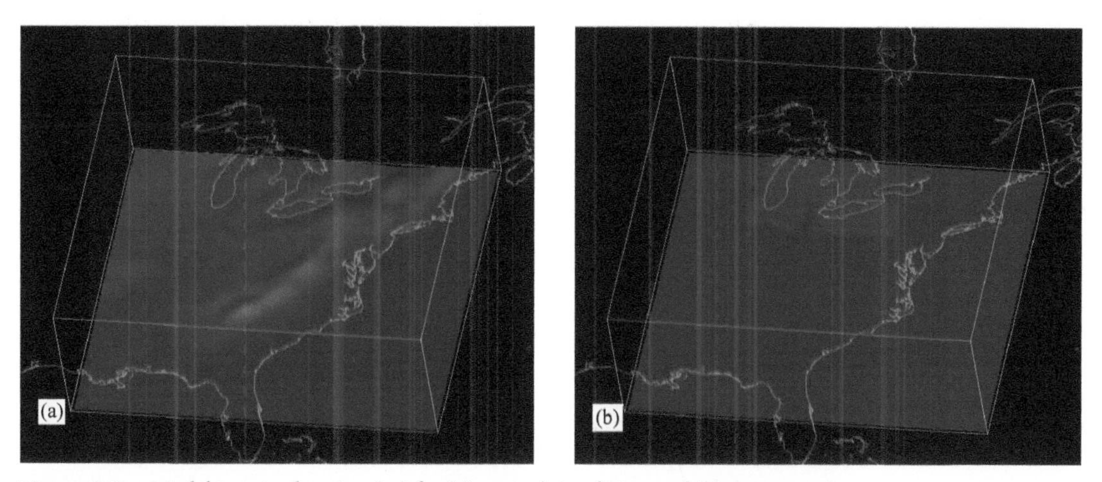


**Figure 8.9.** Model terrain elevation (m) for (a) control simulation and (b) no mountain.

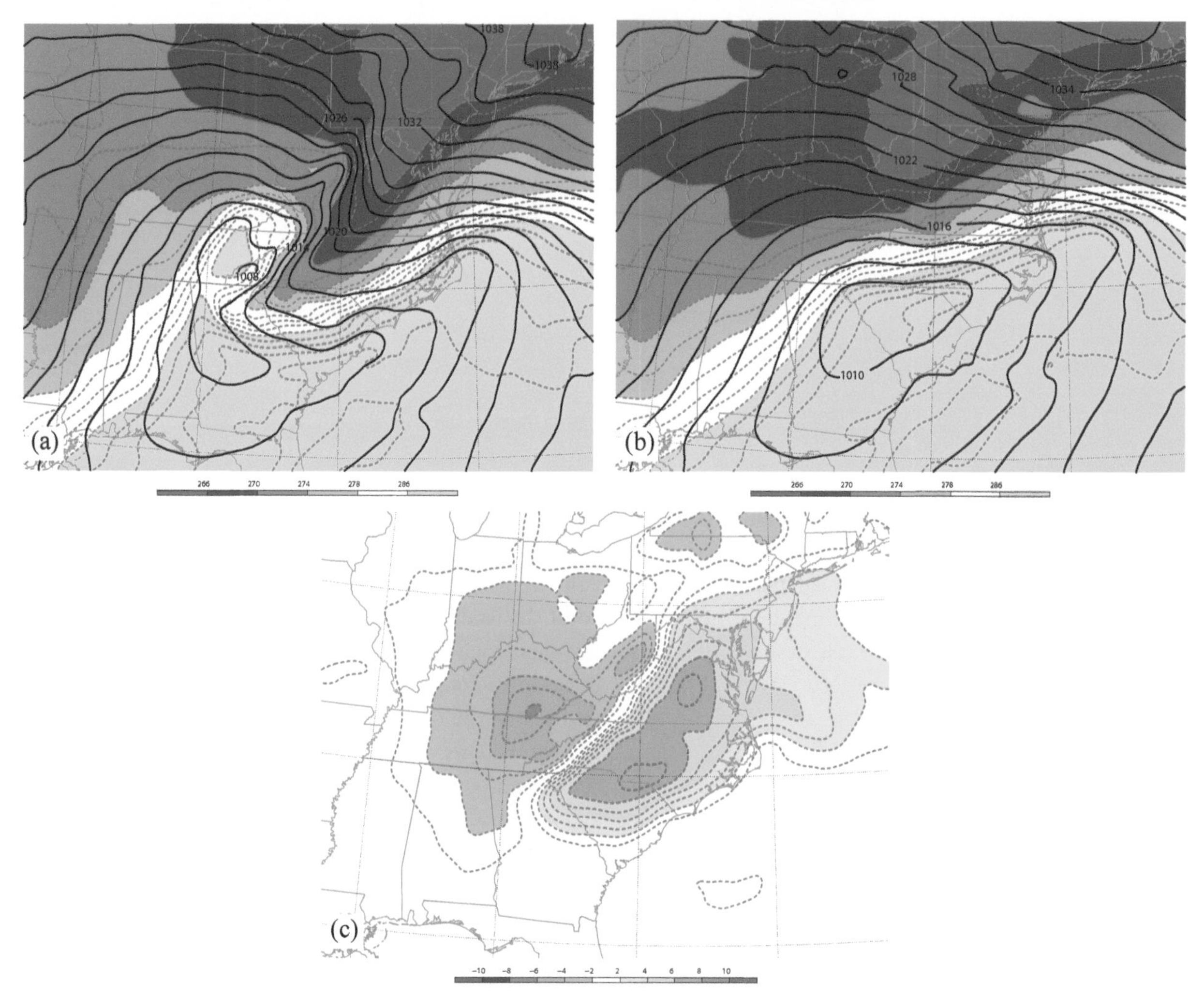


**Figure 8.10.** Sea level pressure (black contours, interval 2 mb) and 950-mb potential temperature (green dashed contours, interval 2 K and shaded as in legend at bottom of panel) valid 0000 UTC 17 Feb 2003: (a) control simulation and (b) no mountain simulation. (c) Difference in sea level pressure, control minus no-mountain simulation (interval 1 mb, zero contour omitted, shaded as in legend at bottom of panel).

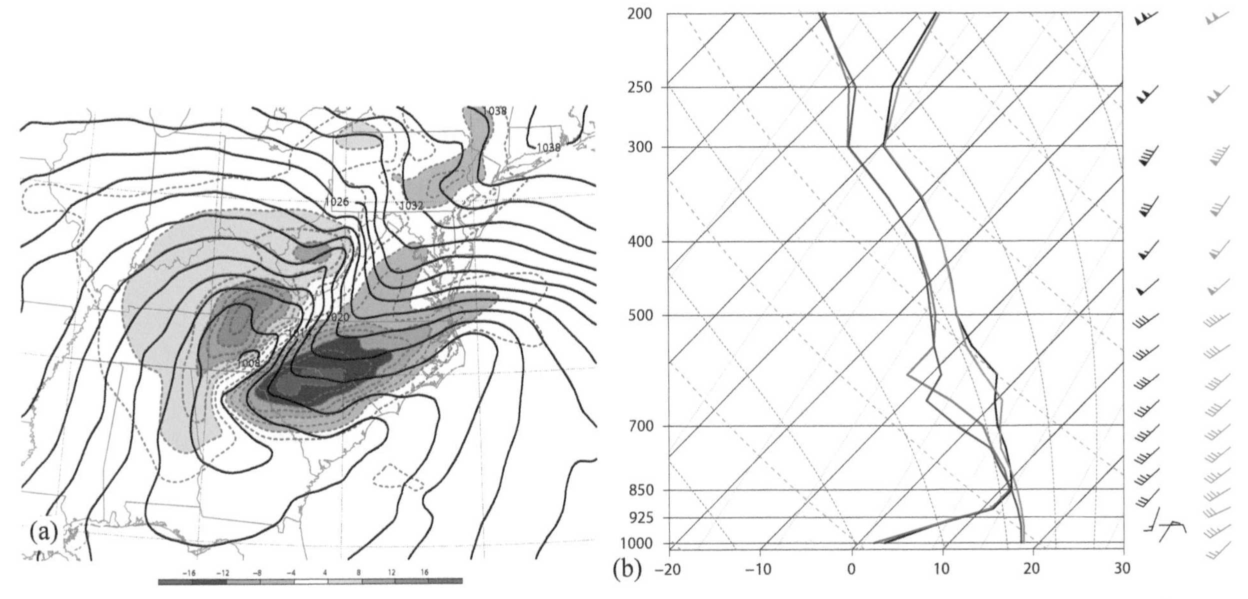


**Figure 8.11.** Simulation comparison valid 0000 UTC 17 Feb 2003. (a) Difference in 950-mb potential temperature, control minus no-mountain simulation (contour interval 3 K, dashed contours and shaded as in legend at bottom of panel, and zero contour omitted), with sea level pressure from control simulation superimposed (black solid contours every 2 mb). (b) Comparison of model forecast soundings in Skew *T*-log*p* format for Columbia, South Carolina, for control simulation (black and green) and no-mountain simulation (red and blue).

consistent with the idealized thought experiments discussed previously (e.g., Fig. 8.3); in addition, another separate trough is evident to the east of the damming region, along the coast of the Carolinas (Figs. 8.8e,f, 8.10a,b). This feature may be due to the presence of a *coastal front* or a synoptic warm front, but it is likely not a coincidence that this feature is present in the same location as the Gulf Stream current. Given that this trough appears to contribute significantly to the U-shaped isobar pattern evident in this CAD event, additional experimental simulations are needed to fully understand the role of surface features in producing the classical southeastern U.S. CAD isobar pattern.

Accordingly, two additional model experiments were performed: one in which the Gulf Stream was removed (the *No-Gulf Stream* simulation) by setting the sea-surface temperatures there equal to a spatially averaged value over the region (Fig. 8.12) and another in which *both* the Gulf Stream *and* Appalachian Mountains (the *no-Gulf Stream no-mountain* simulation) were removed (Fig. 8.13). Omission of the Gulf Stream does in fact have a pronounced impact on the U-shaped isobar pattern in the CAD region (Fig. 8.12). Pressures are more than 7 mb greater in the region occupied by the coastal trough in the control simulation. This demonstrates the importance of surface heat fluxes to the evolution of wind, temperature, and pressure patterns in the coastal zone.

When both the Appalachian Mountains and Gulf Stream are removed, the resulting sea level pressure pattern looks drastically different from that in the control run (cf. Figs. 8.13a, 8.10a). Without the mountains and Gulf Stream, the ridge–trough couplet bracketing the Appalachians is absent, and there is no trough along the coast. Instead, the cyclone centered over upstate South Carolina appears as a classical wave cyclone, with an indication of cold- and warm-frontal features to the west and east of the low center, respectively. This set of model experiments serves to illustrate the significant impact that surface and terrain features can exert on the evolution of synoptic-scale weather systems.

## 8.2. CAD CLIMATOLOGY, CLASSIFICATION, AND IMPACT

### 8.2.1. Global CAD environments

Cold-air damming is observed to take place in many locations around the world, and the list provided here is by no means complete. Along mountainous coasts, a tongue of cool, dense marine air can surge along the barrier; an

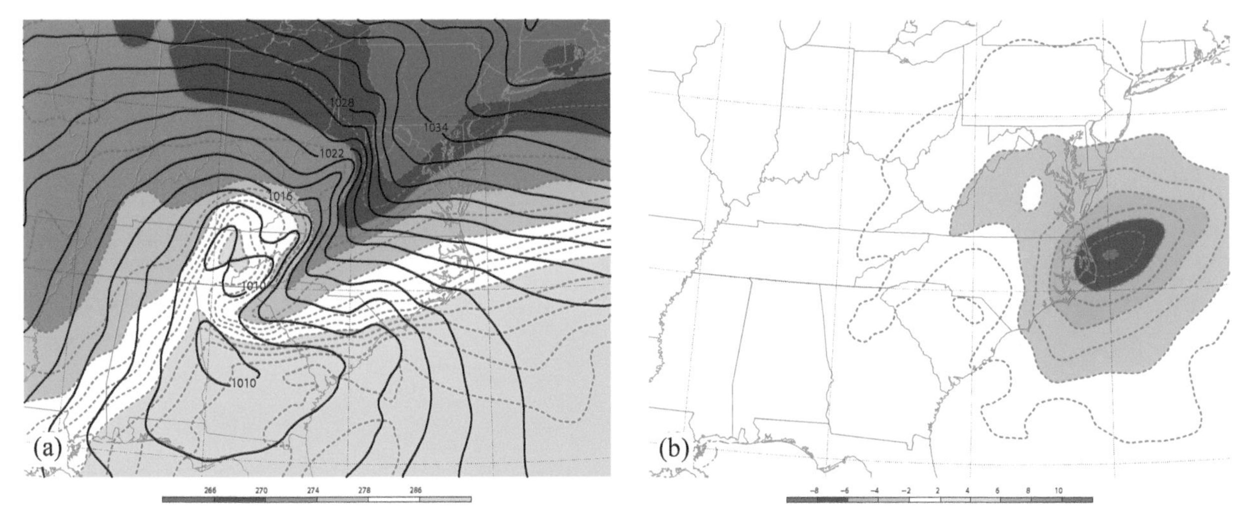


**Figure 8.12.** (a) As in Figs. 8.10a,b, but for no Gulf simulation; (b) difference field of sea level pressure, control minus no-Gulf, contours every 1 mb, shaded as in legend at bottom of panel.

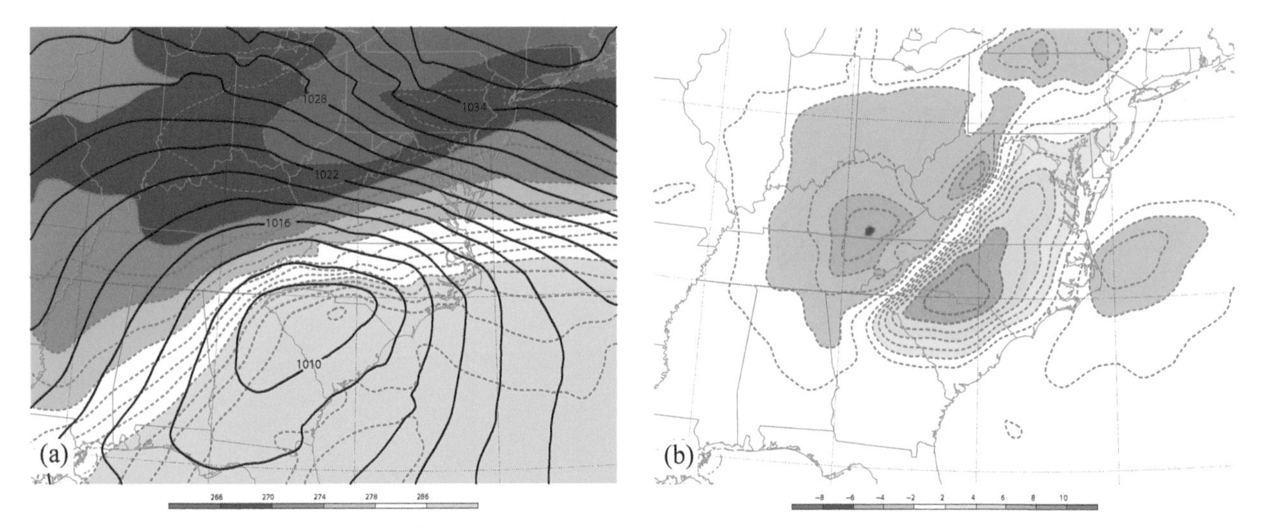


**Figure 8.13.** As in Fig. 8.12, but for no Gulf Stream, no mountain simulation compared to control.

example of this from the U.S. West Coast is presented in Fig. 8.14. These events often mark the end of hot weather episodes and offshore flow in western Washington State. As the cool marine surge propagates northward, upon reaching the Chehalis Gap in western Washington State, where the barrier of the coast range is effectively absent, the cool air surges into the western Washington interior. A similar phenomenon takes place in southeastern Australia, where it is known as a "southerly buster" or "burster" event (e.g., Baines, 1980).

Cold surges along the east slopes of the Rocky Mountains and in eastern Asia are fundamentally similar in character to CAD events described for the Appalachian Mountains, only the presence of a much wider orographic barrier leads to greater separation between trough and ridge. Sloping terrain and the absence of a warm water body to the east also result in pressure and temperature distributions that differ somewhat from those shown previously for the Appalachian region in the southeastern United States. Additional examples are highlighted by Smith (1982) for Iceland and New Zealand; the latter example is shown in Fig. 8.15. Damming-type phenomena have been observed for the Andes Mountains of South America, the Brooks Range in Alaska, and the Alps in Europe. Essentially, whenever stable flow encounters a significant topographic barrier in the middle or high latitudes, a distortion to

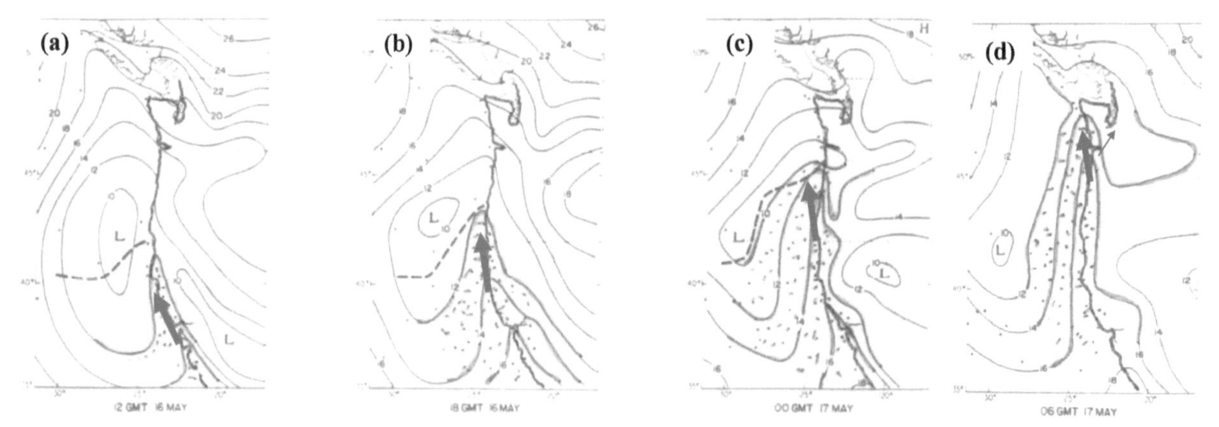


**Figure 8.14.** The CAD-like phenomenon of the alongshore surge near the West Coast of the United States, from Mass and Albright (1987) with modifications to clarify the location of the marine surge: (a) 1200 UTC 16 May 1985; (b) 1800 UTC 16 May; (c) 0000 UTC 17 May; and (d) 0600 UTC 18 May.

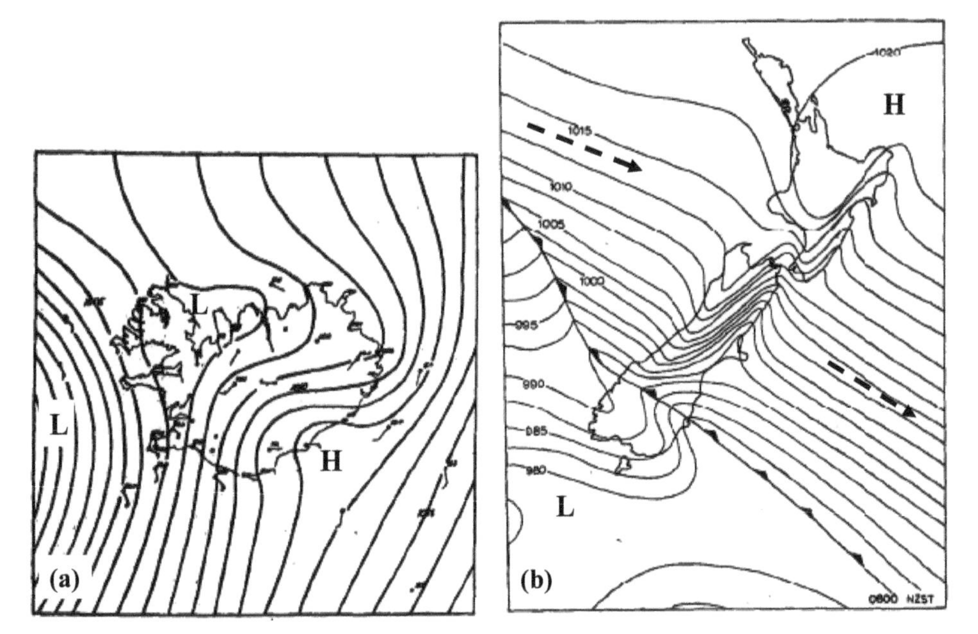


**Figure 8.15.** Example of orographically disturbed sea level pressure (mb) from (a) Iceland, and (b) New Zealand on 1 Aug 1975, contour interval 2.5 hPa. (From Smith 1982).

the wind and pressure fields will result, with geostrophic adjustment taking place.

### 8.2.2. The CAD spectrum for the southeastern United States

In the southeastern United States, because of its impact on regional population centers, CAD has been the subject of extensive research and operational forecasting attention. As mentioned previously, the initiation and maintenance of some CAD events are strongly dependent on diabatic processes such as evaporational cooling and solar sheltering. Other events are driven more strongly by synoptic-scale processes. Because not all CAD events are characterized by the same dominant physical processes, the *predictability* of CAD may also vary; one would expect events driven by synoptic-scale processes to be inherently more predictable than those in which diabatic processes are more important. The sensible weather impacts associated with a given CAD event can also be related to the dominant processes as well. This recognition

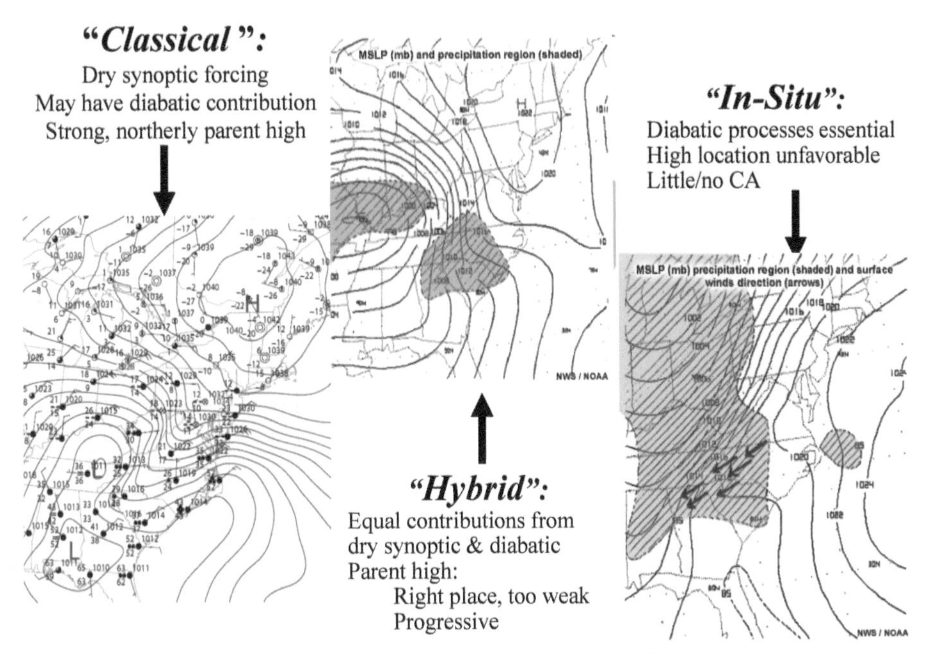


**Figure 8.16.** A classification system for Appalachian CAD events, based on research from the NWS office in Raleigh, NC: (a) sea level pressure (interval 2 hPa) and surface observations thickness for a classical CAD event; (b) sea level pressure (interval 2 hPa) for a hybrid CAD event; and (c) as in (b) but for an in situ CAD event. Examples are representative of each CAD category (graphic modified from G. Hartfield, NOAA/NWS, Raleigh).

led the National Weather Service to categorize CAD events on the basis of the relative strength of synoptic and diabatic contributions. Subsequent research, undertaken in collaboration with regional universities, helped to develop an objective scheme for the classification of CAD events in the southeastern United States. In addition to stronger "classical" CAD events associated with strong arctic high pressure systems, other more subtle events occur. These events can be difficult to forecast, but they are capable of strongly influencing weather conditions in the damming region. The classification scheme is designed to include weaker, more subtle types of damming in addition to the more obvious cases. Another important distinction is the extent of impacts on sensible weather conditions; not all CAD events exert a strong influence on weather parameters.

*Classical* CAD events feature a *strong parent anticyclone*, positioned well to the north of the Appalachian damming region (Fig. 8.16a). Here, "strong" generally corresponds to a central pressure in excess of 1030 mb, and a favorable position for the anticyclone center is over the northeastern United States. In situations where the parent anticyclone is weaker or less favorably located, diabatic process can play a larger role in the development of the event (Fig. 8.16b); these events are dubbed "hybrid" damming events. Finally, subtle, often shorted-lived in situ events may occur in the absence of a favorable synoptic pattern, but when diabatic processes lead to stabilization of an air mass approaching the Appalachians, weak, narrow damming events may occur (Fig. 8.16c). These events often present a challenge to forecasters and numerical forecast models alike.

An objective CAD-detection algorithm was devised and run on a 10-yr sample of surface observations in an attempt to quantify the climatology of CAD for the southeastern United States by Bailey et al. (2003). The CAD detection algorithm for the southern Appalachians identified 374 events over an 11-yr period in the southeastern United States, with the fewest observed during July (Fig. 8.17). A large number of weak CAD events took place from August to October, but most of the strong events were wintertime cases, with maximum occurrence in January and February; the occurrence of strong events is consistent with that documented by Bell and Bosart (1988).

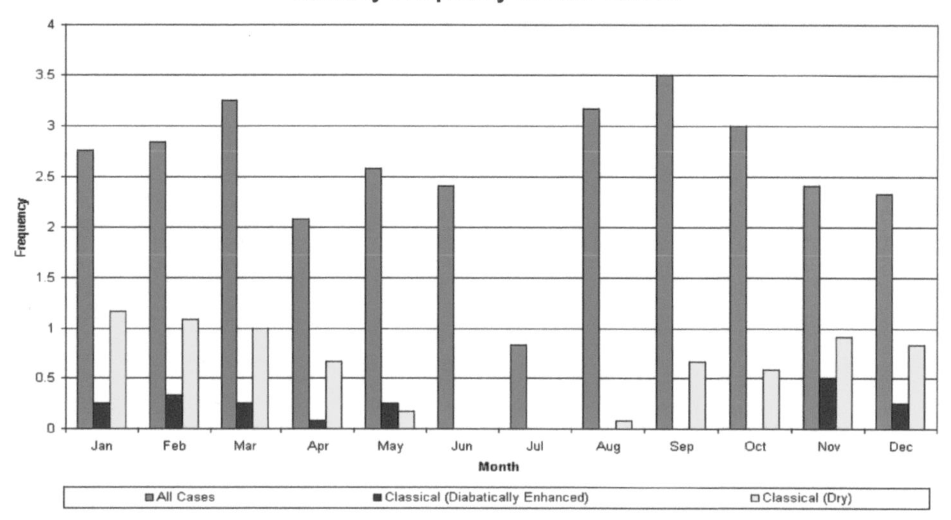


**Figure 8.17.** Monthly frequency of all CAD events (green bars) and classical CAD events (yellow and purple bars). Classical CAD events in which diabatic processes were found to contribute strongly to the event are dubbed "diabatically enhanced" classical CAD events and are shown as purple bars (adapted from Bailey et al. 2003).

By objectively quantifying the strength and duration of CAD events identified by this algorithm, one can devise a "spectrum" for CAD, with intensity plotted on the ordinate and the ratio of diabatic to synoptic forcing along the abscissa. However, when all of the CAD-like events were identified, a number of events that did not fit into the CAD spectrum proposed in Fig. 8.18a appeared. Although it is cumbersome to have additional classifications, the more complete spectrum shown in Fig. 8.18b represents a larger fraction of damming events. However, distinctions based on sensible weather impacts may offer the greatest utility to operational forecasters, as discussed below.

### 8.2.3. Impacts on sensible weather

Using the sample of CAD events identified by the objective detection algorithm, various measures of sensible weather impact during these events were examined. There are many useful sensible weather parameters that could be used in the evaluation of CAD impact. Here, the daily maximum temperature at stations within the southern

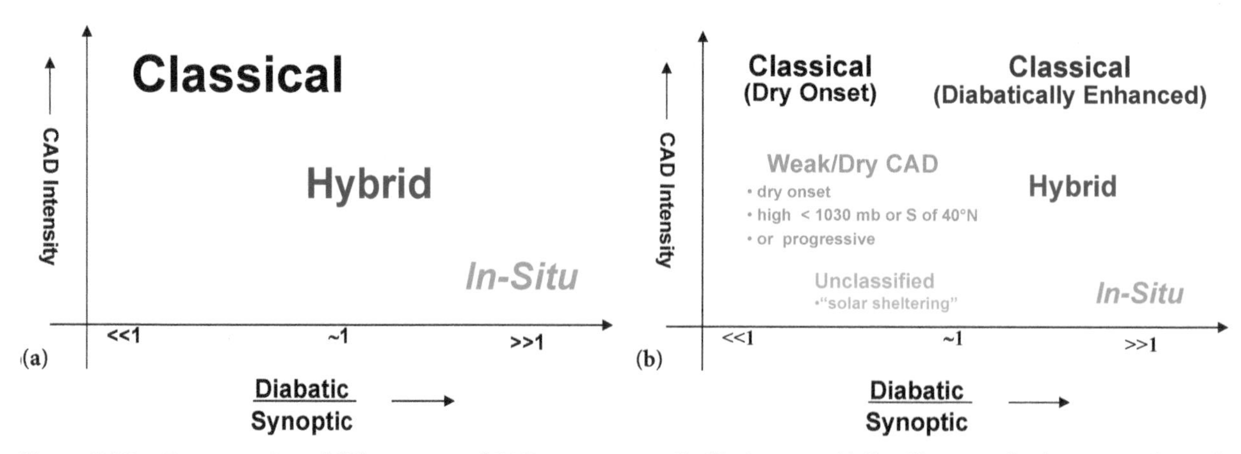


**Figure 8.18.** Representation of different types of CAD events as stratified by intensity (defined by strength of pressure ridge and duration of event) and the ratio of diabatic to synoptic-scale forcing (adapted from Bailey et al. 2003).

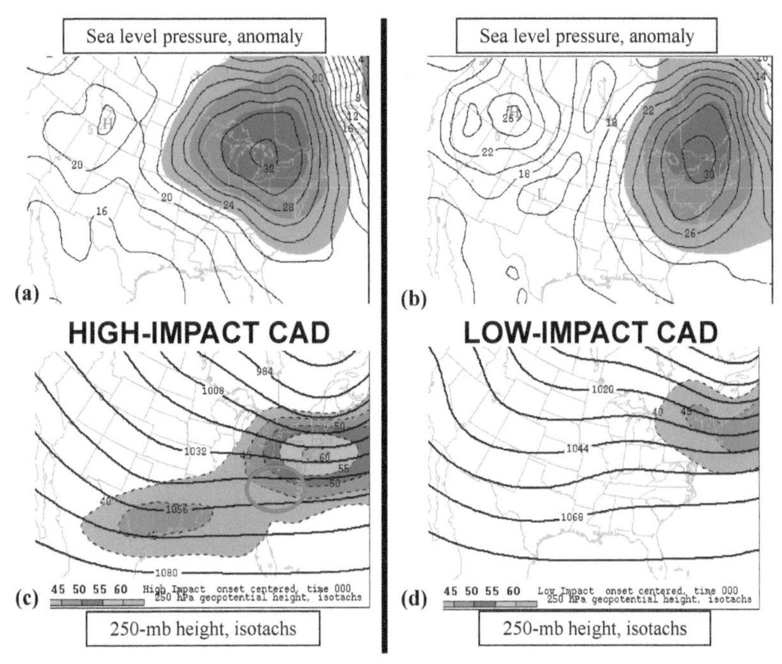


**Figure 8.19.** Composite comparison of (a),(c) high-impact and (b),(d) low-impact CAD events. (a),(b) Composite sea level pressure (contour interval 2 mb) and statistical significance of anomaly determined using two-sided Student's *t* test (green shading for 95% and 99% significance). (c),(d) The 250-mb height (solid contours every 12 dam) and isotachs (kt; shaded beginning 40 m $s^{-1}$ as in legend).

Appalachian damming region, Raleigh-Durham airport (RDU) and GSO were used to stratify the CAD events into high- and low-impact cases. Then, composite maps were computed using a sample of those events exhibiting a strong deviation of the maximum temperature from climatology and another sample of cases that did not.

Although the sea level pressure patterns are fairly similar for high- and low-impact events (Figs. 8.19a,b), more significant differences are evident aloft. High-impact events feature a more pronounced jet-entrance region over the mid-Atlantic region and a secondary jet streak along the Gulf Coast (Fig. 8.19c). These features, combined with weaker ridging aloft in the geopotential height field to the west of the damming region, are consistent with enhanced QG forcing for ascent and are likely associated with more extensive cloud cover and precipitation, which are often critical to the maintenance of CAD in this region. The low-impact composite exhibits a relatively weak jet entrance signature and slightly more pronounced ridging aloft to the west of the damming region (Fig. 8.19d).

Division of sample of classical CAD events into heavily precipitating (Figs. 8.20a–c) and dry (Figs. 8.20d–f) events reveals a more pronounced jet entrance for the wet events relative to dry, similar to that found for the high- and low-impact events defined using the maximum temperature. For the heavily precipitating composite, the presence of warm advection at the 850-mb level is clearly evident over the CAD region, whereas for dry cases the ridge axis at the 850-mb level was centered over the damming region. See Bailey et al. (2003) for additional details.

## 8.3. CAD EROSION

Based on discussions with regional forecasters in the southeastern United States over many years, it is clear that one of the greatest forecasting challenges associated with CAD is not the onset but the demise of the event. Numerical models were historically notorious for underestimating the duration of CAD events, with the model often tending to scour out the cold, shallow air mass too quickly. Before discussing this possible model bias, it is necessary to review the key physical processes and CAD erosion patterns for the southeastern United States. The physical mechanisms responsible for CAD erosion presented here apply to CAD events in any region, although the focus remains somewhat regional.

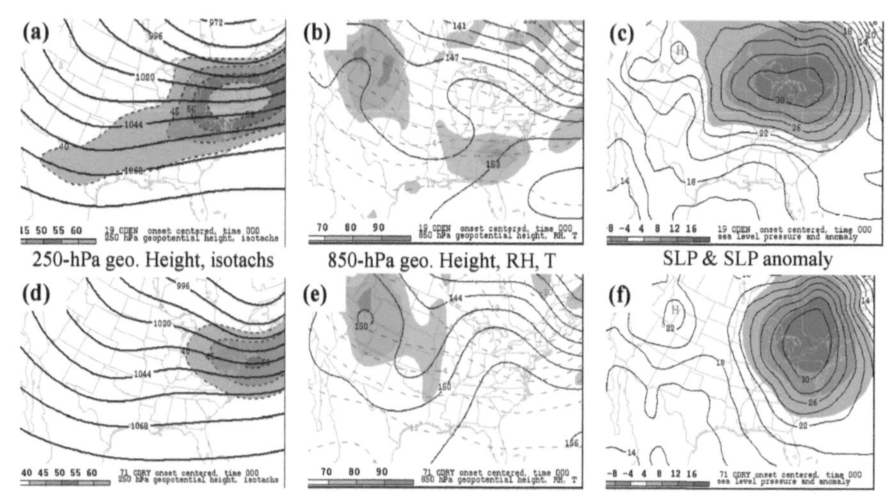


**Figure 8.20.** Composite comparison of (a)–(c) heavily precipitating with (d)–(f) dry: (a),(d) as in Fig. 8.19c; (c),(f) as in Fig. 8.19a; and (b),(e) representing 850-mb height (solid contours every 3 dam), temperature (red dashed contours every 4°C), and relative humidity (percent; shaded as in legend at bottom; adapted from Bailey et al. 2003).

### 8.3.1. Physical erosion mechanisms

Once a CAD air mass is in place, there are at least five physical mechanisms that can act to contribute to its demise; these processes are not independent of one another, and some are more easily handled by numerical weather prediction models than others. These processes may act in different synoptic settings, and specific processes are dominant in certain settings. Consider the vertical stratification of the cold dome during a CAD event as represented in a sounding (e.g., Fig. 8.6b); processes that act to strengthen the capping inversion generally prolong a CAD event, and those that weaken the inversion facilitate mixing and erosion of the cold dome. The bulk Richardson number Ri provides context for discussion of physical processes,

$$Ri = \frac{g\Delta\theta_v/\theta_v}{\left[(\Delta U)^2 + (\Delta V)^2\right]/\Delta z}. \tag{8.5}$$

The numerator in (8.5) represents the stability; in the case of CAD, this corresponds to the strength of the inversion layer separating the CAD cold dome and the ambient atmosphere above. The denominator represents the square of the vertical wind shear across the inversion layer. When the Richardson number becomes sufficiently small, turbulent mixing will occur, and this mixing can weaken the inversion layer and ultimately allow the erosion of the cold dome. For each of the processes discussed below, consider how the process would act to change the Richardson number.

#### 8.3.1.1. Cold advection aloft.

Examination of a large number of CAD events indicates that one of the most effective erosion mechanisms is cold advection aloft. It may seem counterintuitive that an effective way to eliminate a cold dome is through cold advection, but cooling *above* the inversion results in its weakening, eventually leading to mixing and the end of CAD. In section 8.1.2, we noted that warm advection within and above the inversion layer, with cold advection beneath, produced strengthening of the inversion due both to the differential thermal advection and to differential latent heating, especially in cases where dry air and evaporation were present beneath the inversion (Fig. 8.6a). During the *demise* of CAD, the opposite thermal advection pattern is found (Fig. 8.21a). The change in the vertical profile of potential temperature through the cold dome due to cold advection aloft is shown in Figs. 8.21b,c.

With cold advection maximized above the inversion layer, the cooling aloft can effectively weaken the inversion layer, thereby reducing the Richardson number. This may promote mixing, especially if there is strong vertical wind shear present; recall that cold advection is associated with backing geostrophic shear as well. Additionally, there is an indirect effect: the cold advection favors

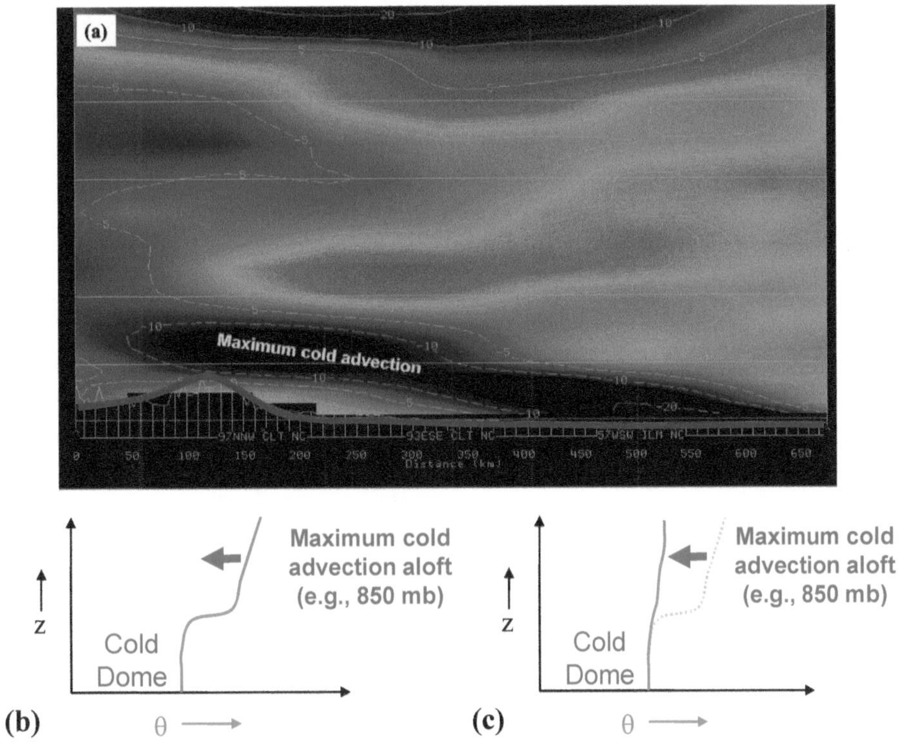


**Figure 8.21.** (a) As in Fig. 8.6a, but 18-h forecast valid 1800 UTC 5 Dec 2002 (courtesy NWS Raleigh). (b),(c) Potential temperature as a function of height before and after the action of cold advection aloft of the type shown in (a).

subsidence and drying, which can effectively strengthen other erosion mechanisms, such as solar heating beneath the inversion. In the presence of westerly cold advection, downslope flow to the east of the Appalachian Mountains may further contribute to subsidence and scouring of the cold layer east of the mountains. The synoptic settings most often characterized by cold advection aloft include cold-frontal passages or cold advection in the wake of a coastal cyclone.

**8.3.1.2. Solar heating.** Barring the presence of a thick overcast within or above the cold dome, solar heating can warm and erode the cold dome from the surface up, eventually resulting in mixing and CAD erosion. If even a shallow stratus layer is present, especially during the cold season and with a strong CAD inversion, this mechanism can be somewhat ineffective. However, with breaks in the overcast and during the warm season, warming due to absorption of solar radiation at the surface often plays a major role in the demise of CAD events (Fig. 8.22). Forecasters must monitor satellite imagery and surface observations carefully to accurately predict this process. As with cold advection aloft, the Richardson number is reduced, and mixing is promoted by this differential heating mechanism.

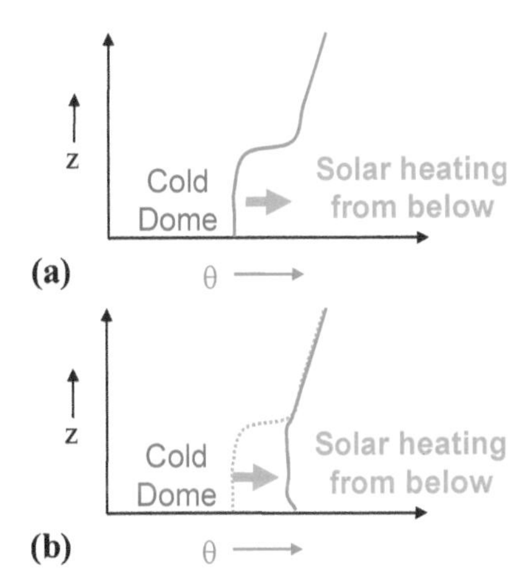


**Figure 8.22.** Idealized depiction of bottom-up CAD erosion due to solar heating. Panels show before and after profiles of potential temperature as a function of height for a location within the damming region.

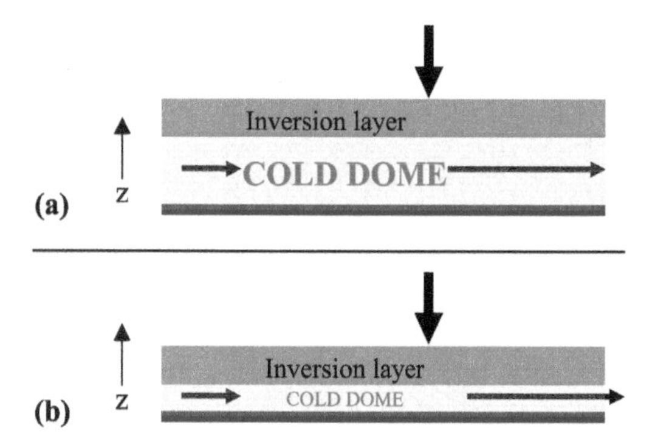


**Figure 8.23.** Idealized cross-sectional depiction of (a) divergence within the CAD cold dome and (b) a resulting decrease in the depth of the cold air layer.

**8.3.1.3. Near-surface divergence.** Frequently, there is divergence within the CAD cold dome during the mature and dissipating stages of an event. This can occur when the parent anticyclone moves eastward, resulting in a cessation of the northerly flow across Virginia and points north. If northeasterly winds persist in the southern portion of the damming region (e.g., over South Carolina), net divergence is implied, reducing the depth of the cold dome (Fig. 8.23). Through continuity, divergence is also associated with subsidence, which may serve to reduce cloud cover and increase the ability of solar heating to warm the cold dome from below.

**8.3.1.4. Shear-induced mixing.** In some cases, strong vertical wind shear can develop across the inversion layer, particularly when a cyclone is located to the west or northwest of the damming region. While the strong static stability of the CAD inversion layer often precludes turbulent mixing even in the presence of shear, if the shear becomes very large or with a weakening of the inversion, the cold dome becomes more susceptible to shear-induced mixing. This "top down" erosion can sometimes be observed in profiler data, such as shown during a CAD erosion event in Maryland on 26 September 2002 (Fig. 8.24). Time advances toward the left on this plot of profiler data from Fort Meade, Maryland; the depth of the northeasterly flow becomes increasingly shallow, and a downward progression of strong southerly flow is evident. In this example, other processes may have contributed to the reduction in the depth of the cold dome, but the top-down progression of winds in the presence of strong southerly shear suggests that this mechanism was at least partially responsible for the reduction in cold-dome depth.

**8.3.1.5. Frontal advance.** During the demise of CAD events, erosion often takes place first in locations where the cold dome is relatively shallow (near the periphery). As mixing progresses and the cold dome begins to erode, the boundary of the cold air, often

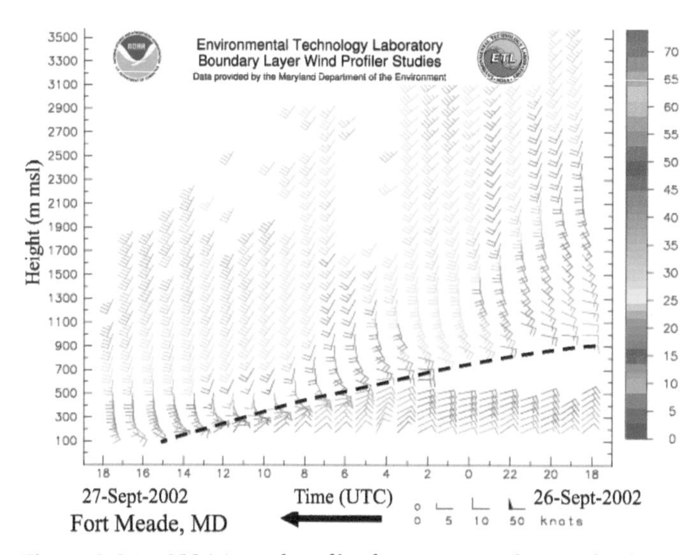


**Figure 8.24.** NOAA wind profiler from Fort Meade, Maryland, on 26 Sep 2002. Wind barbs (kt) are colored by strength, as indicated in legend at right. Time progresses toward the left (courtesy of Jeff Waldstreicher, NOAA/NWS).

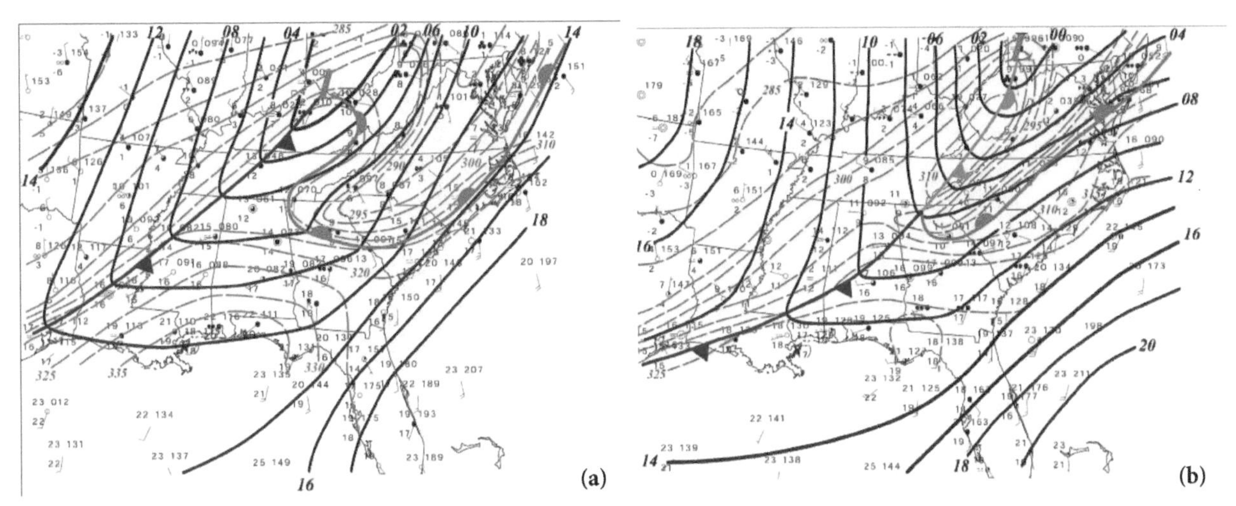


**Figure 8.25.** Surface analyses during the demise of a CAD event on 14 Feb 2000: (a) 0600 and (b) 1200 UTC. Sea level pressure (black contours every 2 mb), surface equivalent potential temperature (red dashed contours every 5 K), frontal analyses, and standard surface observations (from Brennan et al. 2003).

indicated as a coastal or warm front, will move inland as the width of the cold dome diminishes. The mechanisms at work may be similar to those that govern the movement of warm fronts, as discussed in section 6.4.

An example of frontal encroachment on a weakening cold dome from an event on 14 February 2000 is provided in Fig. 8.25. In this case, pressure falls over the cold dome were in part due to latent heat release aloft; the surface cold dome was already saturated, and therefore evaporational cooling did not serve to maintain the cold dome in this event. However, model experiments without latent heating demonstrated that the cold dome was maintained longer in the absence of this process, as discussed by Brennan et al. (2003).

### 8.3.2. Synoptic settings accompanying CAD erosion

Research has shown that there are four primary synoptic settings accompanying CAD erosion for the southern Appalachian region: (i) the passage of a synoptic cold front, (ii) the passage of a low to the northwest of the damming region, (iii) gradual demise as a stagnant residual airmass warms and mixes, and (iv) the passage of a coastal cyclone to the east of the damming region. Each synoptic scenario may be accompanied by one or several of the erosion mechanisms listed above, but certain scenarios feature different mechanisms more prominently. Events that end with the passage of a cold front are often fairly well predicted, because numerical models are capable of accurately representing differential thermal advection (cold advection aloft) and forecasters have a strong signal to work with heralding CAD demise. Erosion during cases in which a cyclone passes to the northwest of the Appalachian damming region but during which a cold front does not pass over the mountains is often premature in numerical model forecasts (Stanton 2004). Without a strong physical mechanism to drive the erosion, the cold air may persist for many hours beyond when the model predicts erosion.

Cases in which stratus cloud cover and saturated conditions have developed within the cold dome exhibit a tendency to persist. To illustrate an example that presented a challenge to numerical model forecasts, an Appalachian CAD event from October 2002 is presented in Fig. 8.26. As indicated by sea level pressure analysis shown in Fig. 8.26a, a relatively narrow tongue of cold air and high pressure was present across the state of Virginia and the Carolinas at 0000 UTC 30 October 2002. A strong stable layer and saturated conditions are present from the surface up to the 700-mb level, evident in the corresponding GSO sounding (not shown). Twenty-four hours later, a coastal cyclone has strengthened to the east of the damming region and a lingering region of flat pressure gradient is evident over the state of North Carolina. Satellite imagery and surface observations demonstrate that a persistent deck of stratus is in place in the damming region (Fig. 8.26c).

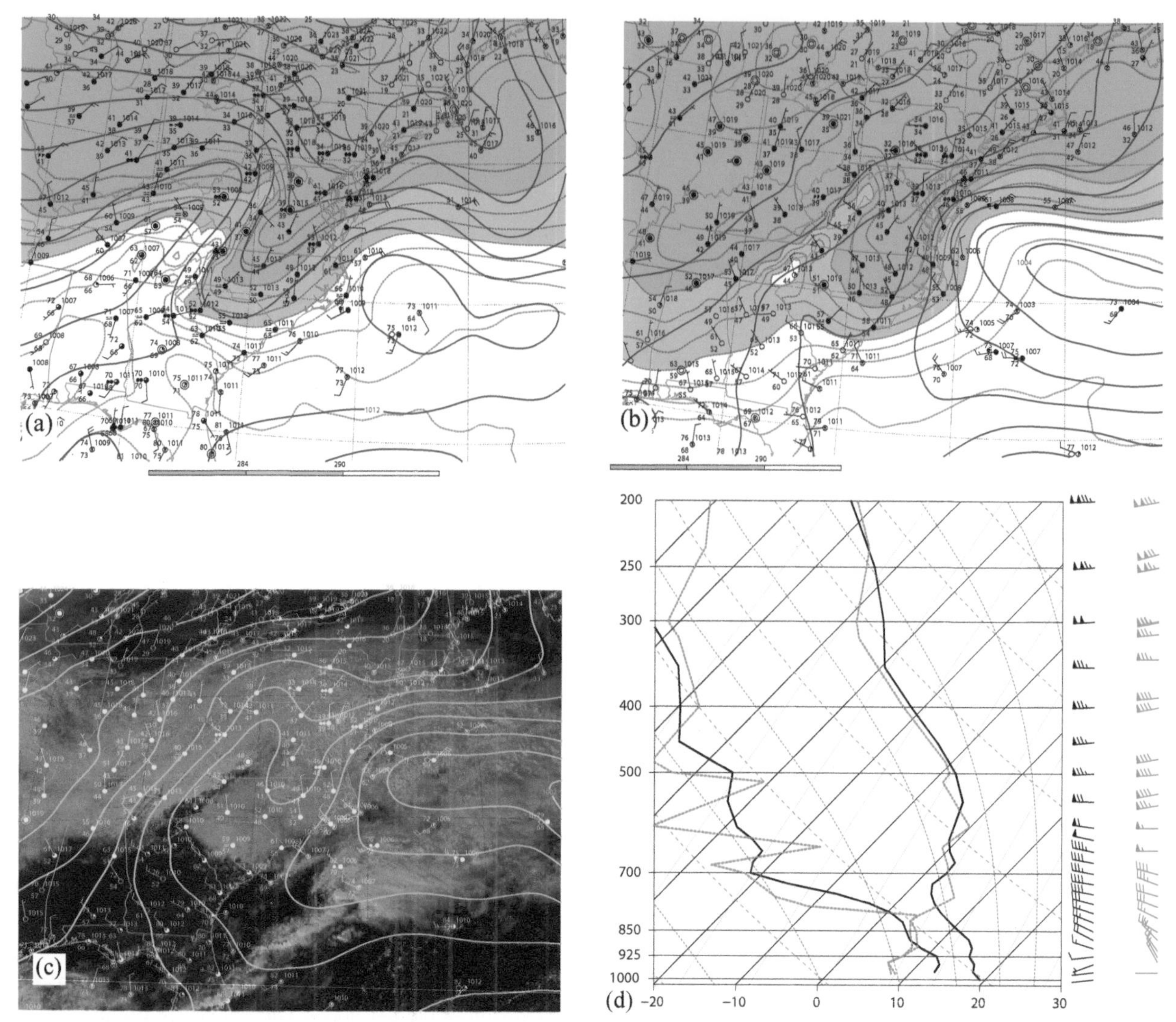


**Figure 8.26.** Summary of CAD event of 30 and 31 Oct 2002: (a),(b) Surface observations and RUC model sea level pressure analyses (dark blue contours every 2 mb) with 2-m isentropes (red dashed contours every 4 K) for 0000 UTC 30 Oct and 0000 UTC 31 Oct 2002. (c) Surface observations superimposed on *GOES-8* visible satellite image for 1800 UTC 30 Oct; and (d) comparison of GSO observed sounding (red dashed contours) and 36-h Eta model forecast sounding (solid black lines) in skew $T$-log$p$ format.

However, even short-term model forecasts had indicated a much more pleasant weather scenario for the Carolinas; the forecast sounding corresponding to this time indicates no stratus layer, and much warmer temperatures than were observed at all vertical levels below 800 mb (Fig. 8.26d). Subsequent analysis of the model forecasts for this case revealed that the cause of the poor model performance was due to a number of factors, including excessive transmission of solar radiation through low cloud layers and shallow mixing promoted by the model convective parameterization scheme. It is important to point out that these model shortcomings have largely been corrected in more recent versions of the NCEP models. We will examine these model components in greater depth in section 10.4.

Cold-air damming is due to the interaction of synoptic-scale pressure systems with topography in the presence of rotation, and the resulting wind and pressure fields can be interpreted as resulting from a geostrophic adjustment process. The varying contribution of synoptic-scale and more subtle diabatic processes, along with differences in the strength of sensible weather impacts, make CAD a difficult forecasting challenge. Regional forecasters in the southeastern United States initiated research that has led to useful classification of CAD events based on the relative importance of synoptic and diabatic processes. Stratification of events by the strength of sensible weather impacts demonstrates that the presence of synoptic-scale forcing for ascent, in the form of jet streak circulations and lower-tropospheric thermal advection, is often a critical factor in determining CAD impacts.

Numerical model representation of CAD will be discussed at length in chapter 10, where the impact of model physics parameterizations are shown to affect the ability of a model to accurately predict the erosion of a CAD event.

## REVIEW AND STUDY QUESTIONS

1. What is meant by the statement "cold-air damming is a geostrophic adjustment process"?
2. The trough–ridge couplet often seen to accompany Appalachian cold-air damming events features a ridge east of the mountains that is stronger than the trough to the west. Explain why the ridge is stronger.
3. Examine a map showing world topography and identify at least three locations that you expect may experience cold-air damming that are not mentioned in the text. Justify the selection of each location based on your knowledge of CAD.
4. Explain how diabatic processes can help to strengthen or maintain a cold-air damming event.
5. What processes are involved in cold-air damming *erosion*, and why are forecasts of CAD erosion difficult?
6. How do the Froude number and the Rossby radius of deformation relate to the evolution and structure of CAD events?

## PROBLEMS

1. Suppose that you are the meteorologist on a team that is in charge of determining the location for a new air field to be located to the east of a major mountain range in a remote, midlatitude, Northern Hemisphere location. You suspect that the region is prone to cold-air damming. Thankfully, there is a nearby upper-air sounding site, and you have access to many years worth of sounding data. Unfortunately, there are few surface stations in the area.
   i. What would you look for in the sounding data to determine the climatological frequency of CAD events? What evidence of CAD could be identified in the soundings?
   ii. Your sounding analysis indeed reveals a significant number of CAD events. List and discuss at least two potential impacts on aviation that might accompany these CAD events.
   iii. How could you use the available data to optimally select the location for the airport, given this situation? What additional data would you need, and how would you use these data?
2. The Rossby radius of deformation is the horizontal *e*-folding length scale for geostrophic adjustment. For a continuously stratified atmosphere (a single air mass), the relevant form of the Rossby radius is $L_R = \frac{NH}{\varsigma + f}$, where H represents the depth of the troposphere and $N = \sqrt{\frac{g}{\bar{\theta}}\frac{d\theta}{dz}}$, in which $\bar{\theta}$ is the average potential temperature of the troposphere (~310 K). This form is known as the *internal* Rossby radius. Using average synoptic-scale values, perform a scale analysis for the internal Rossby radius for the midlatitude atmosphere. Recall that $f = 2\Omega \sin(\varphi)$ and $\Omega = 7.292 \times 10^{-5}\, s^{-1}$.
3. Consider the hypothetical sea level pressure field and mountain barrier pictured here. Given a *Southern Hemisphere* midlatitude location and that the Froude number is small, sketch the 1016-, 1012-, and 1008-mb isobars in the diagram at right as they might look at

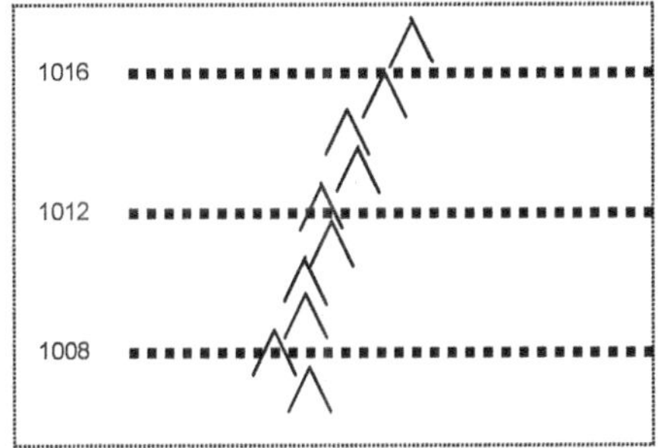


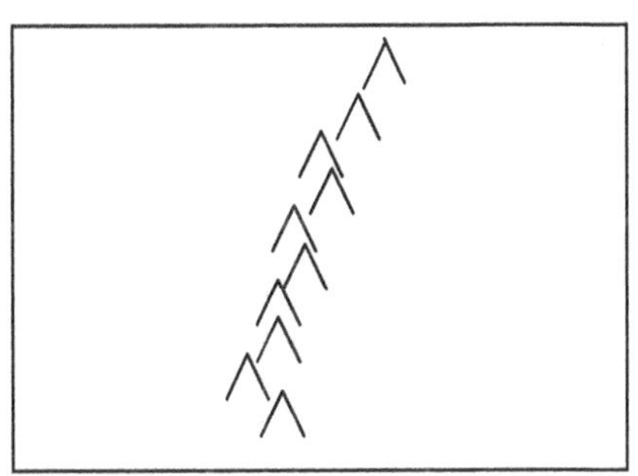

**Figure for Problem 3**

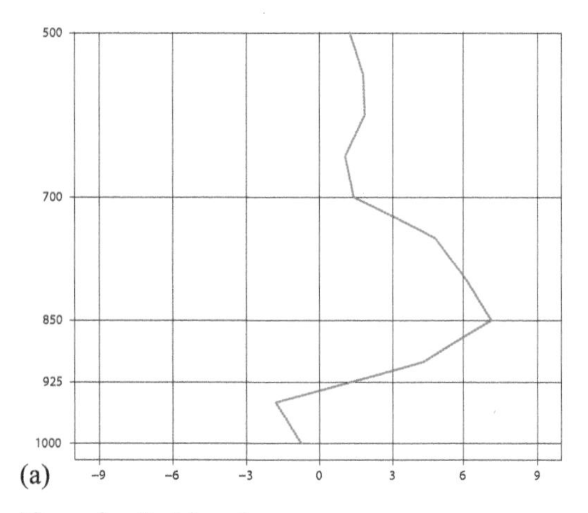


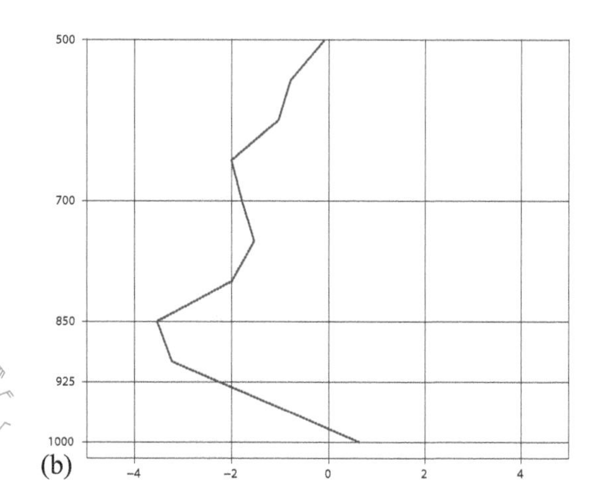


**Figure for Problem 4**

some later time after undergoing some geostrophic adjustment (e.g., 24 h later). Label any ridge and/or trough axes.

4. The two panels below show profiles of temperature advection for central North Carolina. One of these panels corresponds to a period when CAD was first setting up, and the other corresponds to a period of CAD erosion. Which was which? Justify and discuss your answers.
5. The left panel below shows the GFS sea level pressure analysis for 1200 UTC 11 Nov 2009. A large high pressure system to the north, in conjunction with the remnants of Hurricane Ida to the south, has set up a pattern that is seemingly favorable for CAD. The right panel shows the GSO sounding for the same time.

   i. Based on the GSO sounding and pressure analysis, discuss the extent to which CAD is occurring in the Carolinas at the time of this analysis.

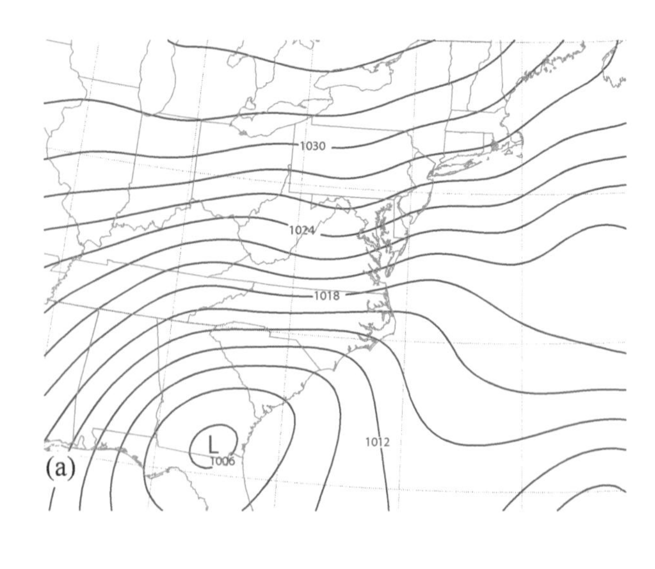


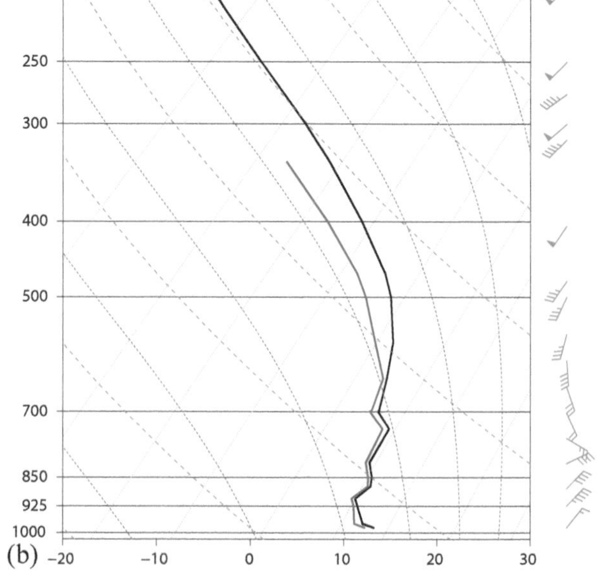


**Figure for Problem 5**

ii. Recall that the Froude number, $F_r = U/(NH)$, can help to ascertain the likelihood of flow blocking and CAD. Use the sounding along with the expression to perform an approximate calculation, and to determine if conditions are generally favorable or unfavorable for CAD. Discuss your findings.

## REFERENCES


Bailey, C. M., G. Hartfield, G. M. Lackmann, K. Keeter, and S. Sharp, 2003: An objective climatology, classification scheme, and assessment of sensible weather impacts for Appalachian cold-air damming. *Wea. Forecasting,* **18,** 641–661.

Baines, P. G., 1980: The dynamics of the southerly buster. *Aust. Meteor. Mag.,* **28,** 175–200.

Bell, G. D., and L. F. Bosart, 1988: Appalachian cold-air damming. *Mon. Wea. Rev.,* **116,** 137–161.

Brennan, M. J., G. M. Lackmann, and S. E. Koch, 2003: An analysis of the impact of a split-front rainband on Appalachian cold-air damming. *Wea. Forecasting,* **18,** 712–731.

Cortinas J. V., B. C. Bernstein, C. C. Robbins, and J. W. Strapp, 2004: An analysis of freezing rain, freezing drizzle, and ice pellets across the United States and Canada: 1976–90. *Wea. Forecasting,* **19,** 377–390.

Mass, C. F., and M. D. Albright, 1987: Coastal southerlies and alongshore surges of the west coast of North America: Evidence of mesoscale topographically trapped response to synoptic forcing. *Mon. Wea Rev.,* **115,** 1707–1738.

Meisinger, C. L., 1920: The precipitation of sleet and the formation of glaze in the eastern United States January 20 to 25, 1920, with remarks on forecasting. *Mon. Wea. Rev.,* **48,** 73–79.

Smith, R. B., 1982: Synoptic observations and theory of orographically disturbed wind and pressure. *J. Atmos. Sci.,* **39,** 60–70.

Stanton, W. M., 2004: An analysis of the physical processes and model representation of cold air damming erosion. M.S. thesis, Marine, Earth and Atmospheric Sciences Dept., North Carolina State University, 207 pp. Available from http://repository.lib.ncsu.edu/ir/handle/1840.16/2493


## FURTHER READING


Colle, B. A., and C. F. Mass, 1995: The structure and evolution of cold surges east of the Rocky Mountains. *Mon. Wea. Rev.,* **123,** 2577–2610.

Colquhoun, J. R., D. J. Shepherd, C. E. Coulman, R. K. Smith, and K. McInnes, 1985: The southerly burster of southeastern Australia: An orographically forced cold front. *Mon. Wea. Rev.,* **113,** 2090–2107.

Dunn, L., 1987: Cold air damming by the Front Range of the Colorado Rockies and its relationship to locally heavy snows. *Wea. Forecasting,* **2,** 177–189.

Forbes, G. S., R. A. Anthes, and D. W. Thomson, 1987: Synoptic and mesoscale aspects of an Appalachian ice storm associated with cold-air damming. *Mon. Wea. Rev.,* **115,** 564–591.

Fritsch, J. M., J. Kapolka, and P. A. Hirschberg, 1992: The effects of subcloud-layer diabatic processes on cold air damming. *J. Atmos. Sci.,* **49,** 49–70.

Hartfield, G., 1998: Cold air damming: An introduction. National Weather Service Eastern Region Training and Evaluation Module 4, 16 pp. [Available online at http://www.erh.noaa.gov/er/hq/ssd/erps/tem/tem4.pdf.]

Keeter, K. K., S. Businger, L. G. Lee, and J. S. Waldstreicher, 1995: Winter weather forecasting throughout the eastern United States. Part III: The effects of topography and the variability of winter weather in the Carolinas and Virginia. *Wea. Forecasting,* **10,** 42–60.

Lackmann, G. M., and J. E. Overland, 1989: Atmospheric structure and momentum balance during a gap-wind event in Shelikof Strait, Alaska. *Mon. Wea. Rev.,* **117,** 1817–1833.

Richwien, B. A., 1980: The damming effect of the southern Appalachians. *Natl. Wea. Dig.,* **5,** 2–12.

Stauffer, D. R., and T. T. Warner, 1987: A numerical study of Appalachian cold-air damming and coastal frontogenesis. *Mon. Wea. Rev.,* **115,** 799–821.

Xu, Q., 1990: A theoretical study of cold air damming. *J. Atmos. Sci.,* **47,** 2969–2985.

——, S. Gao, and B. H. Fiedler, 1996: A theoretical study of cold air damming with upstream cold air inflow. *J. Atmos. Sci.,* **53,** 312–326.

# CHAPTER 9

# Winter Storms

*The meteorological record of February, 1920, at New York, will long be remembered by reason of the remarkable storm.... The maximum depth of snow and sleet was 17.5 inches, of which 8.8 inches were sleet and 8.7 inches were snow.... Flame throwers, a remnant of the war, were tried, as well as another heat device, but these were so local in their results that they were impractical; the steam shovel was tried, with fair success, where the drifts were deep enough; a newly constructed snow digger was tried with great success, but it was impossible to get enough of this type of machine to be of use in the emergency; finally, the fire hose was used with the greatest effect. This experience should warn all large cities where such storms are a possibility, and lead to the provision of adequate equipment for such an emergency.*

—C. L. Meisinger,
"Demoralization of Traffic in New York City by Snow and Sleet" (1920)

A man shovels snow off of a roof in the Tughill Plateau region of New York state, following an intense lake-effect snow event that took place in the vicinity of Lake Ontario in early February, 2007.

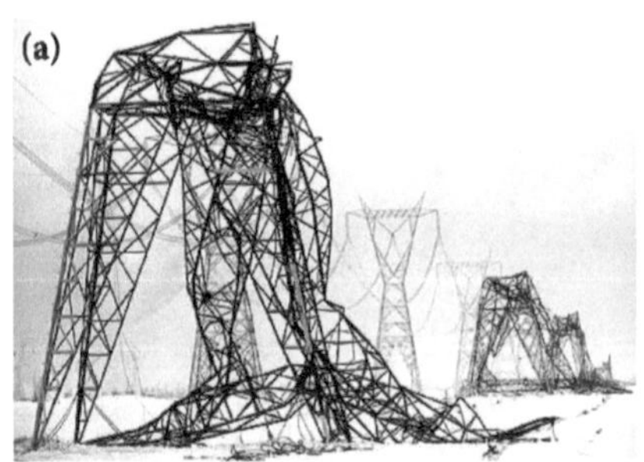

**Figure 9.1.** Photographs of high-impact winter weather: (a) the collapse of transmission line towers near St. Bruno, Quebec, during the January 1997 ice storm that left more than 1.4 million people without power; (b) late morning snow on 19 Jan 2005 that snarled traffic in Raleigh, North Carolina; (c) a major arterial street in Raleigh on 25 Jan 2000 during a record-setting snowstorm (photos courtesy of Jacques Boissinot, Jonathan Blaes, and Neil Jacobs).

Along with severe convective storms, floods, and hurricanes, winter storms rank among weather phenomena most capable of bringing severe societal impacts, in addition to being challenging to forecast. For seasoned winter storm veterans living in cities characterized by appreciable annual snowfall, efficient snow-removal strategies and experienced drivers lessen the impact of winter storms to some extent. However, for major metropolitan areas and in regions that are climatologically less accustomed to wintery weather, snowstorms are often associated with significant economic and societal disruptions (e.g., Fig. 9.1). Travel delays and cancellations, school closings, event rescheduling, and reduced access to emergency services can all occur during a winter storm. Retail losses of up to $10 billion per day have been reported during winter storms. Flight delays cost U.S. air carriers over $3 billion annually, the annual cost of U.S. snow removal exceeds $2 billion, and damage to utilities and infrastructure can exceed $2 billion per storm (NOAA Economics 2008).

In this chapter, we discuss the physical processes at work that determine precipitation type, and we will discuss the ability of numerical forecast models to represent these processes. A variety of winter-weather phenomena are presented, ranging from the mesoscale phenomenon of lake-effect snow (LES) to synoptic-scale winter storms.

## 9.1. GENERAL FORECASTING CONSIDERATIONS AND CLIMATOLOGY

When frozen or freezing precipitation falls, there is a heightened demand for the accurate prediction of precipitation occurrence *and amount*. However, quantitative precipitation forecasting (QPF) is widely regarded as the least accurate aspect of numerical weather prediction. Winter storm impacts are highly sensitive to precipitation amount; for freezing-rain events, the difference between a short-lived glaze and major power outages can hinge on this parameter. This is not to say that relatively light precipitation amounts are not occasionally associated with major impacts. Precipitation intensity can also impact precipitation type, as we will see, further increasing the challenge of winter-weather forecasting.

Accurate prediction of the *impacts* of winter weather can add complication to the winter-storm forecast process. Warm soil and road temperatures could mean that snow accumulation will largely be restricted to grassy or elevated surfaces and travel will not be severely impacted. An otherwise insignificant snowfall that takes place during rush-hour traffic can wreak havoc. Should meteorologists be expected to predict impacts, in addition to what is falling from the sky? In recent years, increasing efforts have been made to link forecasts to societal weather impacts; for instance, National Weather Service (NWS) field offices are now staffed with a warning-coordination meteorologist (WCM), who serves as a liaison between the forecasters and emergency managers and local decision makers.

One of the reasons why winter-weather prediction can be so difficult is that small variations in atmospheric parameters can dictate changes in precipitation type at the surface. Subtle thermodynamic processes, many of which are QPF dependent, are not always represented properly or accurately by numerical weather prediction models. Latent heat release and absorption can alter atmospheric temperature profiles sufficiently to affect precipitation type. These processes and their prediction will be discussed at length in section 9.2.

The geographical distribution of freezing and frozen precipitation is highly variable, ranging from no

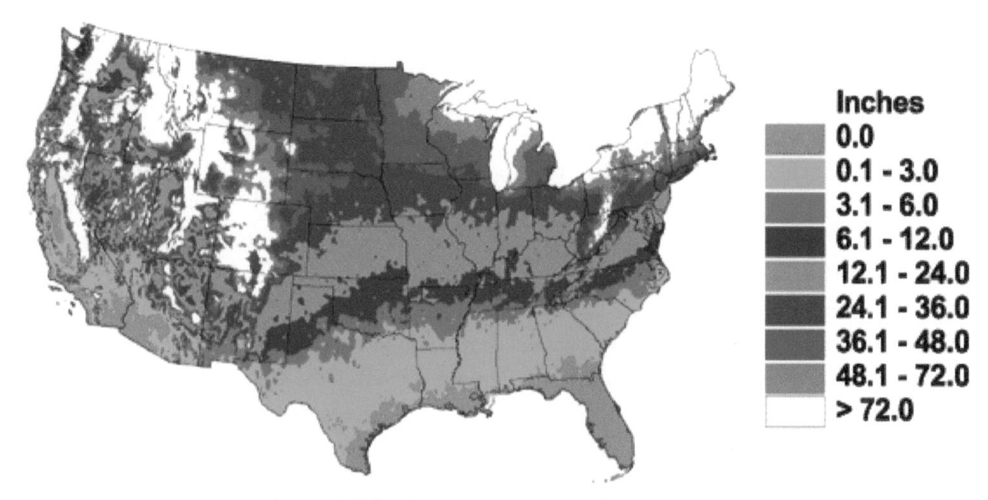


**Figure 9.2.** Mean annual snowfall for the contiguous United States (inches; shaded as in legend at lower right) based on a 30-year average of monthly snowfall data (image source: http://www.ncdc.noaa.gov/oa/about/cdrom/climatls2/info/atlasad.html).

annual snowfall to over 200 in. across the contiguous United States (Fig. 9.2). Different regions of the world are faced by unique winter-weather forecasting challenges, including lake-effect snowstorms, coastal cyclones, freezing rain and sleet linked to cold-air damming, and orographic snowfall. In many parts of the southern United States, winter weather is uncommon, making it difficult for regional forecasters to accrue experience in its prediction. Also, in some regions the public may tend to overreact to predictions of winter weather, and forecasters must choose their wording very carefully when winter weather threatens! In the southeastern United States, *mixed precipitation* is common in the Carolinas and Virginia, in part because of cold-air damming. For example, Moyer (2001) found that 86% of winter precipitation events at Greenville–Spartanburg (GSP) in upstate South Carolina consisted of two or more precipitation types (e.g., a mix of snow, freezing rain, and ice pellets).

Wintry weather can take place in a variety of synoptic settings, and an in-depth discussion of these situations is not presented here. Synoptic patterns conducive to winter weather are strongly regional; for example, Kocin and Uccellini (2004) provide a comprehensive discussion of the patterns and processes associated with snowstorms in the northeastern United States, whereas Ferber et al. (1993) identified a set of distinct synoptic patterns accompanying snowfall in the western lowlands of Washington State. Still different circumstances characterize the synoptic environment for lake-effect snow in the Great Lakes region.

Clearly, wintry weather is less likely in the warm sector of cyclones, and it is favored in the cold regions north of warm fronts, in the vicinity of or to the northwest of cyclone centers, and for anafront-type cold fronts. In the mountain and plains regions of the United States and Canada, upslope flow (orographic lift) represents a significant lifting mechanism that must be considered in winter-weather prediction even in the fall and spring for the high plains. Upper-level cutoff lows can produce copious mountain and high-plains snowfall, often working in concert with orographic processes. In the midwestern United States, inverted troughs to the north of a cyclone center or snowfall along an occluded front can be important. We will examine conditions favoring the occurrence of lake-effect snow in section 9.3. For freezing rain and ice pellets (Fig. 8.1), in addition to topographically favored regions featuring cold-air damming, the region ahead of an approaching warm front is characterized by a vertical temperature structure that is conducive to freezing precipitation (e.g., Fig. 8.2). The emphasis of the discussion here is on the physical processes at work that can determine the precipitation type rather than on the individual synoptic patterns leading to wintry weather.

Mesoscale processes can also be important during winter weather. The importance of gravity waves, local topographic effects, lake-induced convection, and mesoscale snowbands have all been documented to impact snowfall distributions (e.g., Sanders and Bosart 1985; Colle et al. 1999; Niziol 1987; Novak et al. 2004, 2009). Rather than discuss each phenomenon here, lake-effect snow will be presented as an example of such a mesoscale winter storm phenomenon.

## 9.2. PHYSICAL PROCESSES

Although cloud and precipitation physics are not reviewed in this text, it is expected that the reader has some familiarity with the cloud-formation process, and is familiar with terms such as *deposition*, *mixed-phase*, *riming*, *ice nucleation*, as well as the difference between classes of hydrometeors such as ice pellets, freezing rain, hail, and graupel. At the end of the chapter, some cloud and precipitation review questions are provided as a study guide for those desiring additional review.

### 9.2.1. Overview of processes affecting the thermal profile

Consider the Greensboro, North Carolina (GSO), rawinsonde sounding for a winter storm that featured freezing rain and cold-air damming, which took place in December 2002 (Fig. 9.3). It is a worthwhile exercise to think carefully about the large variety of physical processes at work that affect the temperature and dewpoint profiles and also to consider what evidence is available in the soundings to indicate the presence and importance of these processes. In anticipation of high-impact winter weather during this event, personnel at GSO launched sondes every 6 h rather than at the usual 12-h intervals.

At 1800 UTC 4 December 2002, a deep layer of saturation extended from the 925-hPa level upward, with saturation with respect to ice suggested by the gradual spreading of the temperature and dewpoint profiles above the 600-mb level (Fig. 9.3a). If we had plotted frost point rather than dewpoint, this layer of saturation would be more evident. Frequently, subfreezing clouds feature a mixture of ice crystals and liquid droplets, so the dewpoint remains a meaningful quantity in these situations. With a deep cloud layer and parts of the cloud exhibiting temperatures ranging from near freezing to colder than −40°C, it is almost certain that a mixed-phase cloud was present, with active Bergeron–Findeisen snow crystal growth. Furthermore, a veering wind profile in the lower atmosphere implies warm advection and thus quasigeostrophic (QG) forcing for ascent (Fig. 9.3a). The ascent and expansional cooling leads to the development of clouds and precipitation, but ascent in a stable atmosphere such as seen here provides a strong cooling mechanism [see Eqs. (1.27) and (1.28)]. Precipitation was falling aloft at this time (confirmed by radar imagery, not shown), although falling snowflakes were undergoing sublimation in the dry layer between the surface and the 925-mb level.

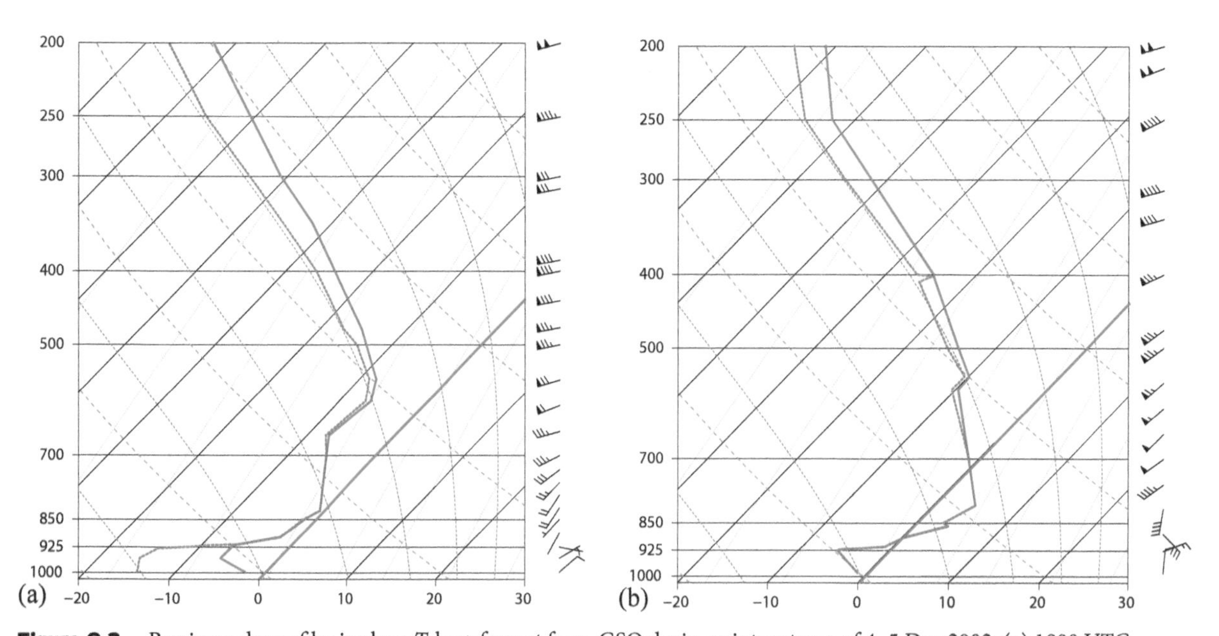


**Figure 9.3.** Rawinsonde profiles in skew *T*-log*p* format from GSO during winter storm of 4–5 Dec 2002: (a) 1800 UTC 4 Dec and (b) 0600 UTC 5 Dec. Zero degree isotherms highlighted red.

Based on evidence in the 1800 UTC sounding, thermodynamic processes releasing heat included condensation and deposition, whereas latent heat absorption and cooling was taking place in the lowest portion of the atmosphere because of sublimation. Additionally, any riming of falling snowflakes with supercooled cloud droplets would release latent heat of freezing.

As discussed in the previous paragraph, warm advection was taking place, but the accompanying ascent partially compensated this warming process through adiabatic expansion. Additional processes that are not clearly evident in this sounding may also be at work. It was stated that this winter storm was accompanied by a CAD event, so it is likely that a shallow layer of cold advection accompanied the northeasterly wind seen in the lowest portion of the GSO sounding (Fig. 9.3a). Radiative processes, including upward infrared (IR) radiation from the ground and downward IR from the cloud base, may have affected the temperature profile as well. Examination of model-derived soil temperatures indicate a warm ground, with temperatures in the uppermost 10 cm of soil above 5°C (not shown). This suggests that an upward heat flux from the ground was present, which would have contributed to warming the lower atmosphere. Isentropic analysis indicates that strong ascent was taking place in the lower troposphere (not shown), and surface analyses support the above contention that cold, dry air was advecting southward across the region with an ageostrophic northerly flow (not shown).

After 12 h (Fig. 9.3b), significant changes have taken place in the profile relative to that shown in Fig. 9.3a. The layer from ~860 to ~710 mb is above freezing at this time, likely the result of pronounced warm advection, which is consistent with the strong, veering wind profile (Fig. 9.3b). Beneath this warm tongue, the atmosphere was saturated, with precipitation reaching the surface. Because of the warmth and depth of the above-freezing layer, snow crystals would not fall through this layer without melting. However, with subfreezing air beneath the warm layer, the raindrops would likely either refreeze before reaching the ground or freeze on contact with the surface. The melting of snow aloft in the above-freezing layer would absorb heat, cooling the air there, whereas the freezing process at the surface results in the release of latent heat there. Surface observations for this time (not shown) indicate freezing rain, and the shallow surface-based mixed layer evident in Fig. 9.3b may be a reflection of the latent heat of freezing released at the surface or of upward heat flux from the warm ground. Unlike the earlier sounding (Fig. 9.3a), which was characterized by sublimational cooling near the surface, at this time little additional cooling potential existed near the surface because of saturated conditions there.

Even though we have considered a large number of processes here, there remain others of unknown importance. For example, sensible heat transport by falling rain could be significant in a sounding such as this, if sufficiently heavy precipitation was falling. This is a process that is typically neglected in numerical models, but justification of this neglect for cases of heavy precipitation may require additional research.

### 9.2.2. Warm snowstorms

The term *warm snowstorm* has been applied to events in which cooling due to melting is a critical factor in determining precipitation type (Gedzelman and Lewis 1990). Dramatic examples of such events include the upstate New York snow of 4 October 1987; a snowstorm on 26 December 1974 in Seattle, Washington; and a snowstorm in Boston in April 1953 to name a few. Early studies of warm snowstorms by Wexler et al. (1954) and Lumb (1960) recognized the role of melting in such events. More recent studies by Bosart and Sanders (1991) and Kain et al. (2000) document the ongoing challenges that these potentially high-impact events present to forecasters.

Cooling (warming) of the atmosphere due to melting (freezing) is often weak relative to temperature changes resulting from condensation or evaporation. This is partly because the latent heat of fusion is nearly an order of magnitude smaller than the latent heat of vaporization ($3.34 \times 10^5$ J kg$^{-1}$ versus $2.5 \times 10^6$ J kg$^{-1}$ at 0°C) and also because of compensating processes that will be discussed below. However, the cooling potential via evaporation is usually bounded; once the lower atmosphere becomes saturated, barring dry-air advection, then no additional cooling below the *wet-bulb temperature* will occur. Furthermore, the amount of vapor required for saturation diminishes rapidly with temperature, meaning that the cooling potential for a given dewpoint depression decreases with temperature (put another way, the wet-bulb temperature, which is the temperature to which air will cool when brought to saturation via evaporation, becomes closer to the air temperature as it gets colder). This is not to say that evaporational cooling is not a critically important

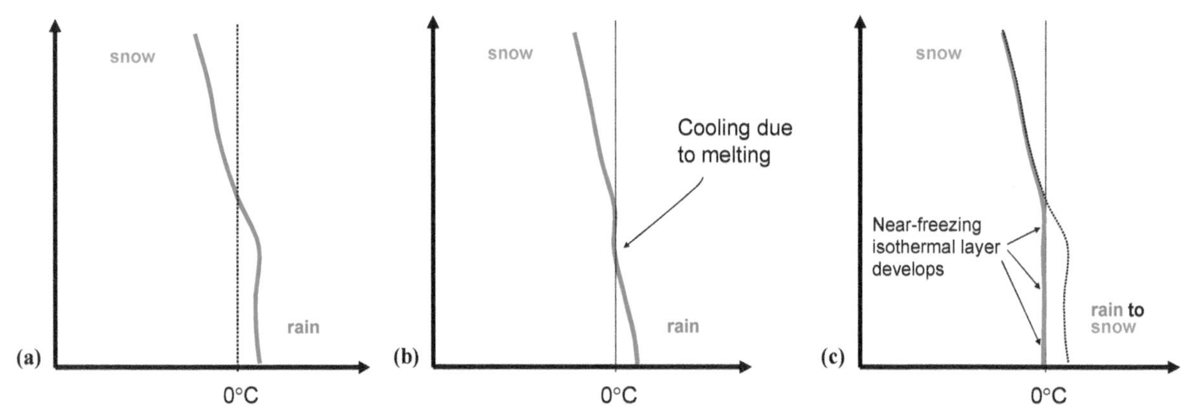


**Figure 9.4.** Idealized sequence of temperature profiles with height (red lines) showing the development of an isothermal melting layer that descends to the surface. The vertical axis represents height and the abscissa represents temperature; 0°C reference line included. Dotted black line in (c) indicates original temperature profile from (a).

process—forecasters must pay careful attention to the wet-bulb temperature in borderline winter-weather situations.

In some meteorological scenarios, cooling due to melting can play an important role. Unlike evaporation, melting-induced cooling will continue as long as snow is falling into an above-freezing layer and melting. An idealized sequence of temperature profiles illustrates the development of an *isothermal melting layer* of the type that can occur in the presence of continued melting of falling snow (Fig. 9.4). As snow falls into an above-freezing layer and melts, the atmosphere will cool, eventually reaching the freezing point in the absence of compensating mechanisms. Once the air has cooled to the freezing point, no additional cooling will take place due to melting. For the above-freezing portion of the air column, melting-induced cooling may result in the freezing level descending all the way to the surface, with precipitation changing from rain to snow as indicated in Fig. 9.4c. Of course, rain may mix with and change to snow before the surface temperature reaches the freezing point.

The melting process can give rise to situations in which precipitation intensity determines precipitation type. Areas with lighter precipitation may observe rain, whereas locations with heavy precipitation, where stronger cooling would occur, may witness a changeover to snow. Even if the ground temperature is above freezing, melting of snow as it hits the ground cools the top layers of the soil; if the snow is sufficiently heavy, this can result in accumulation despite an initially warm surface.

Given the depth and warmth of the above-freezing layer, how can forecasters determine if melting will be sufficient to bring about a warm snowstorm? First, one must assess the relative importance of other processes, such as warm advection, evaporational cooling, and precipitation intensity and amount. Provided that heavy precipitation is expected with marginal compensation from other processes, one can apply a bulk form of the first law of thermodynamics to estimate the amount of precipitation needed to "melt out" a given above-freezing layer. The following is a simplified expression of this type presented by Kain et al. (2000),

$$P_{\text{melt}(cm)} \approx -\frac{\delta T \times \delta p}{193}, \tag{9.1}$$

where $P_{\text{melt}}$ is the amount of liquid-equivalent precipitation needed to melt out the above-freezing layer (in centimeters), $\delta T$ represents the average temperature of the above-freezing layer (in K), and $\delta p$ is the pressure depth of the layer in mb.

The melting process would be more effective in cooling the atmosphere if it were not for an implicit compensating effect from condensation. If a layer is saturated and melting snow is falling, the melting-induced cooling results in condensation, which releases latent heat. Surface conditions during this process are often characterized by fog as a result. Despite the compensation from condensation, cooling over a significant depth can occur; during an event from 1 November 1942 in the United Kingdom, the wet-bulb freezing level descended more

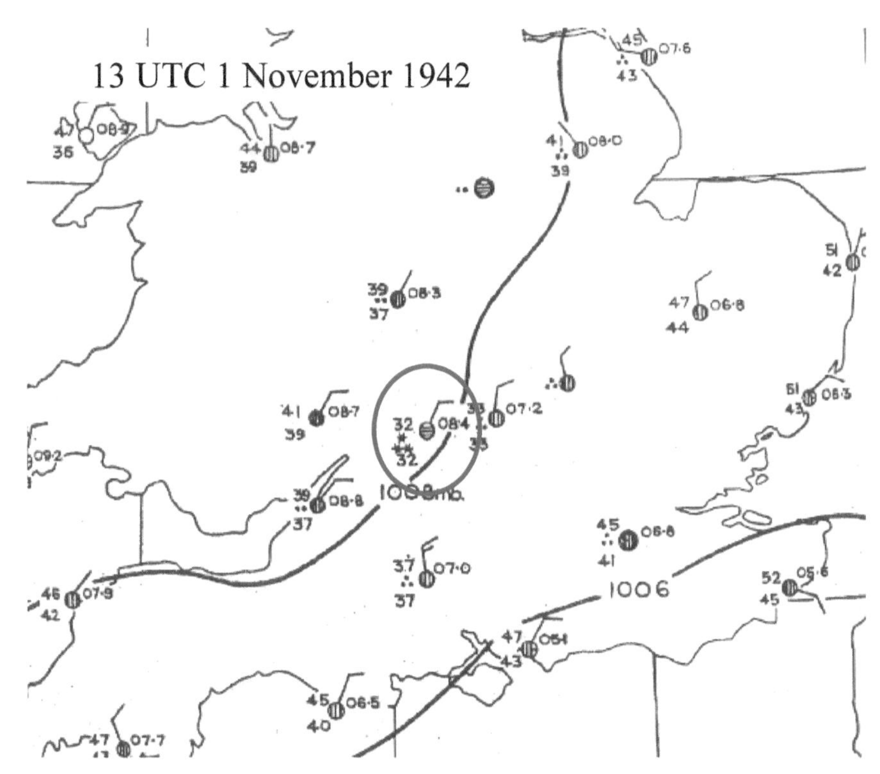


**Figure 9.5.** Surface observations for the United Kingdom for 1300 UTC 1 Nov 1942 (from Lumb 1960). Circle indicates location of station reporting moderate snow.

than 1,500 m (Lumb 1960)! Note the isolated station reporting snowfall, surrounded by areas of rain and above-freezing temperatures in Fig. 9.5.

### 9.2.3. Freezing rain

In certain synoptic situations, such as during the advance of a warm front, and in specific geographical regions, a layer of above-freezing air may become established above a shallow subfreezing layer near the surface. This configuration can result in freezing rain or ice pellets. The term *ice pellets* is preferred to *sleet*, because the latter term is taken to refer to a mixture of rain and snow in some locations. As its name implies, freezing rain is liquid rain that freezes on contact with the surface, creating a glaze that can result in extremely treacherous driving conditions, power outages, and damage to trees and plants. As is the case with snow, accurate prediction of the amount, as well as type of precipitation, is a critical forecast parameter during freezing-rain events. Although ice pellets can also result in significant disruption to travel, the absence of a glaze renders these events less damaging than would be an event characterized by the same liquid-equivalent amount of freezing rain.

In freezing rain and ice pellet events, one must, in addition to the processes discussed previously, consider the latent heat released by the refreezing of liquid rain drops in the lower atmosphere or at the surface. Depending on the circumstances, all or only a fraction of this heat may be imparted to the atmosphere, as discussed below. Consider the hypothetical situation shown in Fig. 9.6a, which could correspond to a cold-air damming event. Temperatures aloft are easily warm enough to support rain falling into a shallow subfreezing layer near the surface. What processes are acting to determine the evolution of temperature in the subfreezing layer? There will be warming that results from the latent heat released during the freezing of falling rain, radiative processes, and perhaps a contribution due to sensible heat transport by the falling rain. Depending on the soil temperatures, there may be cooling or warming due to heat flux into or out of the ground. There may be continuing cold and/or dry-air advection within the subfreezing layer that would act to maintain it. In the absence of cold advection in this layer, however,

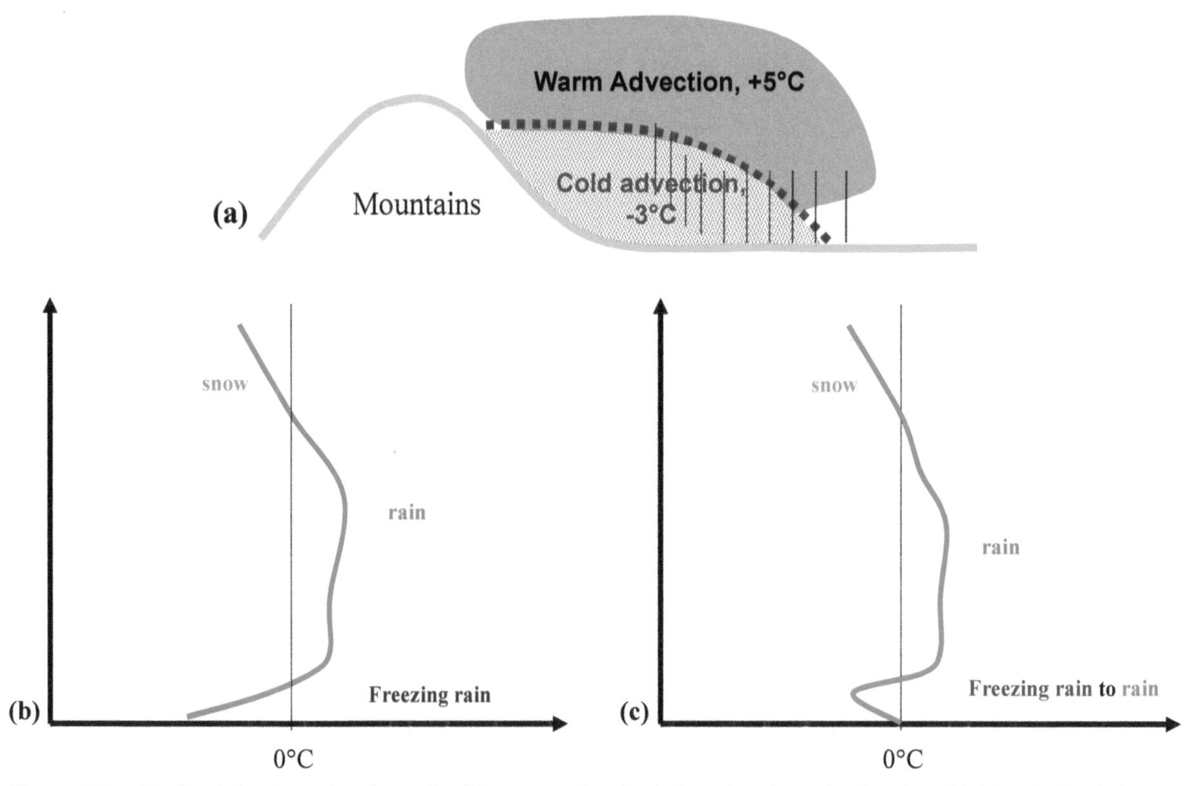


**Figure 9.6.** Idealized freezing-rain schematics: (a) cross section depicting a freezing-rain situation; (b),(c) as in Fig. 9.4.

the processes are generally acting to warm the layer. For cases of heavy freezing rain, the heat released by freezing can increase the surface temperature to the freezing mark, bringing an end to further ice accretion, as depicted in Figs. 9.6b,c.

Freezing-rain events are *self-limiting* in that the freezing process itself warms the surface air temperature, potentially bringing the event to an end. This fact has been recognized by several past studies in the meteorological literature (e.g., Stewart 1984), and the fact that the large majority of freezing-rain events are short lived (less than 6 h in duration) has been attributed to this mechanism (Cortinas et al. 2004). In light of this, major freezing-rain events must either be accompanied by very cold air to begin with or else, more commonly, a mechanism is required to sustain the subfreezing near-surface layer. Such mechanisms include cold or dry advection in the lowest layers, which ventilate the warmed air. If the depth and strength of the subfreezing layer are known, then the first law of thermodynamics allows an accurate estimate of the amount of ice accretion that would "freeze out" the subfreezing layer in the absence of compensating advection.

Another important consideration during freezing-rain events is the soil temperature and surface characteristics. If soil temperatures are above freezing and there is no snow cover, then freezing-rain accretion will tend to be restricted to elevated or insulated surfaces. The large majority of heat released by freezing will be imparted to the atmosphere (let $F_A$ denote the fraction of heat released by freezing that goes to warming the atmosphere), and upward infrared radiation and heat flux from the ground may also contribute significantly to warming the lower levels of the atmosphere. In contrast, in a situation with subfreezing soil temperatures, the ground can act as a heat sink, and some of the heat released by freezing will go to warming the soil instead of the atmosphere, thereby prolonging the freezing rain ($F_A$ will be smaller). However, if a deep snow cover is present, the ground may be effectively insulated and may play a smaller role in determining the evolution of near-surface air temperatures. Surface characteristics, such as land use, can also be important; a forested area may support more icing on elevated surfaces and a larger $F_A$ relative to the case of bare soil with a deep layer of subfreezing ground temperatures. The severity of a given freezing-rain event may depend on several additional

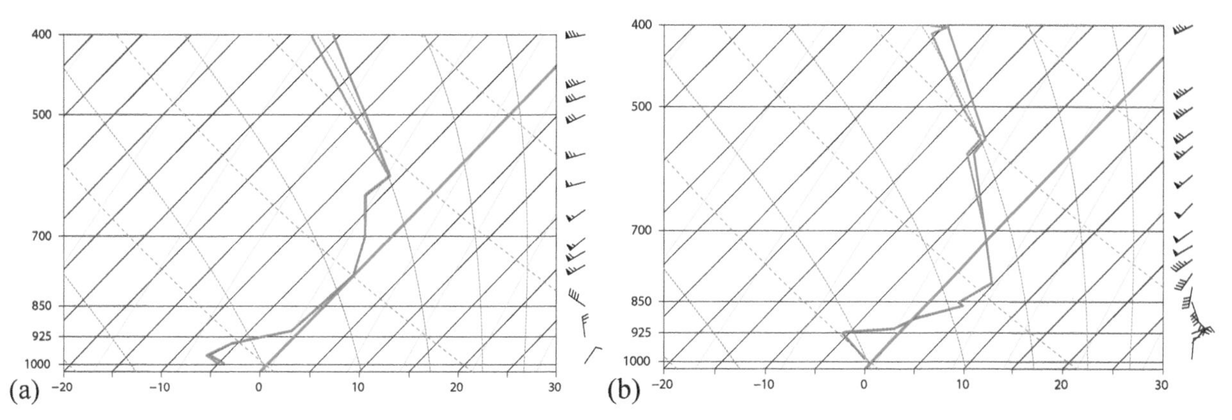


**Figure 9.7.** As in Fig. 9.3, but for (a) 0000 UTC and (b) 0600 UTC 5 Dec 2002. Zero degree isotherms highlighted red.

considerations, such as whether raindrops become supercooled prior to reaching the surface, the temperature of objects at or near the surface, whether leaves are still on deciduous trees, and road temperatures, for example.

The formula below, provided by Lackmann et al. (2002), is a bulk form of the first law of thermodynamics that can provide a useful estimate of freezing rain in situations when horizontal advection and other processes are relatively weak:

$$\Delta T \cong \frac{F_A L_f \rho_\ell R_m}{M\left(C_p + L_v \dfrac{\partial q_s}{\partial T}\right)}. \tag{9.2}$$

In (9.2), $F_A$ again represents the fraction of latent heat released by freezing that goes to the atmosphere, $L_f$ is the latent heat of fusion, $R_m$ is the amount of rainfall in meters, $q_s$ is the saturation specific humidity, and $\rho_\ell$ is the density of liquid water (1,000 kg m$^{-3}$). Inspection of the denominator of (9.2) reveals the compensating effect of evaporation, analogous to the compensating condensation for the case of a melting layer. Here, as freezing of rain warms the layer, it becomes subsaturated, allowing the cooling process of evaporation to take place. If we set $\Delta T$ to be the freezing-point depression of the near-surface subfreezing layer and compute the pressure equivalent of the mass $M$, we can rearrange to obtain an expression for the expected ice accretion that would result in a given situation, assuming that the heat released by freezing is not removed or compensated by other mechanisms. Rearranging (9.2) and grouping constants to provide a more convenient form, we can estimate icing as

$$\text{Icing (mm)} = 0.05 dT(^\circ\text{C})\, dP/F_A, \tag{9.3}$$

where $dT$ is the average freezing-point depression in the near-surface subfreezing layer, and $dP$ is the depth of subfreezing layer in mb. Note that $F_A$ is often ~1 for cases of warm ground or snow cover.

Consider a CAD event accompanied by freezing rain in central North Carolina, which took place on 4–5 December 2002; a sounding from this case is also shown in Fig. 9.3. At 0000 UTC 5 December, the Greensboro, North Carolina (GSO), sounding still supports snow, with a deep isothermal near-freezing layer extending from 925 to ~750 mb (Fig. 9.7a). Strong warm advection in the 925–700-mb layer resulted in a changeover of precipitation to freezing rain shortly after this time, with the development of a deep above-freezing layer by 0600 UTC 5 Dec (Fig. 9.7b).

Based on the 0000 UTC sounding, we estimate that the average temperature of the subfreezing layer is about −4°C and that the layer depth is roughly 75 mb. Warm soil temperatures were present, suggesting a value of $F_A$ of ~1. Using these values, (9.3) yields 15 mm (0.6 in.) of icing. It is important to recognize that this calculation neglects the possible compensating mechanism of cold or dry advection, processes that can be significant in CAD events such as this one.

By 0600 UTC, the near-surface temperature had already warmed to near 0°C, with a significant freezing-rain event underway. Spotter reports to the National Weather Service Forecast Office in Raleigh, North Carolina, were used to construct the analysis shown in Fig. 9.8, which indicates ice accretion between 0.5 and 1 in. for this event. The simple formula (9.3) provided a useful estimate of the accretion here, but in general other processes must be considered before trusting the values obtained from (9.3).

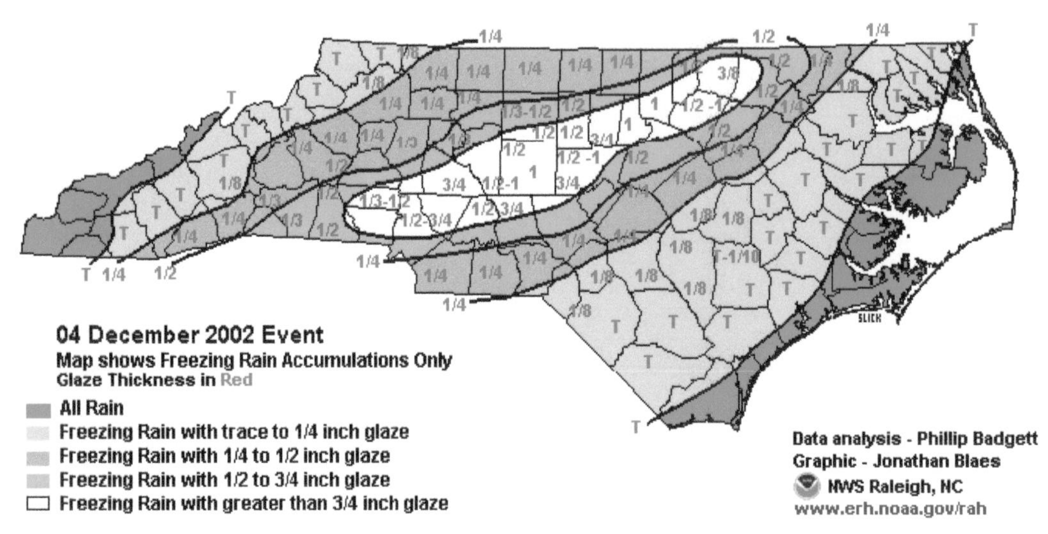


**Figure 9.8.** NWS analysis of spotter reports of freezing-rain accumulations (in.) for the 4–5 Dec 2002 event; T denotes a trace of freezing rain (graphic courtesy Jonathan Blaes and Phil Badgett, NWS Raleigh).

### 9.2.4. Ice pellets

There are two distinct situations in which precipitation may reach the surface in the form of ice pellets. In the presence of a significant layer of above-freezing air aloft and a deep or very cold surface-based subfreezing layer, rain drops may refreeze before reaching the surface. An example sounding from an ice pellet event that coincides with the simulated CAD event presented in chapter 8 (from February 2003) is provided in Fig. 9.9. Whether the rain refreezes in the surface-based cold layer or not is a complicated issue, because refreezing can depend on subtle factors such as the type of condensation nuclei present in the raindrops and the temperature at which these nuclei may activate as freezing nuclei.

A second thermodynamic situation that can result in ice pellets involves partial melting (and subsequent refreezing) of snowflakes, leading to ice pellets even in the absence of a strong surface-based cold layer. In the latter situation, the remaining ice in partially frozen droplets serves as an ice nucleus, allowing for rapid refreezing when the temperature drops below freezing again. Also, contact nucleation of ice is possible in this situation.

### 9.2.5. Process summary

Given the number and complexity of physical processes at work during winter-weather events, forecasters often rely on numerical models to quantify the predicted thermal profile. Provided that the model represents the processes adequately, this can be an acceptable strategy, but a skilled forecaster can often identify and compensate for model biases through a physical understanding of the processes at work. Furthermore, research has shown that some processes are not adequately represented in numerical forecast models: for example, sensible heat transport by falling rain is currently neglected in most NWP models.

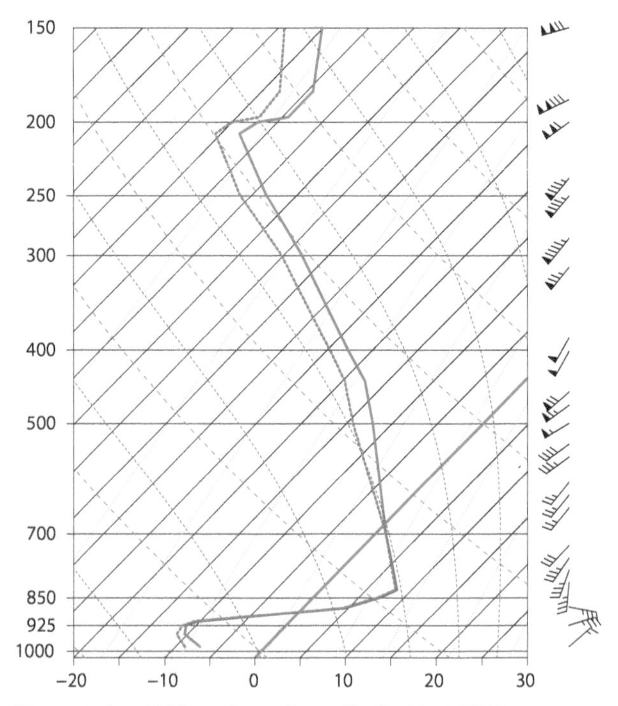


**Figure 9.9.** GSO rawinsonde profile for 0000 UTC 17 Feb 2003. Zero degree isotherm highlighted red.

Cloud–radiation interactions, as well as any QPF-dependent processes such as melting, are often unreliable in numerical model forecasts.

Consider an example scenario in which a human forecaster might add value to a numerical prediction. Suppose that, for a given event, NWP models are predicting a light rain event but with temperatures only slightly above freezing in the lowest layers of the atmosphere. As the event unfolds, radar imagery and surface observations indicate the development of considerably heavier precipitation than that shown by short-term model forecasts. An astute forecaster would be alert to the possibility that the melting process could result in the occurrence of heavy snow; monitoring of observations, including a descending radar bright band, could alert forecasters in advance of impending adverse weather in a situation of this type.

Despite limitations, it should be clear that NWP models provide an extremely useful tool for all types of weather prediction, including winter-weather forecasting. The ability of numerical models to simultaneously represent the main physical processes discussed in this section is far more quantitatively accurate than what a human could do without model guidance. Further discussion of the strengths and limitations of NWP models will be presented in chapter 10. Due to the limiting processes involved with freezing rain and ice pellet events, major ice accretion generally requires one or more of the following: (i) cold or dry advection during the event; (ii) extremely cold and dry air initially; (iii) subfreezing soil temperatures and limited snow cover; or (iv) some other local cooling mechanism, such as adiabatic expansion in upslope flow.

## 9.3. PRECIPITATION-TYPE FORECASTING TECHNIQUES

A systematic forecasting method will be outlined in chapter 11, but some of the tools and techniques specifically relevant to winter precipitation forecasting will be presented here. Owing to the potential for significant societal and economic impacts, forecasters must choose their wording carefully when issuing forecasts for winter-weather situations. Inclusion of the word "snow" in a public forecast, even if the wording is "a slight chance of snow flurries on Wednesday night, no accumulation expected" can transform into a conversational "Did you hear? The forecasters are calling for *snow* on Wednesday night!"; in some parts of the country, this can be sufficient to trigger runs on grocery-store items. In the words of Kermit Keeter, the former National Weather Service science officer in Raleigh, "it is better to be right than first" with a prediction of snowfall. For the majority of winter-weather preparations, 24 h of forecast lead time is sufficient, alleviating the need for dramatic long-range winter-storm predictions.

During wintry weather, there are heightened expectations of an accurate forecast of precipitation amount, or the quantitative precipitation forecast (QPF). Techniques relating to QPF are discussed in chapters 10 and 11, with the focus here being on tools for determining the *type* of precipitation. There are several quantitative techniques available for prediction of precipitation type. With any of these techniques, it is important to remain familiar with the physical processes at work and with the assumptions underlying the technique. It is best not to rely exclusively on any single technique in making a precipitation-type prediction. As with any forecasting, regional and seasonal climatology should be weighted into the forecast process.

### 9.3.1. Partial thickness "trends" technique

Forecasters have long utilized the thickness as a tool in the prediction of precipitation type (Bocchieri and Glahn 1976; Keeter and Cline 1991; Keeter et al. 1995). The thickness, which is proportional to the mean-layer virtual temperature as discussed in section 1.4 and defined by Eq. (1.37), provides a robust measure of the temperature over a layer, given that it is obtained from the difference in the geopotential height field between two pressure surfaces. The cutoff values for rain versus snow for 1000–500-mb thickness, as presented by Bocchieri and Glahn (1976), are presented in Fig. 9.10 and demonstrate that a wide range of cutoff values are found in different geographical locations. This is due to the fact that a given value of 1000–500-mb thickness can be associated with a wide variety of temperature profiles (Fig. 9.11). As a very general first-cut estimate, 5400 m is a ballpark cutoff value of 1000–500-mb thickness for rain versus snow. However, in regions where the lower troposphere is relatively warm, as is often found in maritime regions, we expect that a smaller value of 1000–500-mb thickness would be needed as a cutoff for rain and snow. This is precisely what we observe along the Pacific Northwest coast in Fig. 9.10.

In regions of elevated terrain such as the Rocky Mountains, the fictitious 1000-mb surface is located far below

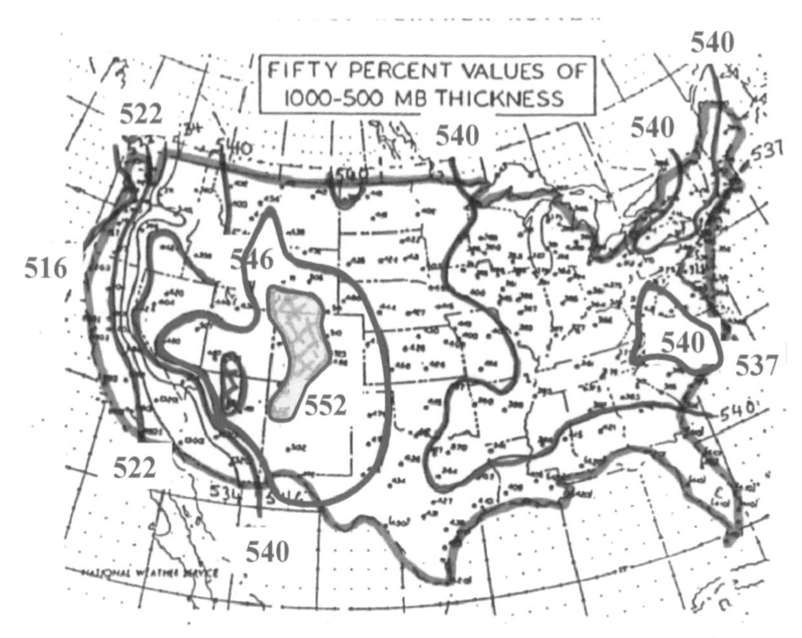


**Figure 9.10.** The 50% values of 1000–500 mb thickness (dam; from Bocchieri and Glahn 1976). Contour labels and colored lines added here for clarity. The thickness contours correspond to values for which 50% of precipitation observations were in the form of snow.

ground, so the 1000–500-mb thickness value only corresponds to actual air temperatures for the upper portion of the layer. In situations where the surface pressure is less than 1000 mb, the 1000-mb height is routinely computed via extrapolation underground using the standard-atmosphere lapse rate. Different hypothetical temperature profiles that correspond to identical thickness values are presented in Fig. 9.11. Snow could be expected in the stable situation depicted in Fig. 9.11a, whereas rain may characterize the relatively unstable situation in Fig. 9.11b, which is somewhat analogous to the situation in the Pacific Northwest mentioned above. Even with the same lapse rate as in 9.11b, a high-elevation station could experience snow (Fig. 9.11c) due to the fact that the warmer portion of the layer does not actually exist.

Clearly, using the 1000–500-mb thickness as a stand-alone technique for prediction of precipitation type does *not work very well!* However, other techniques based on thickness have been devised to provide more useful precipitation-type forecasts. National Weather Service forecasters in the southeastern United States have developed a "partial thickness" approach in which the thickness

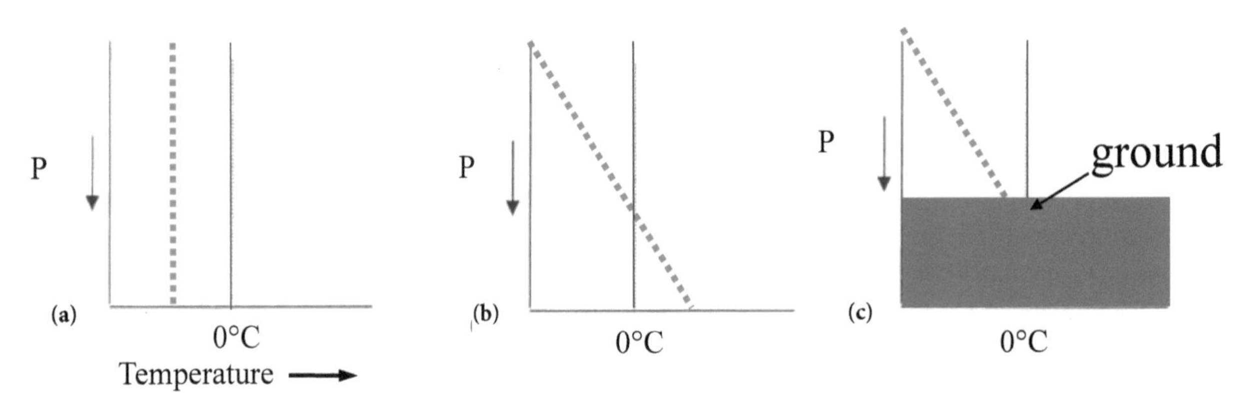


**Figure 9.11.** Idealized plots of temperature as a function of pressure for three hypothetical situations with identical 1000–500-mb thickness.

in two separate layers is used to provide implicit information about the stability of the layer, allowing useful prediction of freezing and frozen precipitation as well as rain versus snow.

Some useful benchmark values can be obtained from rearranging the hypsometric equation to relate a given mean-layer virtual temperature to thickness:

$$\Delta Z = \frac{R_d \overline{T}_v}{g_0} \ln\left(\frac{p_{\text{low}}}{p_{\text{up}}}\right). \tag{9.4}$$

If we ask what value of thickness corresponds to a layer-average virtual temperature of freezing (273 K) in the 1000–850-mb layer, we use (9.4) to find this value, approximately 1,300 m. Thus, if the 1000–850-mb thickness (for locations at which the surface pressure is reasonably close to 1000 mb) is above 1,300 m, we can be certain that some portion of the layer is above freezing, and the likelihood of snow diminishes rapidly. The cutoff value of 1000–850-mb thickness is a somewhat more precise predictor of precipitation type relative to the 1000–500-mb layer, although geographical variations again exist and the stability of the air column must be taken into account. Less stable regions require a smaller value of 1000–850-mb thickness for the midpoint value, following similar arguments to those made previously for the 1000–500-mb thickness.

If the 1000–850-mb thickness is used in conjunction with the 850–700-mb thickness, then a more complete picture of the thermal structure of the lower atmosphere is obtained. If the thickness in the upper layer, 850–700 mb, is sufficiently warm, then liquid precipitation will fall into the lower 1000–850-mb layer. Then, the relative frequency of ice pellets versus freezing rain will depend on the 1000–850-mb thickness.

Forecasters in the southeastern United States utilized rawinsonde data from GSO to develop an empirical technique for use in forecasting the complex winter precipitation events accompanying cold-air damming in that region. By plotting the *observed* precipitation type on a nomogram featuring 1000–850-mb thickness on the ordinate and 850–700-mb thickness on the abscissa, climatologically favored transition points between differing predominant precipitation types could be identified, as shown in Figs. 9.12a,b.

A useful software tool with which to visualize model sounding data is the BUFKIT program, which can be downloaded freely from the National Weather Service BUFKIT home page (available online at http://wdtb.noaa.gov/tools/BUFKIT/index.html). This software package is designed for a variety of operational forecasting applications and includes the ability to plot forecasted points from a sequence of numerical model output times, alongside a skew *T*-log*p* diagram of forecast data, providing a sense of the evolution of the thermal profile with time (Fig. 9.13). Examination of changes in the values of the thickness in these layers also allows assessment of various physical processes: for example, evaporational cooling taking place in the model atmosphere is evident both in the thermodynamic profile and in the sequential right-to-left progression of the points on the partial thickness nomogram in the left portion of Fig. 9.13.

In late January 2010, a winter storm threatened the southeastern United States. Examination of the partial-thickness nomograms from BUFKIT for this event demonstrates how one can relate physical processes to changes in partial thickness in the model atmosphere. In this case, as the model atmosphere becomes saturated from above, the zone of evaporational cooling shifts from the 850–700-mb layer into the 1000–850-mb layer, as evident in Fig. 9.13b. This corresponds to a downward shift in the red points on the nomogram, as temperature changes become small aloft but cool in the lower layer, as represented on the ordinate.

The BUFKIT program also provides the ability to display the wet-bulb temperature on a model-forecast sounding (Fig. 9.13c). Note how the saturated temperature in the 16-h forecast (Fig. 9.13b) is very similar to the wet-bulb temperature profile from the 10-h model forecast, before saturation was reached. Plots of surface wet-bulb temperature can also be useful in identifying threat areas for freezing or frozen precipitation in a variety of winter-weather situations.

### 9.3.2. The area method

Precipitation-type forecasting is largely a thermodynamics problem, and the approach taken by Bourgouin (2000) utilizes this approach via computation of the energy required to melt or freeze falling precipitation and relating this to observed thermodynamic profiles. Recall that the skew T-log*p* and many other thermodynamic diagrams are designed such that area on the chart corresponds to energy. Bourgouin defines "positive area" (PA) as corresponding to portions of the vertical temperature profile in which the temperature is above freezing (Figs. 9.14a-d),

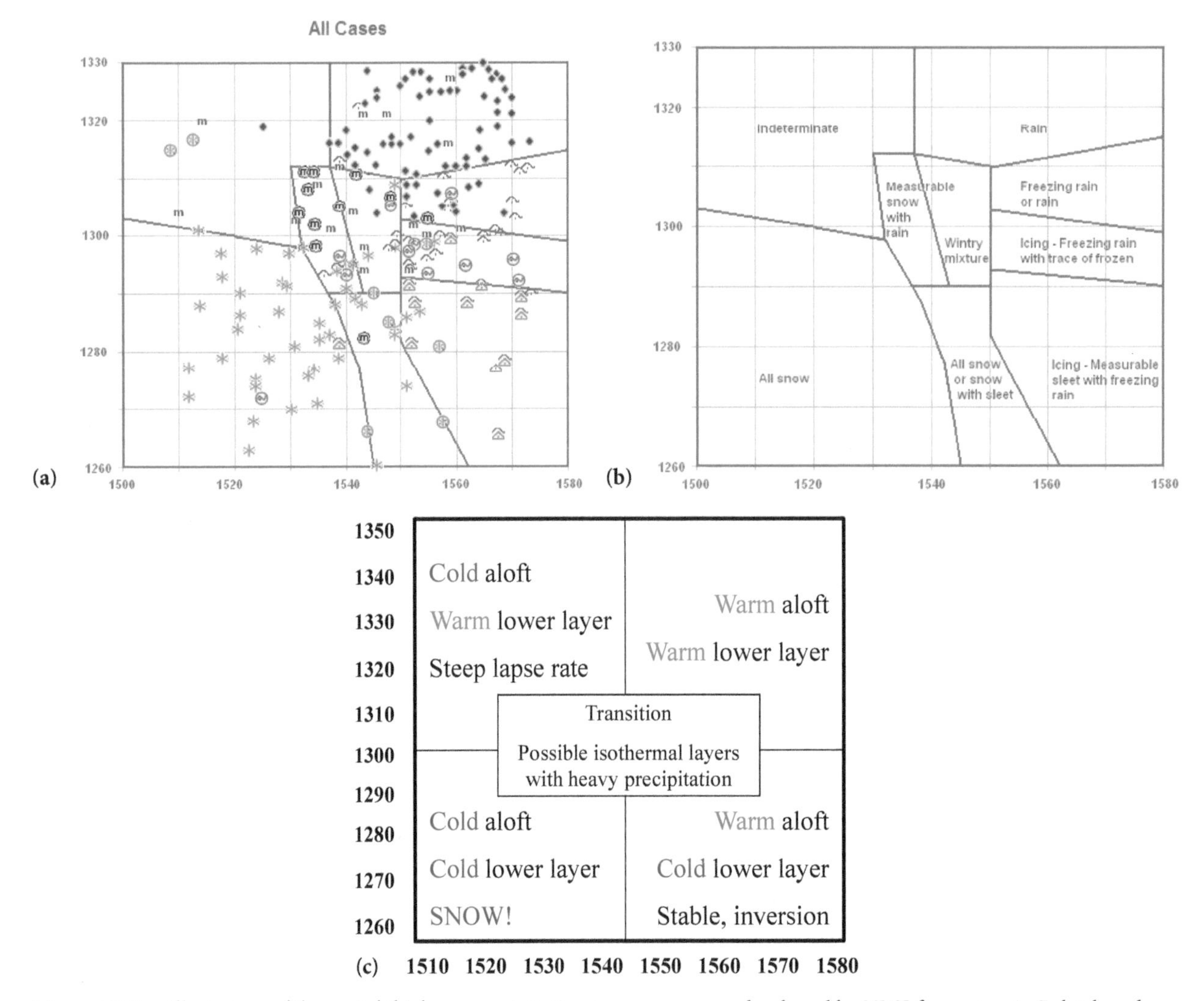


**Figure 9.12.** Illustration of the partial thickness precipitation-type nomogram developed by NWS forecasters in Raleigh and other regional NWS offices in the southeast U.S.: (a) empirical data indicating observed precipitation type corresponding to observed values of 1000–850-mb and 850–700-mb thickness observed from the GSO rawinsonde site and (b) partitioned predominant precipitation types derived from empirical data (image from http://www4.ncsu.edu/~nwsfo/storage/trend/). (c) General atmospheric conditions corresponding to different portions of the nomogram.

with "negative area" (NA) defined analogously as subfreezing regions located beneath above-freezing layers.

By plotting observational data on a diagram with positive and negative area represented on the abscissa and ordinate, respectively, one can derive a useful prediction of precipitation type. For regions of the diagram characterized by small positive area or large negative area, the expected precipitation type would be ice pellets, whereas, with smaller negative area, freezing rain would become more likely. Some software packages, such as the BUFKIT program discussed previously, can display model-forecast sounding data on a diagram of this type.

### 9.3.3. Numerical model output and forecast sounding interpretation

As emphasized earlier in this section, it is wise to avoid reliance upon a single forecasting tool or strategy for any forecast, but this is especially important for high-impact winter-weather forecasts. It is important to remain aware of the vertical thermal structure and to think about the various physical processes responsible for determining it.

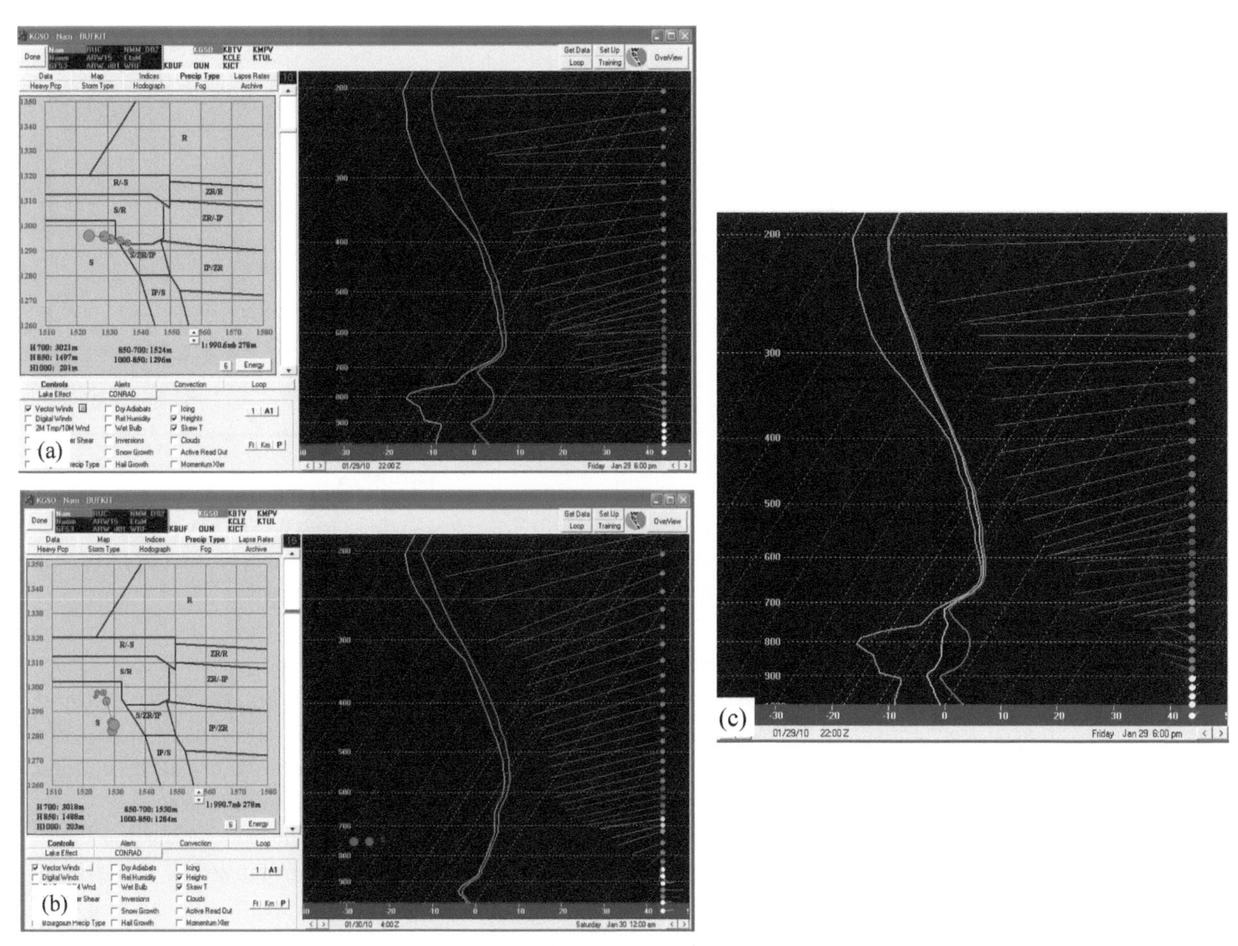


**Figure 9.13.** Examples from the BUFKIT software package for winter-weather forecast from 29 Jan 2010 at GSO. (a)–(c) Model-forecast soundings are indicated in skew *T*-log*p* format at the right, with corresponding partial thickness nomograms at the left. The largest red dot on the nomogram represents data shown in sounding, and progressively smaller dots represent previous times. Only the sounding that corresponds to that in (a) is shown in (c), but with wet-bulb temperature included as a blue line.

Given the intricate balance between processes, some of which are not always accurately represented in numerical weather prediction models, routine examination of observational data is crucial. Observed and forecast soundings, soil temperatures, surface analyses, and satellite and radar imagery can all provide early indications of whether numerical model guidance is trustworthy during a given event.

A "top down" approach is advocated for operational precipitation-type forecasting by the U.S. National Weather Service (Baumgardt 1999; COMET 2005). This method begins by assessing whether the upper portions of the cloud are sufficiently cold for ice crystals to be present. The complexity of ice initiation in clouds necessitates a probabilistic approach, but observational studies have shown that clouds reaching −10°C have a roughly 50% chance of containing ice, while the probability goes up rapidly as the temperature drops below −13°C (Wallace and Hobbs 2004, Chapter 6.5.2). Observationally, rawinsonde data and IR satellite imagery can be used to estimate cloud-top temperatures; model-forecast soundings can also be used in this capacity.

Next, forecasters must assess whether there is or will be an elevated above-freezing layer. It is best to view the profile of wet-bulb temperature to account for potential evaporational cooling. As a rule, if an elevated above-freezing layer is warmer than 3°C, complete melting of falling snow will occur. Finally, the temperature profile in the lowest portions of the troposphere is considered. For cases with a surface-based warm layer, if the altitude of the wet-bulb freezing level exceeds 1,500 m, snow is unlikely, with increasing probability of snow as the depth of a surface-based above-freezing layer decreases to 750 m. If the warm layer is shallower than 750 m, snow is likely

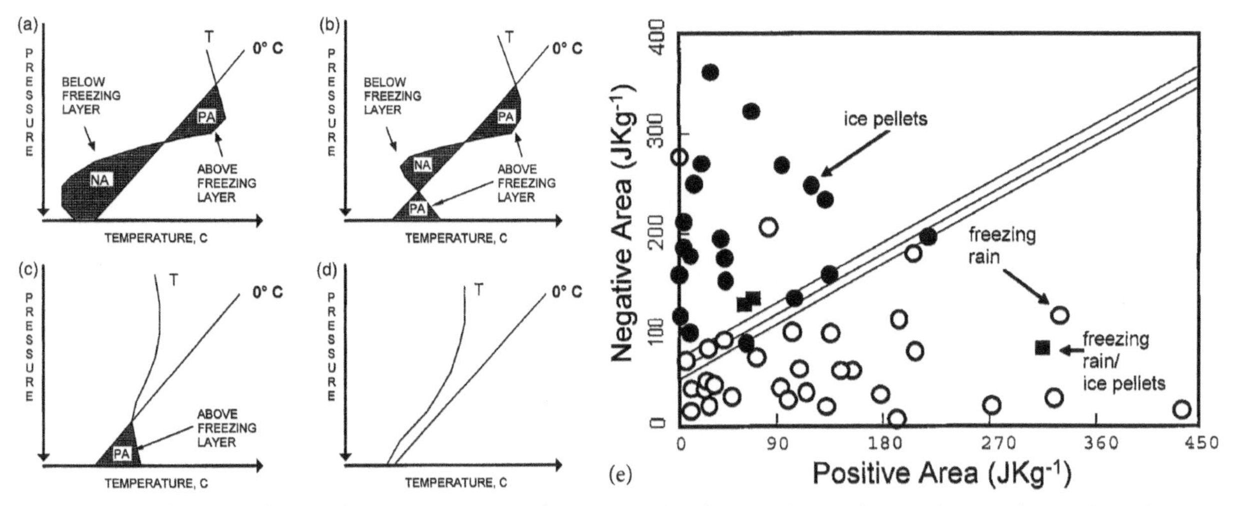


**Figure 9.14.** The area technique for precipitation-type forecasting. (a)–(d) Example soundings with PA and NA indicated. (e) The observed precipitation types on a plot of PA versus NA; solid circles represent ice pellets, open circles correspond to freezing rain, and closed squares correspond to a mix of ice pellets and freezing rain (from Bourgouin 2000).

to reach the surface. These are very general guidelines; as discussed in section 9.2.2, one must consider melting and other processes to accurately anticipate the evolution of the temperature profile.

Even if a forecaster is able to interpret model forecasts perfectly, it is important to recognize the predictability limits in a given forecasting scenario. As will be discussed in section 10.6, *ensemble forecasting systems* provide a quantitative means of assessing forecast confidence. When winter weather threatens, ensemble guidance can alert forecasters to the possibility of differing synoptic evolutions, as well as quantifying uncertainty for key parameters such as QPF.

Forecasters must remain aware of model limitations when viewing model-forecast soundings. In situations where the temperature profile suggests borderline precipitation type, evaporation and melting can result in cooling that may change precipitation type. As we have seen, during heavy precipitation, above-freezing layers can "melt out" and precipitation may change to snow as a result of melting-induced cooling. Such processes are QPF dependent; because QPF is one of the least accurate model-forecast parameters, forecasters must be ready to make adjustments as the event unfolds.

As discussed in the following chapter, most modern numerical model microphysics schemes predict mixing ratios of cloud water, cloud ice, rain water, snow, and graupel. This suggests that a more direct and model-consistent manner for precipitation-type prediction would be to simply examine the model-output hydrometeor fields at the lowest model grid cell in NWP output. Additionally, model-produced snowfall accumulation forecasts will likely gain popularity in the operational forecasting community in the coming years.

A comment should be made concerning the "snow ratio," which is the ratio of snowfall to the liquid-water equivalent. In forecasting snow depth, this can be a critically important consideration, as the ratio can vary from 6 or 8 to 1 for heavy, wet snow, to more than 20 to 1 for lake-effect events with cold temperatures and dendritic snow crystal habits. However, note that some aspects of the snowfall are more strongly tied to the *liquid equivalent* than to the actual snow depth, including the amount of energy required to remove or melt the snow.

## 9.4. LAKE-EFFECT PRECIPITATION

Most of the synoptic-scale precipitation systems discussed in this text are characterized by *stratiform* precipitation: that is, precipitation that arises in airflow that is statically stable to upright convection, with air ascending along sloping isentropes in a stable upglide. Owing to the fact that upright convection is inherently mesoscale in nature, lake-effect snow falls somewhat outside the focus area of this text. However, there are strong linkages between convection and the synoptic environment. For example, synoptic-scale lift and vertical wind shear in the presence of instability can strongly influence the occurrence and organization of convective storms. Despite the synoptic-scale emphasis of this book, an exception will be made in the case of the fascinating mesoscale phenomenon of lake-effect convection.

Differential heating such that the lower portion of an air column warms relative to the upper portion results in a reduction of static stability; a variety of mechanisms can produce such differential temperature changes. Often, solar heating warms the surface and in turn the lower atmosphere. Because solar radiation is only very weakly absorbed by the free tropospheric air, the environmental lapse rate tends to increase with afternoon heating, increasing the risk of convection then. During *lake-effect* precipitation, the surface heat source is not provided by solar heating but by a relatively warm lower boundary in the form of a lake (or ocean, in the case of ocean-effect precipitation). The result can be locally heavy convective snow, and certain geographical regions, such as the areas immediately downwind of the Great Lakes in North America, are prone to significant winter snows through lake-induced convection (Fig. 9.17).

### 9.4.1. Definition and climatology

The American Meteorological Society (AMS) *Glossary of Meteorology* defines the term *lake-effect snow* as "localized, convective snow bands that occur in the lee of lakes when relatively cold air flows over warm water." This definition implicitly accounts for the destabilization process due to heat and moisture transfer processes above the lake surface. More generally, the term *lake-effect precipitation* allows for the fact that rain showers can also arise from this type of destabilization mechanism.

A distinction between lake-effect and *lake-enhanced* precipitation should also be made. The former can be defined as precipitation that would not have occurred without the destabilizing influence of the lake, whereas lake-enhanced precipitation applies to situations in which precipitation would have fallen even without the influence of the lake but additional lake-induced convection enhances precipitation amounts beyond what the synoptic-scale precipitation system alone would have produced.

Before considering the spatial distribution of lake-effect snow, let us examine the annual cycle of the lake influence on the lower atmosphere as depicted in Fig. 9.15. Due to the greater thermal inertia of water relative to land, the annual cycle of lake temperatures lags behind that of the near-surface air temperature. Thus, during the fall and winter, air temperatures will often be colder than the lake

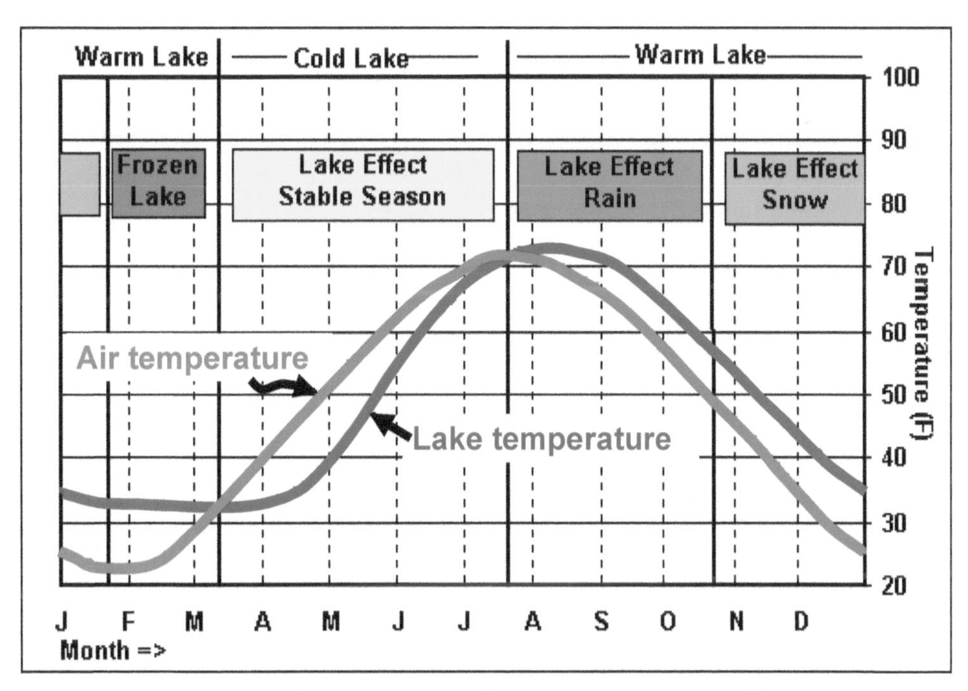


**Figure 9.15.** Average air and lake temperatures throughout the year near Buffalo, NY, with characteristic lake influence indicated (image adapted from Niziol et al. 1995; available online at http://www.erh.noaa.gov/er/buf/lakeffect/lakeclimate.html).

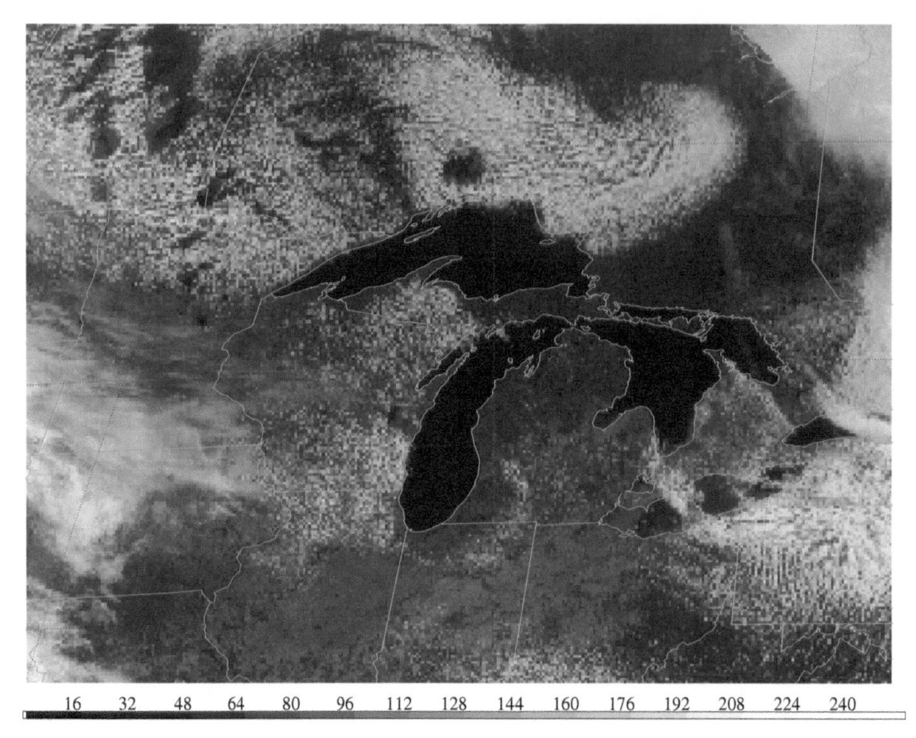


**Figure 9.16.** Example of lake-induced stabilization on 9 May 2010 as evident in *Geostationary Operational Environmental Satellite-13* (*GOES-13*) visible satellite image valid at 1845 UTC.

surface, indicating that the presence of the lake serves as a destabilizing influence during these seasons. However, this is not to say that lake convection will happen continuously during this period: synoptic conditions and other convective ingredients must be present to facilitate the lake convection.

During the late spring and summer months, the air temperature will be relatively warm compared to the lake surface; during these seasons, the lake offers a *stabilizing* influence on the lower atmosphere (lake-effect stable season; Fig. 9.15). As a result, the lakes will serve to reduce or eliminate convection, and this phenomenon can be seen in satellite imagery such as that presented in Fig. 9.16. In this image, note that the absence of convective clouds reveals the location not only of the Great Lakes but also of Lake Winnipeg in Canada, despite the lake outline not being superimposed on the image.

Another consideration that can modulate the lake influence is the extent of ice cover. Lake Erie, which is relatively shallow, generally freezes over at some point during the winter, often before the end of January (Fig. 9.15). The presence of significant ice cover reduces the vigor of turbulent heat and moisture fluxes over the lake to a large extent, with the exception of *leads*, open gaps in the ice surface due to wind-driven ice motion and other processes. Lake Ontario only very rarely experiences significant ice cover, as is the case with Lake Michigan. Forecasters downwind of the lakes must monitor the ice analysis when making lake-effect snow predictions.

### 9.4.2. Mechanisms and environment

**9.4.2.1. Lake-induced instability.** Being an instability-driven phenomenon, the most obvious physical process at work in lake-effect precipitation events is the moist convection itself. A useful measure of the potential for lake-induced instability, long recognized by forecasters in the Great Lakes region, is the temperature difference between the lake surface and the air temperature at the 850-mb level. If this temperature difference exceeds ~13°C, there is typically sufficient instability present to allow lake-effect convection to develop, provided that other necessary ingredients are also present. The "13-degree rule" is partly empirical, but it is also based on the thermodynamics of the situation. Given that

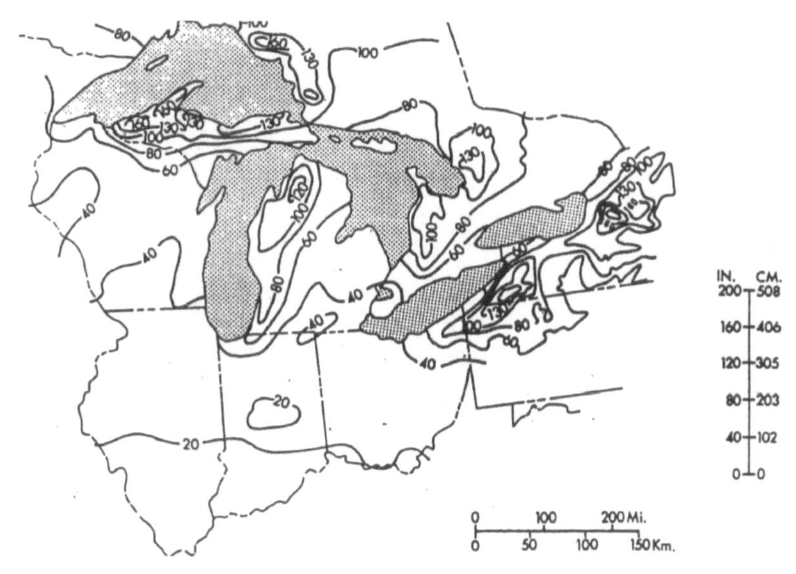

**Figure 9.17.** Annual snowfall in Great Lakes region (contoured in inches, with conversion scale to centimeters provided at right; from Eichenlaub 1970).

the 850-mb height during winter is often near 1,300 m, a lake-surface to 850-mb level temperature difference of 13°C corresponds roughly to the dry adiabatic lapse rate; a difference larger than this value indicates that the potential for unstable lapse rates is present, provided that the lake is able to effectively warm and moisten the lower atmosphere. The height of the Great Lakes above sea level should also be taken into account in this calculation.

An example sounding, taken at Buffalo, New York (BUF), during a complex lake-effect snow event that took place in late November 1996 is provided in Fig. 9.18. This event featured a deep, moist layer near the surface, with lake-surface temperatures near 5°C. The lake-850-mb temperature difference was close to 16°C, indicating more than sufficient potential for instability. The presence of a deep layer of moist air in the lower troposphere, as well as a relatively high altitude of the capping inversion, indicates that surface-based lake convection could easily extend to an altitude of 3 km or more (the 700-mb level); in fact, significant lake-effect snowfall was observed in this event to the south of Lake Ontario in New York State (see Lackmann 2001).

**9.4.2.2. Thermally driven circulations.** Given that a necessary condition for lake-effect snow is the presence of a lake temperature that is substantially warmer than the overlying air temperature, a thermally driven *land-breeze* circulation typically develops. This circulation is similar to the diurnally fluctuating land–sea breeze that often develops near coastlines, only in this situation the direction of the thermal gradient does not typically exhibit a diurnal reversal (although the strength of the circulation may fluctuate diurnally).

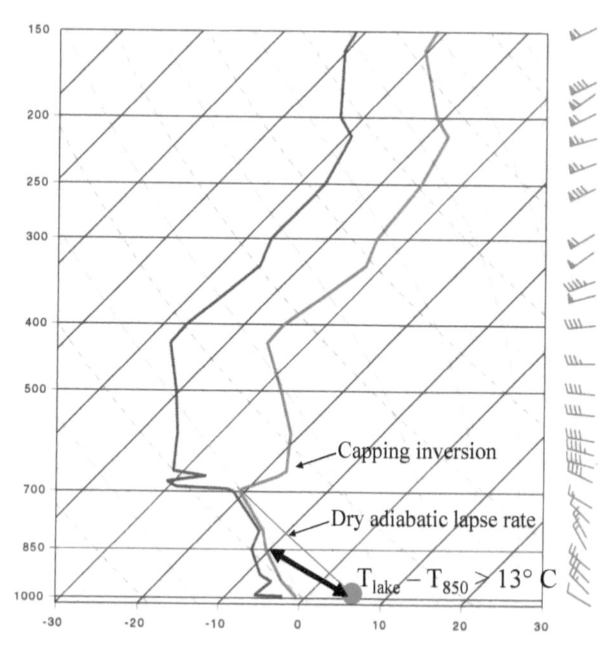


**Figure 9.18.** BUF sounding in skew *T*-log*p* format for an LES event from 0000 UTC 27 Nov 1996 (adapted from Lackmann 2001).

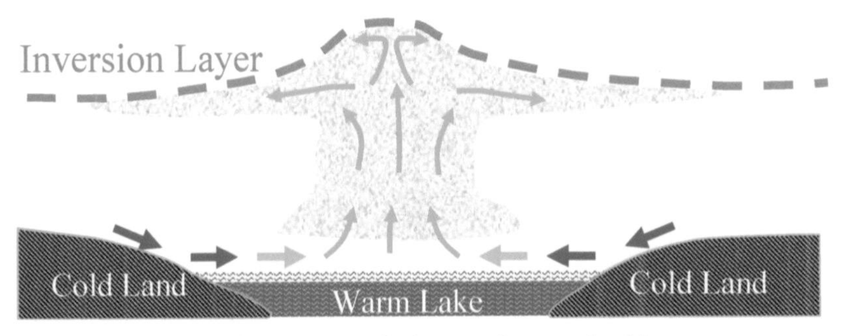


**Figure 9.19.** Idealized cross-sectional schematic depicting land-breeze convergence for a lake-effect convective band (after LaDue 1996).

For sufficiently narrow lakes, the land breezes from opposing shores may converge in the midlake region, with convection organizing as indicated in Fig. 9.19. For cases with vigorous convection, the local height of the capping inversion may be raised by the convective motions and entrainment there.

**9.4.2.3. Frictional convergence and orographic lift.** Consider a hypothetical situation in which flow is crossing a lake (Fig. 9.20). Typically, the effective roughness of the lake surface will be less than that of the surrounding land; as a result, the influence of the frictional force will diminish as air initially moves over the lake and frictional drag will increase again at the downwind shore. The resulting acceleration and deceleration of the flow creates a pattern of divergence and convergence, with associated descent near the upwind shore and ascent downwind (Fig. 9.20).

The straight isobars in Fig. 9.20 are unlikely. This is partly because the warm lower boundary of the Great Lakes induces troughing, which can play an important role in promoting frictional convergence and ascent [recall that cross-isobar frictional flow in the presence of cyclonically (anticyclonically) curved isobars gives rise to ascent (descent)].

In some regions surrounding the Great Lakes, notably in the regions southeast of Lake Erie and Lake Ontario, significant topographic lift accompanies westerly or northwesterly flow across the lakes. In these situations, orographic lifting further enhances precipitation and in some cases can lead to astounding snowfall, such as that pictured at the beginning of the chapter at the eastern end of Lake Ontario. Figure 9.21 provides a summary of the processes discussed so far and again demonstrates how the occurrence of lake convection serves to deepen the mixed layer, resulting in an increase in the altitude of the inversion layer.

**9.4.2.4. Lake-aggregate and far-field lake influences.** The classic paper of Petterssen and Calabrese (1959) was perhaps the first to quantify, albeit with limited observations, the influence of the Great Lakes on the

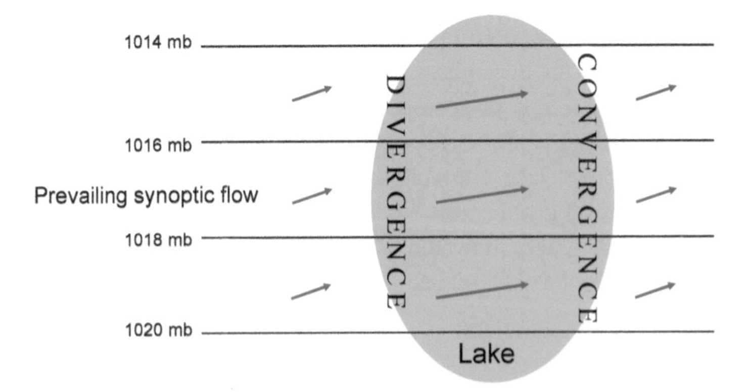


**Figure 9.20.** Idealized schematic of isobars (straight black lines) with flow crossing a lake. Blue arrows indicate acceleration of the flow over the lake, with a pattern of divergence at the upwind shore and convergence at the downwind shore.

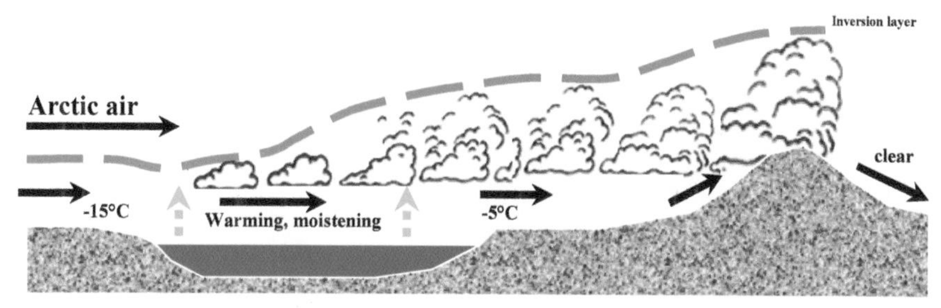


**Figure 9.21.** Idealized cross-sectional schematic of LES development, with downstream orographic lift. Depth of the inversion along the flow shown as a dashed red line.

lower-tropospheric pressure field. During an intense arctic outbreak that took place in February 1958, they found a significant cyclonic circulation attributable to the lakes and computed a lake-induced pressure fall of several millibars; they estimated that the upper limit of such a pressure fall is 6–7 mb. Thus, diabatic troughing, as well as the associated increase in frictional convergence (Ekman pumping), is an indirect frictional effect during lake-effect events.

With the advent of numerical models, experiments can be conducted in which the lakes in the model atmosphere are artificially "frozen over" or removed completely, allowing comparison of what the atmosphere would look like in their absence. Studies by Sousounis and collaborators demonstrate that the influence of the Great Lakes contributes a *synoptic-scale* signature to the surrounding environment.

The far-reaching nature of the Great Lakes influence is also evident in that the signature of lake-effect convection can in some events extend as far south as eastern Tennessee and North Carolina (Fig. 9.22). Note the cloud band extending from Lake Michigan to the Appalachian Mountains in this satellite image.

### 9.4.3. Lake-effect band morphology

Depending on the orientation of the lower-tropospheric airflow relative to the major and minor axes of the lakes, differing lake-convection morphologies can result.

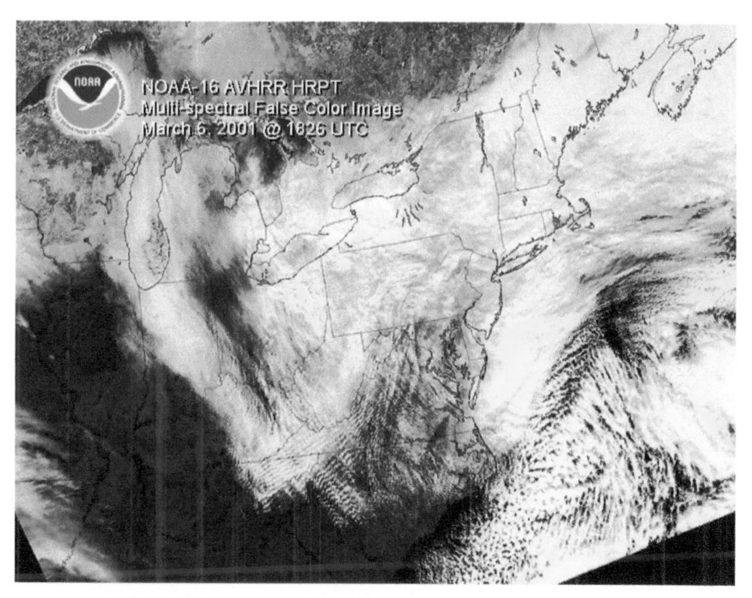


**Figure 9.22.** Advanced Very High Resolution Radiometer (AVHRR) false-color image from *National Oceanic and Atmospheric Administration-16* (*NOAA-16*) polar-orbiting satellite 1826 UTC 6 Mar 2001 [courtesy of National Environmental Satellite, Data, and Information Service (NESDIS); from Keighton et al. 2009].

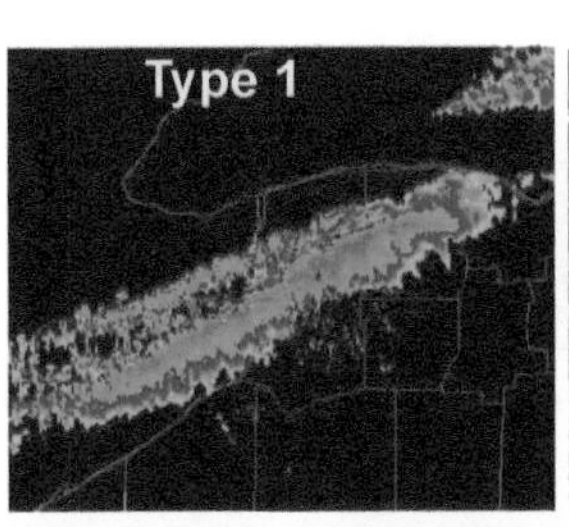


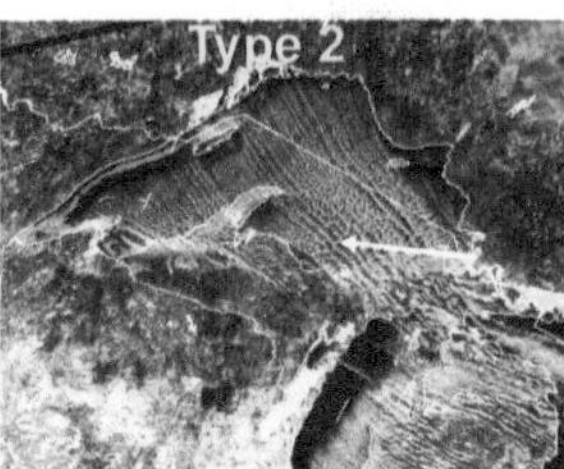


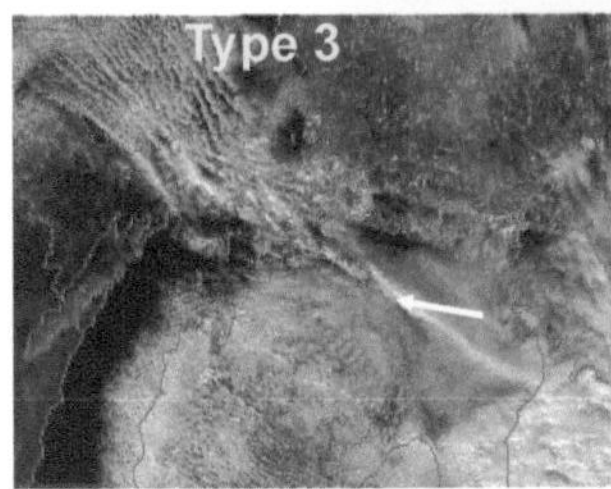


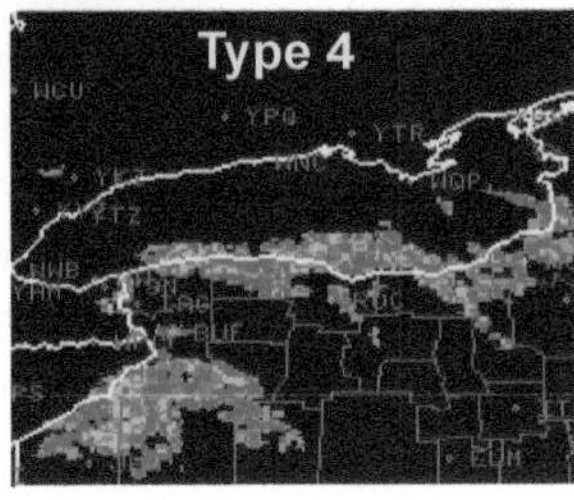


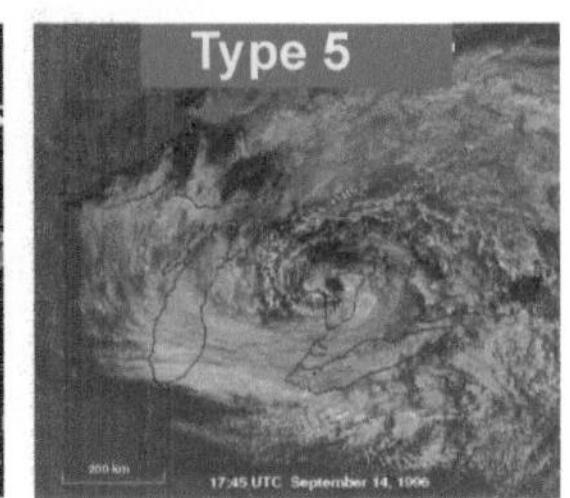


**Figure 9.23.** Radar and satellite examples of Great Lakes band types as classified by Niziol (1987) and Niziol et al. (1995). Image for type 5 from Miner et al. (2000).

The coverage, intensity, and impact of lake-effect snow depends on the type of convective organization that takes place, and forecasters in the Great Lakes region have developed a classification system for different types of bands. These classifications are specific to the U.S. Great Lakes, although they hold relevance for other anisotropic bodies of water. The primary types of lake-effect snow band, as discussed by Niziol (1987), include the following:

Type 1: *Wind-parallel* bands oriented along the *major axis* of the lake, typically 20–50 km wide, 50–200 km long or more. They are often characterized by a single, very intense band with heavy snow, low visibility, and snowfall rates in excess of 1–2 in. $h^{-1}$.

Type 2: *Wind-parallel* bands oriented along the *minor axis* of the lake. Bands are less organized relative to type 1, exhibiting multiple-band structures and generally lighter precipitation rates.

Type 3: As with type 1, except extending across *more than one lake*. These are also known as "superbands."

Type 4: *Shore-parallel* bands that form in conjunction with land breeze and frictional convergence.

Type 5: Mesoscale lake vortices, typically found in light-wind conditions with large air–lake temperature differences.

Examples of the different types of bands are illustrated in Fig. 9.23.

### 9.4.4. Examples, other regions, and other environments

The most extensive research into lake-effect convection has been geographically focused in the U.S. Great Lakes, but there are a plethora of other regions and situations in which similar mechanisms are at work. The Great Salt Lake, the New York Finger Lakes, and Chesapeake Bay are locations in which similar phenomena have been studied and observed (Fig. 9.24). Ocean-effect convection, strongly evident all along the U.S. East Coast in Fig. 9.24a and in Nova Scotia in Fig. 9.24b can also impact Cape Cod, Massachusetts. Globally, parts of Japan, the Aleutian Islands, and northern Europe can witness favorable conditions for ocean-effect snow.

Although ocean-effect snow is quite common, it receives less attention than lake-effect snow due to the fact that often only ships occupy the affected regions (with the exceptions of sufficiently narrow water bodies as mentioned above), and the snow cannot accumulate on the ocean surface. Also, the dynamical mechanisms at work along a large, open coastline during a strong synoptic-scale cold-air outbreak do exhibit some differences with the Great Lakes lake-effect convection. For instance, the

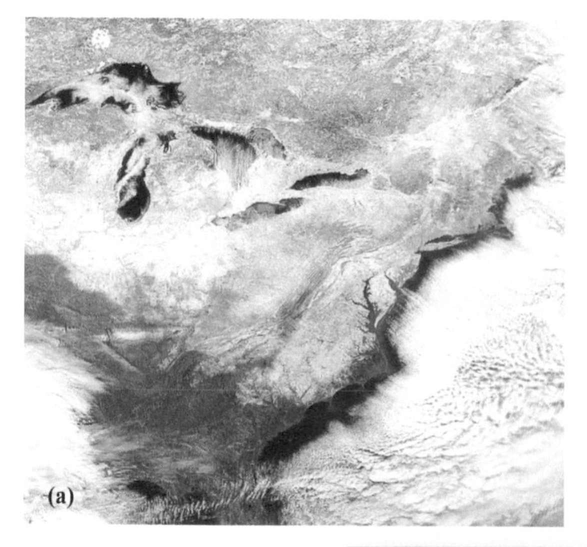

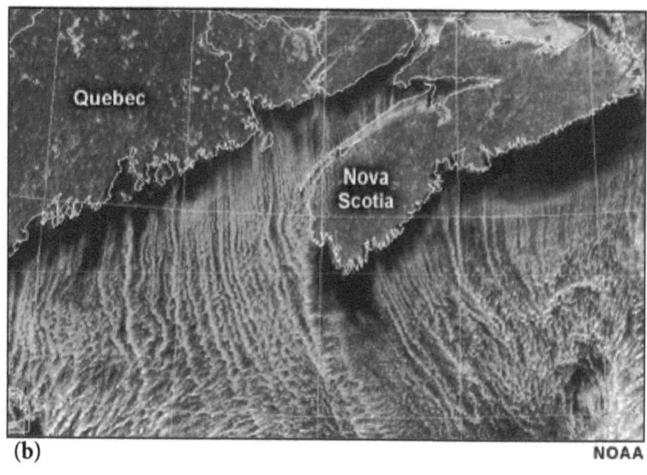


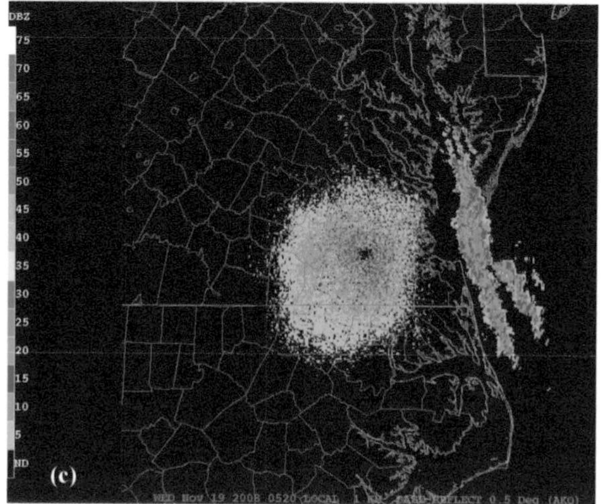

**Figure 9.24.** Examples of ocean- and bay-effect precipitation: (a) false-color visible satellite image from 26 Jan 2000 featuring extensive ocean-effect precipitation along the U.S. East Coast as well as LES; (b) ocean-effect snow affecting Nova Scotia, Canada, at 1400 UTC 18 Jan 2003; (c) base-level reflectivity image showing Chesapeake Bay–effect snow at 0520 UTC 19 Nov 2008 from Wakefield, Virginia, radar site. Images (a) and (b) courtesy NOAA.

roles of converging land breezes, frictional convergence, and orographic lifting are all irrelevant for this phenomenon in open-ocean conditions.

### 9.4.5. Forecasting lake-effect precipitation

Several excellent references on the topic of lake-effect precipitation forecasting have been published in the journal *Weather and Forecasting*, including Niziol (1987) and Niziol et al. (1995). The material here is drawn from these sources, along with the author's experiences living in western New York State in the 1990s.

A necessary condition for lake- or ocean-effect precipitation is the development of sufficient instability. As discussed in section 9.4.2., a convenient and practical measure of lower-tropospheric instability in these situations is the temperature difference between the water surface and the 850-mb level, which generally must exceed 13°C. As with any "rule of thumb," it is important to actually examine the temperature profile, because it is possible that, if the inversion layer were located below the 850-mb level, a shallow layer of instability could exist beneath this level. Prediction of the 850-mb temperature for the synoptic scale by numerical models is generally reliable, and lake-surface temperature data are available from both in situ and remote sensing techniques.

A related consideration is the *depth* of the cold, unstable air, and this is often related to the depth of the planetary boundary layer. Polar or arctic air masses are generally quite stable, and the planetary boundary layer in these air masses is often quite shallow. Mixing and destabilization as stable arctic air moves over a relatively warm water body increases the depth of the mixed layer (e.g., Figs. 9.19, 9.21), but a sufficient distance of over-water trajectory (known as the *fetch*) is required for the turbulent fluxes to deepen the boundary layer enough to allow vigorous convection. A capping inversion usually exists at top of the mixed layer. Any additional lifting mechanisms (e.g., synoptic-scale upper-level troughs or cyclonic curvature of sea level isobars) can serve to increase the altitude of the capping inversion. In general, the deeper the surface-based layer of reduced stability, the greater the potential intensity of snowfall, provided that other conditions needed for lake-effect snow are met. In general terms, the capping inversion height should exceed 1 km for minimal lake-effect snow. The most intense lake-effect events are often accompanied by a capping inversion of 2.5 km or higher.

In the presence of significant directional *vertical wind shear,* observations show that lake-effect bands can be disrupted. Directional shear of greater than 60° between the near-surface flow and that at the 850-mb level tends to favor weaker, multiple-band structures. Small values of directional shear (e.g., less than 30° over the depth of the cold-air layer) favor stronger, single-band events. A synoptic interpretation of this suggests an additional detrimental effect of shear in addition to the "band disruption" aspect; during periods of strong cold advection (backing shear), the associated QG forcing for subsidence also limits the height of the capping inversion, and the associated drying may weaken lake-effect precipitation. Often, the strongest and best organized lake-effect precipitation begins some time *after* a cold frontal passage marks the onset of an arctic outbreak over the lakes. This is consistent with the cessation of strongest QG forcing for descent and a time when the synoptic environment becomes more favorable for lake convection to become organized. Weaker shear is consistent with weaker thermal advection.

The *relative orientation of synoptic-scale flow to lake axes* is important for determination of which type of band morphology will be observed, but also this factor is related to the fetch experienced by air parcels moving over the lake. The required fetch depends on characteristics of the upstream air mass: for example, the depth of the upwind boundary layer. A related factor is the *moisture content of the upstream airmass.* If the air initially reaching the water body is extremely dry, a larger fetch is required to allow moist convection to develop. Numerical modeling experiments have confirmed that the presence of upwind lakes Huron and Superior, along with Georgian Bay, can facilitate lake convection over the lower lakes (Michigan, Erie, and Ontario) via preconditioning of the air mass (e.g., Ballentine et al. 1998).

## REVIEW AND STUDY QUESTIONS

1. Define the terms *coalescence, riming, aggregation, ice nucleation,* and *deposition.*
2. List all phase-change possibilities for water, and discuss which absorb and release heat. Why do these processes absorb or release heat?
3. Discuss the physical mechanisms at work during lake-effect precipitation.
4. Why do we tend to hear more about lake-effect snow than we do about ocean-effect snow?
5. Why is the 1000–500-mb thickness not a very good predictor of precipitation type relative to 1000–850-mb thickness? In what situations would use of the 1000-850-mb thickness for precipitation-type forecasting not be recommended?
6. How can the partial thickness nomogram aid in precipitation-type forecasting?

## PROBLEMS

1. Consider the vertical temperature profile shown below. Suppose that the thermal advection is weak and heavy precipitation is falling, meaning that freezing and melting precipitation can exert a significant influence on the local temperature profile. Sketch a modified (approximate) temperature profile at right below that accounts for latent heat exchange via freezing and melting. (Original profile included as a light gray line.)
2. The sounding below was taken at Greensboro, North Carolina, during a winter-weather event.
   i. What precipitation type do you expect to have been falling at this time? Justify your answer.
   ii. List 3 dynamic or thermodynamic processes that would result in *warming* of the temperature profile at some level. For each, cite evidence of the process in the sounding.

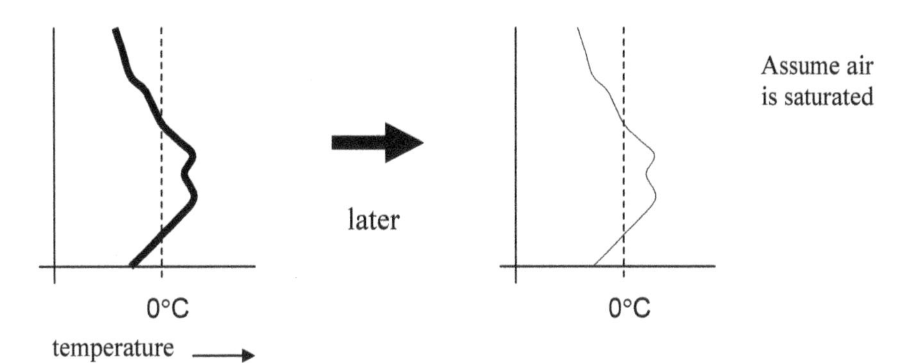


**Figure for Problem 1**

iii. List 3 dynamic or thermodynamic processes that would result in *cooling* of the temperature profile at some level. For each, cite evidence of the process.

3. Consider the four Buffalo, New York, soundings shown below. Assume that the relevant lake temperature was +1°C at the time.
   i. Which *one* of these soundings do you think would *most likely* be associated with significant lake-effect snow at the eastern end of Lake Erie (select from a to d)? Briefly justify your answer.
   ii. Which soundings, if any, would NOT be supportive of LES? Explain why not for any of the soundings you selected.

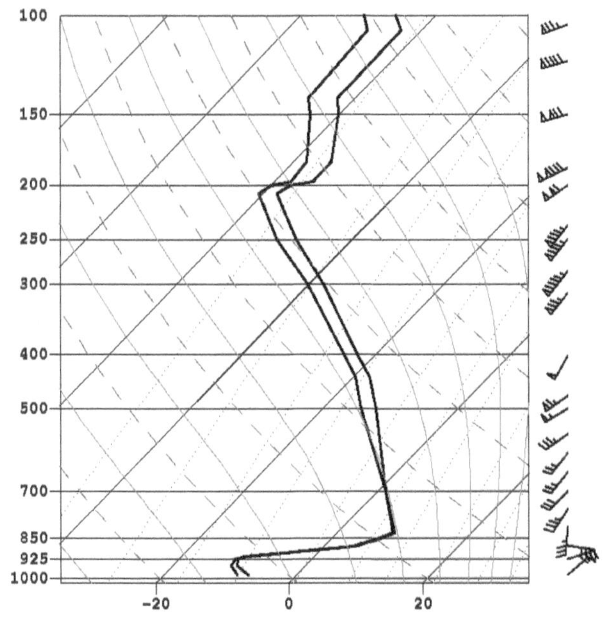


**Figure for Problem 2**

## REFERENCES


Ballentine, R. J., 1982: Numerical simulation of land-breeze-induced snowbands along the western shore of Lake Michigan. *Mon. Wea. Rev.,* **110,** 1544–1553.

Ballentine, R. J., A. J. Stamm, E. E. Chermack, G. P. Byrd, D. Schleede, 1998: Mesoscale model simulation of the 4–5 January 1995 lake-effect snowstorm. *Wea. Forecasting,* **13,** 893–920.

Baumgardt, D., 1999: Wintertime cloud microphysics review. National Weather Service Weather Forecast Office. [Available online at http://www.crh.noaa.gov/arx/micrope.html.]

Bocchieri, J. R., and H. R. Glahn, 1976: Verification and further development of an operational model for forecasting the probability of frozen precipitation. *Mon. Wea. Rev.,* **104,** 691–701.

Bosart, L. F., and F. Sanders, 1991: An early-season coastal storm: Conceptual success and model failure. *Mon. Wea. Rev.,* **119,** 2,831–2,851.

Bourgouin, P., 2000: A method to determine precipitation types. *Wea. Forecasting,* **15,** 583–592.

Byrd, G. P., R. A. Anstett, J. E. Heim, and D. M. Usinski, 1991: Mobile sounding observations of lake-effect snowbands in western and central New York. *Mon. Wea. Rev.,* **119,** 2323–2332.

Colle, B. A., K. J. Westrick, and C. F. Mass, 1999: Evaluation of MM5 and Eta-10 precipitation forecasts over the Pacific Northwest during the cool season. *Wea. Forecasting,* **14,** 137–154.

COMET, cited 2005: Topics in precipitation type forecasting. UCAR COMET Program. [Available online at http://meted.ucar.edu/norlat/snow/preciptype/.]

Cortinas, J. V., Jr., B. C. Bernstein, C. C. Robbins, and J. W. Strapp, 2004: An analysis of freezing rain, freezing drizzle, and ice pellets across the United States and Canada: 1976–90. *Wea. Forecasting,* **19,** 377–390.

Eichenlaub, V. L., 1970: Lake effect snowfall to the lee of the Great Lakes: Its role in Michigan. *Bull. Amer. Meteor. Soc.,* **51,** 402–412.

Ferber, G. K., C. F. Mass, G. M. Lackmann, and M. W. Patnoe, 1993: Snowstorms over the Puget Sound lowlands. *Wea. Forecasting,* **8,** 481–504.

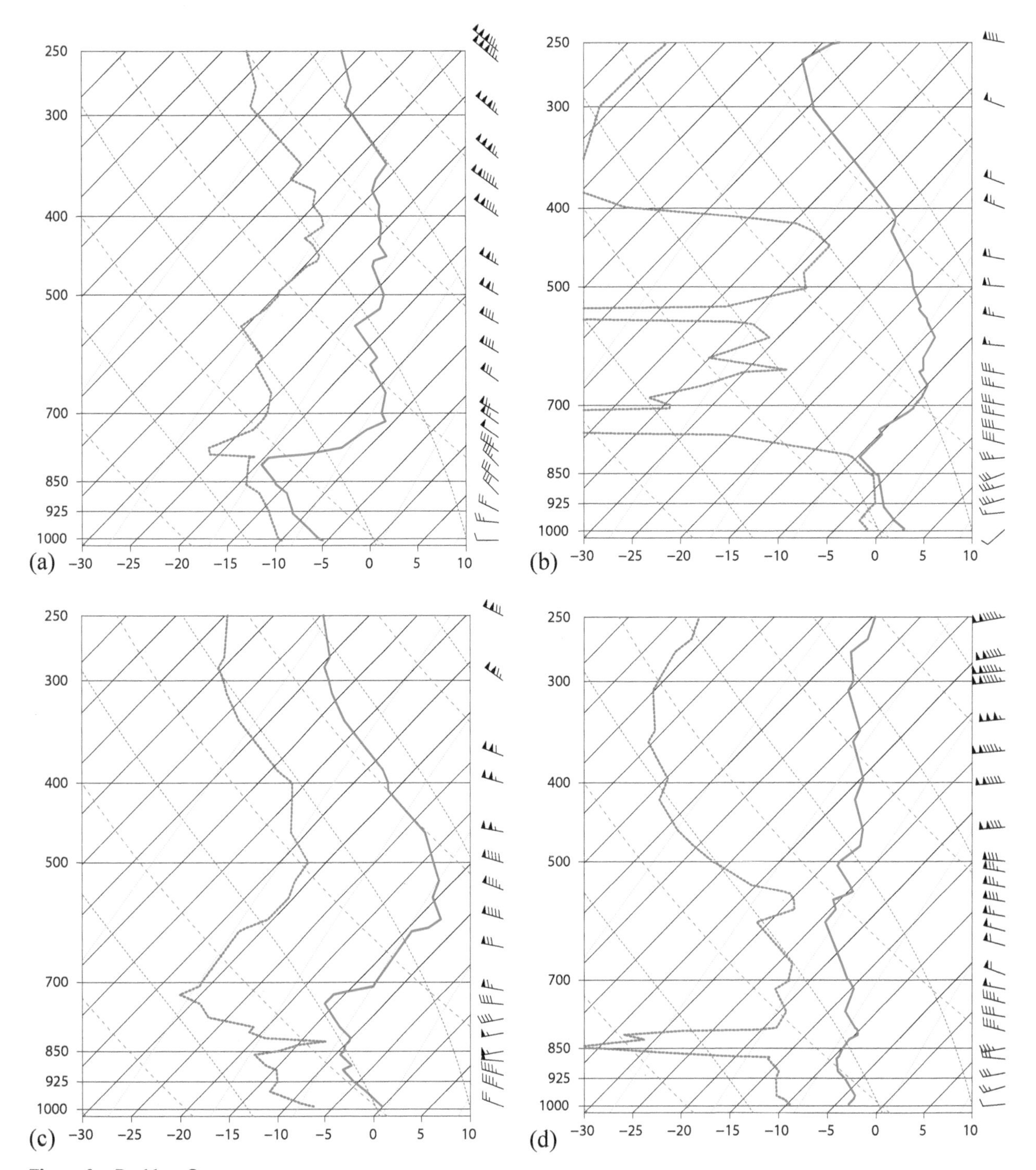


**Figure for Problem 3**

Gedzelman, S. D., and E. Lewis, 1990: Warm snowstorms: A forecaster's dilemma. *Weatherwise,* **43,** 265–270.

Hjelmfelt, M. R., 1990: Numerical study of the influence of environmental conditions on lake-effect snowstorms over Lake Michigan. *Mon. Wea. Rev.,* **118,** 138–150.

Jiusto, J., and M. Kaplan, 1972: Snowfall from lake-effect storms. *Mon. Wea. Rev.,* **100,** 62–66.

Kain, J. S., S. M. Goss, and M. E. Baldwin, 2000: The melting effect as a factor in precipitation-type forecasting. *Wea. Forecasting,* **15,** 700–714.

Keeter, K. K., and J. W. Cline, 1991: The objective use of observed and forecast thickness values to predict precipitation type in North Carolina. *Wea. Forecasting,* **6,** 456–469.

——, S. Businger, L. G. Lee, and J. S. Waldstreicher, 1995: Winter weather forecasting throughout the eastern United States. Part III: The effects of topography and the variability of winter weather in the Carolinas and Virginia. *Wea. Forecasting,* **10,** 42–60.

Keighton, S., and Coauthors, 2009: A collaborative approach to study northwest flow snow in the Southern Appalachians. *Bull. Amer. Meteor. Soc.,* **90,** 979–991.

Kocin, P. J., and L. W. Uccellini, 2004: *Northeast Snowstorms. Meteor. Monogr.,* No. 54, American Meteorological Society 818 pp.

Lackmann, G. M., 2001: Analysis of a surprise western New York snowstorm. *Wea. Forecasting,* **16,** 99–116.

——, K. Keeter, L. G. Lee, and M. B. Ek, 2002: Model representation of freezing and melting precipitation: Implications for winter weather forecasting. *Wea. Forecasting,* **17,** 1,016-1,033.

LaDue, J., 1996: COMET course notes and satellite meteorology modules. [Available from http://www.comet.ucar.edu/class/smfaculty/byrd/sld027.htm]

Lavoie, R. L., 1972: A mesoscale numerical model of lake-effect storms. *J. Atmos. Sci.,* **29,** 1025–1040.

Lumb, F. E., 1960: Cotswolds snowfall of 1 November 1942. *Meteor. Mag.,* **89,** 11–16.

Meisinger, C. L., 1920: The precipitation of sleet and the formation of glaze in the eastern United States January 20 to 25, 1920, with remarks on forecasting. *Mon. Wea. Rev.,* **48,** 73–80.

——, 1920: Demoralization of traffic in New York City by snow and sleet. *Mon. Wea. Rev.,* **48,** 80–80.

Miner, T., P. J. Sousounis, J. Wallman, and G. Mann, 2000: Hurricane Huron. *Bull. Amer. Meteor. Soc.,* **81,** 223–236.

Moyer, B. W., 2001: A climatological analysis of winter precipitation events at Greenville-Spartanburg, SC. U.S. National Weather Service Eastern Region Technical Attachment NO: *NO: 2001-01.*

Niziol, T. A., 1987: Operational forecasting of lake effect snowfall in western and central New York. *Wea. Forecasting,* **2,** 310–321.

——, W. R. Snyder, and J. S. Waldstreicher, 1995: Winter weather forecasting throughout the eastern United States. Part IV: Lake effect snow. *Wea. Forecasting,* **10,** 61–77.

NOAA Economics, cited 2008: Extreme events, snow and ice. NOAA Economics. [Available online at http://www.economics.noaa.gov/?goal=weather&file=events/snow&view=costs.]

Novak, D. R., L. F. Bosart, D. Keyser, and J. S. Waldstreicher, 2004: An observational study of cold season-banded precipitation in northeast U.S. cyclones. Wea. Forecasting, **19,** 993–1,010.

——, B. A. Colle, and R. McTaggart-Cowan, 2009: The role of moist processes in the formation and evolution of mesoscale snowbands in the comma head of northeast U.S. cyclones. *Mon. Wea. Rev.,* **137,** 2,662–2,686.

Passarelli, R. E., and R. R. Braham Jr., 1981: The role of the winter land breeze in the formation of Great Lakes snowstorms. *Bull. Amer. Meteor. Soc.,* **62,** 482–491.

Petterssen, S., and P. A. Calabrese, 1959: On some weather influences due to warming of the air by the Great Lakes in winter. *J. Meteor.,* **16,** 646–652.

Sanders, F., and L. F. Bosart, 1985: Mesoscale structure in the megalopolitan snowstorm of 11–12 February 1983. Part I: Frontogenetical forcing and symmetric instability. *J. Atmos. Sci.,* **42,** 1,050–1,061.

Stewart, R. E., 1984: Deep 0 isothermal layers within precipitation bands over southern Ontario. J. Geo. Res., **89,** 2567–2572.

——, 1985: Precipitation types in winter storms. Pageoph, 123, 597–609.

Wallace, J. M., and P. V. Hobbs, 2006: *Atmospheric Science: An Introductory Survey.* 2nd ed. Academic Press, 483 pp.

Wexler, R., R. J. Reed, and J. Honig, 1954: Atmospheric cooling by melting snow. *Bull. Amer. Meteor. Soc.,* **35,** 48–51.

## FURTHER READING

Rauber, R. M., L. S. Olthoff, M. K. Ramamurthy, D. Miller, and K. E. Kunkel, 2001: A synoptic weather pattern and sounding-based climatology of freezing precipitation in the United States east of the Rocky Mountains. *J. Appl. Meteor.,* **40,** 1,724–1,747.

Robbins, C. C. and J. V. Cortinas 2002: Local and synoptic environments associated with freezing rain in the contiguous United States. Wea. Forecasting, **17,** 47–65.

Sousounis, P. J., 1998: Lake-aggregate mesoscale disturbances. Part IV: Development of a mesoscale aggregate vortex. *Mon. Wea. Rev.,* **126,** 3,169–3,188.

——, and J. M. Fritsch, 1994: Lake-aggregate mesoscale disturbances. Part II: A case study of the effects on regional and synoptic-scale weather systems. *Bull. Amer. Meteor. Soc.,* **75,** 1,793–1,811.

——, and G. E. Mann, 2000: Lake-aggregate mesoscale disturbances. Part V: Impacts on lake-effect precipitation. *Mon. Wea. Rev.,* **128,** 728–745.

Steenburgh, W. J., S. F. Halvorson, and D. J. Onton, 2000: Climatology of lake-effect snowstorms of the Great Salt Lake. *Mon. Wea. Rev,* **128,** 709–727.

# CHAPTER 10

# Numerical Weather Prediction

*What satisfaction is there in being able to calculate tomorrow's weather if it takes us a year to do it? To this I can only reply: I hardly hope to advance even so far as this. I shall be more than happy if I can carry on the work so far that I am able to predict the weather from day to day after many years of calculation. If only the calculation shall agree with the facts, the scientific victory will be won. Meteorology would then have become an exact science, a true physics of the atmosphere. When that point is reached,* then *practical results will soon develop.*

—V. Bjerknes, "Meteorology as an Exact Science", (1914).
Delivered at the University of Leipzig, January 8th, 1913

Atmospheric scientists watch a three-dimensional view of a numerical-model simulation of Hurricane Katrina (Aug 2005): a rainwater isosurface along with 10-m wind vectors are shown as Katrina approaches landfall along the United States Gulf Coast. Courtesy of Steve Chall, Renaissance Computing Institute (RENCI).

A complete treatment of atmospheric modeling would require several textbooks and should be the subject of a sequence of courses. Numerical models are used for a wide variety of applications, and it is important that users of model output understand how models work. However, many users of numerical model output have limited formal training concerning the models, despite the obvious advantages of a thorough understanding of their workings. The Cooperative Program for Operational Meteorology, Education and Training (COMET) program has developed and organized useful, up-to-date information about operational numerical weather prediction (NWP) in a series of professional development modules, and this is one avenue through which users of model output can further their knowledge of operational models.

The material presented in this chapter is not exclusively relevant to short-term NWP; computer models of many types are now utilized extensively in the geosciences. These range from simple one-dimensional (1D) radiation or boundary layer models to complex coupled dynamical earth system models, which represent the atmosphere, oceans, and even biosphere. For climate prediction and projection, newer models include feedbacks from dynamic vegetation and land-use change, in addition to full representation of the atmospheric and oceanic physics and dynamics. Air-quality forecasting now utilizes models complete with chemistry packages, and coupled ocean–atmosphere–wave models are used for diverse applications, including the prediction of oceanic pollutant dispersion and transport, and hurricane storm-surge forecasting, for example. Professionals in many geoscience disciplines are expected to stay abreast of an increasing volume of information concerning the configuration and workings of numerical models.

The overall objective of this chapter is to provide students with background information concerning how models work, with emphasis on physical process representation and weather forecasting. Chapter 11 builds upon this material in discussing weather forecasting techniques and offers suggestions for how to best utilize guidance from numerical models in operational forecasting.

## 10.1. HISTORICAL PERSPECTIVES

As is evident from his quote at the beginning of this chapter, it was the dream of Vilhelm Bjerknes to develop meteorology as an "exact science," rather than maintain the status quo as an ad hoc collection of rules and observations. The basic premise of numerical weather prediction was outlined by Bjerknes (1904), as discussed in section 5.4. Of course, electronic computers did not exist at that time, but Bjerknes explained how the problem could be split into many smaller tasks, and presumably he envisioned groups of people computing future weather conditions by graphical integration of the governing equations forward in time. Although the governing equations (momentum equations, conservation of mass, energy, and water substance, along with the gas law) were known well before 1904, it was evidently not recognized until then that these equations formed a closed set that could be solved to obtain future states of the atmosphere. Bjerknes was evidently the first to organize the equations in this way. Bjerknes (1904) is recommended reading for those wishing to learn more about the fascinating birth of NWP and the emergence of meteorology as an "exact science." The optimism evident in Bjerknes (1904) was naïve in that relations between time step, grid length, and computational instability were not yet known; however, this optimism served to inspire and encourage other investigators to push aggressively toward making NWP a reality.

One who was inspired by V. Bjerknes's ideas was L. F. Richardson, who is believed to be the first person to have actually put Bjerknes's NWP theory into practice and publish the results. It must be mentioned that Felix Exner had undertaken numerical forecasts prior to Richardson's famous work, but these efforts were not as comprehensive as those of Richardson. Richardson's numerical weather forecast was for 20 March 1910 (for which upper-air data were available from Bjerknes, in addition to detailed surface analyses), and he published a detailed account of the method for numerical forecasting (Richardson 1922). Unfortunately, because of noise in the initial data and other factors, Richardson's initial tendency calculations were not accurate. Although Richardson recognized that his results were not reasonable (he attributed the error to bad wind observations), it has been subsequently shown (e.g., Lynch 1992, 1999, 2006) that by filtering the initial data, a drastically better forecast is obtained. The lack of accuracy in his initial tendency calculations does nothing to diminish the enormity of his contribution; in the 1922 text, he presented staggered spatial grids and a time-differencing technique now known as the "leap frog" scheme, and he went so far as to estimate the number of human "computers" needed to keep pace with the weather (approximately 64,000). Richardson envisioned a "forecast factory", which consisted of a vast theatre with each seat representing a spatial location on the globe. A "conductor" shone lights of differing colors on different sections of the room to coordinate the pace of calculations. This fantastic vision is not unlike the structure of modern distributed-memory computing clusters! When the first electronic computers became available in the 1940s, the NWP problem was ripe for solving, partly because of the pioneering efforts of Bjerknes and Richardson.

The Courant–Fredrich–Lewy (CFL) condition, which places a constraint on numerical stability, was published in 1928. The CFL condition can be expressed as

$$\mu = \frac{c\,\Delta t}{\Delta x} \leq 1, \tag{10.1}$$

where $\mu$ is the *Courant number* and $c$ represents the speed of the fastest wave or wind in the model domain. This relates the length of a model grid cell $\Delta x$, the model *time step* (the interval of time $\Delta t$ between sequential forecasts), and the speed of the fastest wave or wind in the model domain. *Smaller grid lengths require smaller time steps*, placing an important constraint on the horizontal resolution of numerical models. Violation of this condition can produce large errors and result in computer model crashes; despite extensive development of techniques to circumvent this problem, CFL errors can still (rarely) occur in modern operational NWP models.

The development of the upper-air rawinsonde network in the 1940s paved the way for routine operational NWP efforts. Radio communications were critical to this development, both for transmission of data from the balloon-borne sonde to ground stations and for broadcasting data to centralized locations where analyses could be assembled.

In the 1940s, Charney, Fjørtoft, von Neumann, Thompson, Phillips, and other pioneers quickly recognized that electronic computers could be applied to the numerical forecasting problem. The development of the electronic computer, headed at Princeton University by John von Neumann, was closely linked with the development of numerical weather prediction. On the meteorological side, Rossby, Charney, Reichelderfer, and members of the U.S. military services worked to set up a suitable model. Charney (1948, 1949) outlined the benefit and means of filtering the governing equations while retaining the important dynamical essence of the synoptic-scale forecast signal. Recognition that useful forecasts could be obtained via integration of the barotropic vorticity equation (see section 1.5.3) led to the first successful test prediction (Charney et al. 1950). The first operational numerical weather forecast was made in Sweden in 1954, and in the United States operational numerical predictions began in 1955.

The transformation of meteorology by numerical weather prediction since that time has been remarkable, evident, for example, in the steady forecast improvements shown in Fig. 10.1. Perhaps even more significant was that scientific computing became widely recognized; the advent of scientific computing spread to several scientific disciplines, and this process was accelerated by the developments in meteorology (for an interesting and thorough discussion of this time period, see Harper 2008).

The use of sophisticated vertical and horizontal grid meshes, optimization of model initial conditions through *data assimilation* (DA) techniques, more complete representation of physical processes involving clouds and land surface characteristics, and exploitation of ensemble forecasting strategies are just some of the advances that characterize the past 50 years. In conjunction with improved models, new observing platforms (including radar, satellite, and in situ measurements) have allowed more accurate model initial conditions, as well as contributing to the formulation of reliable analyses against which to evaluate model forecasts. Increased computing speed, data storage, and transfer rates have conspired to inundate weather forecasters with rapidly increasing volumes of model output. Although this can be daunting, without question the rapid improvement of numerical weather prediction has been a major factor in improved weather forecasts over the past several decades.

Although many measures of forecast skill are available, the trend in forecast improvement shown in Fig. 10.1, for the European Centre for Medium Range Weather Forecasts (ECMWF) global model, provides a continuous history of progress in NWP from 1980 through May 2010. The accuracy of Southern Hemisphere forecasts relative to that in the Northern Hemisphere is now very similar, due in large part to increasing availability of satellite data there, in conjunction with improved data assimilation. Expanding the scale on the ordinate above 90% reveals that the upward trend in forecast accuracy continues even for the day-3 forecast, although at some point further increases will be very difficult to achieve (Fig. 10.1). The impressive trend in forecast accuracy led former U.S. National Weather Service (NWS) director Joseph Friday to state that "The success of numerical weather prediction represents one of the most significant scientific, technological, and societal achievements of the 20th century."

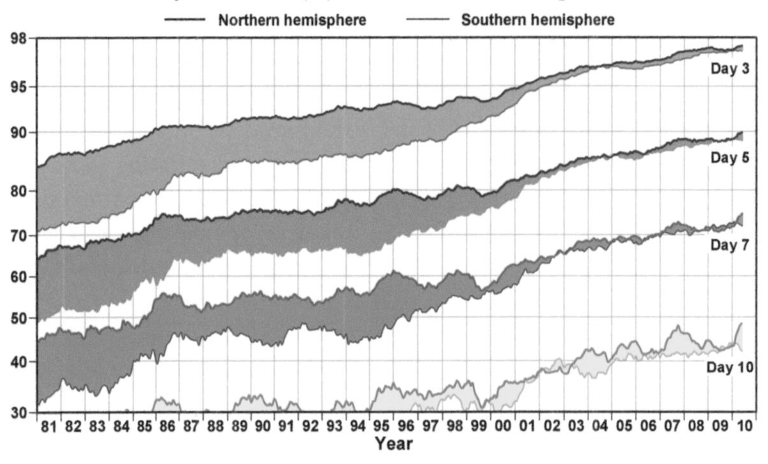


Courtesy of ECMWF. Adapted and extended from Simmons & Hollingsworth (2002)

**Figure 10.1.** Evolution of forecast skill with time for ECMWF global model. Plotted are 500-mb anomaly correlations for 3-, 5-, 7-, and 10-day ECMWF 500-mb height forecasts, plotted as running means for the period of January 1980 through May 2010. Shading shows differences in scores between hemispheres at the forecast ranges indicated. Note nonlinear scale on ordinate (updated image courtesy of Dr. Adrian Simmons, ECMWF; adapted and extended from Simmons and Hollingsworth 2002).

## 10.2. A SIMPLE EXAMPLE

What, exactly, *is* a "model"? The Merriam-Webster dictionary definitions include "a usually miniature representation of something" and "an example for imitation or emulation." *Physical models*, such as rotating dishpans, were used historically in the investigation of dynamical atmospheric phenomena and have enjoyed a recent resurgence (e.g., Illari et al. 2009). The type of model considered here is a *dynamical model*, which is essentially a set of computer programs, typically written in FORTRAN for optimum speed in solution, that are designed to emulate the real atmosphere through the integration of the governing equations. Using suitable initial and boundary conditions, the programs can be run to solve for future atmospheric states, just as Bjerknes had envisioned.

Bjerknes (1904) stated, "If it is true, as every scientist believes, that subsequent atmospheric states develop from the preceding ones according to physical law, then it is apparent that the necessary and sufficient conditions for the rational solution of forecasting problems are the following: a) A sufficiently accurate knowledge of the state of the atmosphere at the initial time, and b) A sufficiently accurate knowledge of the laws according to which one state of the atmosphere develops from another." There are some interesting themes in this statement, for example, the implication that some may not have believed that the atmosphere was constrained to evolve in accordance with physical law and that these people were not to be considered scientists. More important is the idea that one can cast the governing conservation laws into a form that can be solved via the representation of continuous derivatives by numerical approximation. There are many techniques for this, but here we will only describe the simplest, that of centered finite differences.

Rather than converting the entire set of governing equations into finite-difference form, let us consider an example based on the following simplified situation. A cold front has passed St. Louis, Missouri (KSTL), from the northwest at ~1800 UTC 29 November 2009. The initial surface temperature field is known from the surface observing network, and the horizontal temperature gradient is observed to be fairly uniform. The wind behind the front is blowing steadily from the northwest at 10–15 kt (~5–7 m $s^{-1}$). Based on the information provided, how can we utilize our knowledge of the physical processes at work, along with basic equations representing those

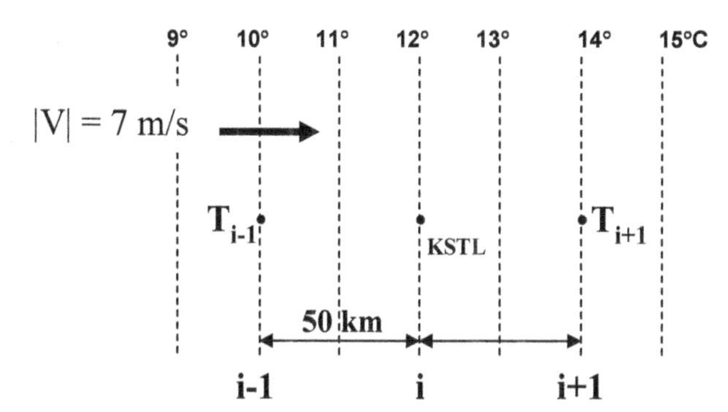


**Figure 10.2.** Idealized schematic of the near-surface temperature field for the advection problem outlined in the text. Dashed contours represent isotherms (as labeled); the bold arrow at top indicates the speed and direction of the prevailing wind.

processes, to *quantitatively* predict the temperature at KSTL, say, 6 h later, at 0000 UTC 30 November? What process(es) do we expect to dominate the temperature-tendency equation in this situation?

This is a straightforward problem if we restrict our attention to horizontal temperature advection, recognizing that the neglect of other processes could lead to large errors. Diurnal heating or cooling, turbulent mixing in the planetary boundary layer (PBL), and processes relating to clouds and precipitation must also be considered. However, if we can assume that horizontal temperature advection is the dominant process at work, we can utilize an *advection model* for the purpose of illustration,

$$\frac{\partial T}{\partial t} = -u\frac{\partial T}{\partial x} - v\frac{\partial T}{\partial y}. \tag{10.2}$$

To further simplify, we can rotate the coordinate system so that the $x$ axis is parallel to the temperature gradient, as pictured in Fig. 10.2, eliminating the second right-hand term in (10.2). Suppose that the isotherms are spaced evenly, with a uniform horizontal gradient as pictured, and that our model grid length is 50 km. We use the indices $i$ and $j$ to represent the gridpoint numbers in the east–west and north–south directions respectively, although here we are only concerned with east–west ($i$) values. The station in question, KSTL, is located at point $i, j$, and we have temperature information at the point $i + 1, j$ and $i - 1, j$ as well as the wind speed at point $i$.

Clearly, we expect it to get colder at KSTL because of the process of horizontal temperature advection. To quantify this, we use (10.2), but first we must convert the partial derivatives to *finite-difference* form,

$$\frac{T_{final} - T_{initial}}{t_{final} - t_{initial}} \approx -u_i \frac{T_{i+1} - T_{i-1}}{(x_{i+1} - x_{i-1})}. \tag{10.3}$$

The changes in time or distance can be written as $\Delta t$ and $\Delta x$, respectively. In the limit where these time and distance increments become infinitesimally small, the approximation (10.3) becomes an exact representation of (10.2). If we then rearrange (10.3) to solve for the temperature at the final time, we have one equation and one unknown, and we can obtain a forecast temperature at KSTL for 0000 UTC,

$$T_{final} = T_{initial} - \Delta t \left[ u_i \frac{T_{i+1} - T_{i-1}}{2\,\Delta x} \right]. \tag{10.4}$$

Plugging in the numerical values using data from Fig. 10.1, we obtain a predicted 0000 UTC temperature of slightly less than 6°C. In this problem, we have a clearly specified "governing equation" and "initial conditions," and several *major* assumptions are implicit in our solution. Aside from neglecting physical processes other than horizontal temperature advection, we took the liberty of a *6-h time step*, assuming that conditions were steady over that time period. In general, this is a very poor assumption.

This simple example represents a method for generating a forecast based on a prognostic equation. A major difference with dynamical models, however, is that, in such a model, the full equations governing the evolution of the wind, temperature, pressure, and moisture fields are solved simultaneously to provide a dynamical update after each time step. In today's operational NWP models, time steps are on the order of 60 s, with even smaller values in high-resolution mesoscale models.

In the preceding example, we assumed that we had a complete representation of the temperature field at all spatial locations (grid cells) in our domain, and this enabled us to determine the needed values at the locations used in our finite-difference calculation. In that case, the isotherm analysis provided the needed data. In general, before computations of this type can be made, the initial data must be interpolated onto an evenly spaced grid

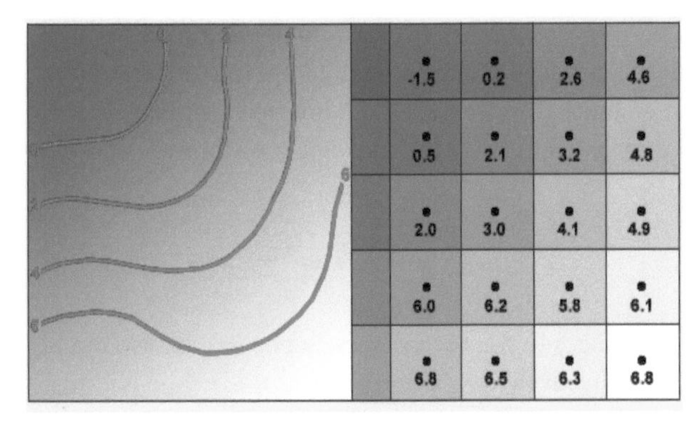


**Figure 10.3.** Hypothetical representation of temperature data in (left) continuous and (right) gridpoint form (courtesy of the COMET program).

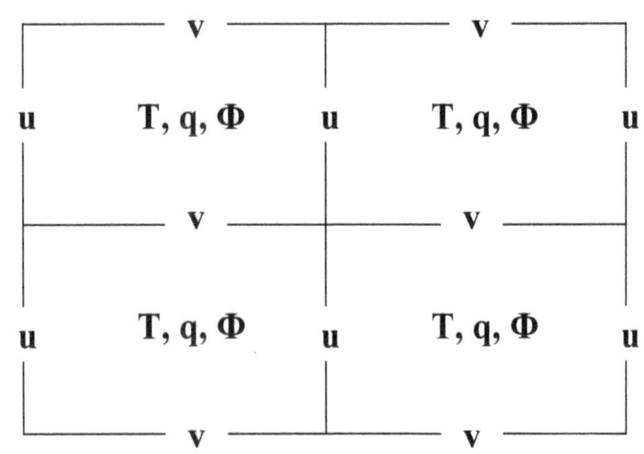


**Figure 10.4.** Horizontal view of staggered Arakawa C grid used in the WRF-ARW.

mesh, to allow finite-difference computations to be made (e.g., Fig. 10.3). Given the data provided on the right side of Fig. 10.3, one could easily compute the spatial gradients of temperature, provided knowledge of the grid spacing.

The starting point for an isotherm analysis of this type is usually a set of unevenly distributed observations. In the past, these were mostly from surface stations and aloft, from rawinsonde measurements. In recent decades, increasing volumes of data are provided by aircraft, satellites, radar, profilers, and other observing platforms. Another extremely valuable source of information that goes into modern analyses is short-term forecasts from previous runs of numerical models. By combining forecast information with all available observations over a specified time window, highly accurate analyses are routinely constructed at operational centers around the world via the process of *data assimilation*. Modern DA systems constantly *cycle* the forecast and blend it with observations, producing initial conditions for operational NWP models; data assimilation will be discussed in section 10.5.

## 10.3. THE DYNAMICAL CORE

The grid architecture, vertical coordinate, time stepping technique, boundary conditions, and other numerical aspects of the system for solving the governing equations are collectively referred to as the *dynamical core* of the model. This aspect is distinct from the *model physics*, which describe the treatment of physical and thermodynamic processes other than advection, including the representation of clouds and precipitation, radiation, turbulence, and convection.

A complete understanding of differencing techniques and grid structures is useful to comprehend the workings of numerical models. However, these topics are only briefly introduced here, as our emphasis is on model physics and to a lesser extent model initial conditions, with emphasis on weather analysis and forecasting.

### 10.3.1. Grid configuration

Modern atmospheric numerical models include sophisticated grid architectures that are designed both for speed of computation and accuracy. Simple grid meshes of the type depicted in Figs. 10.2 and 10.3 are not found in modern models; rather, *staggered grids* of the type depicted in Fig. 10.4 are the norm. This particular grid is used in the Weather Research and Forecasting model (WRF) "Advanced Research WRF" dynamical core (WRF-ARW), developed at the National Center for Atmospheric Research (NCAR).

The grid features wind components at the midpoint boundary along each grid cell, and mass variables such as temperature and specific humidity are computed for the center of the grid cells. Staggering is also performed in the vertical, and the time differencing is partitioned too. This is done for computational efficiency and also because it increases the effective resolution of the model.

The use of the term "grid cell" is preferable to "grid point," because the data values carry information that is relevant to some spatial distance surrounding the center point of the data locations. For some quantities, such as cloud cover, the fractional coverage of a grid cell is useful in accurate computation of the effective albedo, for example.

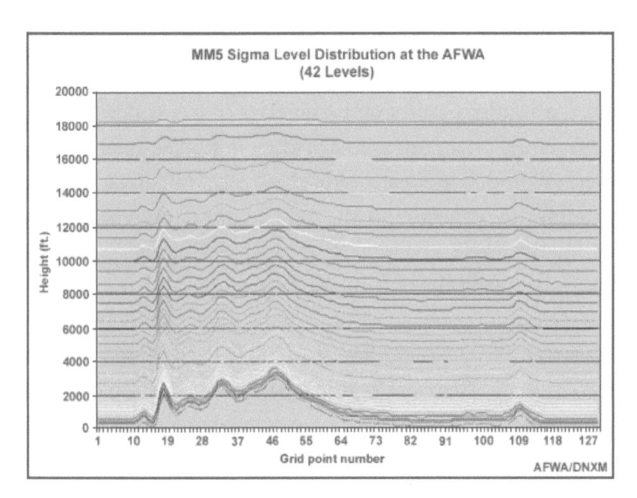


**Figure 10.5.** Cross section of sigma coordinate surfaces in a region of elevated terrain from the Air Force Weather Agency (AFWA) operational model (courtesy of the COMET program).

Despite some advantages of pressure as a vertical coordinate, complications exist when model coordinate surfaces intersect the surface. As a means of eliminating this complication, N. Phillips introduced an ingenious vertical coordinate, denoted sigma ($\sigma$), where $\sigma = p/p_s$ is the pressure at any given point above the surface normalized by the surface pressure $p_s$. Thus, the $\sigma = 1$ coordinate level corresponds to the surface, and $\sigma = 0$ corresponds to the top of the atmosphere. Because most models do not extend vertically to the top of the atmosphere, the following variation on the original sigma coordinate is used in many numerical models:

$$\sigma = \frac{p - p_T}{p_s - p_T}, \tag{10.5}$$

where $p_T$ is the pressure at the model top and now $\sigma = 0$ corresponds to the model top rather than to the top of the atmosphere. The distribution of sigma coordinate surfaces for the U.S. Air Force operational model configuration is presented in Fig. 10.5.

The simplicity and advantages that go with sigma coordinates are not without cost, however. For regions of steep terrain, where the sigma coordinate surfaces may slope strongly, the horizontal pressure-gradient force expressed in sigma coordinates is related to the difference between two terms, and numerical errors in the computation of the pressure-gradient force can become problematic.

The use of staggered grids and novel vertical coordinates is often transparent to the user of model output, because the postprocessing phase of the numerical forecast process includes destaggering of the model output data and vertical interpolation to standard or regularly spaced isobaric or height levels. The model output viewed by forecasters and other users is rarely on the native model grid; in fact, several different standard grid interpolations of a given operational model run are available (see section 10.7.1).

### 10.3.2. Hydrostatic and nonhydrostatic models

As discussed in chapter 1, the hydrostatic approximation is highly accurate for synoptic-scale and some mesoscale weather phenomena. Because the CFL condition (10.1) is dependent on the speed of the fastest waves supported by the model equations, there is a significant advantage to using the hydrostatic approximation in a model rather than solving the full vertical momentum equation. This is because, in a hydrostatic model, sound waves are effectively filtered, because the pressure at any point is then determined by the hydrostatic mass in the overlying air column, and pressure fluctuations due to sound waves are not permitted. Hydrostatic models may have a correspondingly larger time step for a given grid length relative to what a nonhydrostatic model would require, because sound waves are fast, $c \sim 300\ \mathrm{m\ s^{-1}}$.

As of this writing, nearly all *global* climate and weather prediction models are hydrostatic models, along with many regional models. However, nonhydrostatic models have become more popular in recent years because it is desirable to simulate and analyze mesoscale weather systems, such as supercell thunderstorms, that are characterized by nonhydrostatic dynamics, and, as computing power has increased, it is now possible to run models with grid lengths that resolve phenomena for which nonhydrostatic processes are important. In the 1990s, nonhydrostatic models such as the fifth-generation Pennsylvania State University–NCAR Mesoscale Model (MM5) were developed; more recently, the WRF model was developed [both the Nonhydrostatic Mesoscale Model (NMM) core developed at the National Centers for Environmental Prediction (NCEP) and the ARW core from NCAR are nonhydrostatic models].

For nonhydrostatic models, there are several means of circumventing the full restrictions imposed by the CFL condition. These include partitioning the governing equations and integrating those that support fast waves separately or otherwise filtering specific wave types from the model at each time step, thereby removing the wave from the solutions. For hydrostatic models, such as the NCEP Global Forecast System (GFS) model or the ECMWF global model, how is the vertical motion computed if the model has only the hydrostatic relation in place of the vertical momentum equation? A common technique for this is to compute the *kinematic* vertical motion, obtained diagnostically by vertical integration of the continuity equation. During each model time step, once the new horizontal velocity components have been computed, then the divergence field is known, and a vertical integral of the continuity equation can be performed to determine the vertical motion. This is in turn used to compute the condensate and ultimately the precipitation in each grid column.

For most operational forecasting applications, the quantitative difference between hydrostatic and nonhydrostatic models is not significant. However, as limited-area models are run with decreasing grid lengths, the importance of nonhydrostatic processes will increase. Currently, for model grid lengths less than 10 km, especially in the presence of high terrain, nonhydrostatic processes can be important.

### 10.3.3. Spectral and gridpoint models

Representation of spatial derivatives using finite-difference methods, even for very sophisticated higher-order finite-difference schemes, introduces some truncation error. Clever grid designs and higher-order differencing schemes are now able to reduce the truncation error substantially relative to what simple centered differencing would provide, however with considerable added computational expense. A different approach to the truncation problem is to represent the equations as a sum of basis functions, which could be, for example, a sum of sine and cosine functions. Then, spatial derivatives can be computed *analytically*, resulting in an extremely accurate solution. The truncation error when using this approach is then given by the first wave that is neglected. For example, suppose that we were representing a given variable as the sum of 30 waves in the zonal direction. By neglecting the signal associated with the 31st wave, some error would be introduced. These errors are thus related to resolution, with the error being associated with the scale of the neglected waves.

Some processes cannot be represented with spectral techniques: for example, convection, precipitation, and vertical advection; these quantities are represented in gridpoint space. Accordingly, spectral models solve part of the dynamical equations in a transformed spectral space, but the model physics and some dynamical processes are represented on a grid, as in a gridpoint model. Operational spectral models are really hybrid models, utilizing a variety of numerical strategies to integrate the governing equations.

Due to the convenience of representing fields with a cyclic lateral boundary in spectral space and the highly accurate numerical solutions, many *global* models are of the spectral type, including the widely used ECMWF and GFS models. There are regional spectral models, although the majority of limited-area models employ gridpoint architectures. To some extent, the dynamical core of a model is transparent to the end-product user, but it should be recognized that the dynamical core in a spectral model is fundamentally different from that in a gridpoint model.

### 10.3.4. Resolution

The *resolution* of a model is a term that describes what scale of meteorological feature can be adequately represented by the model. Concerning terminology, the effective *resolution* of a numerical model is *not* the same as the model *grid spacing* or *grid length*, despite widespread treatment of these terms as synonymous. The reason for the difference is that it takes several grid cells to truly *resolve* a meteorological feature; therefore, the resolution of a model is typically on the order of 5 times larger than the grid length. The precise relation between grid length and resolution depends on the nature of the specific meteorological feature in question, as well as the specific grid architecture employed by the model. Due to these complications, it is preferable for gridpoint models to discuss the model grid length, or grid spacing, rather than resolution.

For spectral models, additional discussion regarding resolution is required. As mentioned above, the error in spectral models is related to the scale of waves which are neglected in the spectral representation, but also spectral models must perform some computations on a grid. The grid architecture in a spectral model is typically aligned with latitude and longitude lines, and the effective zonal grid spacing therefore may change as a function of latitude. Spectral models only carry information for waves of a certain spatial scale and ignore that associated with smaller-scale waves. For these and other reasons, it is not possible to exactly relate the resolution of gridpoint and spectral models.

However, we can estimate the equivalent grid spacing of a spectral model to make it comparable to a gridpoint model, which can be helpful in determining the ability of a model to represent a given weather system. If we assume that a gridpoint model would require 3 points to minimally represent a given wave, then the approximate grid spacing needed (in units of *km*) to represent the set of *N* waves of a spectral model is given by

$$\Delta x = (111\,km)[360^\circ/(3N)] \qquad (10.6)$$

The most common means of expressing the spectral model truncation error is for "triangular" truncation, and the model resolution is described as T*N* (e.g., T80), where *N* is the number of waves included in the model.

At the time of this writing, the U.S. GFS model, a global spectral model, has T574, which according to (10.6) yields an approximate grid length equivalent of ~23 km (NCEP lists the equivalent grid length as 27 km). On 10 January 2010, the ECMWF global spectral model increased resolution, going to T1279. The documentation provided by ECMWF with this upgrade lists the effective grid spacing as 16 km, which is evidently the length of grid cells in the gridpoint portion of that model. If we instead utilize (10.6), the effective grid spacing equivalent of the ECMWF model is ~10 km. Because model characteristics such as resolution frequently change, users of model output are encouraged to utilize the aforementioned COMET Operational Models Matrix.

Does higher resolution lead to improved forecasts? The answer to this question is more complex than one might think, and it is dependent on the phenomenon in question. In regions of complex terrain and for orographically forced precipitation, a documented benefit has been found in reducing the grid length from 36 to 12 km, but with less improvement as the grid length was further reduced to 4 km (e.g., Colle et al. 2000; Mass et al. 2002). For convective precipitation, a distinct change in the character of the forecast comes about when the grid length becomes small enough to explicitly represent the precipitation systems (e.g., Gallus 1999, 2002); new verification techniques are needed for very-high-resolution model forecast verification (see Gilleland et al. 2010).

## 10.4. PARAMETERIZATION OF PHYSICAL PROCESSES

Whether a given model represents the atmosphere as a set of grid cells or using spectral methods, there are many physical processes that cannot be explicitly resolved by the model. Some processes, such as radiative energy transfer or cloud microphysics, take place in nature on a range of spatial scales extending down to the molecular level. In representing the continuum of atmospheric motions and processes, approximations must therefore be made.

Consider a situation in which there is a convective marine boundary layer, capped by a weak inversion layer, with dry air above the inversion. In nature, buoyant production of turbulence beneath the inversion gives rise to cumulus clouds, some of which carry sufficient buoyancy to penetrate the inversion layer, increasing the depth of the planetary boundary layer with time and moistening the air above it through detrainment and mixing. The horizontal dimension of many of these buoyant cumulus clouds is less than 500 m, well below the resolvable scale for a typical operational NWP model. However, the model must represent the transport of heat, moisture, and momentum accomplished by these small-scale convective motions or else the model forecast would feature, in this example, an overly shallow mixed layer and a dry bias above the inversion layer. To represent the processes described here, a *planetary boundary layer* parameterization would be employed, perhaps with additional contributions from a *convective parameterization* (CP) scheme.

As discussed in chapter 1, we cannot introduce new dependent variables to represent these small-scale turbulent transfer processes, or the system of equations would no longer be closed and the model could not be integrated forward in time. The term *parameterization* refers to a means of expressing unknown or unresolvable quantities in terms of other, existing dependent variables. This may be done out of necessity because the unknown quantity in question is not easily measured, or it may be necessary to *close* a system of equations (reduce the number of unknowns to equal the number of equations). Parameterization involves accounting for unresolved physical processes without introducing additional dependent variables to the model equations.

The objective of this section is to provide an overview of the parameterization schemes used in many operational numerical models, with an emphasis on weather forecasting applications. A description of these parameterization packages is challenging, because the schemes themselves are often modified with subsequent releases of a given model and because a given scheme can be configured in different ways for use in various models and applications. This necessitates a more generic approach.

Parameterization of the *planetary boundary layer*, *land surface* processes, *subgrid-scale convection*, *radiation*, and *cloud microphysics* will each be discussed here.

### 10.4.1. Planetary boundary layer representation

Recall from section 1.6 that the PBL is defined as that layer of the atmosphere in which the direct influence of the surface is felt; this layer is characterized by turbulence, along with generally uniform potential temperature and specific humidity (at least for the daytime boundary layer). The PBL can be further subdivided into the *molecular boundary layer*, which is a very thin layer immediately adjacent to the surface in which molecular diffusion is an important transfer process, the *surface layer* (sometimes also known as the *constant flux layer* because of the property that turbulent fluxes of a given meteorological variable are often relatively constant with height in this 10–30-m-deep layer), the *mixed layer*, and the *inversion layer* (Fig 10.6). The atmospheric PBL is extremely important in that the exchange of heat, moisture, and momentum between the surface and atmosphere must take place within this layer. Model PBL schemes account for these exchanges, requiring inclusion of realistic representation of related processes such as surface properties and vegetation, coastal processes, the surface radiation balance, and cloud cover.

The properties of turbulence present a challenge to its numerical representation: it is chaotic, small-scale turbulent motions are isotropic (fully 3D with no preferred directions), and energy must be supplied to maintain it because of constant turbulent kinetic energy (TKE) dissipation. Turbulence represents a "cascade" of kinetic energy, from larger to smaller scales, with ultimate dissipation to thermal energy at the smallest scales. Turbulence flux divergences also contribute nonnegligible tendencies to the larger-scale dependent variables, and therefore models must represent turbulence in a realistic manner to generate correct forecasts even for the synoptic scale. Another important characteristic of this layer is that it is the layer in which we live! So, when we make forecasts for surface weather conditions, we are forecasting for the PBL, and knowledge of the relevant processes acting there can be very useful in weather forecasting.

It is convenient to separate temporal fluctuations of a given variable into those due to turbulence and to those attributable to more slowly varying processes, such as would be obtained by a time average. Generally, the fluctuations of interest for the variables being predicted are not due to turbulence but correspond to a time average over several minutes. Recall from section 1.6 that turbulent fluxes are related to correlations between turbulent flow components and the perturbation quantities of other variables in the turbulent flow. For instance, the vertical turbulent heat flux involves the quantity

$$\overline{\theta' w'}, \tag{10.7}$$

which is simply the product of turbulent fluctuations in potential temperature and vertical velocity, averaged over some time interval (indicated by the overbar). If fluctuations of these two variables were completely independent of one another, then on average the turbulent heat flux would be zero.

As discussed in section 1.6, oftentimes there are physical reasons for turbulent fluctuations *not* being independent of one another. For example, in the surface layer, where the horizontal wind speed typically increases

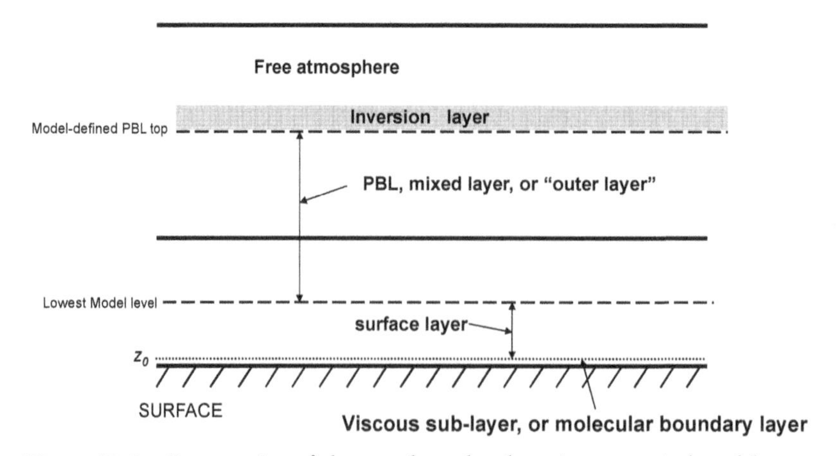


**Figure 10.6.** Cross section of planetary boundary layer in a numerical model.

with height (often approximately linearly with log $z$), we would expect air parcels that are rising in turbulent updrafts to be characterized by smaller horizontal wind speeds than would sinking parcels, because the rising parcels originate in a region of lower average wind speed relative to the sinking parcels. This is consistent with a downward momentum flux, as is often observed in the lower atmosphere. The turbulent heat flux described by (10.7) is a *second-order turbulent moment*, meaning that it involves the product of two turbulent fluctuations. A term involving the product of three turbulent quantities would be a *third-order moment*, etc.

For example, the zonal momentum equation, written to include turbulent transfer processes, demonstrates that it is the turbulent flux *divergence* that is related to the local tendency of a given meteorological variable,

$$\frac{d\bar{u}}{dt} = -\frac{1}{\rho_0}\frac{\partial \bar{p}}{\partial x} + f\bar{v} - \frac{\partial \overline{u'w'}}{\partial z}. \qquad (10.8)$$

An analogous relation is evident in equation (1.58). If we can then devise a means of representing $\overline{u'w'}$ at two different levels, then the rightmost term in (10.8) can be evaluated. The means of representing the turbulent flux is referred to as the *closure problem*, because the turbulent flux must be represented in terms of the other dependent variables to keep the number of unknown quantities equal to the number of independent equations.

**10.4.1.1. PBL closure techniques.** There are several strategies for closure, beginning with the bulk aerodynamic method discussed in section 1.6. However, modern NWP models use more sophisticated strategies than the bulk scheme. One method that represents increased complexity from the bulk scheme draws upon an analogy between turbulent transfer and molecular diffusion, where the diffusive properties of turbulent motions, known as the *eddy diffusivity*, replaces the molecular diffusion. Using this technique, the turbulent fluxes of the zonal velocity components and potential temperature can be represented as

$$\overline{u'w'} = -K_m \frac{\partial \bar{u}}{\partial z}$$
$$\overline{v'w'} = -K_m \frac{\partial \bar{v}}{\partial z}$$
$$\overline{\theta'w'} = -K_h \frac{\partial \bar{\theta}}{\partial z}, \qquad (10.9)$$

where $K_m$ is the *eddy viscosity coefficient* and $K_h$ is the *eddy diffusivity coefficient*; these coefficients differ as a function of the variable in question, as well as other properties of the flow, and are typically determined using empirical data. The turbulent fluctuations are then related to the vertical gradient of the time-averaged quantities, multiplied by the eddy coefficients ($K_m$, $K_h$, etc.). Measurements demonstrate that these coefficients are functions of static stability, and numerous field experiments are required to obtain values that are appropriate for a wide variety of situations. This closure technique is known as *K theory* or *flux-gradient theory*.

The molecular analogy has limitations when applied to the full depth of the PBL but is more relevant in the surface layer. In fact, many numerical models have separate routines for the surface layer and mixed layer, although these routines are closely coupled in the model. Often, an empirically based technique known as *Monin–Obukov similarity theory* is utilized in the model surface layer. With this technique, the K coefficients are determined to be a function of a *mixing length*, which can be interpreted as the vertical distance over which an air parcel in turbulent flow will retain its original properties before mixing with the surroundings. Because this closure technique replaces the second-order turbulent moments, it is a *first-order* closure technique; the order of closure is always one less than the order of the replaced turbulent moments.

Above the surface layer in the mixed layer, the turbulent heat flux can be large despite the fact that the vertical potential temperature profile is neutral (as was shown in chapter 1). The relevance of flux-gradient theory (10.9) is questionable there, and higher-order closure techniques are typically used. One such technique involves solving for the local *turbulent kinetic energy* in the model; by adding a new variable (TKE) and a new equation, the system remains closed. The TKE equation is

$$\frac{d(TKE)}{dt} = -\overline{u'w'}\frac{\partial \bar{u}}{\partial z} - \overline{v'w'}\frac{\partial \bar{v}}{\partial z} + \frac{g\overline{\theta'w'}}{\theta_0} - \varepsilon + (other), \qquad (10.10)$$

where $TKE \equiv (u'^2 + v'^2 + w'^2)/2$, $\varepsilon$ represents turbulent dissipation, and (other) refers to pressure work terms.

The first two right-hand terms in (10.10) represent mechanical production of turbulence, drawing on the shear of the background flow that is analogous to the barotropic energy conversion discussed in section 2.7. Typically $\partial\bar{u}/\partial z > 0$ and $\overline{u'w'} < 0$, meaning that this term often results in *production* of TKE. The third right-hand term in (10.10) accounts for buoyant production (daytime) or destruction (nighttime) of turbulence.

For an NWP model to utilize (10.10), the turbulent quantities must be expressed as the product of the TKE itself, the mixing length, and other parameters determined from the closure scheme. The TKE, a second-order moment, is predicted by the model. This is a "1.5-order closure" technique, referred to as "TKE closure", and it is widely used in operational models. A full second-order technique would have prognostic equations for all second-order turbulent moments.

**10.4.1.2. Local and nonlocal closure.** Different PBL closure techniques can exert a pronounced influence on the model forecast. One fundamental difference concerns whether the turbulence scheme operates based on vertical gradients computed locally in a given grid column, imparting tendencies to the dependent variables only between adjacent model levels (local closure) or whether the model generates changes that simultaneously span the depth of the PBL (nonlocal closure). In the PBL, large eddies can mix air across the depth of the PBL, and turbulence should be envisioned as a superposition of eddies on a broad range of scales, so it can be argued that nonlocal closures more closely mimic what happens in nature (Fig. 10.7). Generally, model analysis studies show a

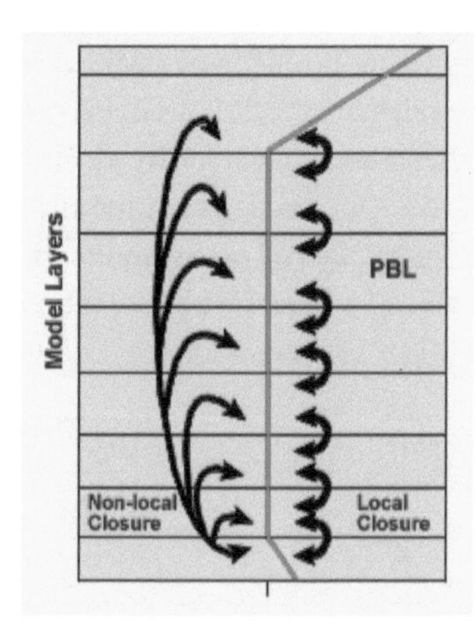


**Figure 10.7.** Comparison of vertical turbulent mixing process in nonlocal versus local formulation of a model PBL scheme. Potential temperature as a function of height (red line) is shown to indicate the depth of the PBL (courtesy of the COMET program).

tendency for nonlocal schemes to produce a deeper PBL than local schemes, although other model components can affect this, making it dangerous to generalize.

Some PBL schemes, such as a popular scheme that has been implemented in the WRF-ARW by researchers at Yonsei University in Korea, include entrainment at the PBL top designed to mimic the action of shallow cumulus and stratocumulus clouds. In some models, the convective parameterization scheme (see section 10.4.4) also includes shallow mixing, meaning that physics options must be selected with care to avoid duplication of process representation.

### 10.4.2. Land surface models

There is no shortage of examples in which the properties of the local land surface exert a pronounced impact on the overlying atmosphere. One dramatic example is a region of corn growing in moist soil. The corn plants can act effectively as moisture pumps, drawing moisture from the soil and transferring it to the overlying atmosphere via evapotranspiration, where turbulence and convection then distribute it vertically. For a NWP model forecast in this situation, the difference between actively evapotranspirating corn and bare soil is critically important. Failure to account for this moisture source could result in clear skies in the model atmosphere, whereas afternoon convective storms may take place in reality. Proper prediction of the boundary layer depth, convective initiation, surface fluxes of heat, moisture, and momentum, and the surface radiation budget all depend on accurate representation of the surface, vegetation, and soil properties.

Today's NWP models include sophisticated schemes to account for evapotranspiration along with many other processes taking place at the land surface, and in the soil beneath. These land surface models (LSMs) include predictive equations for soil temperature and moisture, and they account for local albedo changes and thermodynamic processes due to snow cover, vegetation, and vegetation changes with season, as well as surface evaporation and radiation budgets. The LSMs account for soil properties, including thermal and moisture conductivity of a given soil type. A schematic of the popular Noah LSM that is currently used operationally in the NCEP North American Mesoscale (NAM) and GFS models is provided in Fig. 10.8.

One aspect of the LSM that directly affects the overlying atmosphere is the computation of the *surface*

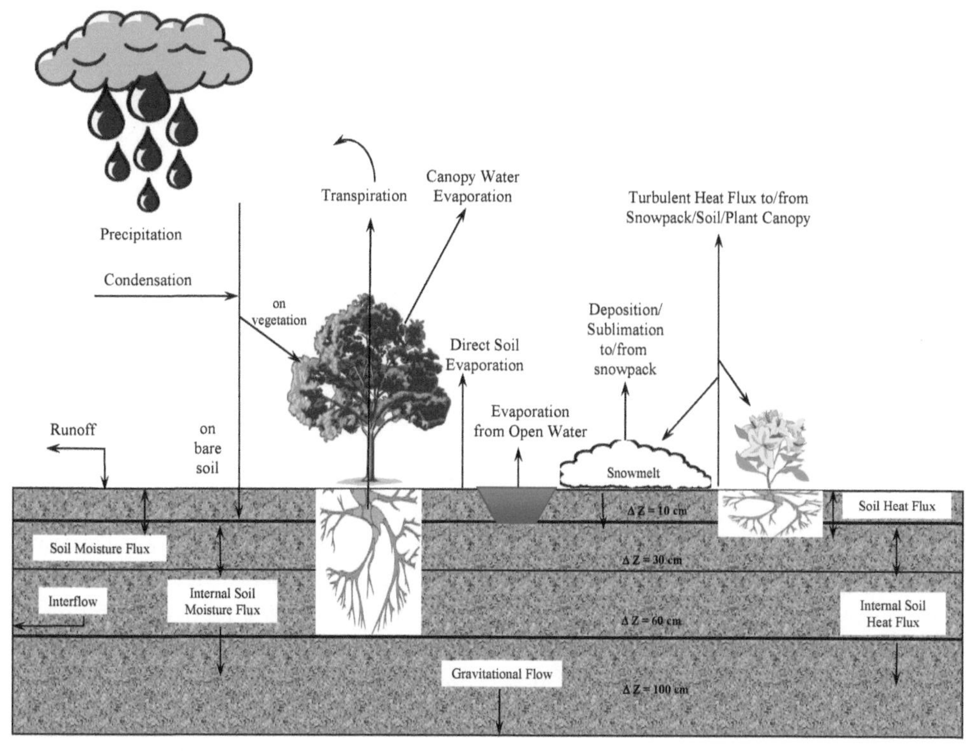


**Figure 10.8.** Schematic representation of processes in cross-sectional view for the Noah land surface model (from http://www.ral.ucar.edu/research/land/technology/lsm.php). © University Corporation for Atmospheric Research, UCAR/RAL.

*skin temperature*. This is not the same as the air temperature near the surface, but this represents the physical temperature of the surface itself, which then determines outgoing infrared radiation, and influences the surface fluxes of heat and moisture. Over water, models typically use other means to determine the sea surface temperature (SST), and so the LSM is only invoked for grid cells that include land.

In chapter 9, it was pointed out that the model treatment of snow and freezing rain is important to correct prediction of the near-surface air temperature. This requires, among other things, that the LSM be initialized with correct snow cover information. Additionally, during the course of the model run, snow that falls in the model atmosphere may accumulate on the surface, and this is accounted for by the LSM. In that case, the surface albedo in the grid cells that experience snow accumulation is modified (increased) to account for the snow cover; if above-freezing temperatures develop at the surface, the LSM will account for the energy absorbed in melting the snow pack.

During spring and fall, an LSM will commonly use climatological values for the growth or fall of leaves in regions of deciduous forest. If a drought or unusually warm or cold period takes place during these seasons, it is possible that the model LSM will not have accurate information concerning vegetation properties in the affected regions, but these situations are relatively infrequent.

### 10.4.3. Gridscale precipitation: The microphysics scheme

Obviously, NWP models must be able to predict the development of clouds and precipitation, which are required as forecast parameters, but also because other aspects of the model forecast may depend on their representation. For instance, when clouds form in the model atmosphere, the radiation scheme must account for the corresponding change in albedo for shortwave radiation, as well as for the emission and absorption of infrared radiation by the cloud. The model must also account for the latent heat released by condensation, which can affect the geopotential height, temperature, wind fields, etc.

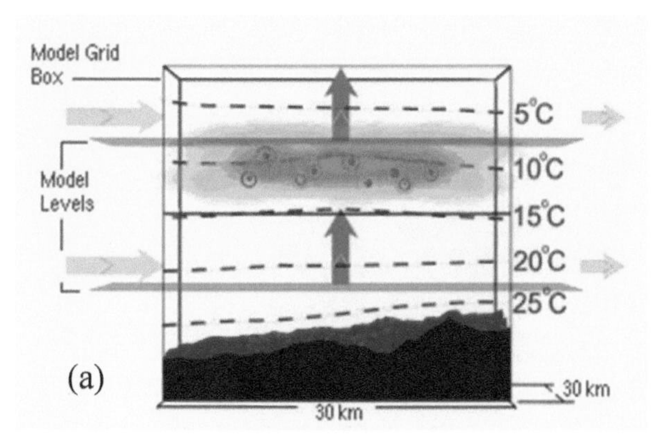

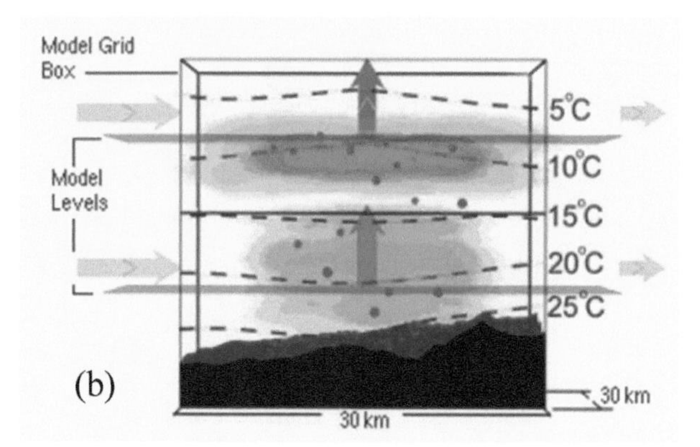


**Figure 10.9.** (a) Idealized representation of condensational heating in a model cross section, (b) followed by cooling beneath as falling precipitation evaporates (graphics from COMET program).

When saturated air rises, condensation (or deposition) takes place. In the model atmosphere, when an entire grid box reaches a specified saturation threshold (near 100% relative humidity), condensation occurs on the grid scale and the model predicts the development of a cloud and/or precipitation. Modern NWP models have predictive equations for cloud water and cloud ice, along with predicted mixing ratios of snow, rain, graupel, and sometimes hail. In the past, many operational models would remove condensate immediately to the surface; however, in the last 10–20 years, because of increases in computational power, the sophistication of microphysics schemes has increased to the point where most now explicitly account for the formation and subsequent evolution of different classes of hydrometeors.

For saturated gridscale ascent, the model microphysics scheme computes the degree of supersaturation, and the corresponding amount of excess water vapor is converted to the appropriate form of condensate (cloud liquid water or cloud ice). The associated latent heat release is computed, and tendency terms for the temperature and water vapor are passed back to the main model routines. The model predicts the 3D distribution of rain, snow, hail, or graupel; some research-grade microphysics schemes include additional classes of hydrometeors. The fall velocity of precipitation dictates when the hydrometeors reach the model surface, and the falling precipitation may evaporate or sublimate en route, depending on the humidity in the intervening layers (Fig. 10.9).

Because the amount of vapor available for condensation is determined by the vertical motion field, the details of the model microphysics scheme will not typically exert a dominant influence on the model forecast for synoptic-scale systems. However, feedbacks can arise that have large impacts on the amount and location of precipitation for some types of weather systems, such as hurricanes and organized convective storms.

Suppose that, at a given instant, we were to capture a unit volume of air from a given cloud as depicted in Fig. 10.10a; the temperature of this portion of the cloud is above freezing, and therefore the volume is assumed to contain only liquid cloud droplets. If an exact count of the number of droplets of any given size could be obtained, we could plot this information in log coordinates as in Fig. 10.10b, with the number of drops of a given size per cubic meter on the ordinate and the droplet diameter on the abscissa.

For a numerical model to represent the temporal evolution of this size distribution, two strategies are available: (i) predict the changes to the number of particles in each size bin (Fig. 10.11a, the bin method) or (ii) approximate the distribution with an analytic function and predict the evolution of the distribution through the parameters of this function (Fig. 10.11b, the bulk method). There is very large computational expense associated with the bin method, so most operational models utilize bulk microphysics schemes.

Microphysics schemes predict the mixing ratio of water vapor, along with the size distributions of several classes of hydrometeor, including cloud water, cloud ice, snow, rain, and graupel (or sometimes hail). An analytic expression is obtained for each hydrometeor class,

$$N_x(D) = N_{0x} D^{\alpha_x} \exp(-\lambda_x D), \qquad (10.11)$$

where $N_x(D)$ is the number of a given hydrometeor class $x$ of a given diameter $D$, $N_{0x}$ is the *intercept value* for that

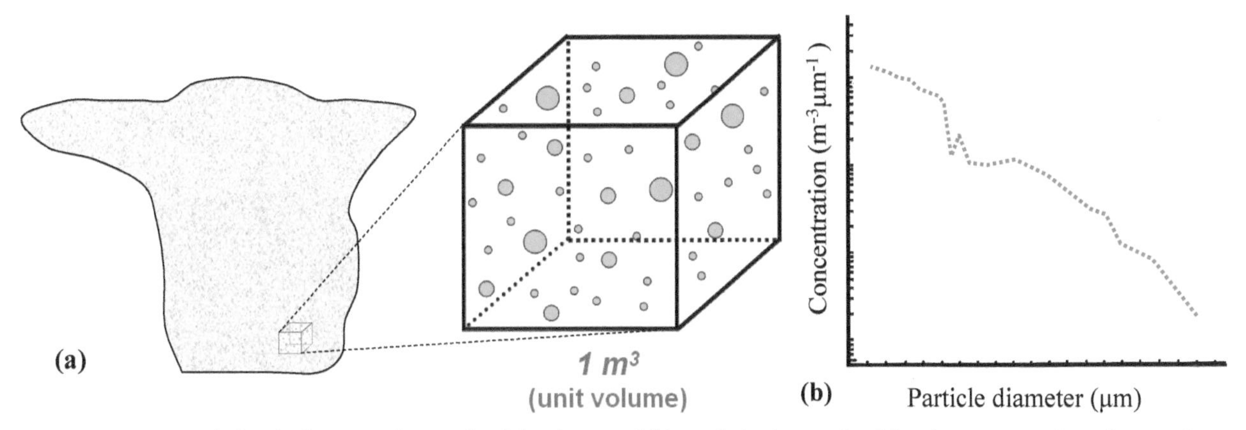


**Figure 10.10.** (a) Idealized schematic of a sample of cloudy air and (b) graph displaying cloud droplet properties from the sample plotted as droplet diameter (abscissa) versus log of concentration (ordinate; figure courtesy Jason Millbrandt, Environment Canada).

class (as depicted in Fig. 10.11b), $\alpha_x$ is a *shape parameter*, and $\lambda$ is the slope of the analytic function. The total number concentration of a given hydrometeor class $N_{Tx}$ can be obtained by integrating with respect to diameter and multiplying this by the density of liquid water or ice; in this way, the mass or mixing ratio $q_x$ of a given class can be obtained for each grid cell.

Representation of the size distribution allows computation of the fall velocity as a function of particle size and changes in the distribution are obtained through prediction of $N_{Tx}$, $q_x$, or both. *Single-moment* bulk microphysics schemes predict either $N_{Tx}$ or $q_x$ but not both. Then, the slope $\lambda$ and intercept $N_x$ are computed as functions of $q_x$ or $N_{Tx}$. *Double-moment* schemes predict *both* $q_x$ and $N_{Tx}$. Then, both the slope and intercept are functions of these quantities; this allows more accurate specification of the size distribution, which in turn allows better representation of the fall speed, collection efficiency, and sedimentation (fall-speed sorting by size). There are now some *triple-moment* schemes that also predict the shape parameter $\alpha$, but these are only recently being implemented into research-grade models.

The computational expense associated with sophisticated microphysics schemes precludes their implementation into operational models; as a result, almost all operational NWP models use single-moment bulk schemes, and sometimes computational demands dictate that additional measures are taken to further reduce

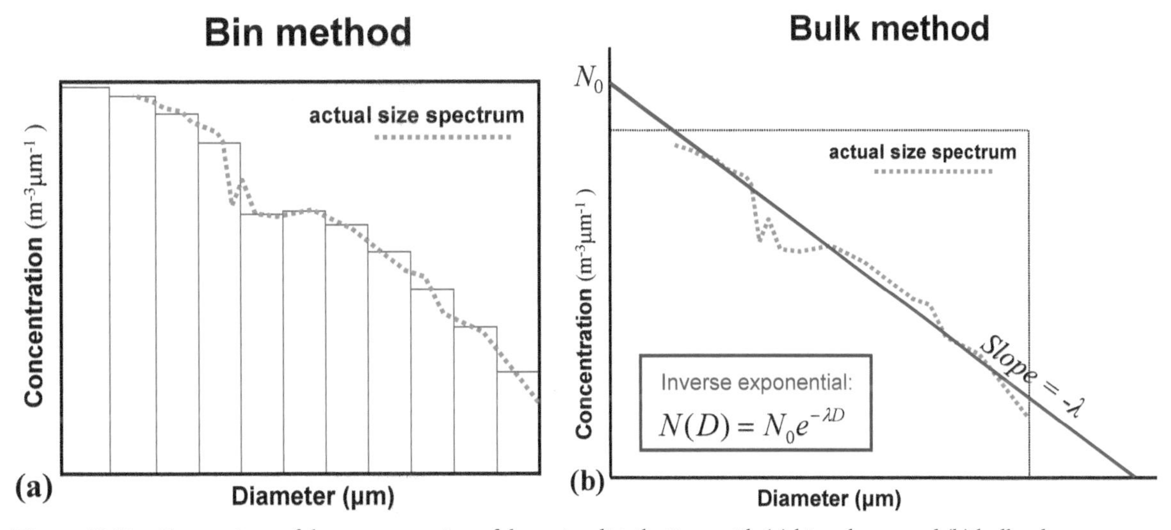


**Figure 10.11.** Comparison of the representation of drop size distributions with (a) bin scheme and (b) bulk scheme (courtesy of Jason Millbrandt, Environment Canada).

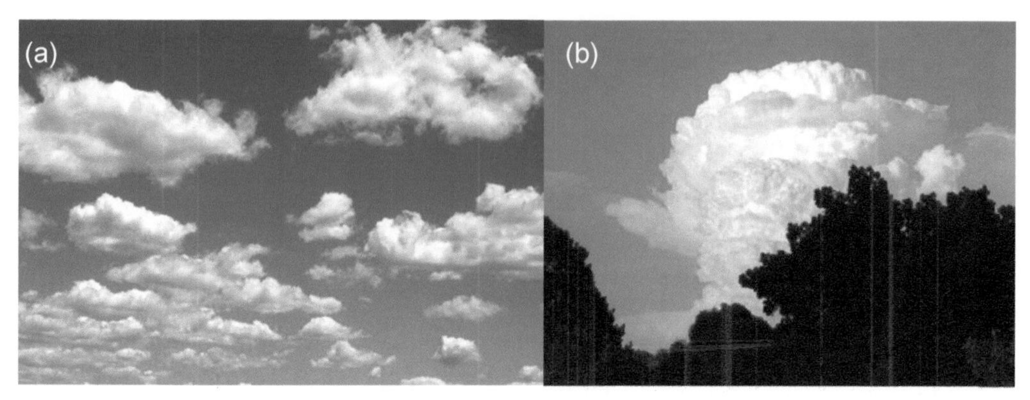


**Figure 10.12.** Photographs of (a) fair-weather cumulus and (b) cumulonimbus.

computational expense: for example, grouping the hydrometeor classes that are passed back to the model advection routines. As computing power continues to increase and model grid lengths continue to diminish, increasing use of double-moment schemes is likely.

Depending on the grid spacing and configuration of a given model, not all model precipitation takes place on the grid scale. For models running with a grid length of about 5 km or more, a *convective parameterization* (CP) scheme should also run, which can account for precipitation that takes place on horizontal scales unresolved by the model grid; the CP scheme can generate precipitation independent of the microphysics scheme.

### 10.4.4. The parameterization of subgrid-scale convection

Some cloud and precipitation systems are much smaller than a model grid cell. Consider a field of fair-weather cumulus clouds (e.g., Fig. 10.12a) or a cumulus congestus cloud that grows into a cumulonimbus (e.g., Fig. 10.12b). In the cumulus case, the model will need to represent the albedo and mixing processes associated with the clouds, but the precipitation forecast may not depend on this. For the cumulonimbus cloud, the model cannot resolve the system, even though it may be a significant rain producer and have strong implications for the evolution of weather in the surrounding environment.

If a model is run with sufficiently small grid length, then precipitation can be handled entirely by the model microphysics scheme as discussed in the previous section. At the time of this writing, most operational models are still run at a grid length that is too large to eliminate the use of a *convective parameterization* scheme. Global models, including general circulation models used to study climate change, will run at grid lengths that require the use of CP schemes for the foreseeable future; however, for regional-scale NWP models, an increasing number of research and forecasting applications now omit the CP scheme by running with grid lengths smaller than ~4–5 km. Several studies have examined the question of when it is acceptable to omit the CP scheme. The answer to this question is partly phenomenon dependent; however, for operational prediction, ~4 km represents a grid length below which it appears safe to turn off the CP scheme (Weisman et al. 1997).

A classic study by Molinari and Dudek (1992) investigated what happens when a model is run with coarse grid length but without a CP scheme. They found an unrealistic delay in the onset of precipitation, because enough instability was required to allow entire grid cells to overturn. When convection did take place, it was in the form of unrealistic "gridcell storms," and precipitation was excessively heavy. The environment in such a model becomes unrealistically unstable, because of the omission of small-scale cumulus convection, which serves to transport heat upward and stabilize the environment. The authors also warn of a "no-man's land" defined in terms of model grid spacing, where the separation between grid-scale and subgrid-scale processes becomes questionable. Some newer CP schemes have pushed the boundary of this range of grid lengths, but the fact remains that some grid lengths, for instance around 10 km, are not well suited for some applications due to this scale-separation issue. There appears to be a "sweet spot" between 1- and 4-km grid length, however, in which an adequately realistic representation of convection is obtainable that is sufficient for numerical prediction, although not for some research applications.

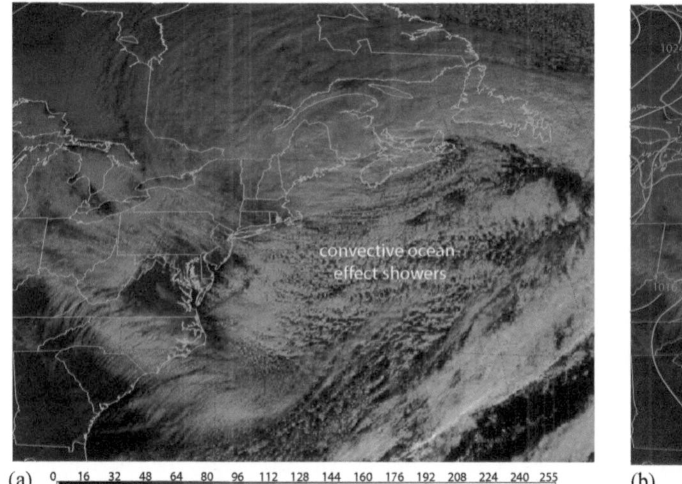


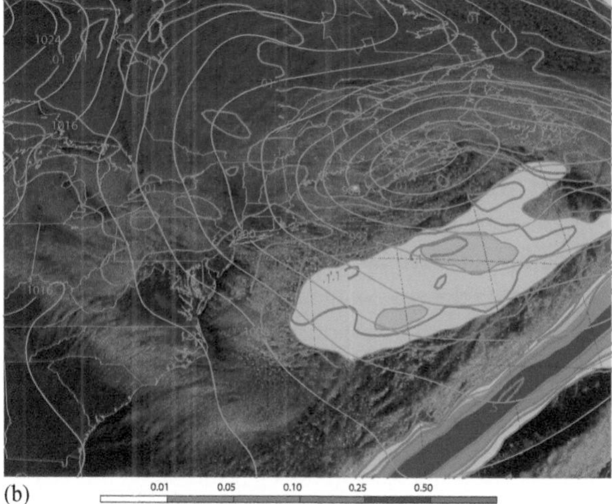


**Figure 10.13.** (a) *GOES-12* Visible satellite image valid 1445 UTC 4 Dec 2007; (b) 12-h NAM forecast of sea level pressure (green contours with 4 mb interval) and 6-h precipitation (magenta contours are total precipitation and shading corresponds to convective precipitation, in.) ending 0000 UTC 5 Dec 2007.

**10.4.4.1. Explicit and implicit precipitation in models.** Before delving into the details of CP schemes, let us consider for a moment the interplay that must take place in the model atmosphere between the CP and microphysics schemes. First of all, for a model running a CP scheme, *there are two fundamentally different "flavors" of precipitation generated by the model.* Forecasters viewing model output can benefit from examination of the different types, because they stem from different processes and are triggered by different conditions in the model. The satellite image provided in Fig. 10.13a corresponds to a situation with cold air moving over warmer ocean waters along the U.S. East Coast. A precipitation forecast from the NCEP NAM model shortly after the time of the satellite image demonstrates that the model CP scheme was accounting for the forecast precipitation in the convective region offshore, whereas farther west, the precipitation was of the grid-scale variety.

In regions where the CP scheme triggers deep convection, the model produces precipitation even in the absence of grid-scale saturation. In viewing model forecast soundings, it may appear inconsistent to have the model producing precipitation in regions where there are no model clouds, but they are being represented implicitly, so the model production of precipitation is in fact entirely consistent with its design. Some CP schemes, when active, can alter the model gridcell albedo to implicitly represent the presence of convective clouds and also pass hydrometeors back to the grid cell to account for convective anvils.

It is beneficial for forecasters and users of model output to understand the basic structure, strengths, and weaknesses of the various convective schemes in the operational models. Experienced forecasters can then recognize signatures of the CP scheme, which helps in interpretation of the model output and may aid in assessing the uncertainty in the forecast. Detailed descriptions of a given CP scheme are not possible, because the schemes are constantly evolving, have altered characteristics in different models, and may even differ between release versions of the same modeling system. Therefore, it is more profitable to summarize some of the basic strategies behind the CP schemes.

Most CP schemes work in single atmospheric grid columns in the model atmosphere. Many operational CP schemes only adjust temperature and moisture profiles and do not adjust momentum (despite the fact that real convection exerts a strong influence on the flow); the reason for this omission is the complexity of storm-scale influences on the wind field. For a typical CP scheme to activate (or "trigger"), several conditions typically must be met. Once triggered, tendencies of the meteorological variables due to the CP scheme (e.g., latent heat released to account for condensation in subgrid-scale convective updrafts and net removal of water vapor to match the amount of precipitation produced) are distributed over a specified time scale, allowing the model atmosphere sufficient time to adjust to the presence of convection.

**10.4.4.2. Adjustment schemes.** One class of convective parameterization schemes, based on observations (an empirical approach), is known as *convective adjustment schemes*. Tropical field experiments were designed to carefully measure atmospheric properties over time; by comparing the environmental conditions prior to and after convection, it was possible to isolate the changes to the large-scale environment that were attributable to the action of convection. One popular CP scheme of this type is the Betts–Miller–Janjic (BMJ) scheme. The BMJ scheme is currently used in the operational NCEP NAM model and is an option in the WRF-ARW.

Essentially, an adjustment scheme considers the model thermodynamic environment in each grid cell. If convection of sufficient depth is possible, then the scheme checks to make sure that the action of convection would lead to a thermodynamically reasonable stabilization of the environment. If these conditions are met, then the scheme adjusts the model temperature and moisture profile toward predetermined "postconvective" profiles that are designed to mimic the changes in the postconvective environment measured in field experiments. Adjustment schemes are not designed to directly represent heating and moistening due to individual convective clouds, and typically they do not account for storm-scale features such as convective outflow, which can be important for convective storms.

Specifically for the popular BMJ scheme, the trigger requires instability in the grid column, a potential convective cloud of sufficient depth, and enough moisture to result in a net warming and moistening of the grid column if the scheme's deep convection were to trigger, as shown schematically in Fig. 10.14.

For the BMJ scheme, first the potential cloud base and cloud top are determined via computation of the level of free convection (LFC) and equilibrium level (EL) in a given grid column. The model computes the properties of hypothetical lifted air parcels from each of the lowest model layers to find the level at which the maximum instability would result from lifting. To save computational expense, the model only lifts parcels from the lower atmosphere, and the depth of this search is a tunable parameter. The exact required cloud depth needed for the scheme to trigger is also adjustable, but it is typically set to 200 mb in depth.

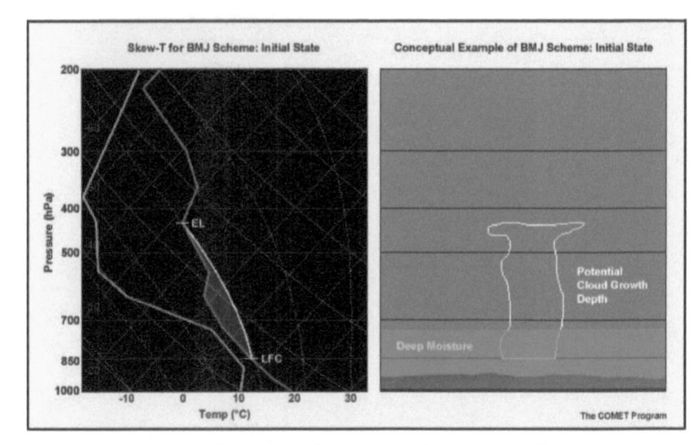


**Figure 10.14.** Schematic diagram of BMJ CP scheme trigger properties: (a) skew $T$–log$p$ diagram showing model sounding and the LFC and EL for a lifted parcel and (b) outline of hypothetical cloud depth and indication of moist layer (courtesy of the COMET program).

If sufficient cloud depth and instability are present, then the model computes postconvective reference profiles, as depicted in Fig. 10.15 in dark blue lines. These profiles are based on points fixed at the cloud base, freezing level, and cloud top. The profile jogs toward cooler temperatures at the freezing level to roughly mimic the effects of melting-induced cooling in convective clouds, as was evident in the empirical data. The reference dewpoint profile is computed from the temperature reference profile along with the specification of the altitude of lifting to the LCL at these three points, measured in pressure. This is a highly adjustable aspect of the scheme; if the postconvective dewpoint profiles are set to very dry values, then the scheme will generally be more active and more vapor will be removed when the convective scheme triggers.

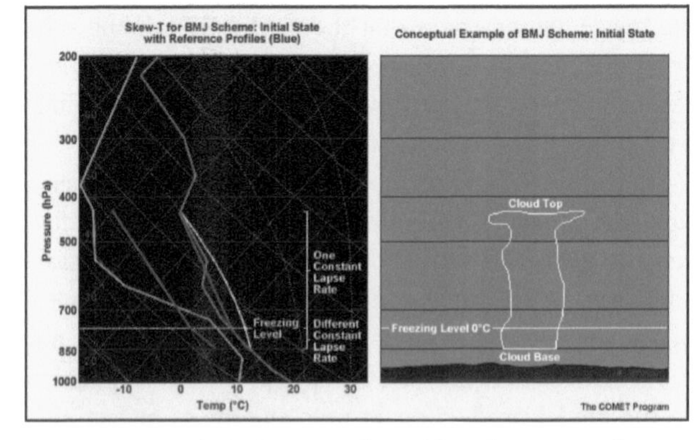


**Figure 10.15.** As in Fig. 10.14, but adding the computed postconvective reference profiles on the skew $T$–log$p$ diagram at left (dark blue lines; courtesy of the COMET program).

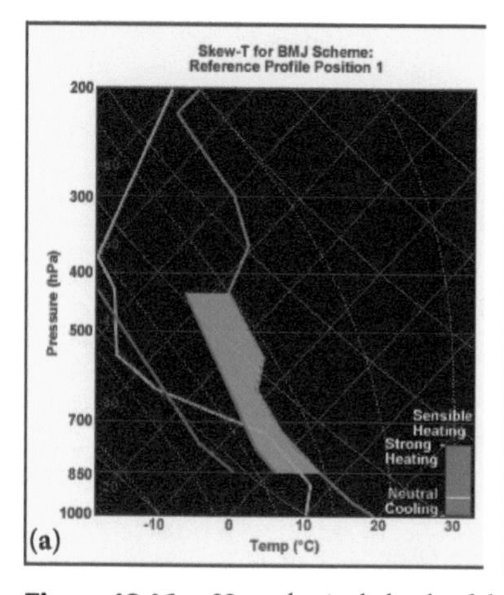


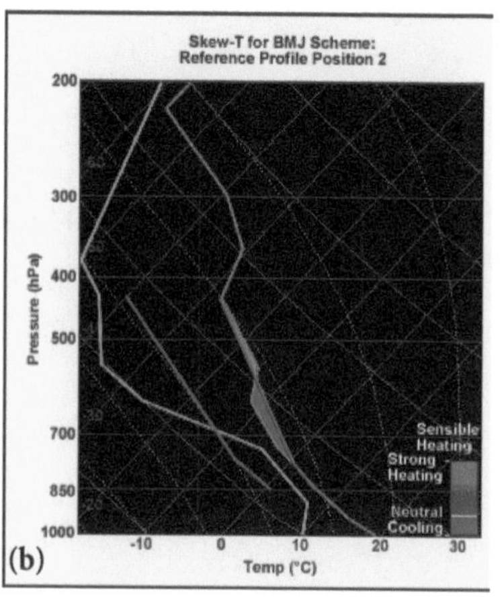


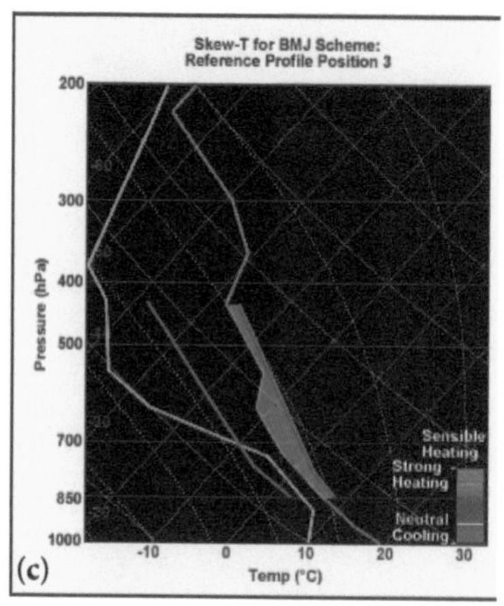


**Figure 10.16.** Hypothetical check of the BMJ reference profiles to determine if net warming and drying can result from adjustment toward the reference profiles: (a) profiles set too cold, where net warming would not result; (b) profiles set to allow net warming and drying; (c) profiles set too warm to allow net drying (courtesy of the COMET program).

However, there is still one other hurdle that must be cleared before the BMJ CP scheme will trigger deep convection. While holding the shape of the temperature and moisture reference profiles fixed, the scheme shifts the profiles to seek a position where *net warming and drying* can be obtained by adjusting the profiles as shown in Fig. 10.16. If no position can be found where a net warming and drying can be achieved with triggering of deep convection, then this portion of the scheme will not trigger, although a separate, nonprecipitating shallow convection component may still activate (discussed below).

Provided that the reference profiles can be defined in a manner that results in net warming and drying when the model gridpoint sounding is adjusted by deep convection, then the corresponding amount of precipitation is generated in that grid cell, with appropriate changes to the temperature and moisture profiles passed back to the model. Thus, the scheme is more easily able to trigger in moist environments, where this *enthalpy constraint* (the requirement for net warming and drying) can more easily be met.

When the BMJ scheme is active for a sufficient period of time, the "signature" of the reference profiles can be evident in model forecast soundings, as shown in Fig. 10.17. The layer over which the scheme has been active is typically characterized by a lack of vertical structure, exhibits cooling near the freezing level, and shows

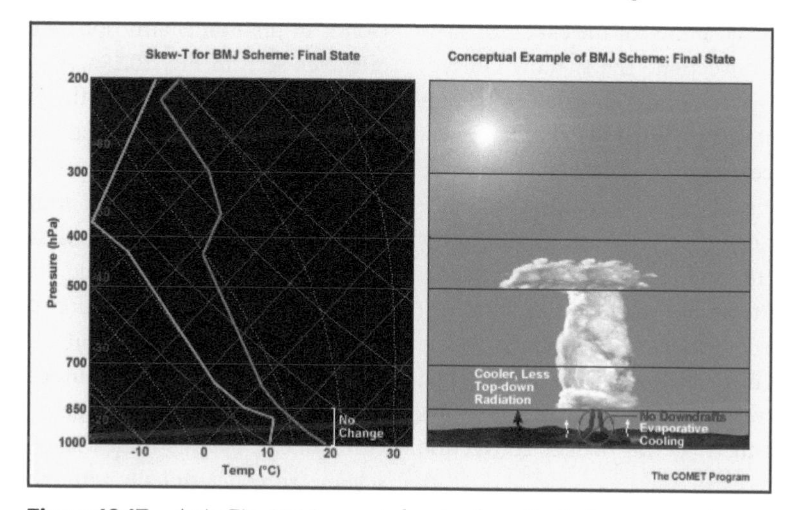


**Figure 10.17.** As in Fig. 10.14, except showing hypothetical postconvective environment after triggering of the BMJ CP scheme (courtesy of the COMET program).

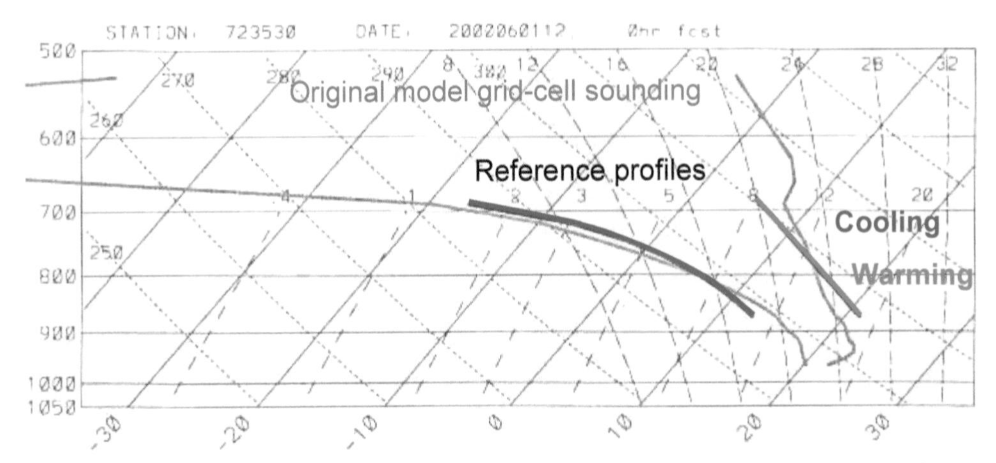


**Figure 10.18.** Model sounding valid 1200 UTC 1 Jun 2000 at KOKC in skew *T*–log*p* format, with BMJ shallow convection reference profiles (red and blue) superimposed (from Baldwin et al. 2002).

little change with time beneath cloud base. The BMJ scheme does not account for convective downdrafts, so the lowest portions of the sounding do not cool as much as one might expect in a postconvective environment. However, when the scheme is active, it can pass this information to other model components such as the radiation scheme to account for partial cloudiness. This can result in surface cooling in the model atmosphere indirectly by reduced surface solar heating. Also, some studies have found that cooling above the surface can produce negative buoyancy there, with cooler air descending to the surface from above (Bukovsky et al. 2006). This only takes place in special circumstances and is not necessarily a realistic representation of convective outflow.

The BMJ scheme also accounts for the effects of shallow, nonprecipitating convection. The shallow convective component can trigger when the enthalpy constraint mentioned above is not met or when the cloud depth is too shallow to allow deep convection to trigger. In these cases, the model scheme will again determine reference profiles but limit their depth to ~200 mb, as shown for an example in Fig. 10.18. The shallow convection component of the BMJ scheme has undergone revision in recent years to reduce its activity; in situations with a capping inversion, the BMJ shallow mixing scheme had been found to aggressively eradicate the cap, sometimes resulting in the unrealistic triggering of deep convection (Baldwin et al. 2002).

One of the issues that numerical models had with the erosion of cold-air damming, as discussed in section 8.3, was that the model would prematurely erode the cold dome through the action of the shallow mixing scheme. For the CAD case study presented in chapter 8, model forecast soundings did not match the observed sounding at Greensboro, North Carolina, very closely (Figs. 8.26d, 10.19a). Comparisons of model simulations using the unaltered BMJ scheme with those using a version in which the shallow mixing was disabled demonstrated that the shallow scheme was in fact responsible for excessive drying and mixing in the inversion layer (Fig. 10.19b).

At the time of the CAD event presented in chapter 8, the operational version of the NCEP Eta Model also had a problem with cloud–radiation interactions in the lowest portion of the atmosphere; a combination of shallow mixing and excessive transmission of solar radiation through clouds in the model atmosphere contributed to the discrepancy seen in Fig. 10.19a. The cloud–radiation issue was addressed in updates to the NCEP NAM model, although the shallow mixing issue may remain. However, it is important to emphasize that, when the shallow mixing scheme is disabled, excessive stratus can develop in the model atmosphere; this scheme is needed for a realistic representation of shallow cloud processes.

The possibility of interaction between the model CP and PBL schemes is apparent for situations like those presented here. As discussed in section 10.4.1, the Yonsei University PBL scheme includes a formulation for moist entrainment, but the local Mellor-Yamada-Janjic (MYJ) scheme, run operationally with the BMJ CP scheme, does not. Thus, one must be careful in selection of compatible physics choices when configuring a numerical model. The BMJ CP scheme works well with the MYJ PBL scheme in

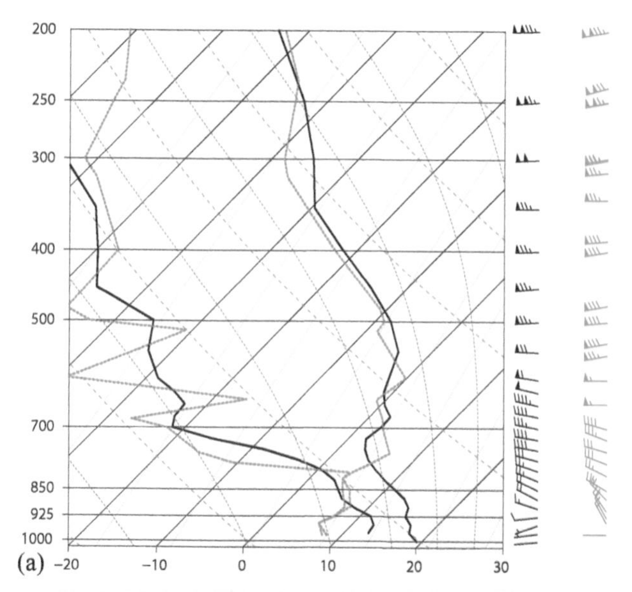


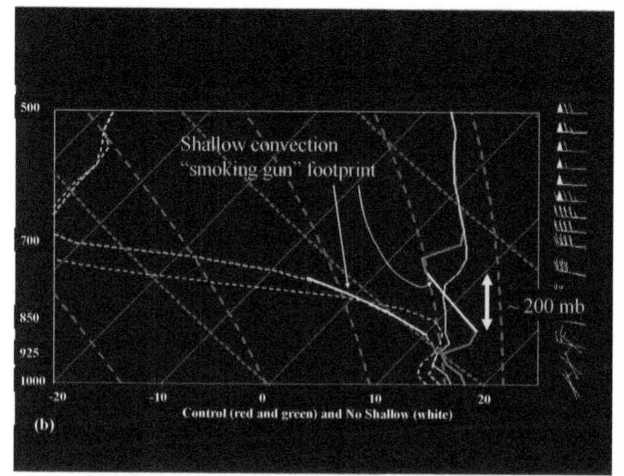


**Figure 10.19.** Soundings for Greensboro, North Carolina (KGSO), during the CAD event presented in chapter 8, valid 0000 UTC 31 Oct 2002: (a) KGSO observed sounding (red) and operational Eta Model 36-h forecast sounding (black) and (b) comparison of retrospective 36-h model forecasts with the full BMJ scheme (red and green) and with the BMJ shallow mixing scheme disabled (white contours).

that shallow mixing is accounted for only by the former scheme.

To summarize, the BMJ scheme has several strengths, including (i) low computational expense due to the straightforward adjustment approach, (ii) good performance in moist environments and for summertime afternoon showers, and (iii) efficient drying and stabilization of the atmosphere to preclude the development of unrealistic grid-scale convection. Disadvantages of this scheme are (i) the neglect of cooling due to downdrafts, (ii) the inability to trigger in relatively dry environments, (iii) difficulty handling capped environments, and (iv) undesirable influences of the shallow convection component in some situations.

**10.4.4.3. Mass-flux schemes.** Another philosophy for convective parameterization is the *mass-flux* approach, in which a one-dimensional cloud model is run within a grid cell to compute updraft and downdraft mass fluxes, allowing inclusion of more complex processes such as entrainment and detrainment. Two popular schemes of this type are the Arakawa–Schubert (AS) scheme, a variation of which is used in the NCEP GFS model, and the Kain–Fritsch (KF) scheme, available in many research models such as the WRF and run in some ensemble members at NCEP.

**Arakawa–Schubert scheme.** There are several different versions of the Arakawa–Schubert scheme, used in a wide variety of models from the operational GFS global model to climate models. Because specific details vary with the model, only general characteristics will be discussed here. For the GFS model, the AS scheme is somewhat unique in that it adjusts momentum in addition to temperature and moisture; in this sense, it differs from many implementations of the BMJ and KF schemes, although the latter can be configured to include momentum adjustment.

To trigger, the AS scheme requires environmental instability [convective available potential energy (CAPE)] and also that the large-scale environment destabilize with time (this could be due to solar heating, cold advection aloft, or other processes). An equilibrium state is assumed in which some instability remains and is consumed by the scheme at a rate proportional to that at which the large-scale environment is destabilizing. This strategy is perhaps best suited to large grid cells and in the tropics. In continental, midlatitude convection, such an equilibrium may not exist, because CAPE can be consumed rapidly by strong, organized convection; the AS scheme is not designed to optimally represent such a situation.

An important aspect of the AS scheme is that it represents convection as an *ensemble of clouds* of different depths (Fig. 10.20). It accounts for entrainment and detrainment, as well as downdraft cooling and compensating subsidence. A one-dimensional cloud model is run for hypothetical clouds of differing depths, and the thermodynamic

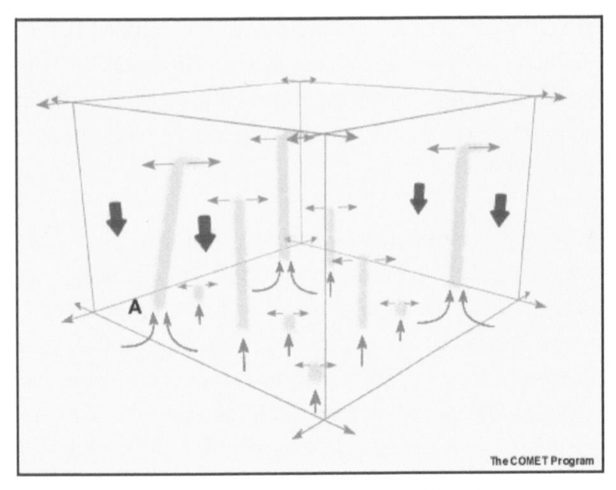


**Figure 10.20.** Schematic representation of the AS CP scheme, showing an ensemble of convective clouds, compensating subsidence, detrainment, and outflow aloft (courtesy of the COMET program).

changes that take place within the entire grid cell are computed accordingly.

Because more explicit information is available as to the source layer of condensate, this scheme is capable of accounting for evaporation in downdrafts, accounting for detrained moisture to the grid-scale environment aloft, and feeding hydrometeors back to the model microphysics scheme (Fig. 10.21). Complex structure from the original model thermodynamic profile can be preserved in the postconvective sounding (Fig. 10.21c).

The GFS model had experienced some problems with excessive development of tropical cyclones, and these warm-core systems were in part due to latent heat release in the AS scheme. So, partly to correct this problem, *momentum adjustment* was added to the AS scheme in the GFS model (Han and Pan 2006). In a warm-core cyclone, vertical momentum adjustment weakens the cyclone and

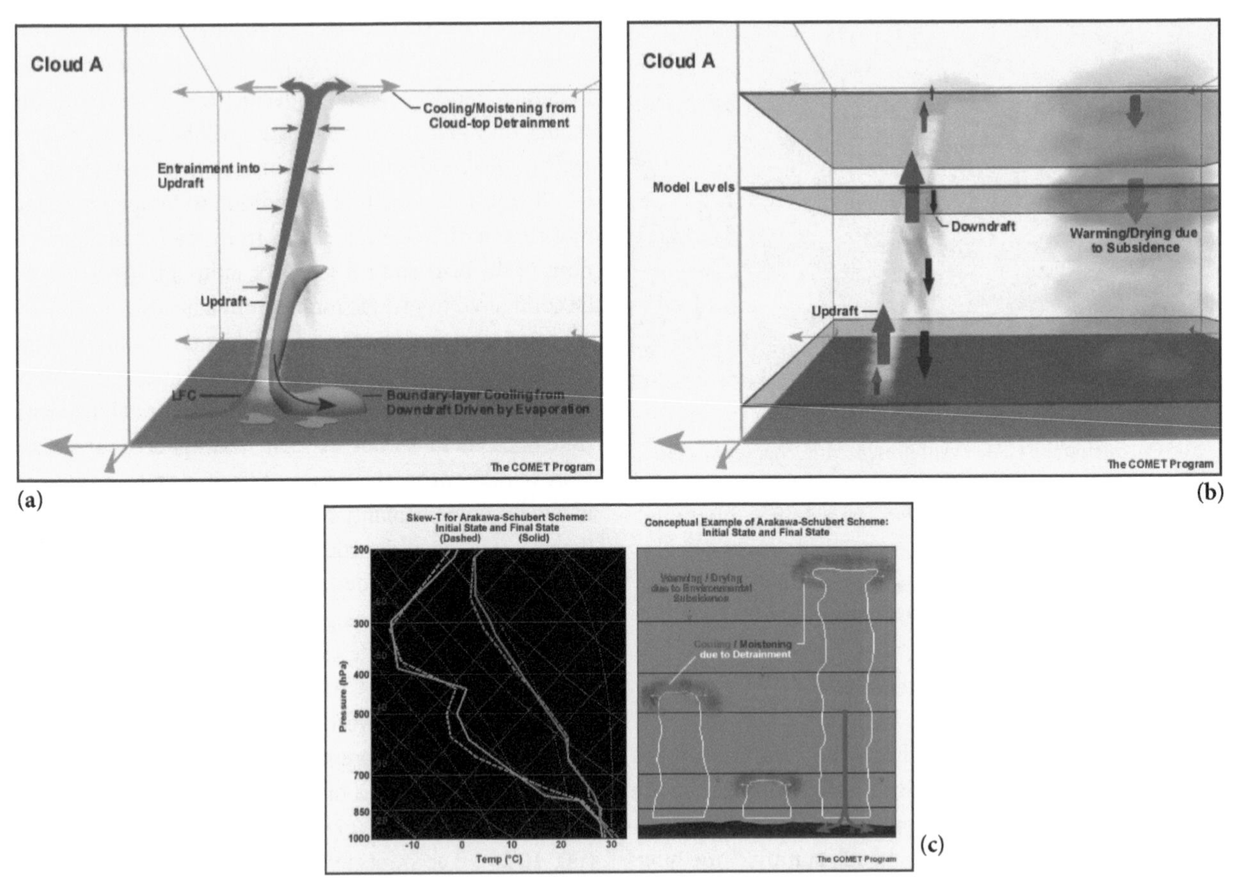


**Figure 10.21.** Schematic description of physical process representation in the AS CP scheme: (a) processes accompanying a deep convective cloud; (b) changes within the grid cell in addition to the convective cloud; and (c) adjustment of model forecast sounding as scheme triggers (courtesy of the COMET program).

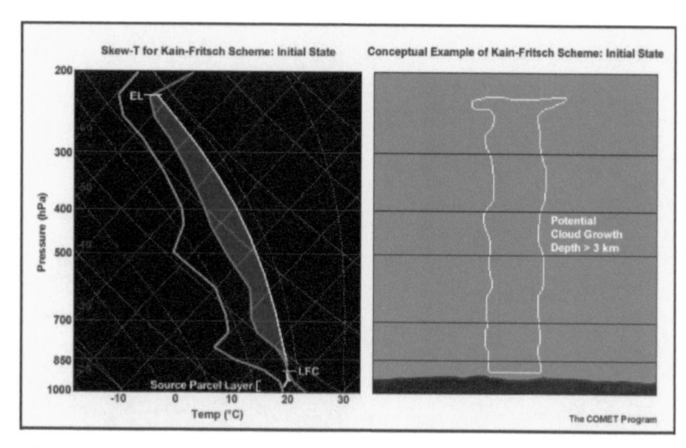


**Figure 10.22.** As in Fig. 10.14, but an idealized representation for the KF CP scheme, indicating hypothetical cloud depth in the preconvective state (courtesy of the COMET program).

resulted in a reduction in the number of spurious tropical cyclones in the GFS model.

Overall, this scheme is capable of handling a variety of complex processes, and it retains pronounced vertical structure in the thermodynamic profiles. The behavior of this scheme is highly implementation dependent, and it is hard to make general statements about its performance. It is more computationally expensive than the BMJ scheme, but it is also capable of representing downdrafts, interacting more directly with the model microphysics scheme, and handling situations with a capping inversion. Disadvantages are that the scheme is not aggressive, and sometimes the model atmosphere can destabilize fast enough to result in grid-scale convection. It is likely better suited to larger grid lengths, as found in global and climate models.

**The Kain–Fritsch scheme.** The KF scheme is designed for smaller grid lengths (e.g., between 10 and 20 km) and for midlatitude, continental convection. This parameterization builds on a predecessor scheme by Fritsch and Chappel that was originally developed for use in higher-resolution mesoscale models, with grid lengths on the order of 20 km. These schemes compute the updraft and downdraft mass flux within a grid cell and account for many storm-scale processes in a more direct fashion than do adjustment schemes, while being more suitable for higher-resolution simulations than the AS scheme. For example, the KF scheme accounts for entrainment in updrafts, storm outflow, and detrainment of hydrometeors to the grid scale.

The KF scheme triggers when there is CAPE and sufficient cloud depth (Fig. 10.22) but also when there is enough lift to potentially elevate air parcels to the LFC. This last condition is based in part on the model grid-scale vertical motion field to see if sufficient environmental ascent is present. If the triggering condition is met, then the scheme initiates deep convection, and the updraft and downdraft properties within the grid cell are computed.

Unlike the AS scheme, the KF scheme assumes clouds of a single depth but specifies entrainment and detrainment at different vertical levels below the cloud top (Fig. 10.23). The scheme can provide hydrometeors to the microphysics scheme, which allows more realistic representation of thunderstorm anvils, for example, than would otherwise be possible. Another difference with the AS scheme is that the time scale for convective adjustment is not tied to the destabilization rate in the large-scale environment but instead to a convective time scale, over which the scheme will effectively eliminate the available CAPE (Fig. 10.24).

The KF scheme is a popular choice for mesoscale models running at grid lengths that are too large to make explicit convective representation feasible. It accounts for many mesoscale processes in a realistic fashion, including convective cold pool generation, detrainment of hydrometeors to the microphysics scheme, compensating subsidence, and mesoscale triggering in capped environments. Because the trigger in the KF scheme uses the grid-scale vertical motion, a more direct linkage between synoptic and mesoscale processes takes place. Drawbacks to this scheme are increased computational expense and

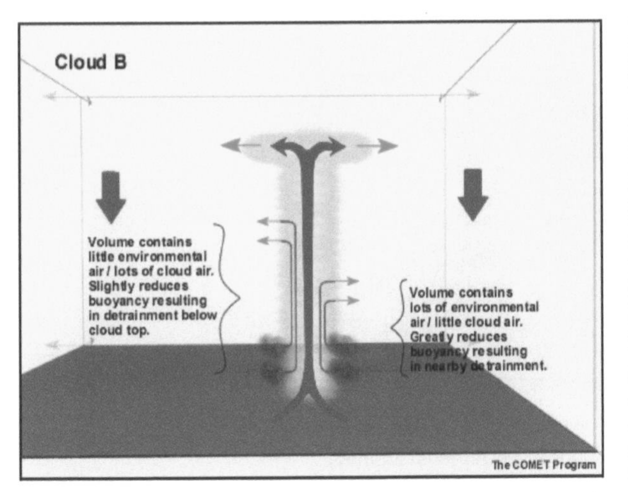


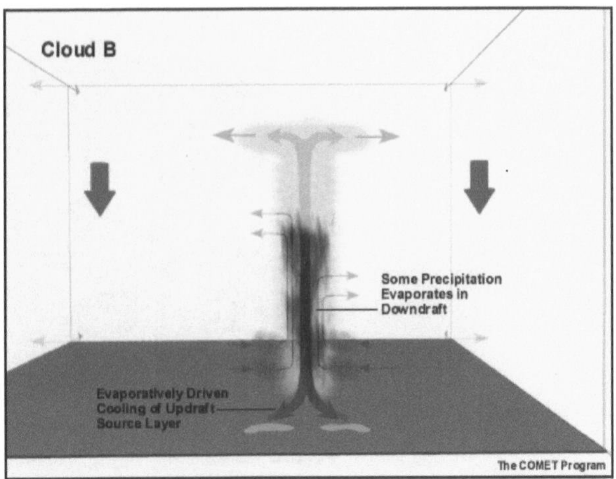


**Figure 10.23.** Schematic representation of processes active during the triggering of the KF CP scheme, as indicated (courtesy of the COMET program).

sometimes unrealistically moist layers in the postconvective environment.

**10.4.4.4. Explicit convection.** Continual increases in computing power will very likely be accompanied by continued reduction in the grid lengths used in operational NWP models. Below approximately 4-km grid spacing, it is advisable to turn off the model CP scheme, at least for deep convection (e.g., Weisman et al. 1997). Several recent and ongoing experiments, such as the Spring Experiment held in Norman, Oklahoma, at the National Severe Storms Lab (NSSL) and Storm Prediction Center (SPC), have shown promising results from high-resolution, explicit convection NWP models (e.g., Kain et al. 2003a,b; Weisman et al. 2008).

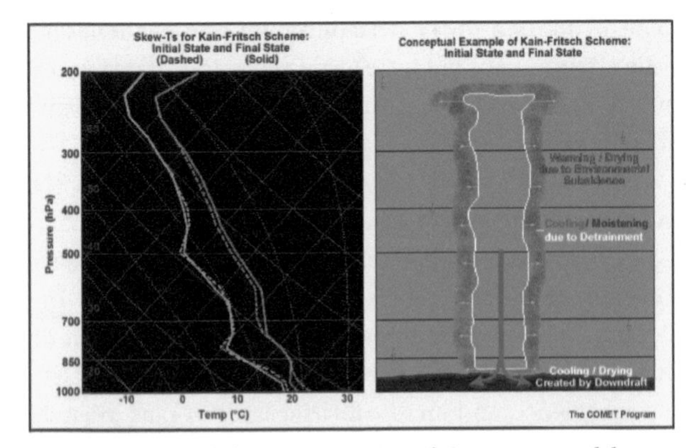


**Figure 10.24.** (left) Representation of changes to model sounding due to KF CP scheme and (right) description of the physical processes leading to the changes (courtesy of the COMET program).

Part of the justification for omitting the CP scheme with sub-4-km grid spacing is because CP schemes have not generally been designed to represent convection with such small grid cells. Another reason is that, for strong convective systems exhibiting some organization, the model at these grid lengths is able to explicitly capture enough of the mesoscale structure of the system to generate a somewhat credible storm in the model atmosphere. However, many caveats are necessary, as discussed below.

Most CP schemes are designed to represent the idealized situation depicted in Fig. 10.25a, where convective clouds are much smaller than the model grid cell, and a clear scale separation exists between these cloud systems and what the model is able to resolve. For organized convection, the situation is more complex, although for a large grid cell the convective complex is still unresolved (Fig. 10.25b). However, as the model grid length decreases to, say, 12 km, some grid cells are contained completely within the convective system, and grid-scale saturation can occur; the system is partially resolved, although not likely to exhibit the same mesoscale structure found in nature (Fig. 10.25c). In this case, the characteristics of the CP scheme would determine the extent to which the model is able to resolve the convective system on the model grid. For a CP scheme such as the BMJ, removal of vapor would likely preclude gridcell saturation, and the system might remain mostly implicit. For the KF scheme, which passes hydrometeors back to the grid scale, it is more likely that grid-scale saturation would take place, possibly leading to the development of

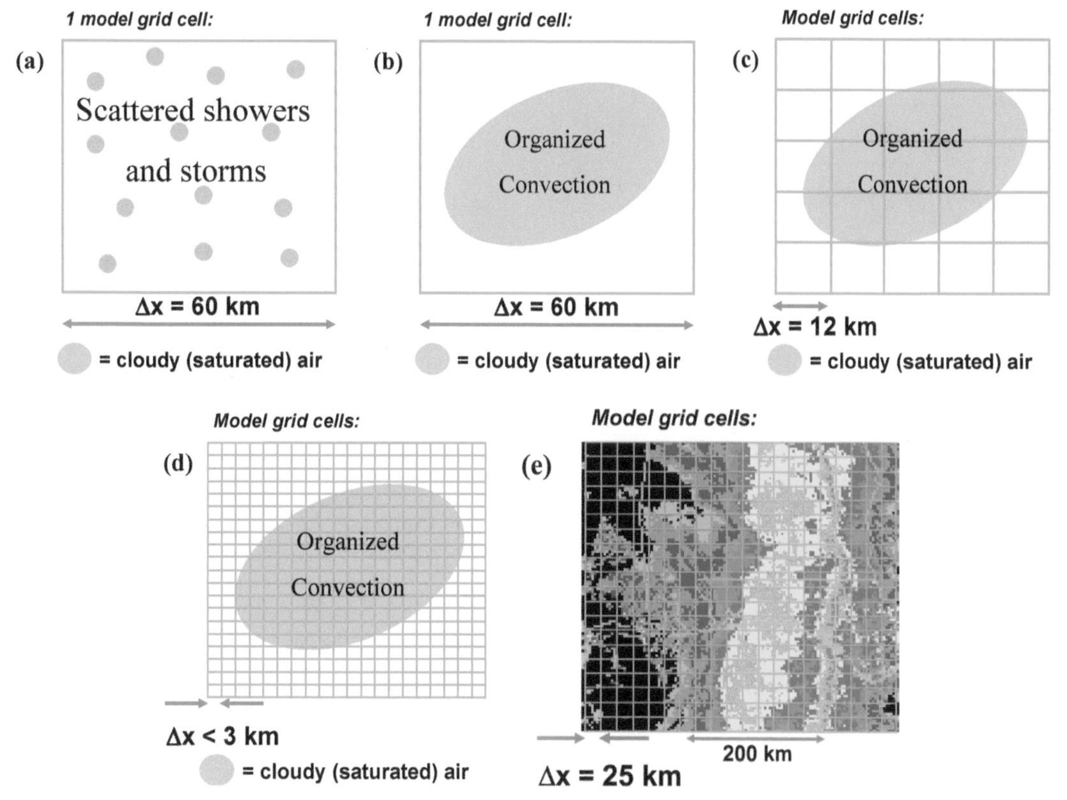


**Figure 10.25.** Idealized schematics of model grids with various spacing and convective systems (graphic adapted from example by Jason Millbrandt, Environment Canada).

a stratiform precipitation region to be represented on the model grid scale.

As the grid length is decreased below 4 km, the model is able to handle the convection entirely on the grid scale, albeit without the small-scale convective motions characterizing natural convection (Fig. 10.25d). For example, the processes of entrainment and detrainment at the cloud boundaries would not be represented unless other model components were active. In this situation, a model diffusion or TKE-based turbulent mixing scheme would be needed to mimic this effect, and some PBL schemes may represent this type of turbulent mixing.

As discussed for the BMJ scheme, most CP schemes include a component to represent shallow mixing, such as might accompany cumulus or stratocumulus clouds. When running without a CP scheme, care must be taken to ensure that this process is somehow represented, either by retaining the shallow component of the CP scheme or by the PBL scheme (for instance, the YSU PBL scheme). An example, from simulations of a MCS event over the southeastern United States on 1 April 2005 with the WRF model, is presented in Fig. 10.26. The areal coverage of cloud water is much greater in the explicit convection simulation relative to a simulation that employed the BMJ CP scheme; satellite imagery for this time (not shown) suggests that the BMJ simulation was more realistic.

**10.4.4.5. Forecast sensitivity to CP scheme choice.** An example will now be provided that illustrates the type of impact that the choice of CP scheme can have on the synoptic-scale forecast. On 17 February 2004, a complex and challenging weather forecasting scenario was present across the southeastern United States and mid-Atlantic region. A 12-h forecast from the NAM model indicated a classical CAD setup, with a vigorous, digging, negatively tilted trough over the southern United States (Fig. 10.27a). As this trough amplified over the south-central United States, coastal cyclogenesis was expected offshore of the Carolinas (Fig. 10.27b).

With subfreezing air in place, the potential existed for a major coastal cyclone and mid-Atlantic snowstorm. Furthermore, comparison of short-term model forecasts with

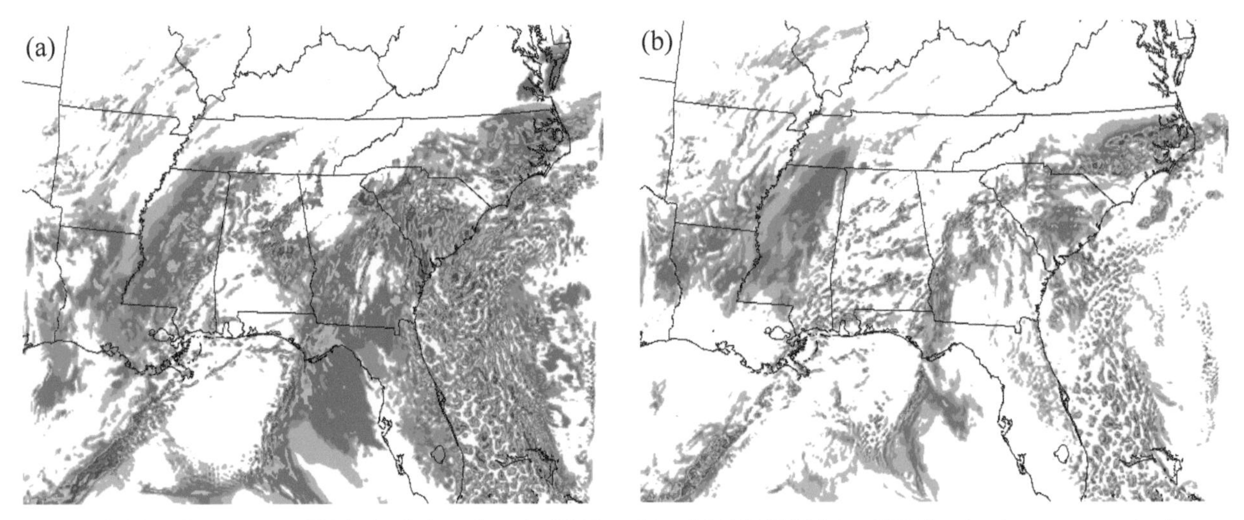


**Figure 10.26.** Comparison of 30-h simulations from the WRF model valid 1800 UTC 1 Apr 2005 showing layer-average cloud water in the 800–950-mb layer: (a) without CP scheme (explicit convection) and (b) with BMJ CP scheme.

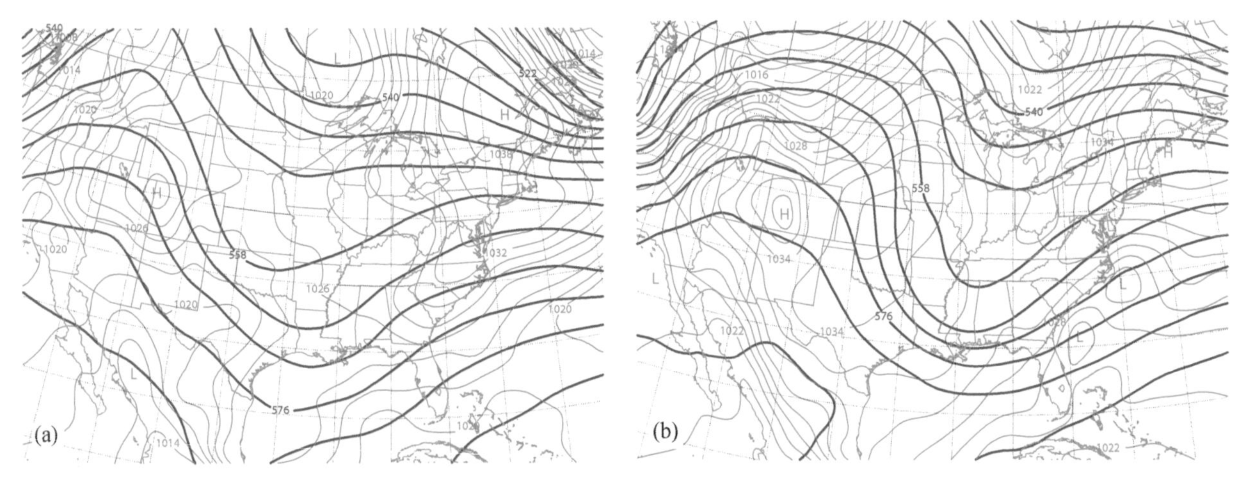


**Figure 10.27.** Forecasts from the NCEP Eta Model depicting 500-mb height (dam; blue contours) and sea level pressure (mb; red contours): (a) 12-h forecast valid 0000 UTC 17 Feb 2004 and (b) 30-h forecast valid 1800 UTC 17 Feb 2004.

upstream upper-air observations indicated that the strength of the jet on the western side of the upper trough was greater than what was indicated in the model initial condition, suggesting a greater potential for trough amplification than what operational models were indicating (see Mahoney and Lackmann 2006 for further discussion of this case).

Examination of model sea level pressure forecasts, superimposed with convective precipitation (from the BMJ CP scheme), revealed that the cyclone centers in the model forecast were colocated with maxima of convective precipitation and were therefore likely to be dependent on details of the model CP scheme (Fig 10.28).

To analyze the CP influence in this case, the forecast was rerun, using the workstation version of the then-operational NCEP Eta Model. The initial conditions, grid spacing, and all model configurations were identical, except that one ran the operational BMJ CP scheme and the other utilized the KF CP scheme. The 24-h forecasts from these two runs are shown in Figs. 10.29a,b.

Rather than the two distinct cyclone centers seen in the run using the BMJ scheme, the KF run maintains an elongated, inverted trough structure, consistent with a coastal front. Precipitation extends farther inland across northeastern South Carolina and North Carolina in the KF run relative to the BMJ run, which was more consistent with the radar mosaic from that time (Fig. 10.29c). However, both model simulations overpredicted inland precipitation, and the event, in the end, produced very

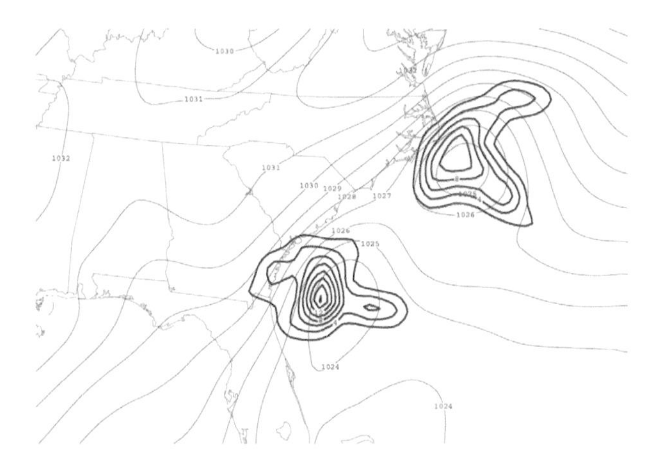

**Figure 10.28.** Operational Eta Model 30-h forecast of sea level pressure (contour interval 1 hPa) and 6-h convective precipitation (thick black contours every 0.1 in.) valid 18 UTC 17 Feb 2004.

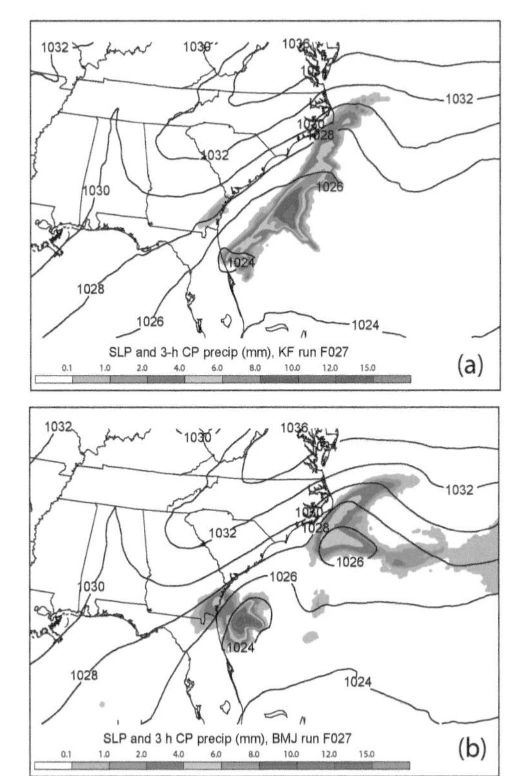


**Figure 10.29.** Comparison of sea level pressure (contour interval 2 mb) and 3-h precipitation [mm; shaded as in legend at bottom of (a) and (b)] for 27-h Workstation Eta Model forecasts valid 1500 UTC 17 Feb 2004: (a) BMJ run and (b) KF run. From Mahoney and Lackmann (2006).

little in the way of snowfall as major cyclogenesis took place well to the northeast of the region. The point of this example is not to evaluate these two simulations against observations, but rather it is to illustrate the type of influence that a CP scheme can exert on synoptic-scale forecasts.

It is evident from Fig. 10.29 that the choice of CP scheme seems to affect the character of the coastal front between these two simulations; this is more clearly evident in Fig. 10.30. The KF run produces a more intense and linear coastal front that is aligned with the coast, whereas the BMJ run produces a more diffuse frontal zone, distorted by the cyclonic flow associated with the two closed cyclone centers (Fig. 10.30).

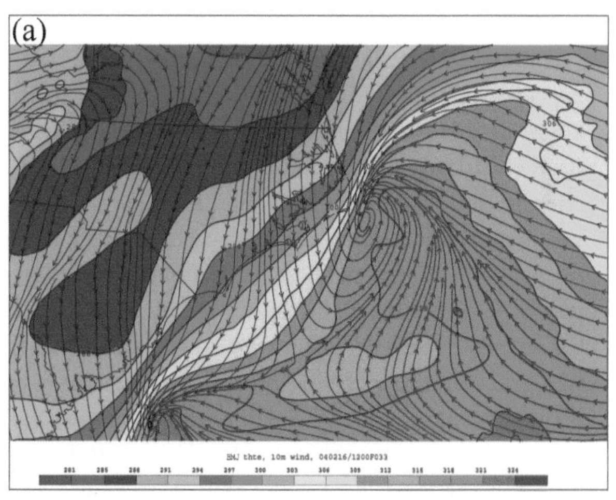


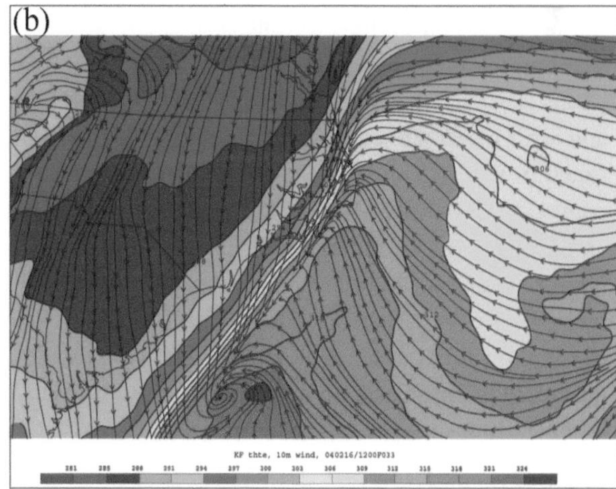


**Figure 10.30.** Comparison of 33-h BMJ and KF simulations for coastal front region of the Carolinas valid 2100 UTC 17 Feb 2004, showing 1000-mb equivalent potential temperature (color shaded as in legend) and 10-m wind (streamlines): (a) BMJ simulation and (b) KF simulation (from Mahoney and Lackmann 2006).

Forecasters examining model output in a situation of this type may not recognize that the surface cyclone and coastal front intensity and location in model forecasts could be so strongly influenced simply by the choice of the model CP scheme. In part to provide more robust guidance in situations of this type, NCEP runs its Short-Range Ensemble Forecasting System (SREF), in which some ensemble members use a variety of different CP schemes and other model and physics configurations. The SREF is designed to quantify uncertainty in features of the model output that are sensitive to physics choices such as CP scheme. When the details of a given forecast are sensitive to the CP scheme choice, an ensemble approach is a useful way to discern the correct level of forecast confidence in numerical forecasts.

Forecasters should also recognize the distinction between grid-scale and convective-scheme precipitation, and plotting these separately can provide useful insight into the workings of the model. The QPF from the grid-scale precipitation scheme is fundamentally different from that of the CP scheme, partly by design. For a model grid cell that is 50 km on a side, it is recognized that active convection within the grid cell would produce some localized regions of heavy rainfall, with other locations in the same grid cell area receiving no precipitation. However, for the model QPF, the convective precipitation is *distributed evenly* across the grid cell, as it must be. This does not mean that the model was really predicting a light, even rainfall across the entire grid cell!

Thus, CP precipitation for summertime showery situations is expected to show wider coverage than observed and with much smoother spatial variation. This is illustrated in Fig. 10.31, which shows summertime afternoon convection over the southern United States in satellite and radar imagery. Cellular structure is evident in the observations, but a uniform coverage of precipitation is produced by the model CP scheme. This is not an error in the model, because it is not possible for the model to distribute the precipitation in any other way, but users of model output should be aware of this distinction.

### 10.4.5. Other physical parameterizations

There are additional parameterization packages in numerical models, including algorithms for radiative transfer, diffusion, gravity wave drag, and atmospheric chemistry. Models used for the prediction of climate are coupled ocean–atmosphere models, and some models run for short-term forecasts are also coupled to wave and ocean models, including the Hurricane WRF (HWRF) model used in short-term hurricane prediction. Today's students will be expected to possess knowledge of numerical models of increasing complexity as the level of model sophistication continues to advance.

It is beyond the scope of this book to cover all of the details of the different model physics schemes, and their interactions. This is a difficult topic because of the constant evolution of these model packages. An excellent text authored by Stensrud (2007) is available for in-depth discussion of the physical parameterizations in models. The COMET MetEd modules, available online, are an outstanding source of information on current operational model configurations as well as background material on how different parameterization schemes operate (http://www.meted.ucar.edu/).

## 10.5. DATA ASSIMILATION

As discussed in sections 5.4 and 10.1, Bjerknes (1904) identifies "sufficiently accurate knowledge of the state of the atmosphere at the initial time" as one of the two fundamental requirements for numerical weather prediction. Indeed, one of the early barriers to numerical prediction was a lack of observations, especially for the upper air, leading Bjerknes and his collaborators to lobby extensively for expansion of meteorological observing and reporting. To test numerical forecasting procedures, special field campaigns were required to secure adequate observations for the atmospheric initial conditions, as well as for post-forecast evaluation. Noise and errors in the initial conditions of Richardson's first published NWP forecast have been shown to be the most important error source in this historic computation, which exhibited a substantial error (e.g., Lynch 1992, 2006).

A number of events that led to the advent of operational NWP took place in rapid succession during the 1930s and 1940s, including the development of the rawinsonde network, explosive growth of the aviation industry, World War II, and the invention of electronic computers. For the earliest numerical predictions, the initial conditions were based on laborious manual analyses used to generate gridded fields that could be used by models. As early as 1951, Charney (1951) recognized the need to automate and improve the means by which initial conditions were obtained. Early *objective analysis* methods were used to generate regularly spaced grids from unevenly

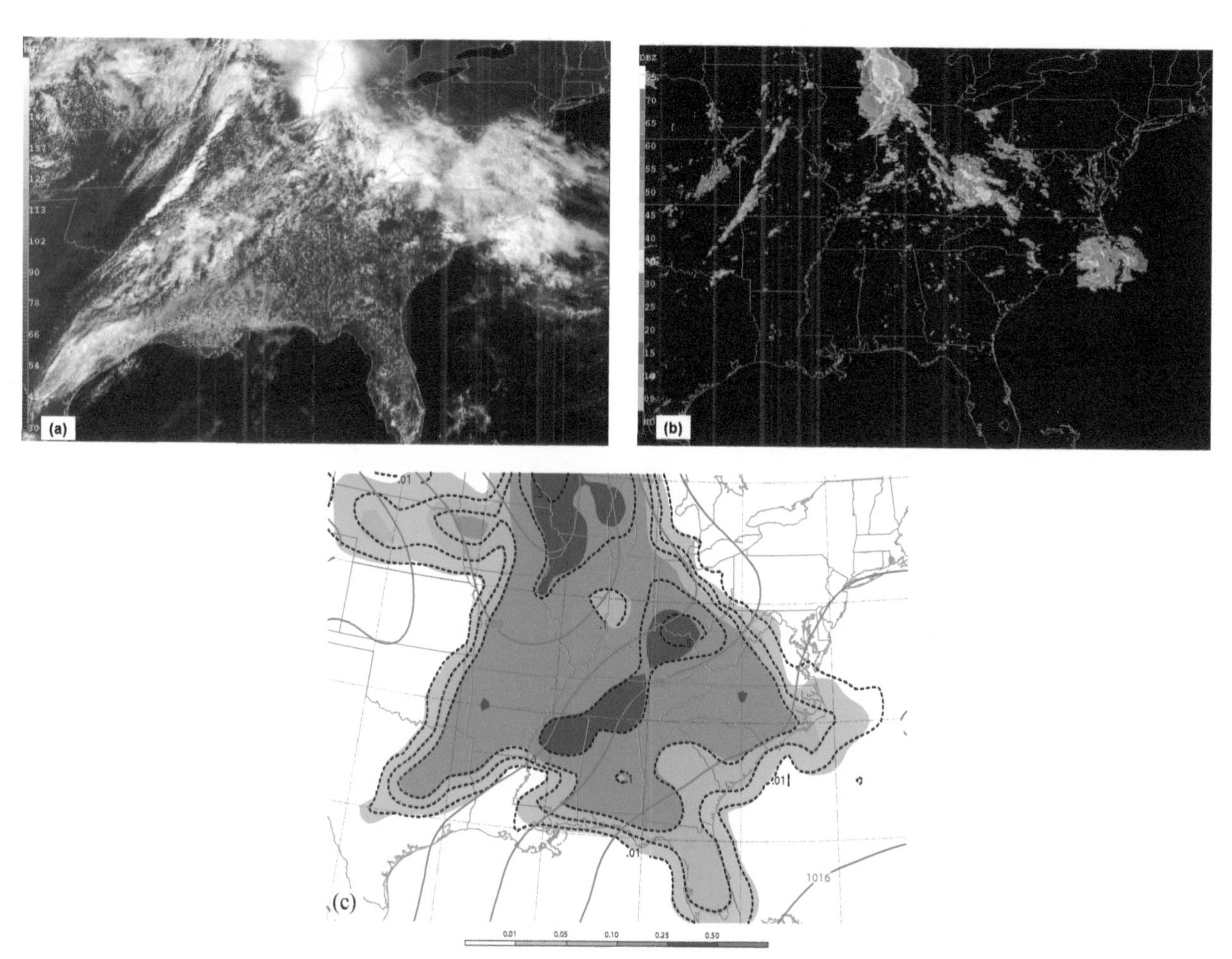

**Figure 10.31.** Examination of model precipitation forecast for 30 May 2004, a day characterized by convective showers: (a) visible satellite image valid 1815 UTC 30 May; (b) radar mosaic valid 1815 UTC 30 May; and (c) 6-h GFS model forecast precipitation (inches, contoured and shaded as in legend at bottom of panel) ending 0000 UTC 31 May 2004. Overlapping blue and green contours indicate that the model precipitation was almost entirely due to the CP scheme.

distributed observations. Due to insufficient data coverage, especially over oceanic regions, the observations were later combined with climatological fields or short-term model forecasts to improve the initial conditions. These early techniques have grown tremendously in sophistication, and today the process of *data assimilation* has developed into an important (and still growing) discipline within the atmospheric and oceanic sciences.

Simply put, modern data assimilation techniques provide *a means of combining all available information to construct the best possible estimate of the state of the atmosphere*. The information provided by the DA process extends beyond the generation of model initial conditions; in fact, understanding of atmospheric behavior and predictability has benefited tremendously from the development of powerful DA techniques. Modern DA methods have been applied retrospectively to archived data to generate relatively consistent long-term analyses of record, known as "reanalyses," which have widespread value in climatological studies as well as for retrospective case studies. The advance of DA techniques has taken place in conjunction with rapid growth in the field of ensemble prediction and, as will be shown subsequently, these topics are closely linked.

The organization of this section will start with some discussion of the observations themselves, in part to demonstrate the diversity and complexity of modern observing systems. Then, a brief overview of some current operational DA techniques is presented, leading to a discussion of ensemble prediction section 10.6.

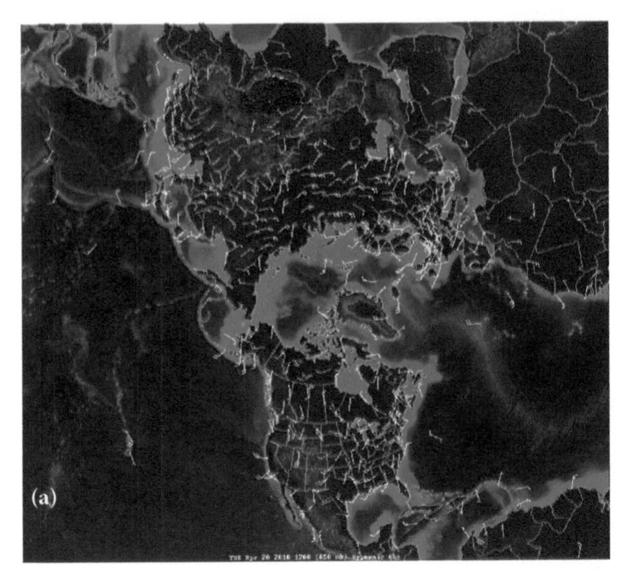


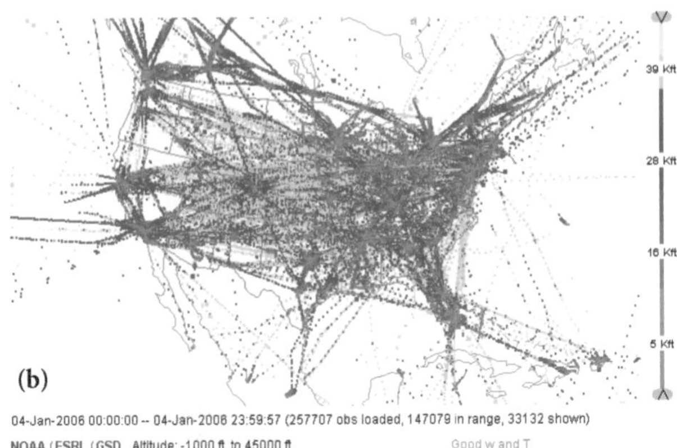


**Figure 10.32.** (a) Sample of 850-mb rawinsonde wind data (barbs) for 1200 UTC 20 Apr 2010; (b) sample of aircraft observations for 24-h period beginning 0000 UTC 4 Jan 2006. The observations are coded by altitude, as indicated in legend at right. Panel (b) courtesy NOAA.

#### 10.5.1. A brief summary of some observational platforms

To illustrate the wide variety of meteorological measurements that are utilized in the data assimilation process, a brief summary of some observational platforms will be provided here. A discussion of the detailed workings of the instruments and their errors is beyond the scope of this text.

**10.5.1.1. In situ surface observations.** In many nations, conventional surface observations are now largely automated and are generally located at airports around the world to support aviation. The automation has some benefits: for example, allowing continuous reporting in remote locations. However, the automation of the observing network has come at a price, as instrument errors sometimes go undetected and some types of measurements, such as snow depth in the United States, are no longer reliably taken.

In the United States, the Automated Surface Observing System (ASOS) data that are routinely collected and transmitted to global centers are often augmented by human observers employed by the Federal Aviation Administration (FAA). Other automated systems, such as the Automated Weather Observing System (AWOS) are similar but are not augmented by human observers and have less detailed information concerning clouds and precipitation. Data for all surface observing systems is coded into METAR[1] format for transmission, consistent with World Meteorological Organization (WMO) standards.

The U.S. ASOS sensors take measurements at high frequency, but the data are time averaged to filter out high-frequency fluctuations associated with turbulence. Official "sustained" wind speeds are averaged over 2 min, with gusts and peak winds reported as the highest 5-s wind average. Temperatures are averaged over a 5-min period, computed as an average of several sub-1-min averages, but the 5-min average temperature is updated every minute. These details can be important, because, when making a forecast, these are the data against which the forecast is evaluated!

**10.5.1.2. In situ upper-air observations.** Rawinsonde measurements remain an important element of the observational network (Fig. 10.32a), as discussed in section 10.5.3 and by Zapotocny et al. (2000). However, in recent years, in situ data collected by commercial aircraft [commercial aircraft reports (ACARS)] have grown in importance and strongly complement the twice-daily sonde measurements with greater temporal coverage and many upper-level measurements over oceanic regions (Fig. 10.32b). Aircraft takeoffs and landings provide a large number of soundings each day at airports; despite

[1] The term METAR is derived from *MÉTéorologique Aviation Régulière*, loosely translated French for routine meteorological aviation report.

uneven distribution, these data are important to the development of initial conditions used by NWP models (see section 10.5.3).

Until recently, a major limitation of ACARS was a lack of humidity measurements. Recently, new sensors have been developed to accurately measure humidity, and these are being mounted on a growing number of commercial aircraft in the United States. Given the importance of accurate moisture analyses to weather forecasting, these new sensors offer promise of improved forecast accuracy in the coming years (e.g., Benjamin et al. 2010; Moninger et al. 2010).

*Dropsondes* are routinely used to sample tropical cyclones, both environmentally and within the core of the storm itself. Given the general paucity of reliable measurements in the vicinity of tropical cyclones, these data are crucial to determining the intensity and structure of these important weather systems. Flight-level in situ observations from hurricane hunter aircraft operated by the U.S. Navy and the National Oceanic and Atmospheric Administration (NOAA), in addition to aircraft-borne radar systems, are valuable in the analysis of tropical cyclones as well. The development of data assimilation techniques that are especially designed to improve tropical cyclone initial conditions is currently the focus of intensive research efforts.

**10.5.1.3. Satellite measurements.** Current versions of geostationary and polar-orbiting meteorological satellites carry an impressive payload of instrumentation, not all of which is directly related to meteorology (e.g., the search-and-rescue communications antenna). For meteorological applications, the two primary observing systems are the *imager*, which includes the ability to construct visible, infrared, and water vapor measurements and images, and the *sounder*, from which vertical profiles of atmospheric temperature and moisture can be obtained.

For both polar-orbiting and geostationary satellites, an orbit is selected so as to allow force balance between the centrifugal and gravitational forces. Geostationary satellites, which are able to remain above a fixed point on the equator (rotating with Earth) in balanced orbit, must be located at a relatively high altitude of ~36,000 km to achieve this. This provides the advantage of a wide field of view but results in lower-resolution data relative to platforms in lower orbits. Polar-orbiting satellites, not constrained to move with a fixed point on Earth's surface, are often placed in orbits less than 1,000 km in altitude. For such orbits, the satellites have smaller fields of view and higher-resolution products. The polar orbits also allow detailed viewing of high-latitude regions that are unavailable from geostationary satellites.

Geostationary satellites have been used for meteorological purposes in the United States since 1974. The first of the modern Geostationary Operational Environmental Satellite (GOES) satellites, *GOES-1*, was launched in 1975. GOES satellites are designated by a letter until launch, at which point the letter changes to a number. For example, GOES-C became *GOES-3* when it was launched in June 1978. During the spring of 2010, *GOES-13*, which had been in storage orbit, assumed operational duties; the former GOES-O, now *GOES-14*, remains in storage orbit as of this writing.

Many other nations have networks of meteorological satellites. A consortium that exists to provide continuous global coverage includes Chinese, Russian, Japanese, Indian, and European Union satellites along with those of the United States and some other nations.

GOES satellites were first equipped with sounders in 1994, with the *GOES-8* satellite, the first of a series of second generation geostationary satellites. The sounder utilizes a 19-channel sampling strategy that includes 1 visible wavelength band and 18 infrared bands for the purposes of constructing vertical thermodynamic profiles, as well as measurement of cloud-top and surface temperature. By carefully selecting sampling bands in the center as well as on the edges of the spectral absorption bands, sampling from various levels of the atmosphere can provide information about the temperature as a function of height. Information derived from the measurement of upwelling radiation from Earth's surface, in conjunction with radiation emitted by atmospheric carbon dioxide, ozone, and water vapor molecules, is used to derive detailed profile information. Water vapor exhibits highly variable concentrations in space and time, whereas carbon dioxide is relatively well mixed. By measuring radiance in the atmospheric *window regions* (wavelength bands in which the atmosphere is a very weak absorber), the surface or cloud-top temperature can be obtained, which is then used in conjunction with the other measurements to compute thermodynamic profiles.

The spectral measurement bands selected with respect to carbon dioxide are used to retrieve the temperature

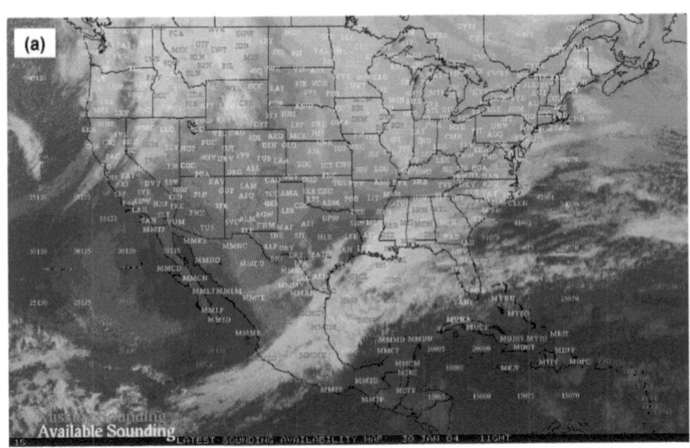

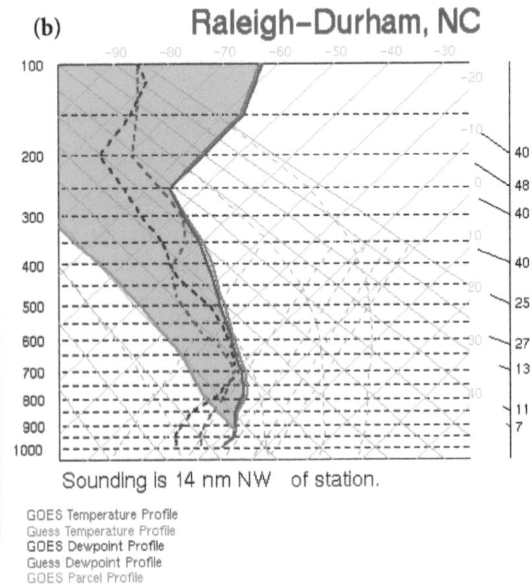


**Figure 10.33.** Sample of GOES sounder-derived temperature and moisture profiles for 19 Apr 2010: (a) plan map showing locations of available (unavailable) soundings in yellow (red) letter identifiers and (b) sample sounding for Raleigh–Durham, North Carolina (KRDU), in skew T–log*p* format. Plot includes both the model "guess" profiles as well as the actual GOES profiles, as indicated in legend at bottom. Graphics obtained from http://www.osdpd.noaa.gov/skewt/html/skewhome.html.

profile, whereas the channels in spectral proximity to the water vapor absorption bands are used to determine the moisture profile. There are several fundamental differences between satellite-derived soundings and those obtained from rawinsondes. The satellite sounder measurements correspond to a finite layer of the atmosphere and a horizontal area of approximately 10 km, whereas rawinsonde measurements are essentially in situ point measurements. These characteristics lead to the satellite soundings exhibiting a relatively smooth vertical profile relative to their rawinsonde counterparts.

Hourly soundings are produced using a combination of sounder-derived measurements with short-term forecasts and analyses produced by the NCEP GFS model (Fig. 10.33b). A major limitation of the current GOES sounder is that soundings are not available in cloudy regions, because of the opacity of clouds in the infrared spectrum where the sounder operates (Fig. 10.33a). Polar-orbiting satellites are equipped with both infrared and microwave sounders and thus can generate soundings in cloudy regions, but polar-orbiting satellites do not provide sufficient coverage to allow continuous soundings and the measurements are restricted to swaths beneath the satellites. Microwave sounders are not yet deployed on GOES systems because of the requirement for very large receiver dishes that would be needed for a comparable geostationary system.

In addition to the traditional satellite measurements discussed above, several specialized satellite sensors and measurement techniques are now utilized in the atmospheric sciences. These include scatterometers, which measure the roughness of the ocean surface to compute the surface wind speed (Fig. 10.34a); radio-occultation sensors, which measure atmospheric water vapor; global positioning system (GPS) sensor measurements of water vapor; and techniques for measuring precipitation rate, lightning, and wind. Cloud-drift winds are a valuable source of upper-air wind measurements (Fig. 10.34b,c); these wind measurements are obtained using a sequence of images and pattern-detection software, which tracks a distinct cloud feature over time to compute the wind vector. The cloud brightness temperature is also used, in combination with numerical model analyses or forecasts, to determine the approximate altitude or pressure level for the wind vectors.

**10.5.1.4. Radar.** Radio detection and ranging (radar) was originally developed for military purposes during the second World War, but wartime radar operators noticed that rain showers, in addition to ships and airplanes,

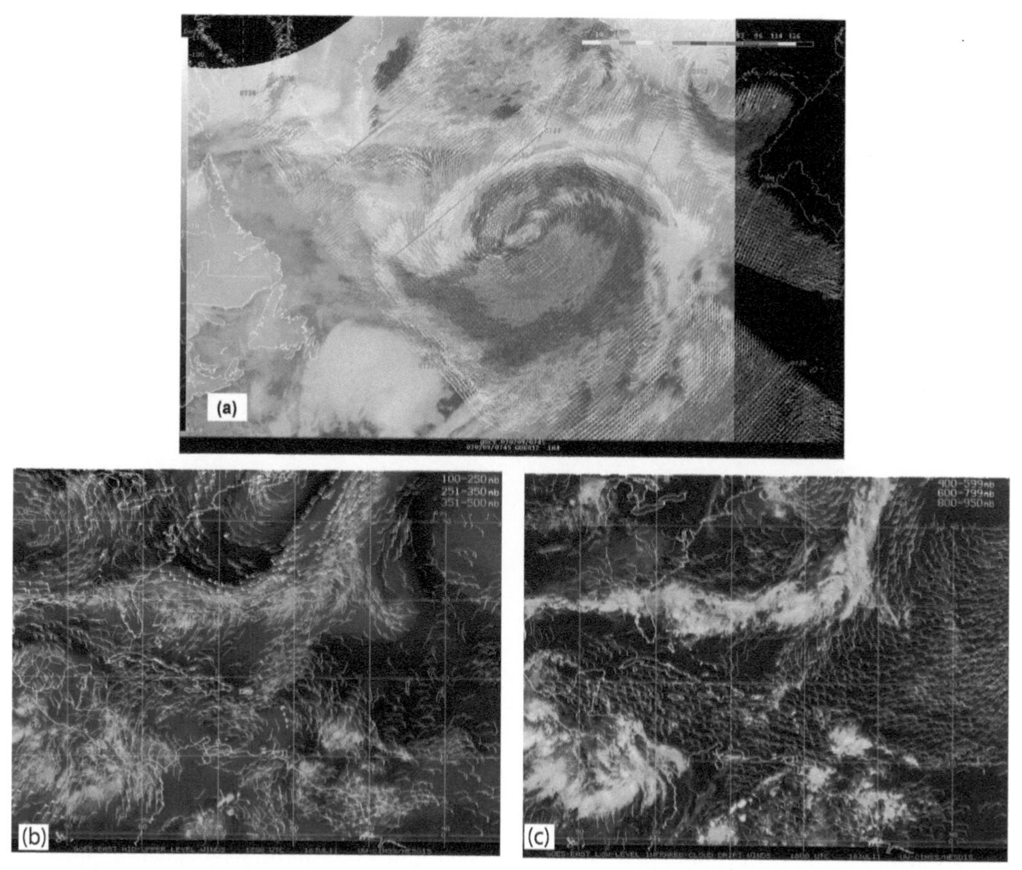


**Figure 10.34.** Examples of satellite-derived products: (a) scatterometer near-surface winds (shown as barbs color coded by wind speed, as indicated in legend at top of graphic); (b) upper-level cloud-drift winds valid 19 Apr 2010, colored by altitude as indicated in legend; (c) infrared lower-level cloud-drift winds, as in (b), but with shifted altitudes for color coding [images from Cooperative Institute for Meteorological Satellite Studies/Space Science and Engineering Center (CIMSS/SSEC) at http://cimss.ssec.wisc.edu/tropic/real-time/atlantic/winds/winds.html and http://cimss.ssec.wisc.edu/tropic2/misc/winds/info.winds.ir.html].

appeared on the radar display. The microwave frequencies used in this active sensor are strongly scattered by raindrops; after the war, meteorological applications were a logical peacetime utilization of the radar units.

In the United States in the early 1990s, the outdated Weather Surveillance Radar (WSR-57) network was upgraded to a Doppler network known as Next Generation Weather Radar (NEXRAD). Improved coverage (Fig. 10.35), in addition to Doppler radial velocity measurement, provides a rich source of raw and derived information, including rainfall estimation, integrated liquid water content, and storm-relative velocity, in addition to reflectivity and radial velocity. These data are now routinely utilized in the development of initial conditions for numerical models, and radar-augmented quantitative precipitation estimates (QPE) are a reliable, high-resolution source of observational data against which NWP QPF can be evaluated.

In North America, radar measurements are available from other networks as well, notably the Canadian meteorological service, and also from a network of FAA radars. Similar networks are available in other locations worldwide, but there is obviously a dearth of ground-based radar information over the oceans.

**10.5.1.5. Other observational platforms.** Profilers, lidar systems, unmanned drone aircraft, and neutrally buoyant balloons are additional sources of meteorological data that will not be discussed here. The point in providing a "laundry list" of observational data sources is

**Figure 10.35.** Locations of NWS Doppler radar sites (white dots) over the CONUS as of spring 2010.

to help build appreciation for the fact that a tremendous variety of data are available and that it is the challenge of data assimilation to utilize all of this observational information, as well as some data that are not observational in nature, to construct the best possible initial analysis to serve as the initial condition for NWP models. Data assimilation extends beyond weather forecasting applications and includes the development of valuable reanalysis datasets that can be used for a variety of research applications.

### 10.5.2. Data assimilation techniques

Early NWP forecasts utilized hand-interpolated data for initial conditions, but clearly this approach was not practical for longer-term real-time forecasting; by the mid 1950s, objective analysis techniques were available for automatic interpolation of observational data onto a regular grid. These techniques typically minimized the least-squares difference between observations and a polynomial defined between the observation locations and the grid points. However, there are not typically enough observations to fully and accurately provide a numerical model with suitable initial conditions. Furthermore, observations contain errors, and there is a need to maintain dynamical consistency in model initial conditions to prevent large imbalances in the initial state from contaminating the forecast. For example, suppose that observations of temperature and moisture are used to compute the 500-mb geopotential height field for a given time and location. If 500-mb wind observations at the corresponding location differ strongly from what gradient-wind balance would yield, the resulting dynamical imbalance could lead to the spurious generation of gravity waves at the beginning of a model forecast. Early in the development of DA systems, there were two steps required in analysis generation: (i) *interpolation* (objective analysis) of unevenly spaced data to a regular grid and (ii) *initialization*, which involved filtering noise and removing imbalances from the initial conditions. These two steps have essentially been combined in modern DA systems.

Utilizing a first-guess field, consisting of either a climatological background state or a short-term model forecast, provides a method of filling in gaps in the observational data network. However, the climatological state is obviously not a very accurate representation of the synoptic pattern on most days, and short-term model forecasts, although superior to climatology, still contain errors. Nevertheless, there is valuable information in these forecasts, and modern DA systems can extract the value provided by each available piece of information, including previous model forecasts.

Consider a modern DA system, such as that for the NCEP GFS model, that is creating an analysis for a given time: for example, 1200 UTC 20 April 2010. Observations

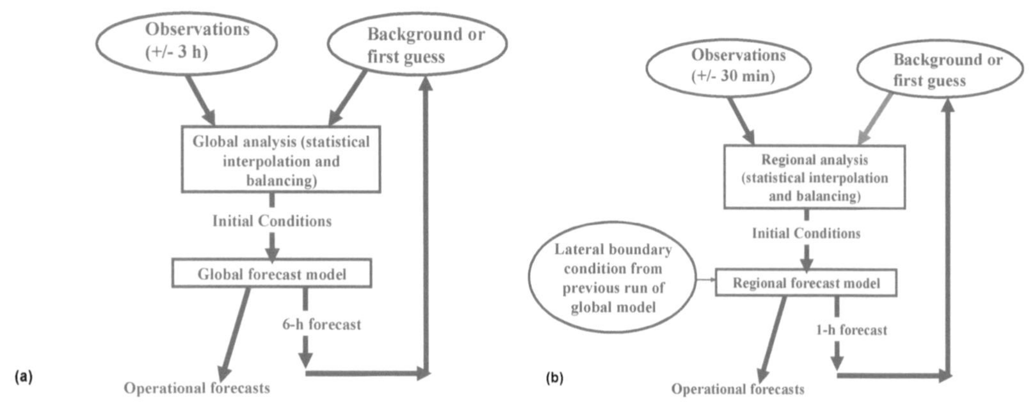


**Figure 10.36.** Flow chart showing the modern data assimilation process for (a) a global model and (b) a regional model using global model data for lateral boundary conditions (from Kalnay 2003).

from before or after this time certainly contain some information about conditions at the analysis time and therefore should be included in the process. The short-term forecast from the 0600 UTC run of the GFS, valid at 1200 UTC, is also likely to be quite accurate in most locations. Figure 10.36 provides flowcharts of the process that is utilized, beginning with observational data from a wide variety of platforms collected over a window of several hours surrounding the analysis time. This information is combined with the "background" or "first guess" provided by the short-term model forecast. The *error statistics* for all of the information utilized represent a critical piece of data, because the error can be used to determine the optimal weighting for all available information. Dynamical consistency can be obtained via the model dynamical equations, and observational quality control is also a part of this process.

Once a valid set of initial conditions have been generated, the operational model forecast is run, with the short-term forecast again being utilized in the development of the subsequent (1800 UTC in this example) analysis. This process of sequential generation of model analyses is known as "cycling" and is performed at all major NWP centers around the world at various time intervals. In the United States, the RUC model generates an hourly analysis, the NAM generates analyses every 3 h, and the GFS model generates 6-hourly analyses as of this writing.

The combined process of objective analysis and initialization is known as *four-dimensional data assimilation* (4DDA), and a variety of techniques are employed for this purpose. The traditional view of an analysis for a given time as an independent "snapshot" does not do justice to the propagation of information in time and space and across the different meteorological variables. The combination of all available information, with knowledge of the error characteristics associated with each piece of information, allows the formulation of a *cost function*, which is essentially a quantitative measure of the "distance" between the true state of the atmosphere and the analysis. By minimizing the cost function, the *most likely estimate* of the true state of the atmosphere is obtained.

**10.5.2.1. An example, following Kalnay (2003).** Following the illustrative example provided by Kalnay (2003), suppose that we have two temperature measurements, $T_1$ and $T_2$. These measurements have errors $\varepsilon_1$ and $\varepsilon_2$, which are equal to the measured value minus the "true" value (e.g., $\varepsilon_1 = T_1 - T_t$). If the measurements are unbiased, then we would expect that averaging over a sufficiently large number of measurements would yield no systematic errors. In other words, the *expected value* E is the same for the analysis and for the true temperature $[E(T_a) = E(T_t)]$. Suppose the measurements are taken from independent data sources, and the errors are known and are *normally distributed* with standard deviation $\sigma_1$ and $\sigma_2$.

Based on the information available, the analysis temperature can be written

$$T_a = a_1 T_1 + a_2 T_2, \qquad (10.11)$$

where $a_{1,2}$ are weights to be applied to the respective temperature measurements. Because the analysis is unbiased in this case, the weights must sum to 1. The temperature analysis will be the best estimate of the true temperature if the weights are chosen in a way that minimizes the mean squared error of $T_a$ ($\sigma_a$) with respect to the observations,

$$\sigma_a^2 = E\left[(T_a - T_t)^2\right] = E[(a_1(T_1 - T_t) + a_2(T_2 - T_t))^2]. \tag{10.12}$$

Because $a_1 + a_2 = 1$, we can substitute to eliminate $a_2$ and minimize (10.12) with respect to $a_1$ to obtain

$$a_1 = \frac{\sigma_2^2}{\sigma_1^2 + \sigma_2^2}\,;\, a_2 = \frac{\sigma_1^2}{\sigma_1^2 + \sigma_2^2}. \tag{10.13}$$

Obtaining (10.13) is left as a problem at the end of the chapter. Using these weights in (10.11) gives an expression for the analysis temperature obtained from the least-squares method,

$$T_a = T_1\left(\frac{\sigma_2^2}{\sigma_2^2 + \sigma_1^2}\right) + T_2\left(\frac{\sigma_1^2}{\sigma_2^2 + \sigma_1^2}\right). \tag{10.14}$$

The approach used here of finding optimal weights is known as an *optimum interpolation* approach.

Optimum interpolation methods are contrasted with *variational* approaches, which are more general and involve defining a *cost function J*. There is more flexibility with this approach; however, if we define $J$ as

$$J(T) = \frac{1}{2}\left[\frac{(T_t - T_1)^2}{\sigma_1^2} + \frac{(T_t - T_2)^2}{\sigma_2^2}\right], \tag{10.15}$$

the expression can be minimized (by taking the partial derivative with respect to $T$ and setting the result equal to zero) to obtain the same result as with the least-squares method, (10.14). In (10.15), the cost function is clearly related to the square of the difference between the actual temperature and the two observations; therefore, minimizing this cost function is consistent with the least-squares approach for this example, although the cost function is more general and could be defined in different ways. It can be shown using Gaussian statistics for normal distributions that the minimization of the cost function allows us to determine the *maximum likelihood* of determining the actual temperature based on the available observations.

The result (10.14) is intuitive if we consider the limiting cases. Suppose that $T_1$ was characterized by very large error (large $\sigma_1$) and $T_2$ was a highly accurate and reliable measurement (small $\sigma_2$). Clearly, the weighting multiplier for $T_1$, $\sigma_2^2/(\sigma_2^2 + \sigma_1^2)$, would then be smaller than that for $T_2$, $\sigma_1^2/(\sigma_2^2 + \sigma_1^2)$, meaning that the reliable measurement is weighted strongly and the unreliable one is less so.

As discussed previously, modern DA systems utilize previous model forecasts as a background or first guess in the analysis process. Integrating this into the same framework used to derive (10.13) yields

$$T_a = T_b + W(T_{obs} - T_b), \tag{10.16}$$

where $T_a$ is the analysis temperature, $T_b$ is the background temperature field, and

$$W = \sigma_b^2/(\sigma_b^2 + \sigma_{obs}^2) \tag{10.17}$$

is the error variance of the background field ($\sigma_b^2$) divided by the total error variance and represents the optimal weight. The difference between the observations and background is called the *innovation* or *observational increment*. The analysis $T_a$ is obtained by adding the innovation, weighted by the optimal weight, to the background field. The error in the analysis can be expressed as

$$\sigma_a^2 = (1 - W)\sigma_b^2, \tag{10.18}$$

meaning that the analysis error variance is equal to that in the background multiplied by the factor 1 minus the optimal weight.

Despite the simplicity of the preceding example, an analogous process is used in modern DA systems, except that the terms in the equation then involve the products of matrices containing large numbers of observations, variables, and gridded data points. The DA techniques differ primarily in *how the background error covariance matrices are obtained*. Numerous strategies are available; for instance, one could compute the "climatology" of error based on past differences between the background field and verifying analysis, as well as between different types of observation and the verifying analysis. However, the errors are functions of space, time, synoptic pattern, etc., and therefore the strategy of applying a uniform, climatological weighting does not yield optimum results. Another strategy for evaluation of the background error covariance matrix is to utilize an *ensemble* of model forecasts, as in the ensemble Kalman filter (EnKF) technique. Then, the spread of the ensemble, which can be computed as a function of space and time for each variable, relevant to the synoptic system in question, represents a more accurate means of determining $W$.

The sequence described above can be "cycled," a process where a model forecast is run to produce values of $T_b$ and $\sigma_b$; this process can be described mathematically as

$$T_b(t_{i+1}) = M(T_a(t_i)), \tag{10.19}$$

where $M$ represents the dynamical forecast model, running with the previous analysis as the initial condition. Of course, we must also estimate $\sigma_b$, which can be accomplished in a variety of ways. A straightforward approach would be to assume linear growth (or another prescribed functional dependence) of the error with time in the model forecast, a strategy known as the *optimum interpolation* approach,

$$\sigma_b(t_i + \Delta t) = a\sigma_a^2(T_i), \tag{10.20}$$

with the value for $a$ being greater than one and the optimum weight then being computed using (10.17).

#### 10.5.2.2. Extension to operational dimensions.

The example from the previous section can be extended to the full multivariate problem by replacing the temperature analysis with a very large vector array representing all of the model variables and all grid cells in the model domain $\vec{x}_a$ and replacing the background temperature with $\vec{x}_b$. The length of these arrays equals $n = i \cdot j \cdot k \cdot \#\text{variables}$, where $i$, $j$, and $k$ represent the dimensions of the model domain. The observations are represented as a vector $\vec{y}_o$ to indicate a different array length from that of the model analysis and background vectors. The length of the observation vector is $p$, the number of observations.

Another important aspect of modern DA systems is that the observations *do not necessarily correspond to the same variables as the dependent variables in the model.* For example, an observation of radiance measured by a satellite sensor is indirectly related to the model variables; however, with knowledge of the properties of the relation, a transformation can be made. To transform the background field into a form that is compatible with the observations, an *observation operator* $H(\vec{y}_o)$ is utilized. This operator handles both spatial interpolation, as well as physical transformation if needed, of the observed and model variables to the same units. The operator $H$ transforms the background to the observations; the transpose of this operator $H^T$ is used to transform the observations back to the model space.

Initial DA efforts to include some types of remotely sensed observations relied on assimilation of observations that had already been transformed into the same variables and units as the model variables: for example, satellite-derived temperatures. However, subsequent research revealed that better results could be obtained by building this transformation process into the DA system. This was because some of the assumptions in the retrieval algorithms were inferior or inconsistent with the model and also because by building this process into the DA system, the full volume of information from the model and DA system, including other observations, could provide added consistency and essentially a more accurate retrieval. Furthermore, the error of the pre-retrieved data was greater than that of the raw measurement, which could be related directly to the sensor specifications if the raw measurements were utilized.

The innovation in observation space is written $\vec{d} = (\vec{y}_{obs} - H(\vec{x}_b))$. Using this and the other notation introduced above, the analogous expression to (10.16) is written

$$\vec{x}_t = \vec{x}_b + W(\vec{y}_{obs} - H(\vec{x}_b)) - \vec{\varepsilon}_a, \tag{10.21}$$

where

$$\vec{\varepsilon}_a = \vec{x}_a - \vec{x}_t \tag{10.22}$$

is the analysis error, and the optimal weight matrix $W$ is of dimension $n \times p$. We could equivalently write (10.21) in a form more directly aligned with (10.16),

$$\vec{x}_a = \vec{x}_b + W(\vec{y}_{obs} - H(\vec{x}_b)), \quad \text{or} \tag{10.23}$$

$$\vec{x}_a = \vec{x}_b + W\vec{d}. \tag{10.24}$$

In expressing $W$, we utilize $B$, an $n \times n$ matrix known as the *background error covariance matrix*, and the observation error matrix $R$ of dimension $p \times p$,

$$W = BH^T(R + HBH^T)^{-1}. \tag{10.25}$$

Comparing (10.17) to (10.25), we see that, as before, the optimal weight is equal to the background error covariance (in observation space) divided by the total error (that of the observations summed with that of the background).

A promising strategy that is currently the topic of much research is the EnKF approach. With this strategy, the background error is computed from an ensemble of model forecasts as a function of space and time, allowing a much more accurate depiction of where the error in the background field is larger or smaller. In (10.21), the optimal weight matrix $W$ is referred to as the *Kalman gain matrix K* when this technique is employed.

#### 10.5.2.3. Variational assimilation (3DVAR and 4DVAR).

A result similar to (10.23) can be obtained by minimizing a cost function, as discussed for the case of a simple scalar variable,

$$J(x) = \frac{1}{2}\left[(\vec{x}-\vec{x}_b)^T B^{-1}(\vec{x}-\vec{x}_b)+(\vec{y}_o-H(x))^T R^{-1}(\vec{y}_o-H(x))\right]. \quad (10.26)$$

It is not an easy task, but it is possible to determine $\nabla_x J(\vec{x}_a)=0$ and solve to obtain an expression for the analysis,

$$\vec{x}_a=\vec{x}_b+(B^{-1}+H^T R^{-1} H)^{-1} H^T R^{-1}(\vec{y}_o-H(\vec{x}_b)), \quad (10.27)$$

which is equivalent to (10.23).

Additional constraints can be added to the cost function, such as a dynamical balance condition. For example, one could impose a constraint of thermal wind balance and allow a certain error in the balance assumption that adjusts the strength of the constraint.

Relative to some other methods, 3DVAR has a relatively low computational expense and is used at NCEP for the GFS and NAM models. Research models such as MM5 and WRF are also now equipped with 3DVAR systems; the flowchart in Fig. 10.37 illustrates the DA procedure for the MM5 model, as described by Barker et al. (2004).

The ECMWF model has outperformed other global models for many years, and this is likely due in part to the use of a 4DVAR DA system in this model. Without going into details, the concept of 4DVAR utilizes the model itself more heavily in the DA process to incorporate information from *different times* more strongly into the analysis. The 4DVAR counterpart of (10.27) is

$$J(\vec{x}_{t0}) = \frac{1}{2}\left[\left(\vec{x}_{t0}-\vec{x}^b_{t0}\right)^T B_0^{-1}\left(\vec{x}_{t0}-\vec{x}^b_{t0}\right)+\frac{1}{2}\sum_{i=0}^{N}\left(H(\vec{x}_i)-\vec{y}^o_i\right)^T R_i^{-1}\left(H(\vec{x}_i)-\vec{y}^o_i\right)\right]. \quad (10.28)$$

Here, the cost function includes information from the model integration to the time of the observations, which may differ from the analysis time. The minimization is conducted in a way that minimizes the cost function over an increment of time, rather than at a single time, hence the "4" in 4DVAR.

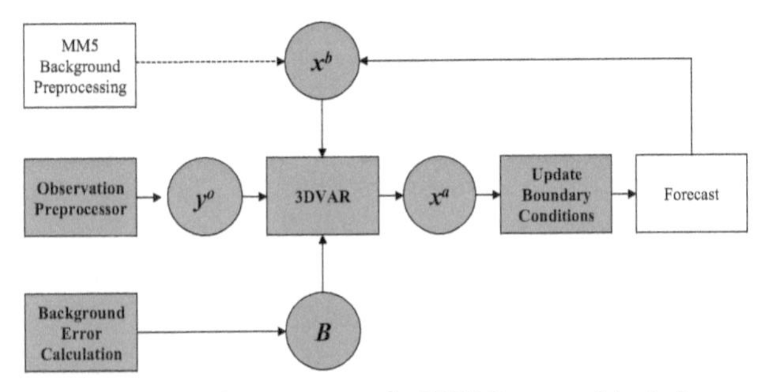


**Figure 10.37.** Components of a 3DVAR system (shaded) used in the MM5 modeling system (from Barker et al. 2004).

### 10.5.3. Data source sensitivity: Which data matter most?

What are the most important aspects of DA for forecasters and users of model forecasts and analyses to understand? The answer to this question will naturally depend on the specific user applications in question, but it can be difficult for end-product users to appreciate the value and importance of the DA process, even if the data they are using for research or forecasting were derived from a complex DA system. For someone using data provided by the NCEP–NCAR reanalysis, the ECMWF reanalysis (e.g., ERA40), or the North American Regional Reanalysis (NARR), it can be important to understand what data went into the analysis, how the inclusion of data from a given platform may have changed over time, and to what extent the analysis is reliable at different times and locations. For weather forecasters, it should be recognized that model initial conditions are the product of a highly complex process, and a simple comparison of in situ surface or upper-air observations to the model analysis may not provide much, if any, insight into the subsequent model forecast accuracy.

In an effort to illustrate the importance and sensitivity of DA in a meteorologically relevant context, we turn to the outstanding study of Zapotocny et al. (2000). Although conducted with a DA system that is no longer widely used, that of the since-retired NCEP Eta Model, this is the DA system that was used to generate the NARR, and the data-denial experiments presented in this paper provide an excellent illustration of the sensitivity of analyses and model forecasts to different data sources. Techniques change with time, and a single case study is insufficient to

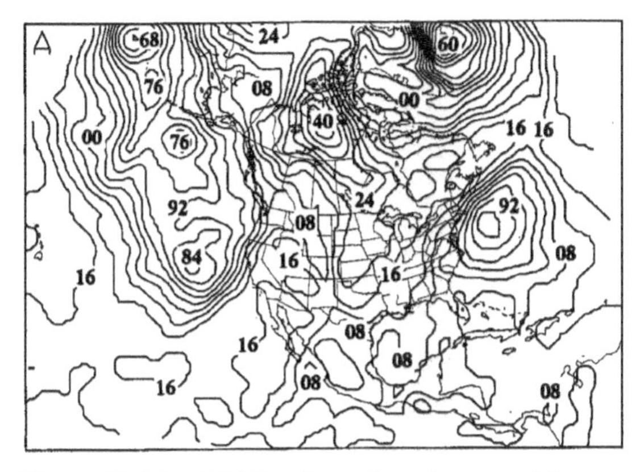

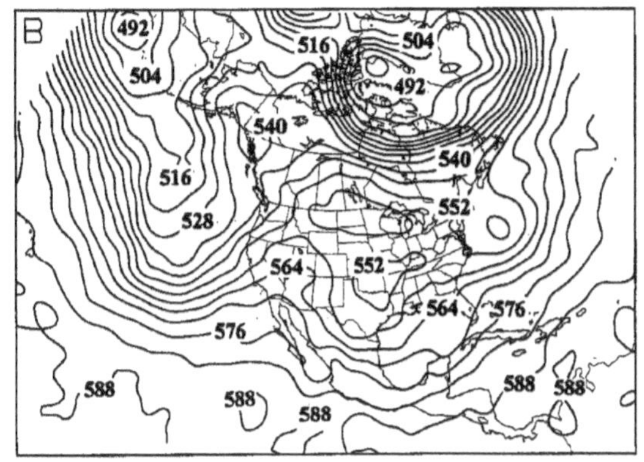


**Figure 10.38.** EDAS analyses from the control forecast run, valid 0000 UTC 6 Feb 1998: (a) sea level pressure (contoured every 4 mb) and (b) 500-mb height (contoured every 6 dam; from Zapotocny et al. 2000).

provide a complete interpretation of how the sensitivity to a given data type will appear in different synoptic settings; however, an in-depth case-study analysis can still yield an insightful appreciation for the DA process.

Using the Eta Model Data Assimilation System (EDAS), the Zapotocny et al. (2000) study ran an Eta Model forecast for a typical day in early February 1998 out to 48 h using initial conditions derived from the full suite of 34 then-operational data sources. This forecast was accompanied by 34 experimental model forecasts, each using initial conditions that were derived with one of the 34 data sources withheld. Comparison of the differing model results in these data-denial experiments allows for diagnosis of both initial condition (analysis) sensitivity to a given data source and examination of the model *forecast* sensitivity to a given data platform.

The synoptic pattern for the study period is shown in Fig. 10.38. For each of the 34 observational data platforms, errors are assigned, taken here to be functions of height (pressure level) and data type. The influence of a given data platform on the final analysis is related both to the error assigned to it, the error assigned to the background, as well as to the coverage and distribution of the data source relative to other sources. Differences in the distribution of rawinsonde data, ACARS, and satellite cloud-drift wind observations are apparent in Fig. 10.39. For densely packed observations, the EDAS system used in this study utilized the median observational value in a given grid cell and ignored others, to reduce data volume and computational expense. Separate designations for mass ("1") and wind ("2") variables are used to further differentiate the types of observations. Some observation platforms (e.g., rawinsondes) collect both mass and wind information. A 12-h assimilation cycle was used, the end of which corresponds to the analysis time, 0000 UTC 6 February 1998.

The sensitivity of the analysis to different data platforms is computed for different variables and at different vertical levels, for both the entire model domain as well as for the conterminous United States (CONUS). For the analysis of the zonal wind component over the entire domain, strong sensitivity to rawinsonde data is evident, with the sensitivity to ACARS and GOES cloud-drift winds also playing major roles (Fig. 10.40a).

When the domain is restricted to the CONUS, the ACARS sensitivity is the largest of any data type for upper-level (300 mb) zonal winds. For other variables, such as humidity or temperature, the ACARS data play a smaller but still important role. Even though at this time the humidity was not a reported variable in the ACARS, influence is still exerted on the humidity analysis because of the dynamical relations between model variables, which illustrates the level of sophistication inherent in modern DA systems (Fig. 10.41). In other words, because the ACARS data include direct information concerning the wind field, the geopotential height, temperature, and humidity fields were all influenced as well, albeit in smaller ways.

The discussion to this point has centered on sensitivity of the *analysis* to different data sources, but it is also important to consider the impact that these data have on the forecast itself. As a means of comparison with Fig. 10.40b, Fig. 10.42 shows the 48-h zonal wind forecast sensitivity over the eastern portion of the analysis domain. It is clear that the influence of ACARS remains strong in the forecast as well.

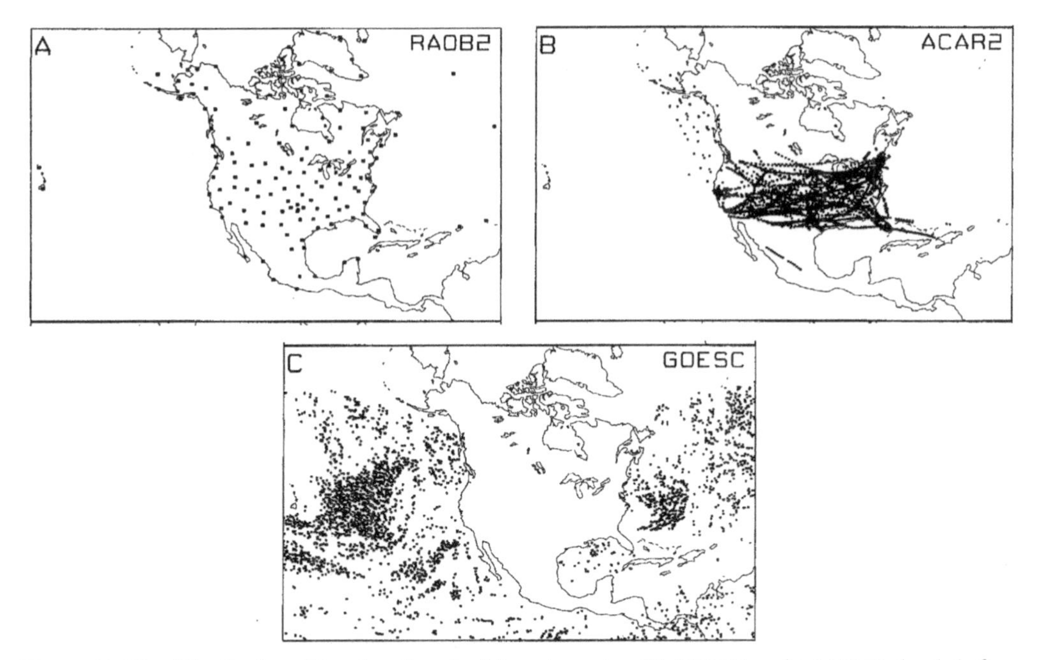


**Figure 10.39.** Distribution of data locations for (a) rawinsondes, (b) ACARS, and (c) GOES cloud-drift wind observations used in the analysis of 0000 UTC 6 Feb 1998 (from Zapotocny et al. 2000).

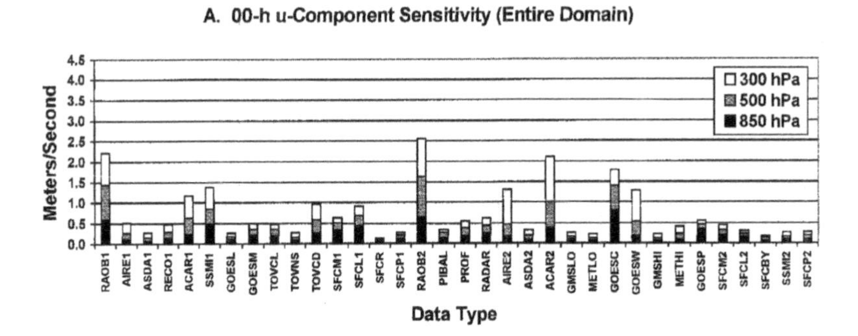


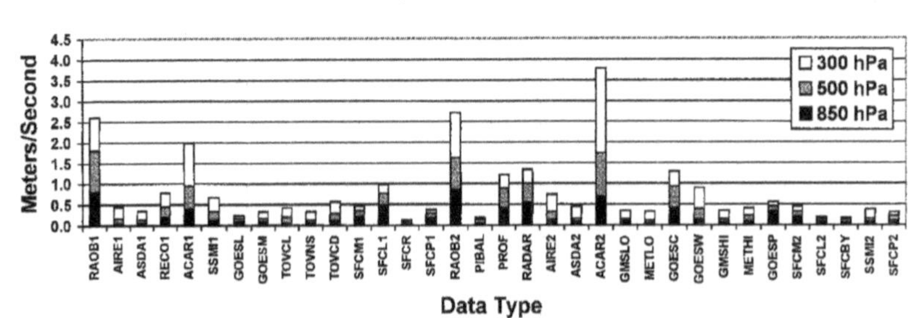


**Figure 10.40.** Sensitivity (m s$^{-1}$) of the final analysis to data platform: (a) for entire analysis domain and (b) for CONUS only. Bars include contributions from 850-, 500-, and 300-mb levels, as indicated in the legends.

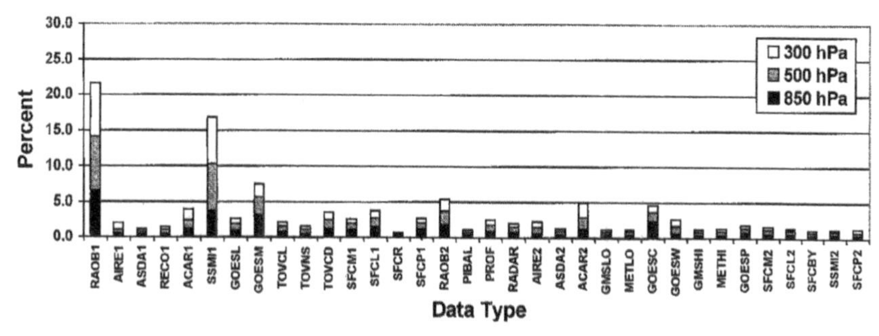


**Figure 10.41.** As in Fig. 10.40, but for relative humidity analysis for CONUS (from Zapotocny et al. 2000).

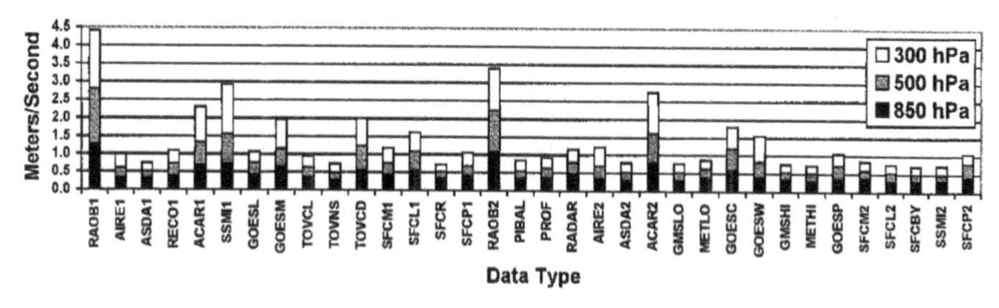


**Figure 10.42.** As in Fig. 10.40, but for sensitivity of 48-h zonal wind forecast to data source and for only a domain subsection at the eastern (downstream) end of the model domain (from Zapotocny et al. 2000).

An important message from the Zapotocny study is that a huge volume of information goes into the analysis, including both model forecast (the background field) and a large variety of observational data. Since the time of the Zapotocny (2000) article, additional data platforms have been incorporated into the DA process, and the means of computing the error covariance has evolved as well. In fact, a completely different operational DA system is now used at NCEP relative to the one described here.

Recognition of the importance of a given data platform can also be useful. During the period from 15 to 23 April 2010, the Eyjafjallajökull volcano eruption in Iceland severely curtailed air travel across much of Europe. Studies from this unintentional data-denial experiment could be performed to determine the extent to which the omission of the normally large volume of ACARS data affected model forecasts over and downstream of Europe during and in the days following the air travel shutdown.

## 10.6. ENSEMBLE FORECASTING

An *ensemble forecast* is a prediction based on not just a single (deterministic) forecast but on a suite of several individual forecasts. In the United States, the field of ensemble prediction has expanded since NCEP first implemented an operational ensemble system in 1992, with applications now for short-range (1–3 days) and extended-range (3–15 days) weather prediction, climate prediction, and air-quality forecasting. *Superensemble* forecasting strategies, in which different ensemble members are weighted according to past skill, have shown value in predicting the track of tropical cyclones, and ensembles form the basis of climate projection at decadal and longer time scales. In addition to improved forecast accuracy, ensembles provide quantitative information about the degree of forecast confidence associated with a given prediction, along with other information that deterministic prediction alone cannot provide. In this section, our objectives are to summarize some of the basic strategies behind ensemble prediction, to discuss some of the information that ensembles provide and how it can be presented, and to illustrate strengths and limitations for some operational ensemble forecasting products.

The ideas that form the basis for ensemble forecasting stem from the work of the late Professor Edward Lorenz, who had been studying nonlinear dynamical systems at the Massachusetts Institute of Technology (MIT) when he was a graduate student. At one point, he attempted to reproduce a forecast from a nonlinear model of his creation. When he examined the model output for some point in the forecast sequence well into the evolution of the forecast, he discovered that it was completely different from that in the original model run. Later, he recognized that the model runs had started out the same but that a very small round-off error in the initial conditions in the second model run had amplified with time, eventually resulting in an entirely different solution. He realized that if the atmosphere behaved in this respect as the simple model did, then numerical weather prediction would suffer the same limitations due to the inevitable small errors in initial conditions. This is the so-called "butterfly effect"; even a miniscule difference in the initial atmospheric state will eventually amplify and at some point result in a completely different forecast.

Subsequent work led Lorenz to estimate that the limit for deterministic prediction of the atmosphere was approximately two weeks, but he also discovered that the length of useful predictability was a function of the atmospheric state. In some situations, the forecast would break down quickly, whereas in others the error growth was much slower. Ensemble prediction provides a means to distinguish these situations, as discussed below.

### 10.6.1. Sources of model error

All NWP forecasts contain error, as do the model initial conditions (analyses), as well as observations used in the DA process. In recent years, the variety of methods by which forecasts are evaluated has become more complex, especially for high-resolution model forecasts, and there is increased appreciation for different types of error as well as for different error sources. For instance, there are new measures of forecast skill that offer a better means of accounting for correct model forecasts of the presence of a feature in, for example, a convective precipitation system, even if the predicted location of the feature differs from what was observed. In this case, a *displacement error* is evident, but there may still be considerable value to operational forecasters in the model having predicted the occurrence of such a feature in the vicinity. However, rather than a focus on novel techniques for forecast verification, the focus here is related to errors in the more traditional sense. Model errors can be classified as stemming from (i) errors in the model initial conditions, (ii) errors in the model physics or dynamics, and (iii) the intrinsic predictability limit for deterministic forecasts of a chaotic system.

**10.6.1.1. Initial condition errors.** Errors in the model initial conditions are unavoidable, whether due to errors in the observations themselves, errors in the background field (model) used in the DA system, or other aspects of the DA system (e.g., faulty quality control algorithms). The contribution of observation errors to the model initial condition can be further subdivided into *instrument errors*, *errors of representativeness*, or *errors in the transformation process* from observation to model space. Errors resulting from uncertainty due to the sensors used to collect the measurements themselves are relatively well known. Observation errors also include those due to insufficient temporal and/or spatial coverage or a lack of measurements of certain variables. Suppose that a rawinsonde is by chance launched into a mesoscale cumulonimbus cloud, when the remainder of the surrounding region is clear. Because the spatial resolution of the rawinsonde network is relatively coarse, an analysis based only on this data platform would artificially amplify the spatial influence of this measurement, an error of representativeness. Initial condition errors are highly variable in time, space, and as a function of phenomenon. For example, it is extremely difficult to obtain accurate initial condition data for tropical cyclones because of their often remote maritime locations, the hazards they pose to in situ observing platforms, and their complex multiscale structure.

**10.6.1.2. Model errors.** Model errors are highly dependent on the synoptic setting and the character of the weather phenomenon being predicted. Model physical parameterizations are often designed using measurements over a relatively small range of conditions, but then are applied over a broad range of conditions when models are run in diverse circumstances. For example, model formulation of turbulent heat, moisture, and momentum fluxes over water involves the computation of surface turbulent exchange coefficients, which are functions of wind speed. Experimental measurements taken over a range of relatively light wind speeds were used to derive formulas that were subsequently incorporated into numerical model surface layer and PBL schemes. However, when a strong tropical (or extratropical) cyclone develops in the model atmosphere, surface conditions in the model atmosphere can fall well outside of the range of conditions (parameter space) over which the measurements used to derive the coefficients were taken and over which the corresponding formulas are valid. Recent field experiments have allowed for improved representation of the turbulent exchange coefficients at high wind speeds over water, and these improved formulations are now available in some models, such as the WRF-ARW. However, this is just one example of an error that is attributable to the model itself.

Errors due to inadequate model vertical and horizontal resolution, representation of topography, or land and ocean surface interactions can also be important. For regional model domains, an additional error source arises because of lateral boundary conditions. There are also model errors due to processes that are typically neglected in the model governing equations, such as the transfer of sensible heat and momentum by falling precipitation or the heat released by a lightning discharge. The impacts of volcanic ash plumes, air pollution, dust storms, or forest fires are not currently represented in most operational model equations, although some of these processes will likely be included in models in the near future. Air-quality forecast models already account for many processes that are often ignored by synoptic weather forecasters, and some of these may be important in weather forecasting.

**10.6.1.3. Intrinsic predictability limits.** Suppose that we could someday develop a "perfect" NWP model, with complete representation of all known physical processes, and that we could capture the initial state of the atmosphere with incredible accuracy and precision to within the measurement error limit from very accurate, high-density sensors. Even if these things were possible, we might be disappointed if we ran the model and found that the predictive skill vanished with lead times of less than a month (probably considerably less). The so-called butterfly effect, the idea that air motions as small as those resulting from the flap of a butterfly's wings, could eventually alter the global-scale forecast is a consequence of the chaotic nature of the atmosphere.

How can we know how long it will take for infinitesimal differences in initial condition to amplify sufficiently to eliminate forecast skill? Knowing that the predictability limit varies with the character of the flow, what can be done to quantify the forecast error in a given forecast situation? Lorenz (1965) studied the behavior of an "ensemble" of forecasts: that is, a series of model runs with slightly different initial conditions. Each model run or ensemble member is an equally probable outcome. By examining the statistical spread among the ensemble members, valuable information is gleaned concerning the predictability of the atmosphere as a function of time and location. Furthermore, this technique provides quantitative information concerning the level of confidence in the particular model or forecast cycle. The third type of model error leads us to a discussion of the concept of *ensemble prediction*.

### 10.6.2. Ensemble forecasting strategies

Several distinct strategies for generating an ensemble of model forecasts are available. These include (i) imposing variations on the model initial condition and (ii) perturbing aspects of the model, such as the choice of model physics, or (iii) using a suite of different models, with the same initial conditions. Each of these categories includes additional variations, only some of which will be described here.

Early ensemble systems involved a *control run*, based on the operational analysis, and ensemble members in which the initial conditions were randomly perturbed with errors falling within the range of the computed analysis error (e.g., equation 10.16). Another efficient strategy is *lagged-average* forecasting, in which operational model runs initialized at different times are compared for the same verification time. The forecasts are weighted according to their expected error to produce a convenient ensemble prediction. The lagged-average ensemble was characterized by perturbations that developed naturally within the model forecasts rather than those applied to the initial conditions. This led to the recognition that dynamically meaningful perturbations might add value beyond what random perturbations to the initial conditions would yield. A disadvantage of this strategy is that a limited number of forecasts are available, and large ensembles require the use of forecasts with very large lead time, which add little skill to the ensemble. A variation on this strategy, scaled lagged-average forecasting, computes the errors between the previous model runs and the latest analysis, and uses these as perturbations in the model, doubling the number of runs available.

In light of the results for lagged-average ensembles, it was recognized that model error growth corresponding to dynamical instabilities was more rapid than that due to random perturbations. In nature, it is dynamical instabilities of this type that amplify to limit predictability. At NCEP, a method was devised to "breed" the most rapidly growing disturbances in the atmosphere and to use these to perturb the initial condition in ensemble generation. Random perturbations are distributed across the model initial condition in a perturbed run, while at the same time an unperturbed control run is conducted. By computing the difference between the control and perturbation run, a set of "bred vectors" is obtained; these are rescaled back to the initial random perturbation amplitude and used to perturb the model initial conditions in an ensemble (while assigning both positive and negative values to the vector amplitude).

An entirely different type of ensemble can be generated by varying the physical parameterizations in a given model (or even by running a different NWP model with the same initial conditions). Several investigators have found that ensemble systems that vary model characteristics can outperform single-model initial condition-based ensembles (e.g., Hamill and Colucci 1997; Stensrud et al. 1999; Fritsch et al. 2000). An example of this type of ensemble system is the NCEP SREF, which uses several different models [the NAM, WRF-ARW, and until recently, the Regional Spectral Model (RSM)], as well as different physics choices within the same models. These variations can be based on a different choice of convective parameterization, precipitation microphysics, boundary layer, or radiation scheme.

Some ensemble systems are based on a set of differing initial conditions (e.g., the analyses produced by different operational NWP centers around the world) in conjunction with a combination of models and model physics. These *mixed ensembles* have the advantage of convenience and perhaps of capturing a broader variety of potential error growth. By combining currently available operational information, The Observing System Research and Predictability Experiment (THORPEX) Interactive Grand Global Experiment (TIGGE) is drawing on the wealth of operationally produced model runs (including ensembles) to generate a massive dataset for ensemble research; this is a global-scale mixed-model ensemble experiment.

### 10.6.3. Advantages of ensemble forecasting

There are four primary advantages to ensemble prediction beyond what a deterministic forecast can provide:

1. The ensemble mean (based on an average or weighted average of the individual ensemble members) exhibits more skill than do the individual ensemble members;
2. The ensemble provides a quantitative measure of forecast confidence as a function of lead time and spatial location;
3. A *probabilistic* forecast is immediately available from the ensemble; and
4. The ensemble system, with the application of additional techniques, provides information regarding the optimum location for additional observations to improve the forecast, which can be utilized for *targeted observations.*

**10.6.3.1. The ensemble mean.** The ensemble mean, even when computed as a simple average of a set of individual forecasts, exhibits, on average, smaller error than any of the ensemble members. This is in part because the elements of the forecast that are least predictable (and which lead to growth of the disturbances that cause the ensemble members to diverge) tend to cancel between the ensemble members; the more predictable aspects remain in the ensemble mean. A similar behavior is seen in human forecasting exercises, where the consensus forecast, computed as the average of individual forecasts, often outperforms the majority of the individual forecasters. This is perhaps due to cancellation of errors between the forecasts of group members who "gamble" on a certain outcome one way or the other.

A *superensemble* technique, developed by Professor Krishnamurti and collaborators at Florida State University, provides a weighted ensemble strategy in which the more historically skillful members of an ensemble are weighted more heavily, providing even further improvement over the traditional ensemble mean. An example from a 10-member NCEP ensemble, based on data from 1997 to 1998, demonstrates the improvement in skill of the ensemble mean over both the individual members and the control run (Fig. 10.43). The operational control

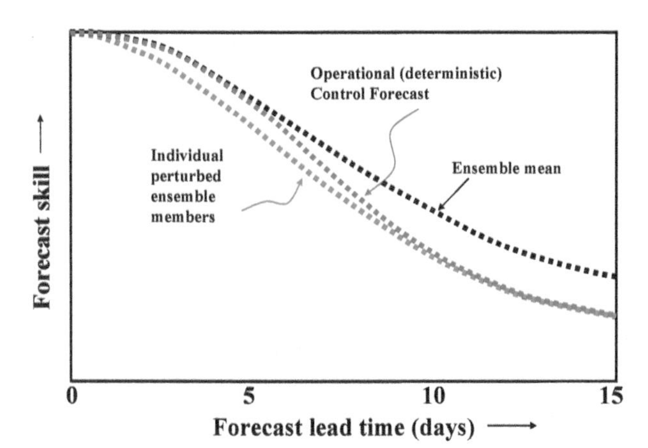


**Figure 10.43.** Anomaly correlation of NCEP GFS model forecasts averaged for the winter of 1997/98. Skill of the ensemble mean (black dashed line), the individual ensemble members (red dashed line), and the operational deterministic model run (green dashed line) is shown [adapted from Kalnay (2003), with credit for data to Jae Schemm of NCEP].

run is typically conducted with higher spatial resolution than the perturbation runs, which explains why the latter exhibits lower skill during the earlier portion of the forecast cycle.

**10.6.3.2. Quantifying forecast confidence.** If a series of multiday forecasts were based only on a single deterministic model forecast, there would be some days where the forecast proved skillful out to several days and others where the forecast accuracy plummeted within a few days. In modern forecasting, operational forecasters typically consult several models when formulating a public weather forecast; in doing so, they are utilizing a sort of multimodel ensemble. When the models being used [e.g., the NAM, GFS, Canadian Global Environmental Multiscale (GEM), Met Office (UKMO), and ECMWF models] all agree on the timing, location, and intensity of a particular weather system, forecasters recognize that the degree of confidence in that particular forecast is high relative to a situation in which the models produce widely divergent solutions. This type of qualitative forecast confidence is common knowledge within the forecasting community.

In addition to informal ensembles as discussed above, operational forecasters are increasingly drawing information from ensemble forecast systems, including the SREF and GFS Ensemble Prediction System (GEFS) systems at NCEP, and others. A full operational ensemble is able to provide *quantitative* information concerning forecast confidence, including specific information about when and where the forecast is more or less reliable.

There are several useful ways to display and communicate forecast uncertainty information derived from ensemble forecasts; "spaghetti plots" display one or a few contours from each individual ensemble member. Alternately, the standard deviation of the ensemble for a given lead time, location, and variable may be plotted. For example, on 22 April 2010, the NCEP GFS model ensemble-mean 500-mb-height forecast revealed a ridge over western Canada and closed lows over the southwestern United States, eastern Canada, and Alaska (Fig. 10.44a). This ensemble consists of 23 model runs, including 11 perturbation runs each from the 0000 and 1200 UTC cycles, plus the control GFS forecast. A spaghetti plot indicates that, for the initial time, as expected, the ensemble members are in close agreement, with the differences being due to the perturbation of the initial conditions (Fig. 10.44b). The standard deviation at this time is likewise small and is largest in regions of strong horizontal height gradient (Fig. 10.44c).

Examining the same sequence of plots for the 72-h forecast, the trough that had previously been centered over the southwestern United States is forecasted to be over the Midwest (Fig. 10.45a). There remains fairly strong agreement in the location of this feature amongst the ensemble members, as evident in the spaghetti plot (Fig. 10.45b) and the modestly small standard deviation there (Fig. 10.45c). The large ridge, now predicted to reside over central Canada, appears to be a stable feature of the ensemble as well, although there is some spread among the ensemble members south of the center of this feature. Evidently, some of the ensemble members reestablish zonal flow to the south of the ridge, leaving a cutoff feature, whereas others maintain a jet stream diversion around this feature. The large standard deviation to the east of the ridge, near the west coast of Greenland, is a reflection both of the large horizontal height gradient there and the spread among ensemble members (Fig. 10.45c).

The plots provided here demonstrate that different information concerning the forecast confidence can be gleaned from spaghetti charts and the ensemble standard deviation. The spaghetti contours do not always provide a complete picture of important meteorological features because of the limitation of a few selected contours. Furthermore, the contours tend to exhibit larger spatial deviations in areas of weak horizontal gradient. However, the

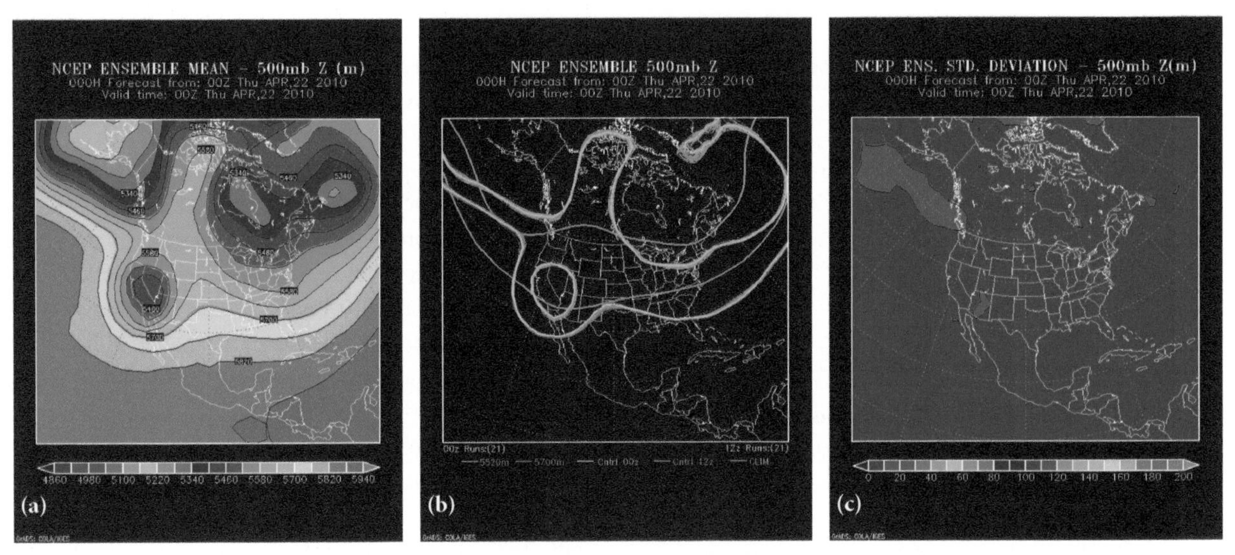


**Fig. 10.44.** NCEP GFS ensemble forecast at hour zero, valid 0000 UTC 22 Apr 2010: (a) ensemble-mean 500-mb height; (b) spaghetti plot showing 5520- and 5700-m contours from individual members (as in legend), along with climatological values (green contours); and (c) standard deviation (m), shaded as in legend (graphic from NCEP ensemble Web page).

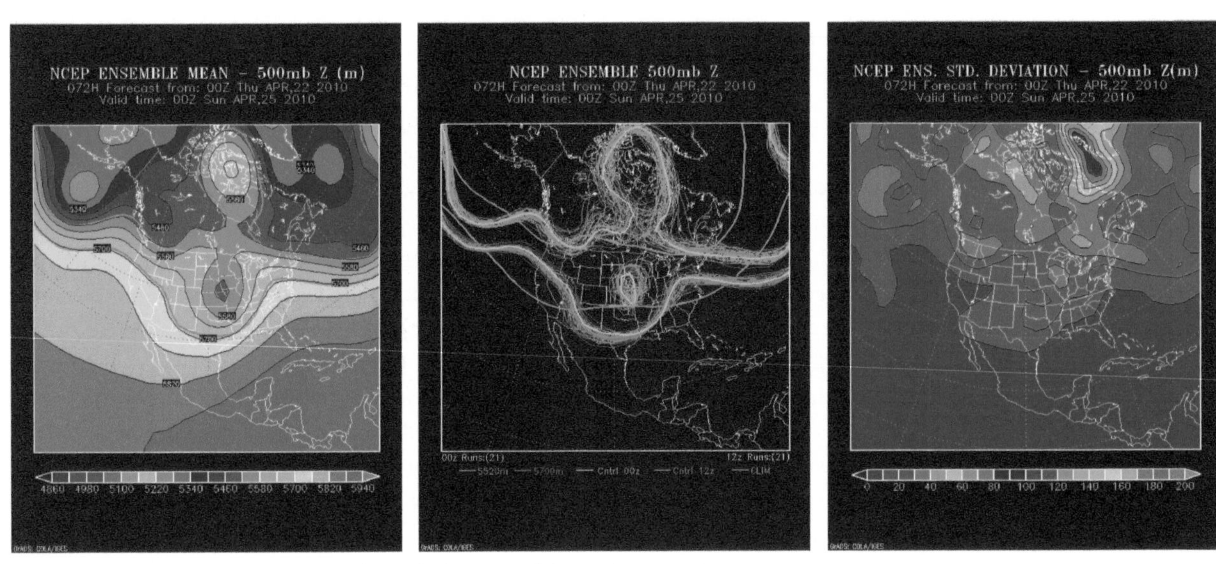


**Figure 10.45.** As in Fig. 10.44, but for 72-h forecast valid 0000 UTC 25 Apr 2010.

spaghetti plot has the advantage of summarizing the behavior of all ensemble members in a single plot. Including the control GFS forecast in a different color is a clever way of providing forecasters who use the GFS with quick context for the operational deterministic model run, which they are likely to have scrutinized closely. The standard deviation plots summarize information from the ensemble but tend to show large values in regions of strong gradient. By using spaghetti and standard deviation plots in combination, along with the ensemble mean, a complete picture of the forecast uncertainty can be efficiently assembled. Viewing these plots (and others) on the NCEP ensemble Web page on a daily basis is suggested as a means of gaining a feel for the information carried in the ensemble.

The ensemble spread not only provides a useful measure of forecast confidence, but this information can also be used to define the background error covariance matrix ($B$ in equation 10.23) in a data assimilation system.

When $B$ is computed in this fashion, it is referred to as the *Kalman gain matrix*, and this technique is known as the ensemble Kalman filter, as discussed in section 10.5.

**10.6.3.3. Probabilistic forecasting.** Given the inherent limits on predictability and the advent of ensemble prediction, in some cases it makes little sense to make a deterministic forecast for a given meteorological quantity. Furthermore, such a forecast does not take advantage of all information available from an ensemble prediction system. Consider a hypothetical example in which a 3-day forecast from a deterministic model indicated that a major winter storm would affect the East Coast of the United States. A hypothetical ensemble forecast demonstrated that the operational, deterministic run was one member of a group of 6 in the ensemble which predicted a coastal cyclone, but another cluster of 17 ensemble members exhibited cyclone tracks farther west, tracks that would bring only rain to the eastern seaboard. Clearly, forecasters would wish to examine the situation more carefully before sounding the alarm for a major snowstorm! Furthermore, the basis for a probabilistic forecast is already available, because, out of a total of 23 ensemble members, 6 predicted a cyclone track that would be conducive to snow in the mid-Atlantic region, corresponding to a roughly 25% chance of this occurrence.

As operational ensembles continue to grow in size and resolution (the European Center currently runs a 50-member ensemble at the time of this writing), it will become increasingly common to derive probabilistic forecast information from such ensembles. In the United States, the Storm Prediction Center has undertaken specialized product development based on the NCEP SREF. For example, based on this ensemble, probabilistic forecasts of specific convective ingredients are now generated. Figure 10.46 displays the SREF-based probability that the most unstable convective available potential energy will exceed 2000 J $kg^{-1}$ for the 30-h forecast valid 21 UTC 23 Apr 2010.

**10.6.3.4. Targeted observations.** When examining an ensemble forecast, one may be able to discern a particular region in which the ensemble members begin to diverge. It is possible using various techniques (essentially running a linear version of the model backward in time) to identify the source region of the forecast divergence early in the forecast cycle. Then, if mobile observational

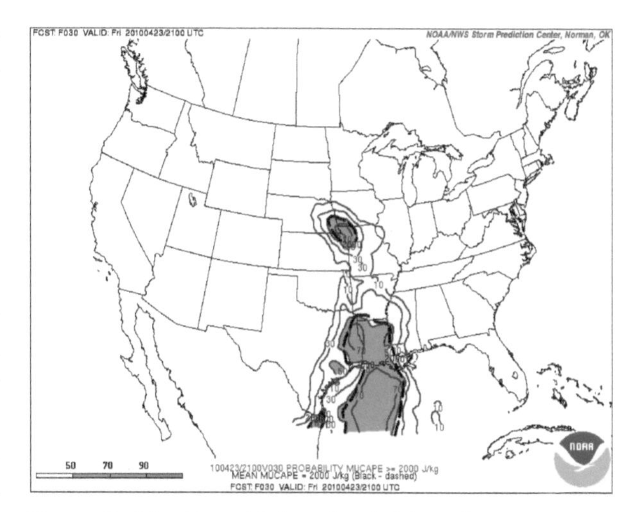


**Figure 10.46.** Ensemble-based probability of the most unstable CAPE exceeding 2000 J $kg^{-1}$ at 2100 UTC 23 Apr 2010, based on 21-member NCEP SREF (for background, see http://www.meted.ucar.edu/nwp/pcu2/ens_matrix/; forecast graphic from SPC experimental Web page).

resources are available, such as dropsondes from a P-3 Orion aircraft, the forecast uncertainty can be reduced and the forecast improved by sampling within the critical region. This strategy was tested over the Atlantic during a field experiment known as the Fronts and Atlantic Storm Track Experiment (FASTEX), and targeted observations have been obtained from aircraft over the Pacific Ocean during several recent winter seasons.

### 10.6.4. Limitations of ensemble prediction

The advantages outlined in the previous section make it clear that ensemble prediction is now a well established approach to a wide range of forecasting problems, with additional applications and expansion likely in the future. Increased emphasis on forecast confidence and uncertainty will likely alter the manner in which forecast information is conveyed to the general public as well as decision makers. However, this is not to say that ensemble forecasting is a "silver bullet" that can solve all of our forecasting challenges. In some situations, the atmosphere diverges outside the ensemble envelope, as schematically represented in Fig. 10.47b.

Large errors in the initial condition (e.g., a poorly analyzed tropical cyclone) can result in rapid error growth in a model forecast. In such situations, the small perturbations or model physics variations in an ensemble system do not accurately reflect the large uncertainties in the model initial conditions. In other situations, model deficiencies

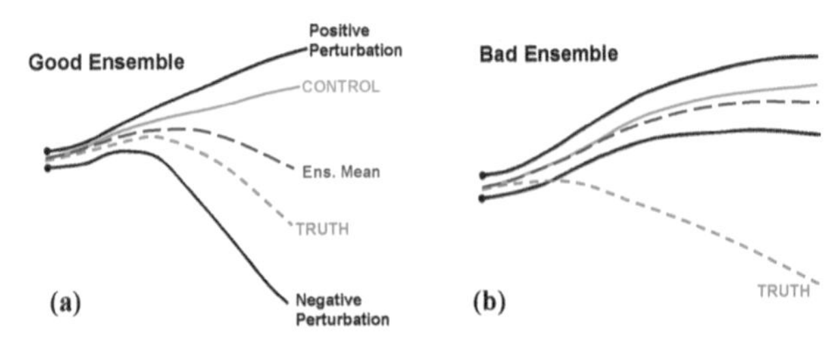


**Figure 10.47.** Hypothetical examples of (a) desirable and (b) undesirable ensemble characteristics. Graphs represent the forecast evolution with time of an arbitrary model output variable (from Kalnay 2003).

may lead to a solution that falls outside of the ensemble envelope.

Another current challenge to ensemble prediction lies with the physical and social science communities to develop more effective means to educate the general public about the value of probabilistic forecasts and forecast uncertainty. Concerning the general public, new means must be devised for presenting forecast uncertainty in useful, straightforward ways. Often, binary decisions must be made (e.g., the school district superintendent must either announce that schools are closed or not the evening before an anticipated winter storm), and decision makers will need guidance for how to optimally utilize probabilistic information to inform such decisions. Despite these challenges, continued development of ensemble prediction represents a growth area that spans the atmospheric sciences, including applications in climate projection and air-quality forecasting as well as in many different areas of synoptic-scale and mesoscale weather prediction.

## 10.7. MODEL CONFIGURATION AND OUTPUT STATISTICS

It is an exercise in futility to present details of current operational NWP model configurations because of the rapid pace of model changes, computer system upgrades, and evolution of physical parameterization packages. However, some basic aspects of the modeling systems are relatively stable and can be usefully summarized. For an up-to-date account of upgrades and changes to the suite of operational NWP systems, the reader is referred to the outstanding COMET Operational Models Matrix Web site. This site includes change information for individual models as well as general educational information about the modeling systems (available online at http://www.meted.ucar.edu/nwp/pcu2/).

### 10.7.1. Some current operational models

In the interest of brevity, this summary is restricted to the most popular NWP models currently run at the U.S. NCEP:

1. NAM model (i.e., NMM)
2. GFS model (and ensemble system)
3. Rapid Update Cycle (RUC) model
4. SREF (a combination of models)
5. HWRF (which is NMM, specially configured)

**10.7.1.1. NAM model.** At times, model naming and renaming can be confusing. The North American Mesoscale model used to be another name for the NCEP Eta Model, but the Eta Model was replaced in June 2006 by the Weather Research and Forecasting Nonhydrostatic Mesoscale Model (WRF-NMM) *while retaining the NAM label.* There is also a separate dynamical core for the WRF model, developed at NCAR, the "Advanced Research WRF", or WRF-ARW. The WRF-ARW is distinct from the WRF-NMM (i.e., NAM), although both models are part of a similar modeling framework. NCEP does run the WRF-ARW for some applications: for instance, at the time of this writing, some members of the SREF ensemble are based on the WRF-ARW.

The NAM is a state-of-the-art *nonhydrostatic* mesoscale model, using a gridpoint architecture and run over a limited-area domain centered over North America. Recall that, by *nonhydrostatic*, we mean that this model solves the vertical equation of motion, rather than diagnosing the vertical motion from vertical integration of the continuity equation. This aspect matters most for organized convective storms and during strong flow in the vicinity of steep terrain. However, the current operational NAM, run with 12-km grid spacing, may still not be sufficiently high in resolution to fully realize the benefits of the nonhydrostatic configuration.

The NAM is a *regional* or a *limited-area* model, meaning that, during the forecast cycle, it needs to be supplied with forecast information at the *lateral boundaries.* This is especially critical at the western (upstream) boundary of the model domain. The lateral boundary conditions used in the operational NAM are provided by the *previous run* of the GFS model. So, near the western edge of the domain, *the NAM forecast is influenced by an older GFS model forecast.* The operational NAM was scheduled to

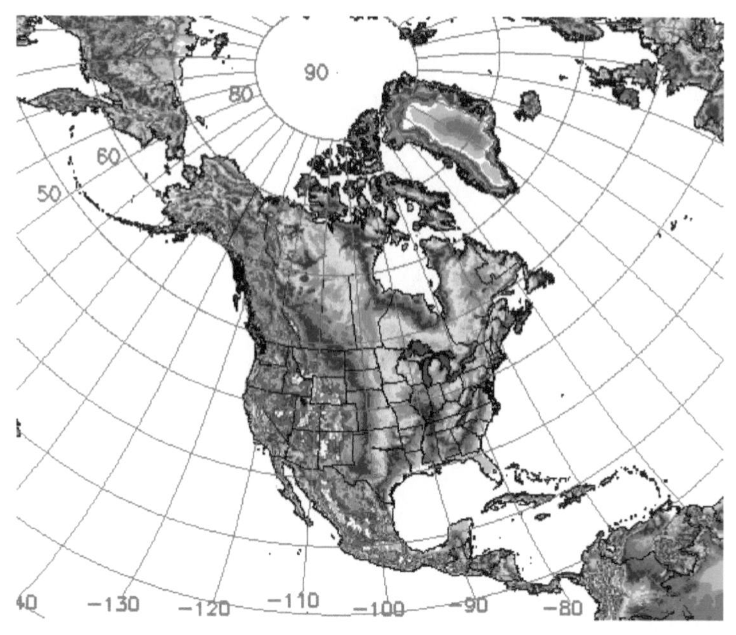


**Figure 10.48.** NAM topography (m, shaded as in legend at right; courtesy of the COMET program, Operational Models Matrix Web page).

add higher-resolution nested domains within the 12-km domain during fall 2011.

The data assimilation system associated with the NAM is now the Global Statistical Interpolation (GSI) system, cycled every 3 h, with NAM forecasts run out to 84 h four times per day, at 0000, 0600, 1200, and 1800 UTC. The NAM is designed for efficiency and rapid dissemination to the forecasting community; therefore, the DA system has a somewhat early *data cutoff* relative to the GFS model. The data cutoff is essentially how long the system waits to collect observations for incorporation into the model initial condition via the DA system. At the time of this writing, the NAM data cutoff time was 1 h, 15 min.

The grid spacing of the NAM is currently 12 km, although there are "high-resolution window" (HRW) NAM runs available with 8- or 5-km grid spacing, and operational versions of the NAM run at 4-km grid spacing are available as of 2011. The output from the NAM (and other models) is interpolated to a variety of different grid projections for distribution, including the "WMO 218" grid featuring 12-km grid spacing, a 20-km grid (the WMO 215 grid), a 40-km grid (the WMO 212 grid), an 80-km grid (the WMO 211 grid), and other specialized grid projections.

Representation of topography is a strength of this model, in part because of the high resolution relative to the GFS model (Fig. 10.48). Also, being a nonhydrostatic model, potential benefits exist for the prediction of organized convection and topographically forced flows such as downslope windstorms. The high resolution is better able to represent small-scale features related to lakes (e.g., lake-effect snow), topography (e.g., gap wind events), and coastlines.

Because the operational NAM still utilizes a CP scheme, the full benefit of its nonhydrostatic configuration and higher resolution relative to the GFS is not yet realized.

**10.7.1.2. GFS model.** Prior to fall 2002, the NCEP global spectral model was known as the Aviation model (AVN), but the name was changed to Global Forecast System (GFS). This global spectral model provides forecasts for the entire world out to 16 days. This is a hydrostatic model that originally came online in August 1980. Currently, the GFS runs 4 times daily out to 384 hours (16 days). The resolution of the model currently decreases beyond forecast hour 192. The resolution degradation is evident in model output graphics, but it should be noted that there is little forecast skill remaining toward the end of the 16-day forecasts anyway. The basic model is the same beyond 192 h, but with decreased resolution.

Being a global model, the GFS serves a distinct purpose relative to the NAM. It can provide forecasts for remote regions and can also be used as initial or boundary

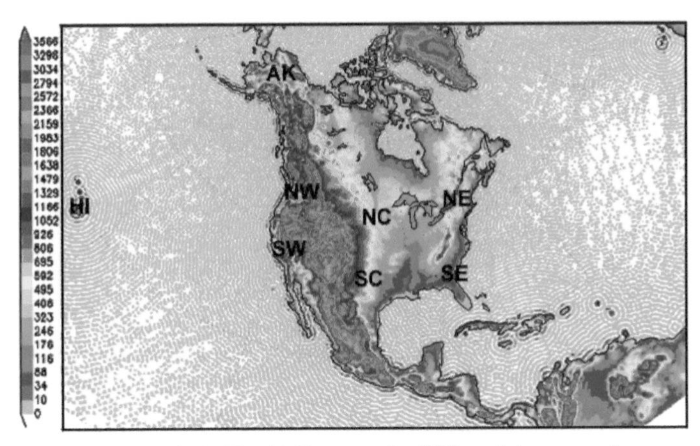


**Figure 10.49.** As in Fig. 10.48 except for GFS model topography (source: COMET MetEd Web page, Operational Models Matrix).

condition data for other regional models run in other parts of the world, including the NAM.

Due to global coverage, the GFS has effectively coarser resolution relative to the NAM. It is a *spectral* model, meaning that it has a fundamentally different way of solving the primitive equations, as discussed in section 10.3.3. The GFS resolution is roughly equivalent to that of a gridpoint model with a ~27-km grid.

One important advantage of the GFS over the NAM is that it has a *later data cutoff*: 2 h, 45 min versus 1 h, 15 min for the NAM. This means that additional data are incorporated into the GFS initial condition, and this can provide an advantage in some situations, especially in maritime regions, or when oceanic storms move ashore.

The GFS model topography is coarser than for the NAM (Fig. 10.49), as expected because of the relatively low resolution. This is most critical in regions of complex terrain or where mesoscale surface features lead to specific atmospheric phenomena.

The GFS model has its own ensemble forecast system, GEFS, as discussed in section 10.6. Recently, this ensemble system has been combined with the Canadian ensemble system to produce a larger ensemble, known as the North American Ensemble Forecast System (NAEFS).

**10.7.1.3. RUC model.** The RUC model, which first came online in 1994, is designed for short-range forecasting applications, especially those related to aviation. It includes its own analysis system and also a research version known as the Mesoscale Analysis and Prediction System (MAPS). As of this writing, the RUC is due to be replaced by a version of the WRF model, which will be called the "Rapid Refresh." The RUC model was developed at the former NOAA Forecast Systems Lab (FSL) in Boulder, Colorado, but this division is called the Earth Systems Research Lab (ESRL) in the Global Systems Division (GSD) at the time of this writing.

The RUC model is run *hourly*, out to 18-h lead time beginning in March 2010; previously, the run length had been 3 or 12 h, depending on the run. This model is therefore useful for short-term forecasts, especially for high-impact weather. The RUC runs at the highest frequency of any forecast model at NCEP, assimilating recent observations to provide high-frequency updates of current conditions and short-range forecasts.

The grid spacing of the RUC was lowered from 20 to 13 km on 28 June 2005. The RUC is excellent for short-term analysis and forecasting, and it includes sophisticated model physics and terrain.

The RUC data assimilation system makes excellent use of ACARS data, and it appears to more strongly represent surface observations in the initial conditions than the NAM or GFS. Because of the rapid cycling and short lead times for the RUC model, a diabatic digital filter system, implemented in 2006, allows the use of radar reflectivity while filtering in a way that allows the maintenance of dynamical balance.

The RUC, with hourly analyses, provides an objective means for the evaluation of short-term forecasts from other models, as well as monitoring current weather conditions. The RUC can be used in conjunction with satellite, radar, and other observational data to assemble a picture of the current forecasting situation.

**10.7.1.4. Operational ensemble prediction systems.** Two ensemble forecasting systems are run at NCEP, each with a different purpose. The GFS model ensemble system, known as the Medium Range Ensemble Forecast (MREF) system was the original NCEP ensemble system, designed as an *initial condition* ensemble as previously discussed in conjunction with the GFS model. The NCEP SREF became operational in 2001 and is a *multimodel* ensemble with physics perturbations of different models.

As of this writing, the SREF included 21 members and ran 4 times daily, at 0300, 0900, 1500, and 2100 UTC, out to 87 h. These times are chosen so as to minimize overlap with other model runs on the busy NCEP computer schedule. To maintain a reasonable level of computational expense, the members are run at relatively coarse grid spacing, currently 32 or 35 km. A broad variety of model architectures, convective parameterization schemes, and initial conditions are used in the SREF. The GFS control run and GFS ensemble members are used for lateral boundary conditions for this limited-area model ensemble. The gridded output from the SREF is broadcast to the community, allowing additional value-added development, such as that shown in Fig. 10.46 from the Storm Prediction Center.

Another groundbreaking, experimental ensemble system has been developed by the National Severe Storms Laboratory (NSSL) and SPC in Norman, Oklahoma. The *Spring Experiment* is run annually during severe weather season in the United States, and it features a rotating team of visiting experts working with a dedicated group of scientists from NSSL/SPC to evaluate the ability of a very-high-resolution ensemble in the prediction of severe convection. The ensemble members are all run with sufficiently small grid length to allow *explicit* treatment of convection (no CP scheme), with grid lengths typically ranging from 1 to 4 km. Model runs are from the WRF-ARW and WRF-NMM, and are not operationally available outside of the experiment; nevertheless, it is important to mention this type of ensemble system, as it represents what the future holds in ensemble prediction.

Several different groups at various locations within and outside of the United States are running high-resolution ensemble forecasts, and many National Weather Service offices are running workstation versions of the WRF model in house. In North Carolina, a suite of deterministic model runs were organized into an ensemble system via a data sharing and model coordination activity. These developments are likely to continue and grow in number in the coming years, which is an indication that it will be increasingly beneficial for students of the atmospheric sciences to educate themselves in the area of numerical modeling and ensemble prediction.

### 10.7.2. Model output statistics

Raw output from numerical models, even when interpolated to common sensor altitudes such as 2 or 10 m, may still contain biases and exhibit other limitations. Examples of limitations could include unresolved mesoscale features that are climatologically common, such as land–sea or mountain–valley circulations. Additionally, it is desirable to obtain numerical forecast guidance for parameters that are not explicitly forecasted by NWP models, such as the probability of thunder, severe thunderstorms, or hail. Using statistical methods, select NWP output parameters can be combined with climatological information and past observations to improve the prediction of sensible weather parameters and to generate new forecast variables. These statistical methods are known as *model output statistics* (MOS). The accuracy of MOS has steadily improved since it was first initiated in the 1970s (Fig. 10.50), because of improved models and also because of improved statistical regression equations (which are partly the result of a constantly growing observational database with which to establish the statistical relationships).

Using multiple linear regression equations for each forecasted variable (predictand), statistical relations between past observations of that variable and a set of NWP output variables and climatological data (predictors) are determined. There are separate regression equations for warm and cold seasons, and there are regional differences in the equations as well. The MOS equations for maximum temperature (predictand) may include as predictors the NWP output temperatures at various vertical levels, measures of vertical motion, the stability of the lower atmosphere, humidity, wind speed and direction, and other parameters computed directly from the model output. Climatological information for that station, such as the long-term climatological average maximum temperature for that date, would also be included in the statistical relationship. Once the statistical relationships between past

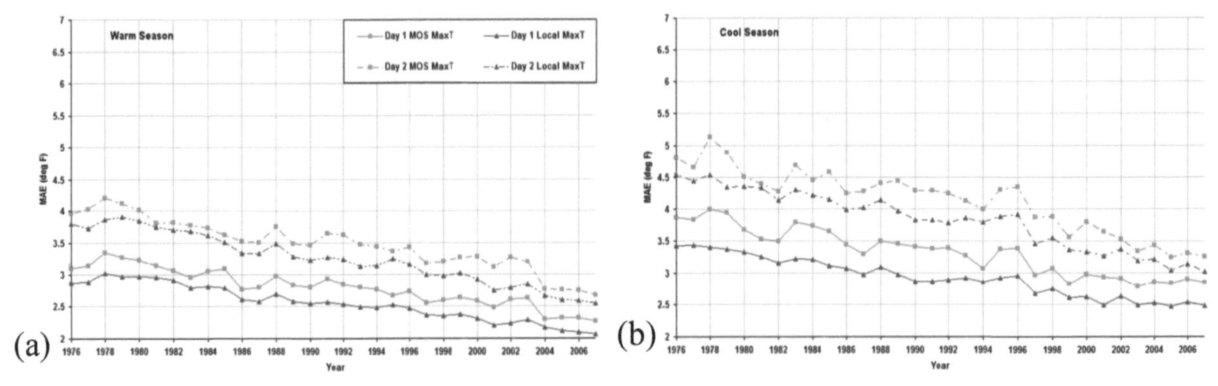


**Figure 10.50.** Day 1 and 2 mean absolute error in MOS and local NWS maximum temperature forecasts for 80 stations in the United States by year: (a) warm season (April–September) and (b) cool season (October–March). These are based on 0000 UTC guidance, and the year corresponds to when season began (from Ruth et al. 2009).

observed maximum temperatures at that station and the past model and climatological values are established, the prediction equations are available for use with future model output data.

The MOS approach can improve the representation of some (but not all) effects not resolved by the parent model. For example, suppose that we are considering the 3-hourly and maximum temperature forecasts for Atlantic City, New Jersey, based on the GFS forecast. The GFS does not have the horizontal resolution to adequately resolve the sea breeze circulation, which often limits the maximum temperature on warm spring days with light synoptic-scale flow. However, MOS, using wind speed as a predictor, would capture this effect due to the repeated occurrence of the sea breeze in the past during light-wind conditions on warm days; the regression equations would show that, with light or onshore flow and very warm temperatures in the raw GFS 2-m temperature forecast, the temperature would tend to be clipped during the afternoon hours (relative to locations farther inland), because of past occurrences of the sea breeze.

Another advantage of MOS is the ability to develop predictions for quantities not explicitly forecasted by the model, such as the probability of thunderstorms, obstructions to visibility, and cloud-ceiling height. An example of a MOS product in tabular form is included in Fig. 10.51, along with some decoding information. The set of stations for which MOS forecasts are available is shown in Fig. 10.52.

The statistical equations that form the basis of MOS must be developed over a considerable period of time to "train" the system with sufficient data. Thus, a new NWP

**Sample Message**

```
KDEN   GFS MOS GUIDANCE    3/04/2010  1200 UTC
DT /MAR    4/MAR   5                /MAR   6                /MAR   7
HR    18 21 00 03 06 09 12 15 18 21 00 03 06 09 12 15 18 21 00 06 12
N/X                     27          53          28          51    24
TMP   48 52 51 40 35 32 30 35 47 50 48 40 38 35 32 36 45 48 45 31 27
DPT   25 25 27 29 29 27 25 25 21 17 18 20 21 21 20 21 21 22 22 23 21
CLD   BK SC SC SC SC OV OV OV BK BK BK SC BK SC SC SC SC SC SC SC SC
WDR   19 14 13 16 20 20 25 25 28 28 28 22 23 23 22 21 12 10 10 17 21
WSP   05 12 11 08 08 07 07 07 08 10 11 08 08 07 07 07 07 10 10 08 07
P06             1     6    14    20    14    25     8     4     2  3  2
P12                     15          27          26           4     3
Q06             0     0     0     0     0     0     0     0     0  0  0
Q12                      0           0           0           0     0
T06          5/11  1/ 2  0/ 1  1/ 2  8/ 3  1/ 1  0/ 0  0/ 0  2/ 8  0/ 0
T12                7/11        1/ 2        8/ 4        0/ 1     2/ 8
POZ    1  1  3  2  2  7  8  3  4  2  3  0  3  3  4  6  3  1  2  0  3
POS   27 16 18 24 28 65 74 72 31 20 37 57 94 87 95 65 41 18 28 50 91
TYP    R  R  R  R  R  S  S  S  R  R  R  S  S  S  S  S  R  R  R  R  S
SNW                      0                       0                 0
CIG    8  8  8  8  8  8  8  7  8  8  7  8  8  8  8  8  8  8  8  8  8
VIS    7  7  7  7  7  7  7  1  7  7  7  7  7  7  7  7  7  7  7  7  7
OBV    N  N  N  N  N  N  N  N  N  N  N  N  N  N  N  N  N  N  N  N  N
```

- **DT** = The day of the month, denoted by the standard three or four letter abbreviation
- **HR** = Hour of the day in UTC time. This is the hour at which the forecast is valid, or if the forecast is valid for a period, the end of the forecast period.
- **N/X** = nighttime minimum/daytime maximum surface temperatures.
- **TMP** = surface temperature valid at that hour.
- **DPT** = surface dewpoint valid at that hour.
- **CLD** = forecast categories of total sky cover valid at that hour.
- **WDR** = forecasts of the 10-meter wind direction at the hour, given in tens of degrees.
- **WSP** = forecasts of the 10-meter wind speed at the hour, given in knots.
- **P06** = probability of precipitation (PoP) during a 6-h period ending at that time.
- **P12** = PoP during a 12-h period ending at that time.
- **Q06** = quantitative precipitation forecast (QPF) category for liquid equivalent precipitation amount during a 6-h period ending at that time.
- **Q12** = QPF category for liquid equivalent precipitation amount during a 12-h period ending at the indicated time.
- **SNW** = snowfall categorical forecasts during a 24-h period ending at the indicated time.
- **T06** = probability of thunderstorms/conditional probability of severe thunderstorms during the 6-hr period ending at the indicated time.
- **T12** = probability of thunderstorms/conditional probability of severe thunderstorms during the 12-hr period ending at the indicated time.
- **POZ** = conditional probability of freezing pcp occurring at the hour.
- **POS** = conditional probability of snow occurring at the hour.
- **TYP** = conditional precipitation type at the hour.
- **CIG** = ceiling height categorical forecasts at the hour.
- **VIS** = visibility categorical forecasts at the hour.
- **OBV** = obstruction to vision categorical forecasts at the hour.

**Figure 10.51.** Tabular example of sample MOS output from GFS model, with decoding information included.

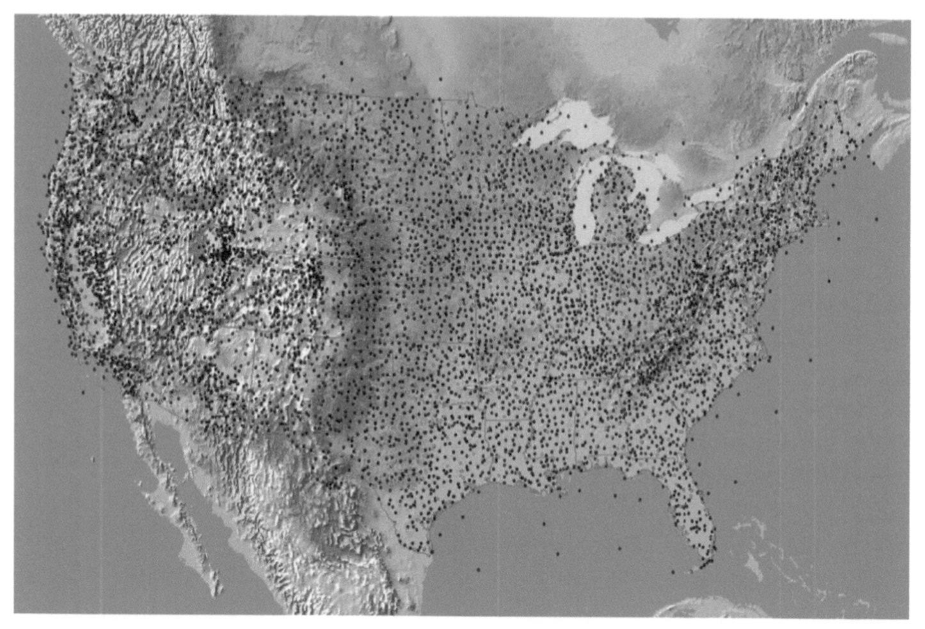

**Figure 10.52.** Locations of 1,200 MOS stations used for verification of NWS gridded forecasts (red dots) and all MOS stations (black dots; graphic from Glahn et al. 2009)

model or changes to an existing one will introduce error into the MOS equations, because the statistical relations between the NWP model output and the observations could change because of the model changes. Another related limitation of MOS is that, during extreme weather conditions, the regression equations may not have sufficient data to properly represent these situations, potentially reducing accuracy.

Rather than provide MOS forecasts in tabular format for selected cities, now the MOS data are available in gridded form, and continuous fields of the forecast variables are available (Fig. 10.53a). This is compatible with the official NWS forecasts, which are now also produced on a grid (with 5-km horizontal grid spacing).

Chapter 11 explores weather forecasting in more detail, and additional suggestions for the use of MOS in the

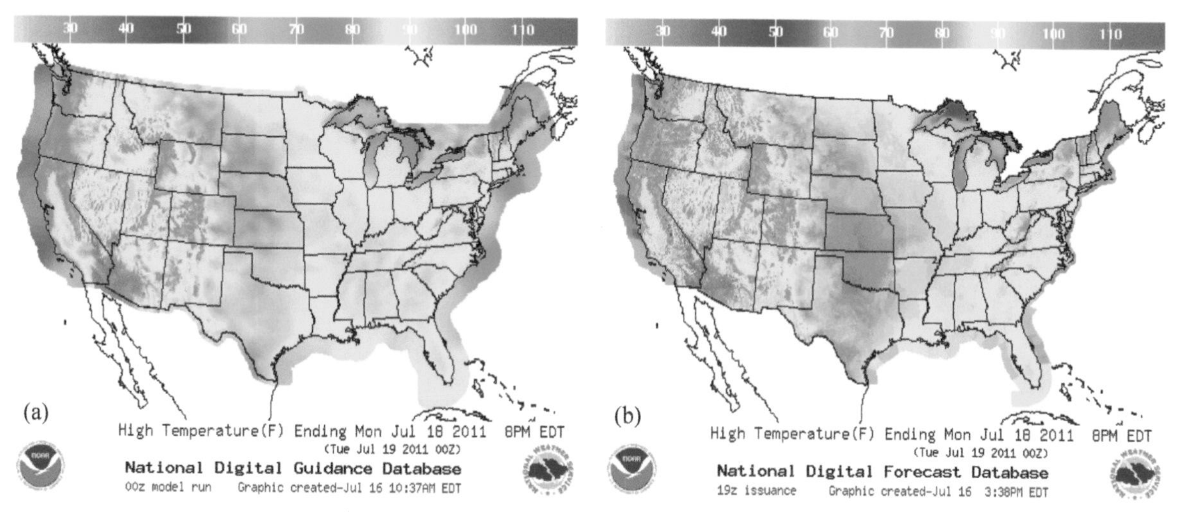


**Figure 10.53.** Comparison of gridded MOS and the National Weather Service Digital Forecast Database valid 0000 UTC 19 July 2011.

forecast process are offered there. Some summary points concerning MOS are provided below:

1. MOS is not direct model output. It is a statistically generated forecast based on multiple linear regressions between past observations and various model output and climatological parameters.
2. MOS can actually account for some (but not all!) processes that are not resolved by models, such as a land/sea breezes, mountain/valley breezes, gap winds, or other features that are climatologically persistent in certain synoptic regimes.
3. MOS is currently available for the NAM and GFS; also, an ensemble MOS product is available for the GEFS.
4. Users of MOS should exercise caution around 1 October and 31 March; there are two different sets of seasonally defined MOS equations, and they change on these dates.

## REVIEW AND STUDY QUESTIONS

1. Why are there two different "flavors" of precipitation in some NWP models? Why is it advantageous for forecasters or others using the model output to distinguish between the two?
2. Physically, how might a model convective parameterization (CP) scheme alter a model forecast? Why are CP schemes necessary in some model runs? What characterizes model configurations in which they are not necessary?
3. Suppose that, in comparing the model analysis (initial condition) to observed rawinsonde data, there is an obvious discrepancy in some parameter. Does this necessarily imply that the model analysis is flawed? Explain.
4. Briefly summarize the data assimilation process. What kinds of information are used, and how is this information combined to form the best possible analysis?
5. What specific types of information are available from ensemble prediction systems that are not available from a single deterministic model run?
6. What are the different ways that ensemble forecasts are made? What methods are used to perturb the forecasts? What are the differences between the NCEP GEFS and SREF?
7. For each of the following forecasting tasks, state whether you would be better off using the NAM or the GFS model and briefly explain why.
    i. You are tasked with forecasting for the western Aleutian Islands in Alaska.
    ii. You are forecasting for a windsurfing competition near the coast of Florida. You expect local sea breezes to be important.
    iii. You are forecasting for Charleston, South Carolina, during a time when a hurricane is approaching the region from the southeast.
    iv. You are forecasting an event with the potential for severe downslope winds in Colorado.

## PROBLEMS

1. Evaluate the partial derivative (10.12) to obtain (10.13).
2. Obtain (10.14) by minimizing (10.15).

## REFERENCES


Baldwin, M. E., J. S. Kain, and M. P. Kay, 2002: Properties of the convection scheme in NCEP's Eta Model that affect forecast sounding interpretation. *Wea. Forecasting,* **17,** 1063–1079.

Barker, D. M., W. Huang, Y. R. Guo, and Q. N. Xiao, 2004: A three-dimensional (3DVAR) data assimilation system for use with MM5: Implementation and initial results. *Mon. Wea. Rev.,* **132,** 897–914.

Benjamin, S. G., B. D. Jamison, W. R. Moninger, S. R. Sahm, B. E. Schwartz, and T. W. Schlatter, 2010: Relative short-range forecast impact from aircraft, profiler, radiosonde, VAD, GPS-PW, METAR, and mesonet observations via the RUC hourly assimilation cycle. *Mon. Wea. Rev.,* **138,** 1319–1343.

Bjerknes, V., 1904: The problem of weather forecasting as a problem in mechanics and physics. *Meteor. Z.* **21,** 1–7.

Bjerknes, V., 1914: Meteorology as an exact science. *Mon. Wea. Rev.,* **42,** 11–14.

Bukovsky, M. S., J. S. Kain, and M. E. Baldwin, 2006: Bowing convective systems in a popular operational model: Are they for real? *Wea. Forecasting,* **21,** 307–324.

Charney, J. G., 1948: On the scale of atmospheric motions. *Geofys. Publ.,* **17,** 3–17.

——, 1949: On a physical basis for numerical prediction of large-scale motions in the atmosphere. *J. Meteor.,* **6,** 371–385.

——, 1951: Dynamic forecasting by numerical process. *Compendium of Meteorology,* T. F. Malone, Ed., American Meteorological Society, 470–482.

——, R. Fjørtoft, and J. von Neumann, 1950: Numerical integration of the barotropic vorticity equation. *Tellus,* **2,** 237–254.

Colle, B. A., C. F. Mass, and K. J. Westrick, 2000: MM5 precipitation verification over the Pacific Northwest during the 1997–99 cool seasons. *Wea. Forecasting,* **15,** 730–744.

Courant, R., K. Friedrichs, and H. Lewy, 1928: On the partial difference equations of mathematical physics. *Mathematische Annalen,* **100,** 32–74.

Fritsch, J. M., J. Hilliker, J. Ross, and R. L. Vislocky, 2000: Model consensus. *Wea. Forecasting,* **15,** 571–582.

Gallus, W. A., 1999: Eta simulations of three extreme precipitation events: Sensitivity to resolution and convective parameterization. *Wea. Forecasting,* **14,** 405–426.

——, 2002: Impact of verification grid-box size on warm-season QPF skill measures. *Wea. Forecasting,* **17,** 1296–1302.

Gilleland, E., D. A. Ahijevych, B. G. Brown, and E. E. Ebert, 2010: Verifying forecasts spatially. *Bull. Amer. Meteor. Soc.,* in press.

Glahn, B., K. Gilbert, R. Cosgrove, D. P. Ruth, and K. Sheets, 2009: The gridding of MOS. *Wea. Forecasting,* **24,** 520–529.

Hamill, T. M., and S. J. Colucci, 1997: Verification of Eta-RSM short-range ensemble forecasts. *Mon. Wea. Rev.,* **125,** 1312–1327.

Han, J., and H.-L. Pan, 2006: Sensitivity of hurricane intensity forecast to convective momentum transport parameterization. *Mon. Wea. Rev.,* **134,** 664–674.

Harper, K. C., 2008: *Weather by the Numbers: The Genesis of Modern Meteorology.* MIT Press, 328 pp.

Illari, L., 2009: "Weather in a tank" exploiting laboratory experiments in the teaching of meteorology, oceanography, and climate. *Bull. Amer. Meteor. Soc.,* **90,** 1619–1632.

Kain, J. S., M. E. Baldwin, and S. J. Weiss, P. R. Janish, M. P. Kay, and G. Carbin, 2003a: Subjective verification of numerical models as a component of a broader interaction between research and operations. *Wea. Forecasting,* **18,** 847–860.

——, P. R. Janish, S. J. Weiss, M. E. Baldwin, R. S. Schneider, and H. E. Brooks, 2003b: Collaboration between forecasters and research scientists at the NSSL and SPC: The spring program. *Bull. Amer. Meteor. Soc.,* **84,** 1797–1806.

Kalnay, E, 2003: *Atmospheric Modeling, Data Assimilation, and Predictability.* Cambridge University Press, 341 pp.

Lorenz, E. N., 1965: A study of the predictability of a 28-variable atmospheric model. *Tellus,* **17,** 321–333.

Lynch, P., 1992: Richardson's barotropic forecast: A reappraisal. *Bull. Amer. Meteor. Soc.,* **73,** 35–47.

——, 1999: Richardson's marvelous forecast. . *Extratropical Cyclones: The Erik Palmén Memorial Volume,* C. W. Newton and E. O. Holopainen, Eds., Amer. Meteor. Soc., 61–73.

——, 2006: *The Emergence of Numerical Weather Prediction: Richardson's Dream.* Cambridge University Press, 290 pp.

Mahoney, K. M., and G. M. Lackmann, 2006: The sensitivity of numerical forecasts to convective parameterization: A case study of the 17 February 2004 East Coast cyclone. *Wea. Forecasting,* **21,** 465–488.

Mass, C. F., D. Ovens, K. Westrick, and B. A. Colle, 2002: Does increasing horizontal resolution produce more skillful forecasts? The results of two years of real-time numerical weather prediction over the Pacific Northwest. *Bull. Amer. Meteor. Soc.,* **83,** 407–430.

Molinari, J., and M. Dudek, 1992: Cumulus parameterization in mesoscale numerical models: A critical review. *Mon. Wea. Rev.,* **120,** 326–344.

Moninger, W. R., S. G. Benjamin, B. D. Jamison, T. W. Schlatter, T. L. Smith, and E. J. Szoke, 2010: Evaluation of regional observations using TAMDAR. *Wea. Forecasting,* **25,** 627–645.

Richardson, L. F., 1922: *Weather Prediction by Numerical Process,* 2nd ed., Cambridge, 236 pp.

Ruth, D. P., B. Glahn, V. Dagostaro, and K. Gilbert, 2009: The performance of MOS in the digital age. *Wea. Forecasting,* **24,** 504–519.

Simmons, A. J., and A. Hollingsworth, 2002: Some aspects of the improvement in skill of numerical weather prediction *Q. J. R. Meteorol. Soc.,* **128,** 647–677.

Stensrud, D. J., 2007: *Parameterization Schemes: Keys to Understanding Numerical Weather Prediction Models.* Cambridge University Press, 459 pp.

——, H. E. Brooks, J. Du, M. S. Tracton, and E. Rogers, 1999: Using ensembles for short-range forecasting. *Mon. Wea. Rev.,* **127,** 433–446.

Weisman, M. L., W. C. Skamarock, and J. B. Klemp, 1997: The resolution dependence of explicitly modeled convective systems. *Mon. Wea. Rev.,* **125,** 527–548.

——, C. A. Davis, W. Wang, and K. Manning, 2008: Experiences with 0–36-h explicit convective forecasts with the WRF-ARW model. *Wea. Forecasting,* **23,** 407–437.

Zapotocny, T. H., and Coauthors, 2000: A case study of the sensitivity of the Eta Data Assimilation System. *Wea. Forecasting,* **15,** 603–615.

## FURTHER READING

Benjamin, S. G., G. A. Grell, J. M. Brown, T. G. Smirnova and R. Bleck, 2004: Mesoscale weather prediction with the RUC hybrid isentropic–terrain-following coordinate model. *Mon. Wea. Rev.,* **132,** 473–494.

Chen, F. and J. Dudhia, 2001: Coupling an advanced land surface–hydrology model with the Penn State–NCAR MM5 modeling system. Part I: Model implementation and sensitivity. *Mon. Wea. Rev.,* **129,** 569–585.

Coniglio, M. C., K. L. Elmore, J. S. Kain, S. J. Weiss, M. Xue, and M. Weisman, 2010: Evaluation of WRF model output for severe weather forecasting from the 2008 NOAA Hazardous Weather Testbed Spring Experiment. *Wea. Forecasting,* **25,** 408–427.

Ek, M. B., K. E. Mitchell, Y. Lin, E. Rogers, P. Grummann, V. Koren, G. Gayno, and J. D. Tarpley, 2003: Implementation of Noah land surface model advances in the National Centers for Environmental Prediction operational mesoscale Eta model. *J. Geophys. Res.*, **108**, 8851, doi:10.1029/2002JD003296.

Jacks, E., J. B. Bower, D. J. Dagostaro, J. P. Dallavalle, M. C. Erickson, and J. C. Su, 1990: New NGM-based MOS guidance for maximum/minimum temperature, probability of precipitation, cloud amount, and surface wind. *Wea. Forecasting*, **5**, 128–138.

Kain, J. S., 2004: The Kain–Fritsch convective parameterization: An update. *J. Appl. Meteor.*, **43,** 170–181.

——, and Coauthors, 2008: Some practical considerations regarding horizontal resolution in the first generation of operational convection-allowing NWP. *Wea. Forecasting,* **23,** 931–952.

Roebber, P. J., D. M. Schultz, B. A. Colle and D. J. Stensrud, 2004: Toward improved prediction: High-resolution and ensemble modeling systems in operations. *Wea. Forecasting,* **19,** 936–949.

Thompson, P. D., 1977: How to improve accuracy by combining independent forecasts. *Mon. Wea. Rev.,* **105,** 228–229.

CHAPTER 11

# Weather Forecasting

*It is possible that more than one solution may be found to the problem of satisfactory and practical weather forecasting. It is also possible that among these we may be able to find methods which obviate the necessity of a complete understanding of the phenomenon whose development we are to forecast. Personally, I have no interest in such methods. I am interested in only that method which is based upon a full understanding of the phenomenon involved.*

—V. Bjerknes, *Weather Forecasting.* Delivered at the meeting of Scandinavian geophysicists, Gothenburg, 28 August 1918.

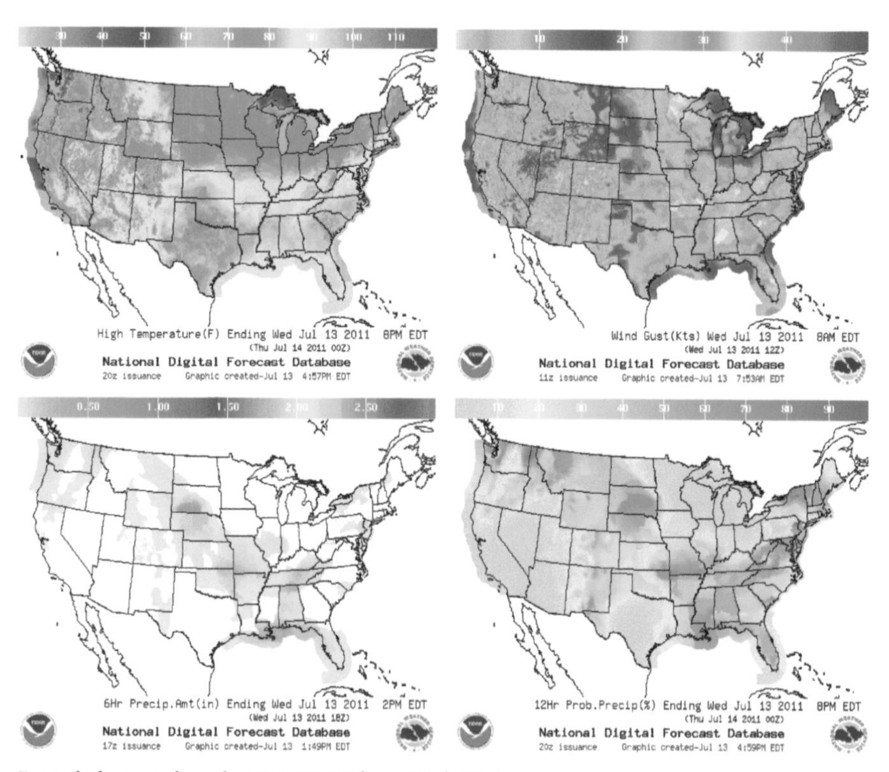


Example forecasts from the NWS NDFD from 13 July 2011.

## 11.1. PHILOSOPHICAL CONSIDERATIONS

The role of humans in weather forecasting has changed in response to technological, scientific, and societal factors. However, in recent years the pace of change has accelerated, and some have gone so far as to suggest that the human element in forecasting should be redefined (e.g., Mass 2003; Doswell 2003b, 2004; Stuart et al. 2006). One driver of change has been the recognition of the importance of weather information in a number of nontraditional applications; another factor has been improved science and technology, particularly the advent of the Internet and other forms of electronic communication, and the increasing speed of computers.

As the field of weather forecasting has diversified, the applications for which weather information is used have grown in number and scope. Weather commodities are traded in some financial markets, some retail chains employ forecasters to optimize weather-related sales, and many insurance and reinsurance companies either have meteorologists on staff or hire meteorological consulting services; this is also the case for the majority of energy companies. Customized transportation forecasts for road conditions, in addition to forecasts tailored for shipping and aviation, are routinely delivered by commercial forecasting services. In many of these examples, meteorologists are directly involved with nonmeteorological applications, and a multidisciplinary background can provide benefit. Other examples include air-quality forecasting, which requires combined knowledge in meteorology and atmospheric chemistry, and fire-weather forecasting benefits from knowledge of forestry as well as the command structure for fire response. Understanding of the financial and economic applications of weather information requires training outside of the traditional physical science disciplines.

Although some forecasting occupations, such as broadcast meteorology, have evolved relatively slowly, the role of forecasters in National Weather Service (NWS) weather forecast offices (WFOs) changed radically in 2003 (after a period of testing). Rather than issue text-based forecasts for a specified county warning area (CWA), forecasters now generate gridded digital forecast products from which text forecasts are derived [the National Digital Forecast Database (NDFD; Glahn and Ruth 2003)]. Both the means of making the forecast, and the ways that forecast information is delivered to the user community have changed as a result (Glahn and Ruth 2003; Mass 2003; Stuart et al. 2006). Some other operational centers that have developed grid-editing forecasting capabilities include the United Kingdom Met Office (e.g., Carroll and Hewson 2005).

All forecasting disciplines are increasingly reliant upon numerical models of various types, along with advancing observational technologies. Within the complex and evolving requirements for forecaster training, a strong understanding of fundamental physical processes remains a requisite foundation. With the emergence of ensemble prediction and probabilistic forecasting, a background in statistics is also helpful. What can today's meteorology students do to make sure that they remain relevant in a rapidly changing field? A diverse background, with strength in computer science, open-minded awareness of cross-disciplinary opportunities, and understanding of the ways that weather impacts society are all advantageous.

On a more personal note, weather prediction has been a source of scientific challenge and enjoyment to me for about 25 yr, dating back to a weather forecasting course I took as a student at the University of Washington in the 1980s (see Bond and Mass 2009 for a description of this course). Over the years, my experience with weather forecasting has come informally in forecasting competitions, as an instructor for classroom activities, and professionally, as a paid consultant. I view each forecast as a scientific hypothesis; by examining all available data, one can ascertain which physical processes are likely to be operating and then consider how these processes will affect a given forecast parameter. For example, a deeper than average planetary boundary layer (PBL) in the presence of strong winds aloft may favor strong and gusty afternoon winds due to vertical mixing of horizontal momentum. In predicting the minimum temperature, clear, calm conditions with a very dry atmosphere can favor strong radiational cooling, provided that local site characteristics and small-scale terrain features favor drainage and collection of cool air in the location of the site in question. Of course, knowing that a given process will be active does not automatically provide a quantitative forecast, but understanding how different processes will affect the weather is an essential element of weather forecasting.

Good forecasters make one or more hypotheses with each forecast, and the joy of short-range weather forecasting is that nature tests the hypothesis automatically as the weather unfolds! It is exciting to call up observations the next day to see how the weather that is taking place in a given location compares to one's prediction. It can bring disappointment or elation, depending on the outcome; whatever happens, there is always a lesson to be learned. I often tell my students that my forecast skills have developed to where they are because I have "busted" so many forecasts over the years; the trick is to expend the requisite effort to learn something from each one.

The primary goal in this chapter is to outline a systematic approach to weather forecasting that is focused upon physical processes. Current and future changes in the field of weather forecasting are also discussed. There may never be unanimous agreement on how the field of weather forecasting should evolve, but it is important that students in the atmospheric sciences understand how their profession has evolved and what changes are being discussed for the future. In developing this material, I have solicited comments from a number of professional forecasters at various stages in their careers to broaden the perspective. This chapter focuses on the U.S. NWS, partly because it would be difficult to inclusively represent the broad spectrum of forecasting activities in the private sector.

The topic of severe weather prediction would require a large volume of background material from mesoscale meteorology, and therefore this important aspect of forecasting is not treated here. Another important area is forecast verification, a necessary counterpart to weather forecasting. A recent and comprehensive text edited by Jolliffe and Stephenson (2003) is available to cover this topic, and the reader is referred to this collection for a complete discussion of verification practices.

### 11.1.1. The NWS NDFD and the human role in forecasting

The blend of computer model output and human meteorological skill in forecasting has been a subject of debate and concern for many years (e.g., Doswell 1986). Concern over the lack of meteorological reasoning employed by some forecasters, a phenomenon dubbed "meteorological cancer," was discussed by Snellman (1977). Many of the concerns raised in Snellman's 1977 article remain relevant today, and some significant steps have been taken by the U.S. NWS to sustain a vital role for humans in the forecast process; one of the most important of these was a restructuring of NWS forecast offices to include a Science and Operations Officer (SOO). The SOO has the responsibility to identify and facilitate the transition of research into forecasting operations. Had NWS forecast offices been structured in 1920 as they are now, perhaps the Bergen School frontal-cyclone model would have been more rapidly adopted by the U.S. Weather Bureau (see section 5.4)!

When the *National Digital Forecast Database* system was officially implemented in 2003, forecasters began to create gridded fields of forecast elements out to 7-day lead times. The advantage to this approach is that the digital forecast is available for use in a variety of derived products and services, both produced by the NWS and other creative partners outside of the NWS. The ability to utilize the forecast grids in geographical information system (GIS) applications and the versatility afforded by visualization and quantification are major advantages of the NDFD. Local text forecasts are generated at the individual WFOs from the gridded fields using automated software. Manipulation of the gridded forecast fields is facilitated by the use of "smart tools," which are numerical algorithms designed to apply corrections and modifications to the fields based on physical processes. The role of humans in this process is to diagnose the meteorology, manipulate the gridded forecast, establish continuity at the boundaries between forecast office county warning areas, and ensure consistency and accuracy.

The primary mission of the NWS is to protect life and property, and the issuance of warnings by local NWS forecast offices remains a top priority for forecasters. Warnings are essentially short-term forecasts for high-impact weather. This aspect has also undergone a recent change, from a county-based system to a more versatile area/polygon-based product.

Methods differ between NWS offices and have evolved over time, but many forecasters adjust or blend the previous forecast with new observations or model output. Forecasters also have the option to populate the gridded forecast straight with model output or the new gridded model output statistics (MOS) products (discussed at the end of chapter 10). In light of this, some view the advent of the NDFD as a step toward forecast automation, with eventual elimination of humans from the forecast process. Articles written about the NDFD pointed to the need for an evolving human role in weather forecasting (e.g., Mass 2003; Doswell 2004; Stuart et al. 2006).

It is essential that humans "add value" to numerical model-forecast output to remain viable participants in the forecast process. Forecasters benefit from knowing how to distinguish situations in which they *can* add value from those in which they cannot. For example, in fair-weather forecasting situations in which NWP forecasts are turning out to be accurate, is it a waste of time for humans to try and improve upon some of the automated forecast elements? The danger is that if human forecasters become removed from day to day forecasting, then, in situations when the model forecasts are not reliable, they will be ill prepared to contribute. How can forecasters play an optimal role in forecasting, especially during threatening weather?

Forecasters who possess a comprehensive knowledge of atmospheric processes, understanding of how these processes are represented in NWP models and the limitations therein, and the ability to effectively communicate forecast information to others will be best equipped to contribute to the forecasting enterprise. It is important that forecasters maintain hands-on involvement with observations so that they understand *how the real atmosphere works* and what physical processes are likely to be important on a given day. It is not likely that these skills can be maintained merely by watching over ("babysitting") automatic forecast generation systems on most days and then jumping in suddenly when high-impact weather threatens. The forecaster must maintain both a linkage and a separation between the atmosphere of computer models and the real one for which they are forecasting.

Other areas of potential involvement for humans in the forecast process include improved communication of NWP output to emergency managers and decision makers, exclusive focus on short-term forecasting, and communication of forecast confidence to users. The U.S. NWS is actively examining *decision support systems* to understand how to optimize the flow of weather information to local emergency managers and other external agencies that may need to act in the event of high-impact weather. In addition, the way in which members of the general public interpret weather forecast information is receiving increased attention (e.g., Morss et al. 2008; Broad et al. 2007).

Personally, I am of the opinion that humans will remain an important component of the forecast process until the year 2025 or beyond but that eventually numerical models will be sufficiently accurate that humans will be able to add little value in terms of deterministic forecast accuracy in most situations. The trend toward probabilistic forecasting [beyond the probability of precipitation (PoP)] is another pronounced change that should be embraced by modern forecasters, provided that weather forecasts can be communicated in a way that meets the needs of the public and decision makers.

There is no question that NWP models have improved our ability to accurately predict the weather (e.g., Figs. 10.1 and 10.48). Several published studies demonstrate that the average errors in various forecast parameters have decreased with time and that forecasts demonstrate statistical skill out to longer lead times than ever before, in large part because of improvements in the numerical models. However, studies also show that humans continue to improve on model forecasts and that a skilled human forecaster can still consistently outperform even the best numerical forecast model when it comes to *specific parameters at specific locations* [e.g., quantitative precipitation forecasts (QPF) for days 1 or 2], especially in high-impact weather situations. There are documented reasons why humans are able to add value beyond what the NWP models alone can provide:

1. Humans can logically anticipate, based on both understanding of atmospheric processes and past experience, how weather conditions will likely evolve with time.
2. Humans can mentally synthesize observational data, conceptual models of atmospheric systems, and numerical forecast output in the development of a forecast that describes the expected "weather" in a given location. The forecast can be tailored and delivered in terms that the general public, emergency managers, policy makers, or other constituents may specifically desire.
3. Humans are able to interpret observational data to formulate accurate short-term forecasts of severe local storms much more effectively than automated techniques. Our "on the fly" pattern-recognition skills have yet to be duplicated by computers.
4. Experienced forecasters are able to utilize observations to evaluate short-term model forecasts, recognize when the models are going wrong, and subjectively correct the forecast. Essentially, humans are able to learn from past experience (and inaccurate forecasts); although NWP output can have bias removed, correcting the models themselves involves active intervention from model developers and researchers.

The "meteorological cancer" of which Snellman (1977) warned sets in when forecasters or researchers treat models as "black boxes" and stop thinking about physical processes taking place in the real atmosphere. As models continue to improve, the challenge of adding value will become greater, and new ways of improving the forecast must be developed for the human role to remain viable. This task is made even more challenging by the frequent and substantial changes made to operational NWP systems. It is difficult for forecasters to remain aware of the model changes let alone to understand their implications in terms of forecast performance.

The emergence of some ensemble forecasting techniques relates to the points above; for example, the superensemble technique is a strategy in which individual ensemble members are weighted by past performance, allowing optimization of a weighted ensemble mean. The ensemble also carries information concerning the overall state of predictability and thus can inform human forecasters when they are most likely to be able to add value to the forecast.

The previous chapter discussed the advantages of ensemble prediction and the importance of recognizing and communicating forecast uncertainty. Given ensemble information, a range of forecast values with a confidence interval can be provided. However, evening weathercasts still frequently present single-number forecasts (when a range would be more appropriate) for maximum and minimum temperature each day out to a week! It may be that the public is not yet ready to embrace probabilistic predictions, at least for some parameters. Or perhaps more effective and easily understood graphics are needed to convey probabilistic forecasts. A comprehensive survey of National Weather Service forecasters in the United States indicates that while there is user demand for information concerning forecast confidence, there are several obstacles that are currently limiting development in this area (Novak et al. 2008). This survey found that limited data access, and limitations in the available ensemble forecasts themselves are two barriers to effective delivery of uncertainty guidance.

For precipitation, it is common to express the probability of precipitation, with which the public is familiar but which may not always be correctly interpreted. With the advent of ensemble forecasting and quantitative information now available concerning forecast confidence, it would be appropriate to show forecasts with error bars or some other easily understood measure of probability and uncertainty. The adoption of such graphical means for expressing these aspects of forecasting is an emerging area in our field (e.g., Morss et al. 2008); new means of communicating uncertainty in tropical cyclone forecasts, winter storm predictions, and severe weather outbreaks are the subject of current study.

### 11.1.2. General approach to forecasting

Weather forecasting involves the application and integration of many of the concepts discussed previously in this book, in addition to some processes that have not been discussed. The goal of forecasting is to identify all processes and to determine how these will affect the forecast parameters at the location and time in question. Not all processes are equally important to forecasting; for example, in many forecasting applications, emphasis is placed upon processes that are related to upward vertical motion and that determine the availability of moisture. In the most basic terms, weather forecasting requires answers to the following questions:

*What physical processes and which weather systems will affect the forecast location during the forecast period?*

*Where were the key weather systems in the recent past, and where are they currently? How are they expected to evolve leading up to the forecast period?*

*Were the expected processes operating at the forecast location in the recent past, and are they operating now? Why or why not?*

*In what way will the expected processes and weather systems specifically affect weather conditions in the forecast location?*

What types of processes and weather systems must we consider? Obviously, the atmosphere is characterized by a complex combination of many processes, often acting simultaneously, and not all of which are important and/or accurately measurable. The ability to identify and prioritize the important mechanisms requires knowledge of the local orography, climatology, and of the current synoptic and mesoscale setting of the forecast site. It can be daunting for beginning forecasters to take into account such a broad spectrum of information. Fortunately, there are techniques available to systematize and prioritize information in developing a forecast.

## 11.2. THE FORECAST PROCESS

It is difficult to describe a systematic forecasting process that is relevant to all types of weather prediction, so we will adopt a flexible and somewhat generic approach that applies to a variety of forecasting situations.

### 11.2.1. Background information

When forecasting for a new or unfamiliar location, background information should be collected prior to making a forecast. If one is making many forecasts over a period of months or years for a given location, a review of relevant published literature is also useful. For shorter-term periods, the following information should be considered:

1. What is the geographical setting of the forecast location? What topographic and physiographic features might influence the weather there? What flow regimes would allow these features to most strongly affect conditions at the forecast site? What types of vegetation and ground cover characterize the area?
2. What sensible weather parameters are being forecasted, and what are their *climatological values* at the location and on the day in question? This could include maximum and minimum temperature, wind speed, probability of thunderstorms, precipitation amount or probability, cloud cover, visibility, etc. The National Climatic Data Center (NCDC) Local Climatological Data (LCD) publication can be helpful in answering these questions.
3. At what times (in UTC) are the local sunrise and sunset for the location and date?
4. What is the station elevation, and what is the approximate *surface pressure* there?
5. Where is the nearest radar site? Where is the nearest rawinsonde site? What is the coverage of surface reporting stations in the vicinity? For U.S. locations, which NWS forecast office is responsible for issuing forecasts for the site?
6. Given the setting and season, what *processes* and circulation systems may be active at the site? These could include a land–sea or land–lake breeze circulations, mountain–valley circulations, or synoptic-scale upslope or downslope flow due to nearby terrain.
7. Are there any mesoscale site characteristics that could influence weather conditions? Where is the sensor location relative to major urban centers? Might an urban heat island influence the site? On the microscale, are there large expanses of blacktop pavement in the climatologically upwind vicinity? For this, the capabilities of Google Earth are quite helpful, along with a National Oceanic and Atmospheric Administration (NOAA) Web site that details station characteristics for most U.S. stations.
8. Finally, some pseudoclimatological data may be needed. For locations near water bodies, is there any ice cover? What is the current surface water temperature? What air temperature regimes would favor a land or sea breeze? Is there deep snow cover in the vicinity? If the surrounding area features agricultural activity, what types of crop or farming are present? What is the sign of soil moisture anomalies? Is the region in the midst of a drought? Has it been abnormally cold, warm, wet, or dry in the preceding weeks or months? In the United States, the NOAA Drought Monitor publication can be helpful in this regard.

This may seem like a lot of information but, if one is to ascertain the important processes, it is necessary to be thorough and diligent in its collection, although obviously only some of this information may be relevant to a given location. It is helpful to organize information of this type in a spreadsheet, along with forecast and verification data.

### 11.2.2. Setting the planetary, synoptic, and mesoscale context

A useful and systematic strategy for organizing forecast information by spatial and temporal scale was presented by L. Snellman. This approach has become known as the "Snellman funnel," and it is summarized in Fig. 11.1.

When forecasting for a specific location, the temptation exists to start analyzing conditions in the immediate vicinity of the forecast location.[1] Such a myopic view will eventually lead to errors, because the weather systems that are likely to affect the location are often situated a considerable distance from the forecast site (or have not yet formed) at the time the forecast is made. Assembling a complete picture of the atmosphere, including account of interactions among weather systems of different scale, requires a look back in time and out to larger spatial scales.

[1] The temptation also exists to jump immediately to model output, without regard for current observations and assessment of how well the current situation is being handled by the short-term model forecasts.

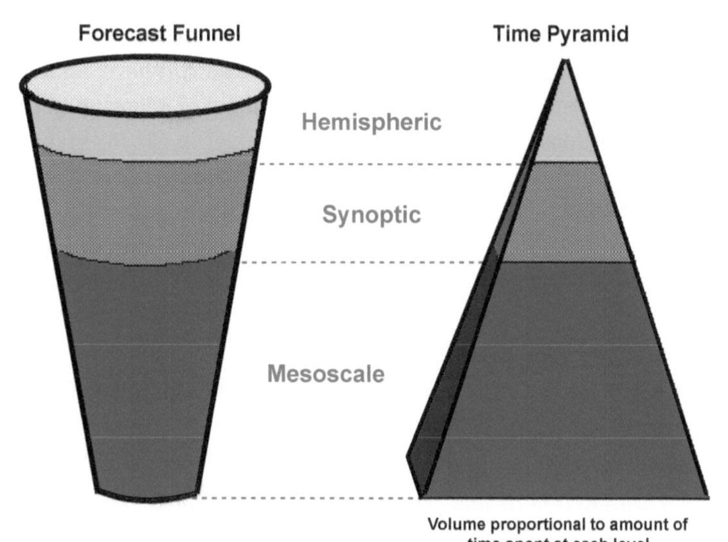


**Figure 11.1.** The Snellman forecast funnel. On the left the spatial scale of the forecast information shows a narrowing toward the mesoscale, while on the right, a pyramid describes the amount of time a forecaster might spend analyzing information for the corresponding scale (adapted from Snellman and Thaler 1993).

However, forecasters are often faced with deadlines, and time constraints do not typically permit a detailed treatment of all scales or a look back too far in time. Accordingly, the Snellman funnel suggests that a relatively small amount of time be spent on these largest scales, with increasing time and attention as one hones in spatially on the forecast site and temporally on current and future short-term forecast information. There is also a need for flexibility. The reality of time constraints in many forecasting environments may necessitate a rapid focus on the most critical forecast problems, and these problems may not occur for only the shortest lead times. Nevertheless, the basic premise outlined here can be tailored to different forecasting environments and forecast lead times.

Another caveat relating to the forecast funnel approach is that, for lead times beyond 2 or 3 days, the larger-scale pattern should be emphasized more, because of increased uncertainty for the smaller-scale details of the pattern. It would not be wise spend a lot of time on mesoscale details in a 7-day forecast, for example.

**11.2.2.1. The hemispheric pattern.** *What characteristics of the planetary-scale pattern are relevant to the forecast?*

In sections 1.5.3 and 2.1.7, we discussed Rossby waves and the concept of downstream development. The eastward propagation of wave energy from upstream locations of amplified wave activity is prevalent in the atmosphere, and to a large extent the waves in the upper jet stream determine the location of synoptic-scale cyclones and anticyclones and their attendant frontal systems. Taking a hemispheric look at weather conditions going back 3–5 days in time serves to establish the planetary-scale context for the forecast. Viewing large-scale GFS analyses at 12-h intervals is sufficient, with an overlay of 500- or 250-mb geopotential height with sea level pressure.

Where has the main westerly jet stream been located relative to the forecast site? Where are the main centers of amplified Rossby wave activity? How are these evolving in time, and where is current development taking place? Where is future development expected? Sometimes, plotting height *anomalies* is useful to set context relative to the long-term average conditions. Online tools are available to compute and plot anomalies of various parameters, defined by subtracting off the longer-term averages (often 10- or 30-yr averages) from the full analyzed or forecasted fields.

**11.2.2.2. The synoptic-scale pattern.** Knowing the planetary-scale setting leading up to the forecast period also provides information relevant to the *synoptic-scale* pattern. A careful look at current satellite and radar imagery, surface and upper-air analyses (e.g., an overlay of sea level pressure and 500-mb height), and a sequence of synoptic-scale analyses from the preceding 12–24 hours is often sufficient to answer these questions:

> *Which synoptic systems* ***have been*** *affecting the forecast location, which ones* ***are*** *presently affecting it, and which ones* ***will be*** *affecting it during the forecast period?*
>
> *Where are these systems now, and what are the associated weather conditions? To what extent will those conditions persist as the system affects the forecast location?*
>
> *Are the systems of importance currently located over relatively data-sparse regions, such as over oceanic regions with limited in situ observations?*

In some situations, the systems that will determine future weather conditions at the forecast location have not yet formed at the time when the forecast is made. However, knowledge of the upper-level pattern should alert the forecaster to the potential for development: for example, the rapid formation of a coastal cyclone. Once the forecaster has a feel for which systems will most strongly influence the forecast, we can further ask the following:

> *What will be the prevailing synoptic-scale wind speed and direction, and how will this bring regional topographic and physiographic features into play?*
>
> *What are the strength, location, and timing of synoptic-scale forcing for vertical air motions at the forecast site?*
>
> *Do short-term forecasts from the operational NWP models adequately represent what observations are currently showing for these systems?*
>
> *How will the weather system(s) in question affect the forecast parameters? Physically, how will the presence of these systems impact conditions?*

The possible passage or close proximity of frontal systems, upper-level vorticity maxima, presence of regions of static instability, availability of moisture, and strength and timing of lifting mechanisms are all under consideration at this point. In addition, although not specifically mentioned above, *air-parcel trajectories* are useful in gauging the characteristics of an incoming air mass.

Having collected sufficient background information and assessed past and present conditions, we are now in a position to examine NWP guidance. However, given a thorough analysis of past and present conditions, the forecaster should have a good feel for what will happen even without looking at model output.

**11.2.2.3. Mesoscale and microscale processes.** Given the planetary and synoptic-scale settings, how will mesoscale processes impact conditions at the forecast location? There is a large variety of processes that are potentially important, but the climatological background should have already alerted us to those upon which we must focus. Local thermally forced circulations (e.g., land–sea breeze circulations), orographic and physiographic effects (e.g., orographic ascent or descent, urban effects), and the precise timing of synoptic features such as frontal passages are examples of mesoscale systems that must be considered.

Mesoscale convection is another important consideration. For instance, if the upstream environment is conducive to mesoscale convective systems, there may be direct and indirect effects on the forecast. Remnant *mesoscale convective vortices* (e.g., Bosart and Sanders 1981) can trigger subsequent convection and are often poorly initialized in NWP analyses (with associated poor forecast performance; see Davis and Trier 2002; Hawblitzel *et al.* 2007). Upstream convection can alter moisture transport, affecting downstream precipitation. Water-vapor imagery from geostationary satellites can be useful in identifying such features observationally, and this can aid forecasters in improving upon NWP output in these situations.

Time management is critical; there is not enough time to examine every possibility. However the preceding analysis serves to narrow the list. In many operational forecasting environments, time constraints can severely limit the ability to approach a forecast in this manner. In the university setting, students sometimes feel overwhelmed by the volume of data, and they may be compelled to examine huge numbers of graphics. This process can be streamlined by gaining experience with which specific graphics to use in a given situation. For instance, in forecasting maximum temperature and peak wind speed, examination of current and forecast soundings provide valuable information concerning mixed-layer depth. If the possibility of lake-effect convection exists, we should evaluate the lake surface to 850-mb-level temperature difference, and the length of the upstream over-lake fetch, in addition to the mixed-layer depth, forecast instability over the water, and synoptic-scale forcing for ascent.

### 11.2.3. Forecasting strategies

A *persistence* forecast is simply a prediction that future conditions will be a repeat of past or present conditions. This strategy can be applied on seasonal as well as short-term time scales but is generally not very accurate and should be used with caution. In general, persistence is less accurate with increasing forecast lead time. The following is an example: *Your forecast area is under the influence of a summertime maritime tropical air mass with weak gradients of temperature and moisture in the surrounding regions. The jet stream is well removed from the area, and you do not expect any significant change in airmass properties in the near future. The current dewpoint temperature is 65°F, and haze and shallow fair-weather cumulus have been the rule for the past several days. At night, clear skies with patchy fog have formed, with a low temperature near 60°F each of the past two nights.* Even the best statistical (MOS) guidance will have a hard time beating a persistence forecast in this situation; a forecast minimum temperature of ~60°F should be excellent. The following day's high temperature, as well as the occurrence of fog and cloud cover, could also be predicted using this method.

Another simple forecast technique is to predict *climatological* conditions for a given location and date. Like persistence, this method works best during benign synoptic patterns, but it is not to be relied upon as a viable forecast strategy in of itself. For example, *in Los Angeles during the month of July, it rains roughly one out of 100 days. A precipitation forecast of no rain (based on climatology) would be correct about 99% of the time.*

*Analog forecasting* has enjoyed some resurgence of popularity recently. This technique involves identification of similar weather patterns in the past and basing the current forecast on the sequence of events that occurred at that time. This strategy is of limited use because it is difficult to find past events that are sufficiently similar for many situations; no two weather patterns are exactly alike! Also, unless computer algorithms are used, it is prohibitively time consuming to search past databases for similar examples. New capabilities, such as the Cooperative Institute for Precipitation Systems (CIPS) winter weather analog site (available online at http://www.eas.slu.edu/CIPS/ANALOG/analog.php) are now available. However, as demonstrated by Roebber and Bosart (1998), precipitation patterns are highly sensitive to details of the synoptic flow pattern, casting some doubt on the viability of analog precipitation forecasting.

A *consensus forecast* involves combining or averaging different independent forecasts to arrive at a blended prediction. As discussed in section 10.6 for ensemble prediction, it is often true that the average of several different individual forecasts (human and/or numerical) is superior to any of the individual forecasts (when performance is averaged over many forecasts). Although there are statistical benefits associated with the consensus strategy, it increases the chances of missing extreme events.

The techniques discussed above should not be used in isolation. Typically, forecasters incorporate elements from several or all of these strategies into their forecasts, in addition to careful synoptic interpretation of the meteorological situation and guidance from numerical weather prediction models.

There is clearly a large volume of information to be organized and accounted for in forecasting. As forecasters gain experience, they find ways to streamline the forecast process and improve time efficiency. In many ways, success in forecasting demands that the forecaster filter out information that is tangential or irrelevant and extract the essential information that is critical to the forecast. Some forecasters use a strategy of identifying the "problem of the day" as a means of sharpening the focus on a given weather system or particularly challenging aspect of the forecast. At times, this can be useful; however, when utilizing this strategy one must be careful not to put on "blinders" and neglect other important processes or weather systems.

## 11.3. SPECIFIC FORECAST PARAMETERS

Bearing the previous discussion in mind, we are now in a position to delve into more specific information concerning the prediction of particular weather elements. To some extent, each forecaster will develop a unique set of methods that will define his or her forecast process.

In forecasting individual weather elements, a critical emphasis is physical consistency. *The sensible weather parameters being predicted are strongly related to one another*, and good forecasts feature consistency. For example, strong winds and thick cloud cover work against the decoupling of the surface layer, promoting the maintenance of an overnight mixed layer. We would not want to forecast a large diurnal temperature range or expect significant radiational cooling in this circumstance, so the reasoning behind our minimum temperature forecast should reflect our expectations for cloud cover and wind speed during the nighttime hours. In light of the interdependence of the forecast elements, the discussion of the forecast elements below should not be viewed as being in order of priority.

An important venue for discussing and defending forecasts is the *weather briefing*, during which consistency and physical reasoning in the forecast process are emphasized. Forecasters who are able to assemble a coherent picture of atmospheric processes are in the best position to make a good forecast. In the United States, the NWS issues a text product known as the Area Forecast Discussion (AFD), which explains the meteorological reasoning behind their forecasts. Typically, I find myself reading these as much or more than the actual forecasts, because the reasoning is where the scientific discussion takes place. A well-crafted AFD can be a masterful demonstration of scientific knowledge; poor AFDs are often characterized by discussion of model selection in the absence of scientific reasoning. Students in my synoptic and forecasting course are asked to write an AFD to accompany their weather briefings, because doing so forces them to organize and justify their forecast in terms of physical processes and this helps to ensure consistency.

Forecast verification should be considered in the forecast process. For example, if one is predicting maximum temperature, wind speed, or cloud cover, how will the forecast be evaluated? It is essential to understand the observing equipment being used to "verify" the forecast. The maximum temperature is a 5-min average for U.S. Automated Surface Observing System (ASOS) stations; if we were forecasting the maximum 5-s average temperature, we would predict a higher value. For wind speed, the ASOS units report 5-s gusts, but the "sustained" wind speed is a 2-min average. For cloud cover, the past ceilometer used in ASOS stations often had an upper limit of 4,000 m (~12,000 ft) AGL; however, if a human observer was supplementing the observations or if aircraft reports or satellite data were used, this limitation might not come into play. New ceilometers, with cloud observing capabilities of up to 8,000 m (~24,000 ft), are now replacing the older units in U.S. ASOS stations. These are just a few examples of the importance of observations and sensor specifications in weather forecasting.

### 11.3.1. Temperature

Prediction of temperature at a given location and time or the maximum or minimum temperature during a particular interval of time requires consideration of expected cloud cover, precipitation, seasonality and diurnal timing, temperature advection, the possible occurrence of frontal passages, orographic effects (e.g., downslope vs. upslope flow), wind speed and direction, boundary layer depth, surface character (e.g., snow cover, soil moisture, urban effects, or local vegetation properties), the overall temperature of the air mass (e.g., lower-tropospheric thickness), and climatology, in addition to other factors. Clearly, these processes are also not independent of one another; if a forecaster were to consider all of these things in isolation, it would take a long time to make the forecast!

**11.3.1.1. Maximum temperature.** One variation of a forecast process for predicting the maximum temperature at a given location involves determination of the following:

1. What are the relevant *climatological* and local surface conditions?
2. What are *current* and *recent-past* surface observations at the site?
3. What synoptic and mesoscale systems will be affecting the forecast location? Are frontal passages expected during the forecast period? Might there be an off-hour maximum (i.e., a deviation from the expected diurnal pattern) due to a warm-frontal passage after sunset? Will the late-morning passage of a cold front or the development of a sea breeze "clip" the high temperature?
4. Assuming a typical diurnal temperature cycle, *how deep of an afternoon planetary boundary layer will develop*? This involves evaluation of sounding data, expected synoptic-scale wind speed and cloud cover, orographic effects such as downslope flow, availability of moisture, and seasonality. Model-forecast soundings can be useful to estimate this as well.
5. Select an isobaric level at or near the top of the mixed layer as determined from (4), typically the 900-, 850-, or 800-mb level for stations near sea level. View the model-predicted temperatures at this isobaric level for the time of expected maximum temperature. An ensemble approach is useful for establishing a range of expected values, with hourly frequency if possible.
6. Using a skew $T$-log$p$ diagram, adiabatically lower an air parcel using the pressure level [obtained as discussed in (5) above] dry adiabatically to the *surface* pressure (not the sea level pressure!). Do this for the range of temperatures aloft.

Working from this range of temperatures, other processes that may be important are weighed in, if needed:

- For example, during peak solar heating in spring or summer, especially with a dry surface, add 1°–2°F for superadiabatic conditions in the surface layer (in which the verifying 2-m temperature is measured).
- *Wind direction and speed* are critical. With nearby topography, wind direction dictates adiabatic modifications. Strong winds favor a deeper mixed layer and warming. Light winds in winter mean reduced mixing and cool weather, whereas light winds in summer can result in superadiabatic conditions and warming. Nocturnal inversions may not mix out with light winds or clouds.
- What is the expected extent of *cloud cover*? Will precipitation occur? Because of evaporative cooling in conjunction with cloud cover, precipitation can significantly reduce the maximum temperature below the adiabatic value. Even rain early in the day, before the expected maximum temperature occurrence, will require energy to evaporate, serving to reduce the maximum temperature.
- What is the recent observational history? Have recent model forecasts done an adequate job of predicting the pressure-level temperatures in recent forecast cycles? Have recent maximum temperature forecasts (including those of the forecaster, NWS, MOS, or model predictions) been accurate? If not, why not?

After these considerations, and arriving at a numerical forecast value, consult MOS and the local weather service forecast to see how they compare. If a good reason for altering the forecast is evident, make adjustments based on this additional information.

**11.3.1.2. Minimum temperature.** Some of the considerations here are the same as for maximum temperature prediction, although perhaps more attention should be paid to local terrain features and station site. What conditions will favor boundary layer decoupling and radiational cooling? Will the minimum temperature be radiationally controlled or dictated by temperature advection (e.g., following a cold frontal passage late in the forecast period)? Assuming a typical diurnal minimum controlled by radiational cooling, considerations include:

1. *Cloud cover:* Downwelling infrared (IR) radiation can prevent the formation of a nocturnal inversion and keep minimum temperatures much higher than they would otherwise be. Will the clouds be thick and low, with a warm base, or thin and high, with a cold temperature? The latter would have a much smaller retarding impact on nocturnal cooling. Careful examination of animated satellite imagery is required.
2. *Humidity, dew, and frost:* Given no airmass change and clear, calm conditions, the expected overnight dewpoint must be considered. During warm-season conditions, especially if the dewpoint temperature is above roughly 15°C (59°F), the dewpoint is often a good lower bound on the minimum temperature. There are several physical processes at work. Water vapor is a greenhouse gas, and high concentrations of water vapor, even in the absence of clouds, can reduce radiational cooling. As dew begins to form, the condensation releases latent heat, directly impacting the temperature. Even if clouds or fog do not form, dew on grass or nearby surfaces means that condensational heating has taken place.

   Because colder temperatures feature a lower saturation vapor pressure, the dewpoint is a less reliable lower bound on temperature during cold weather. The vapor content is very low at temperatures below freezing, and so, even though frost formation releases latent heat, this effect is weak relative to the heat released by heavy dew in warmer conditions. Fog formation is also important; mixing fog can form at sunrise as near-saturated air is mixed vertically with the onset of solar heating. This can impact the maximum temperature later in the day.
3. *Wind speed:* In general, the stronger the overnight winds, the warmer the minimum temperature because of enhanced vertical mixing by mechanical turbulence. Of course, if strong winds are associated with cold advection, cooling can still take place through that mechanism.
4. *Wet-bulb effect:* With the onset of precipitation and little or no temperature advection, the wet-bulb temperature is often a good estimate of what the temperature will be once steady precipitation has begun.

5. *Local effects:* Pay attention to the orientation between the synoptic wind direction and local topography. Drainage flows can cause cool air to collect in some valley stations, and downslope warming/upslope cooling can be important. The urban heat island effect can warm local temperatures by several degrees Fahrenheit.

**11.3.1.3. Temperature and dewpoint.** Some applications call for prediction of daily average or hourly temperature or humidity forecasts. Knowledge of the climatological diurnal cycle of temperature is generally straightforward, although local effects can complicate this near complex terrain and coastlines. The diurnal cycle of dewpoint temperature is more complicated, because it is controlled by the evolution of the turbulent mixed layer, local surface properties, phase changes of water substance, and season.

In the presence of surface vegetation and a reasonably moist environment, one would expect *two* minima in the diurnal dewpoint cycle. One minimum is expected in the immediate predawn hours when the amount of condensation on the surface would be maximized, with the near-surface relative humidity close to saturation; this would be followed by a morning increase as warming temperatures and insolation lead to evaporation of the condensate (dew). As the mixed layer deepens during the morning and early afternoon hours, entrainment of drier air from aloft can result in a decreased dewpoint temperature through the depth of the boundary layer, and a second minimum in dewpoint is often observed during the late afternoon hours. This is, of course, a mechanism that depends on the humidity above the PBL, which varies with synoptic pattern. After sunset, with reduced mixing and given a sufficiently moist surface, evaporation into the lower portion of the boundary layer may result in a dewpoint increase, because this moisture is distributed over a smaller amount of atmospheric mass. Only when the temperature cools sufficiently for dew to form will the dewpoint begin to decrease.

Sensitivity to the type and activity of surface vegetation, soil moisture, and other surface properties can be important. A study using data from the Oklahoma Mesonet demonstrates strong sensitivity of the dewpoint cycle to land use while illustrating the basic mechanisms discussed above (Fig. 11.2).

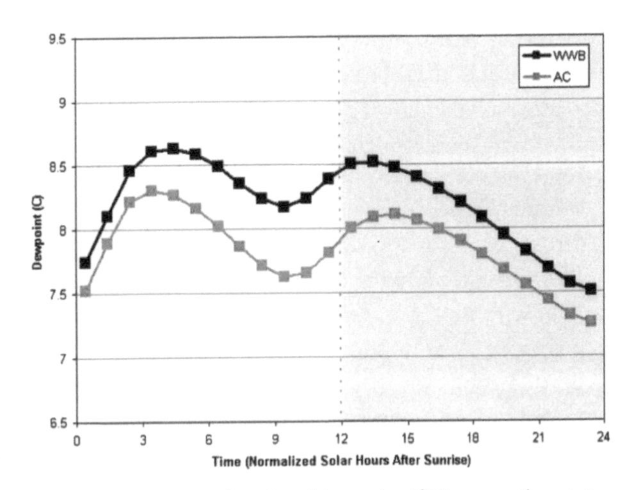


**Figure 11.2.** Diurnal cycle of dewpoint (°C) across the winter wheat belt (WWB) and in adjacent counties (AC) in the state of Oklahoma, averaged from 1994 to 2002 (from Haugland and Crawford 2005).

### 11.3.2. Precipitation

Two common types of precipitation prediction are the *quantitative precipitation forecast* and *probability of precipitation.* Although the general public may sometimes have less use for QPF than PoP forecasts, for agricultural applications, flood prediction, and during winter precipitation, the QPF is a critically important parameter. Forecasts of PoP have traditionally been available to the public, even if there is often confusion as to the actual meaning of such predictions. By definition, *the PoP is the chance that a measureable amount (0.01 in. or more) of precipitation will fall at a single, specific location in the forecast area during a specified time interval.* However, Winkler and Murphy (1976) and Murphy (1978) demonstrated that there is a correspondence between the PoP and the areal coverage of precipitation within the forecast area, provided that the PoP itself is spatially uniform over the forecast area. Thus, this quantity can also be interpreted as the *average PoP in a given region* and is related to the *areal coverage of precipitation in the forecast area.* Whether developing a QPF or PoP forecast, it is useful to first consider whether the precipitation will be *convective* or *stratiform* in nature.

If convective precipitation is expected, our attention must be focused on the strength of static instability and the potential for destabilization, availability of moisture, and expected mechanisms for convective triggering. Although the convection itself is a mesoscale phenomenon, the synoptic environment exerts a strong control on convective-scale processes. The importance of reviewing past and current conditions can be highly beneficial, especially in summertime convective regimes where persistence can be a difficult forecast to beat. The character of the vertical wind shear, which is in large part determined by synoptic weather systems such as fronts, upper troughs, and jets, will influence the most likely *mode* of convection (e.g., supercells, squall lines, multicell storms, or isolated pulse-type airmass convective storms).

A deterministic QPF during convective showers is of little value, because some locations within the forecast area will likely observe heavy rain, whereas others will receive nothing. Some of the issues associated with the use and interpretation of model QPF for convective precipitation were discussed in chapter 10; even when a model convective precipitation scheme is active, the precipitation produced by the scheme must be spread evenly over the grid cell, overestimating coverage and homogenizing the amounts relative to what would be observed. For convection, a high-resolution ensemble prediction would likely yield more useful results, because a measure of the precipitation variability and localized intensity would be better represented. Note that MOS is able to develop a useful PoP for convection even if the parent NWP model does not explicitly represent it.

Precipitation during a given forecasting period is not necessarily restricted to either convective or stratiform, because hybrid examples can occur, such as stratiform warm-frontal precipitation with embedded convective storms. Consideration of the static stability, either through examination of convective available potential energy (CAPE) or measures of potential instability, is important. Regardless of the character of the precipitation, the expected strength, location, and duration of *synoptic-scale ascent* and the *availability of lower-tropospheric water vapor* must be carefully considered.

Tools for the diagnosis of synoptic-scale ascent were presented in chapters 2 and 3, including the quasigeostrophic (QG) omega equation and analysis of storm-relative airflow on isentropic surfaces. Orographic contributions to ascent and descent must be considered as well, because this is not accounted for in the traditional QG omega equation. In keeping with the overall forecast process, review of past and present conditions is critical: Is expected precipitation associated with the movement of an existing weather system? Do analyses clearly reveal synoptic-scale ascent in the upstream region at the present time or in the recent past? Will any current areas of synoptic-scale ascent move over the forecast area? Will they strengthen, weaken, or remain similar between the present and the forecast period?

A number of techniques are available for assessing moisture content and transport, including surface observations of wind and dewpoint; analyses of lower-tropospheric relative humidity, specific humidity, or mixing ratio; and model forecasts of these quantities. The dewpoint temperature, mixing ratio, and specific humidity are measures of the absolute water-vapor content, whereas relative humidity is best suited for locating regions of cloud cover. Observational sources of moisture data are available from upper-air soundings, satellite or aircraft measurements, or other techniques such as from global positioning systems (GPS) and satellite occultation methods. For precipitation associated with midlatitude cyclones, plotting the lower-tropospheric moisture flux or the moisture flux on an isentropic surface (see chapter 3) is an excellent means of diagnosing the character and strength of the *warm conveyor belt*. This moist airstream is often accompanied by a low-level jet. Areas of *moisture-flux convergence*, which are often found at the terminus of the warm conveyor belt, correspond to locations of expected rainfall.

For my own forecast process, after assessing the presence of synoptic lift and available moisture, I consider the thermodynamic profile. Do observed or model-forecast soundings indicate the presence of a deep, mixed-phase cloud? If the layer of strong ascent is relatively shallow and the cloud-top temperature is above freezing, one should limit the QPF to reflect less efficient warm-rain processes, especially for continental locations. Exceptions to this are for orographic lift in marine air masses, where warm-rain processes acting alone have been shown to produce heavy rainfall.

An example is provided below from a forecast of heavy precipitation from the National Centers for Environmental Prediction (NCEP) North American Mesoscale (NAM) model over eastern Virginia and northern and central North Carolina from 13 November 2009 (Fig. 11.3a). Inspection of the grid scale and convective parameterization (CP) precipitation reveals that, along the axis of heavy rainfall from eastern Virginia into northern North Carolina, the precipitation is not convective in the model atmosphere (meaning that the grid-scale microphysics scheme was producing the precipitation). Model-forecast soundings taken in central North Carolina

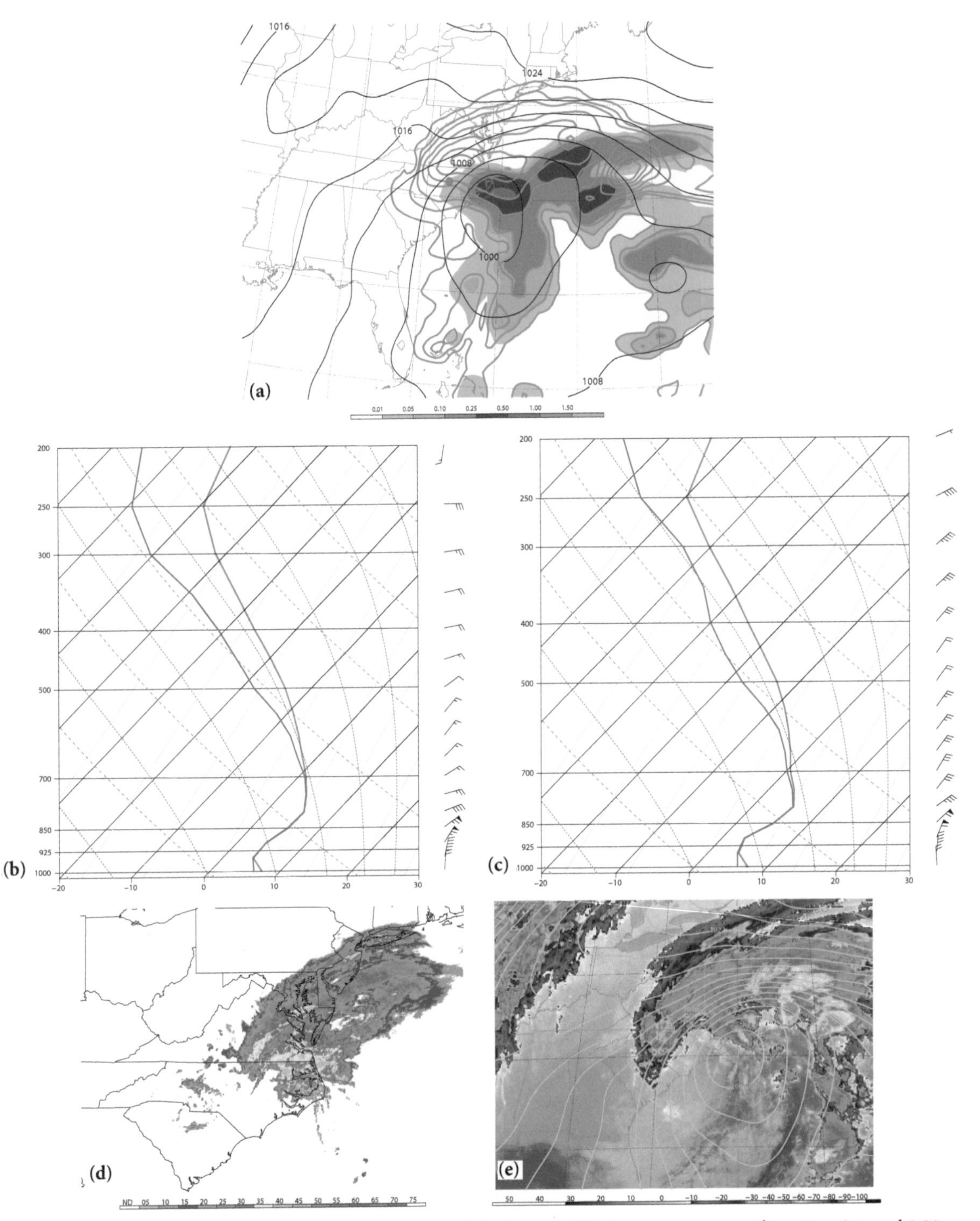


**Figure 11.3.** NAM forecasts valid 0000 UTC 13 Nov 2009: (a) 6-h total QPF (green contours, with contour interval 0.01, 0.05, 0.1, 0.2., 0.3, 0.4, 0.5, 0.75, and 1 in.) and convective QPF (shaded as in legend at bottom of panel); (b) 12-h model-forecast sounding for KRDU, skew *T*-log*p* format valid 1800 UTC 12 Nov; (c) as in (b), but for 18-h forecast valid 0000 UTC 13 Nov; (d) composite radar mosaic valid 1608 UTC 12 Nov; and (e) *Geostationary Operational Environmental Satellite-12* (*GOES-12*) enhanced IR satellite image valid 2345 UTC 12 Nov.

at the location of Raleigh-Durham International Airport (KRDU) indicate that the strongest forcing for ascent, evident in the veering wind profile, is found below the 700-mb level (Figs. 11.3b, c).

Although these soundings indicate layers that may be close to saturation with respect to ice, the dewpoint depression above the 700-mb level, which is the approximate location of the freezing level, suggests subsaturated conditions aloft (Figs. 11.3b,c). Satellite imagery (e.g., Fig. 11.3e) indicates an area of relatively shallow cloud over southeastern North Carolina, consistent with a "dry slot" signature and subsaturated conditions aloft.

In this event, the model forecast provided an accurate overall depiction of the thermodynamic situation and QPF, although a large positive bias in the model QPF was observed for the specific location of KRDU. However, the sharp gradient in the QPF field, along with model-forecast soundings that suggested unfavorable conditions for mixed-phase precipitation processes, were useful guidance to forecasters. Although the *deterministic* QPF for the exact location of KRDU was not accurate, the model forecast of the overall synoptic and thermodynamic setting provided the needed information for a forecaster to make a skilled precipitation forecast. During the forecast period from 1800 UTC 12 Nov through 0000 UTC 13 Nov 2009, only light precipitation fell from the location of KRDU to points south and east (not shown). A snapshot of composite radar mosaic demonstrates that heavy rainfall was restricted to regions of cold cloud tops and deeper moisture (Figs. 11.3d, e).

The utility of ensemble forecasting is apparent in situations of the type described above. A deterministic forecast provides valuable information, but the QPF should not be taken literally for any specific location, particularly those found in regions of large horizontal QPF gradient.

With these diagnostics and data sources in mind, for QPF, we should consider the duration and intensity of precipitation. If a relatively short period of favorable ascent, favorable moisture transport, and favorable thermodynamic profile is expected, such as might accompany a fast-moving cold front, the QPF should reflect this with diminished amounts. It is difficult for even experienced forecasters to formulate a quantitative precipitation forecast just by an eyeball inspection of atmospheric data. Despite the fact that model QPF is one of the least accurately predicted parameters, it provides valuable guidance to forecasters. Experienced human forecasters have consistently been able to improve upon model QPF, and it must be considered a valuable component of the forecast. That is not to say that forecasters should blindly accept the model QPF and use it as their forecast. In my own forecast process, I frequently view the NCEP Short-Range Ensemble Forecasting System (SREF) QPF as a final step in my forecast process. By considering both the range and average, along with the mode, a more complete picture of the most likely QPF is obtained.

Some considerations that can optimize the value of NWP guidance in precipitation forecasting include the following:

1. Is the model QPF of the grid-scale variety or from the model CP scheme?
2. If the precipitation is grid scale in nature, are model-forecast soundings for the location and time of precipitation consistent with physical precipitation processes? In other words, is the altitude range at which the temperature is most favorable for operation of the Bergeron process experiencing saturation and ascent?
3. Is there a clear-cut physical mechanism for ascent (e.g., as diagnosed with QG or isentropic analysis) at the corresponding location and time of heaviest model QPF?
4. Is there a sharp gradient in model QPF in the vicinity of the forecast location? If so, be reluctant to accept the model output at face value and be sure to utilize ensemble information.
5. Examine model ensemble forecasts, particularly output from the NCEP SREF or other high-resolution ensemble systems. Are the operational runs outliers? What is the ensemble mean, mode, and spread?
6. If the precipitation system in question exists at the present time, how well are short-term predictions from NWP models verifying against surface observations, radar, and satellite data?
7. MOS predictions for both QPF and PoP, although based on the assumption of a perfect model forecast, are useful for establishing context and uncertainty. MOS PoP forecasts of 100% are somewhat rare, but they are seldom wrong. If the model setup looks favorable for guaranteed rainfall but the MOS PoP value is only 55%, consider what you might be missing. Using MOS in this way can serve as a "reality check" on a human forecast.

### 11.3.3. Wind speed and direction

In the United States, wind energy generation is expected to exhibit rapid growth, with the projection that 20% of nation's energy needs will be provided by this source by the year 2020. This will clearly elevate the importance of wind prediction on a variety of time scales, as well as motivate research to better understand the boundary layer processes that will ultimately affect the efficiency and longevity of wind turbines.

The challenges of predicting wind speed and direction are exacerbated by local effects on scales from those in the immediate vicinity of the instrument site (e.g., obstructions or channeling due to buildings) to those of local orographic features. Standard ASOS wind measurements include the sustained wind, which is a 2-min average that is constructed from the mean of 24 5-s wind averages, and the peak 5-s wind gust. Currently, ASOS winds are measured at the 10-m level, although at some sites the sensor is located slightly below this level. In predicting the sustained wind speed, one must consider many physical processes, including the following:

1. What is the overall synoptic-scale pressure gradient in the region? The geostrophic wind speed will overestimate the winds but can provide a good starting point for the maximum winds.
2. To what extent will *vertical mixing* in the planetary boundary layer allow transport of momentum from aloft toward the surface? This requires consideration of stability and the depth of the mixed layer, as well as consideration of wind speed predictions at or near the top of the expected PBL.
3. Are there local orographic influences that could result in channeling or downgradient accelerations, such as found during gap wind events?
4. What are local surface characteristics? An upstream fetch of water (lake or ocean) or bare cropland is consistent with greater wind speeds relative to forested or mountainous regions, with all else being equal.
5. For rapidly changing synoptic conditions, isallobaric wind effects can be important. The isallobaric wind is essentially the acceleration in response to *changes* in the pressure field.
6. Are convective storms expected in the vicinity? If organized convection occurs, it can produce very strong winds. Outflow from even weak convective storms has the potential to produce gusty surface winds, even in the presence of weak synoptic-scale flow. The environmental moisture profile can factor in here; if dry air is present in the lower or middle troposphere, the possibility of strong convective outflow increases.

### 11.3.4. Sky cover

Prediction of cloud ceiling and sky cover is of particular importance to aviation forecasting, but it also impacts recreational and agricultural activities. In a future in which solar energy generation becomes increasingly important, cloud-cover forecasts would assume additional relevance. Sky-cover forecasting holds implicit importance for many other forecast parameters, including maximum and minimum temperature, but also the occurrence of convective storms. For instance, the sky-cover forecast was critically important during a major tornado outbreak on 3 May 1999 in the U.S. Midwest, as clearing and subsequent heating allowed for the generation of instability in a volatile synoptic setting (e.g., Thompson and Edwards 2000).

Verification of cloud-cover forecasts can be accomplished by satellite imagery and also through surface observations. As mentioned previously, the standard U.S. surface observing system has recently been replacing laser ceilometers with new units able to detect clouds to twice the altitude (8,000 m) of its predecessor. Even with these improved sensors, satellite images, human-supplemented observations, and aircraft reports may be needed to provide a complete depiction of the actual sky cover. There is a certain element of subjectivity in verifying sky-cover forecasts: for example, when very thin cloud layers are present. The level of subjectivity is in part dependent on the observation methods used.

Predictions of sky cover should consider the cloud altitude, as well as the cloud type and coverage. Surface-based cumulus, cumulonimbus, and some stratocumulus clouds arise from buoyant thermals and are likely to exhibit a strong diurnal character, whereas higher-altitude cloud types, such as altostratus or altocumulus, are less likely to do so. The occurrence of surface-based diurnal cloud cover is a strong function of the depth and humidity of the planetary boundary layer. If the surface dewpoint depression is sufficiently large that the lifting condensation level (LCL) would be above the top of the PBL, then the formation of fair-weather cumulus or stratocumulus at the top of the PBL is not likely. Surface characteristics, including soil moisture and vegetation,

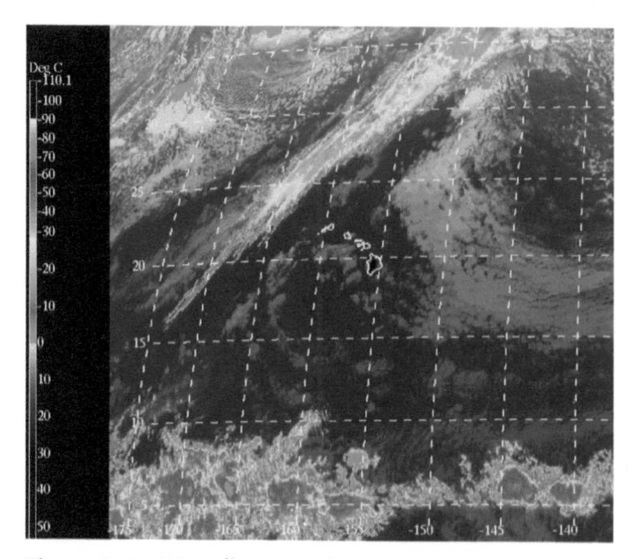


**Figure 11.4.** IR satellite image from 2330 UTC 22 Jan 2010.

are important considerations when making such predictions. For instance, after a recent soaking rain, strong winds with morning sunshine may give rise to cumulus clouds within the boundary layer; with sufficient moisture, these clouds may spread laterally upon reaching the base of the capping PBL inversion, creating overcast or broken sky conditions with stratocumulus. For coastal regions, the passage of an afternoon sea breeze front can be characterized by a reduction in convective cloud cover as stable marine air moves inland. In mountainous terrain, cumulus clouds often first form over east-facing slopes because of the development of morning (upslope) valley breezes.

Prediction of cloud formations that are not surface based is often related to synoptic-scale forcing for ascent; for high cloud types such as cirrus and cirrostratus, conditions in the upper troposphere must be considered. Tropical moisture plumes are often persistent features (e.g., Fig. 11.4) that can be monitored using loops of infrared or water-vapor satellite imagery. Examination of model-forecast soundings can be very helpful in prediction of cloud coverage and altitude, but care must be taken to consider the possibility of saturation with respect to ice in subfreezing regions when predicting high clouds.

### 11.3.5. Other meteorological parameters

Weather forecasts tailored to the needs of agriculture, aviation, shipping and transportation, the energy industry, recreation, and public safety are each associated with a unique set of considerations. Rather than provide an in-depth discussion of each, a suggested general strategy that involves consistency, consideration of a spectrum of physical processes, and attention to observational data as well as output from numerical or statistical models is advocated.

## 11.4. USE OF AUTOMATED GUIDANCE

As presented in chapter 10, numerical modeling, as a research tool and for weather and climate prediction, has emerged as a major subdiscipline within the atmospheric sciences. With continued increases in computing power, there is no question that the importance of modeling will continue to grow. Students of the atmospheric sciences will witness increasing expectations regarding their knowledge and ability in this area.

In addition to the advantages afforded by numerical modeling, there are dangers. If forecasters and researchers cease to think carefully about the physical processes taking place in the atmosphere and come to rely too heavily on output from numerical models, then our collective ability to think critically about limitations in the models will be diminished. What, then, is the optimal manner in which models can be used in the forecast process? How will this evolve in the coming years, as hardware and software advances open up even greater possibilities?

The forecast process discussed in section 11.2 emphasizes beginning with past and present conditions and large spatial scales; thus, observational data (and model analyses, which are a hybrid of observational and model data) can be utilized to identify which phenomena and processes are important to the evolving forecast situation. When considering model output, to the largest extent possible, the physical processes recognized in the observations should be linked to analysis of the *model* atmosphere. Users of model output must at all times remain cognizant of the fact that the model atmosphere is an *inexact replica* of the *real* atmosphere, and the error sources discussed in section 10.6.1 must be borne in mind. A balanced forecasting approach that utilizes observational data, theoretical and conceptual understanding of weather systems and atmospheric processes, and numerical model output can serve to minimize any disadvantages of numerical models while optimizing their value. A similar balance is advantageous to research and analysis in general (Fig. 11.5). Using conceptual and theoretical knowledge

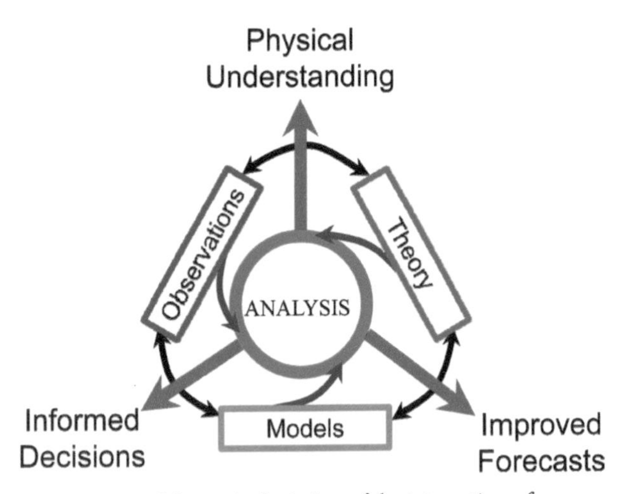


**Figure 11.5.** Schematic depiction of the integration of models, theory, and observations in the analysis process to yield improved forecasts, educated decisions, and physical understanding.

in the interpretation of observational and numerical model data represents an integral component of the analysis process.

### 11.4.1. Deterministic NWP output

In the 1980s and early 1990s, prior to the development of ensemble NWP systems, forecasters would typically dedicate considerable time and attention to the output from individual operational model runs, such as those from the now-defunct NCEP Nested Grid Model (NGM), Eta model, or Aviation model [AVN; now known as the Global Forecast System (GFS) model]. There is still much that can be gained from examination of individual model runs, in part because of the increasing accuracy of these models.

However, beginning in the mid 1990s, an increased appreciation for the role of *uncertainty* in weather forecasting emerged, in large part because of advances in ensemble prediction systems. Research and analysis of output from ensemble systems, discussed in section 10.6, suggests that deterministic approaches are limited by their inability to take advantage of quantitative information concerning forecast uncertainty. Atmospheric predictability is known to vary in time and space, whereas deterministic forecasting does not use all available information concerning forecast confidence. Users of forecast information benefit from understanding the *confidence* behind a certain forecast, in addition to knowledge of the most likely outcome in a given forecast situation.

National Weather Service field offices and operational centers are carefully considering how to best communicate forecast uncertainty to those who utilize their forecasts. This is not to say that deterministic forecasting is a thing of the past! In fact, the general public and other users of weather forecast information still seem to prefer deterministic forecasts, although this is likely in large part because of a lack of exposure to the benefits and format of probabilistic forecasting.

An advantage of the use of a small number of deterministic numerical model forecasts in weather prediction is the relative ease with which specific processes and phenomena can be diagnosed in comparison to ensemble forecast systems. Comparison of short-term model forecasts to observations is also facilitated by examination of deterministic forecasts. However, ensemble prediction systems can also be used to infer the presence of specific processes, but also these systems can elucidate the contribution of specific processes to uncertainty in the forecast and provide information about the dynamical relations between variables. In addition, understanding the range of possible or probable outcomes is of great value.

### 11.4.2. Ensemble guidance

In light of the preceding discussion, and of the advantages of ensemble prediction discussed in section 10.6.3, there are compelling reasons for utilizing ensemble forecasts in all types of weather prediction. Even if deterministic model runs are heavily utilized in the development of a given forecast, examination of ensemble information can provide valuable context and help the forecaster quantify uncertainty in a given prediction. In making a forecast, examination of output from an ensemble may alert forecasters to the possibility that there are two distinct scenarios of likely synoptic evolution or that there is very little forecast confidence beyond a given lead time (evident from strongly divergent ensemble members). In other situations, the operational deterministic forecasts may be outliers from the majority of ensemble members, and knowing this would be very useful to the forecaster.

Certain forecast parameters benefit more strongly from probabilistic treatment than others. For example, convective precipitation typically exhibits random behavior, and deterministic QPF are of limited value in these situations; a probabilistic forecast generated from a high-resolution ensemble provides the mean, extremes, and chances of occurrence of a given precipitation amount threshold.

### 11.4.3. Use of MOS

MOS, as discussed in section 10.7.2, by design, incorporates historical observations and climatology into the forecast, along with numerical model output. Probabilistic forecasts of a variety of parameters, including some that are not directly available from NWP output, such as thunderstorm probability, are available from MOS. Ensemble-based MOS products are also available, and in the future this area will likely grow in importance. Emerging techniques are beginning to employ more sophisticated statistical methods, such as Bayesian statistics (e.g., Raftery et al. 2005), to extend the earlier methods that were often based on multiple linear regression (e.g., Jacks et al. 1990). The availability of gridded MOS products, discussed in chapter 10, also marks a significant advance.

Regardless of the source or technique used as the basis for a MOS forecast, the question again arises regarding the optimal way for humans to utilize this information in the forecast process. In my own experience, I typically formulate a forecast without looking at MOS until I have a complete scenario in mind. Then, if there are large discrepancies between the MOS prediction and my preliminary forecast, I think about what factors are responsible for the differences. Sometimes I identify a possibility that I had overlooked; other times, I "stick to my guns" and discount the MOS prediction (sometimes to my peril!).

There will likely come a time when the combination of improved model physics and resolution, advanced ensemble prediction techniques, and the use of statistical postprocessing techniques such as MOS will produce forecasts upon which human forecasters have little chance to improve. At that point in time, new roles will have been identified for operational forecasters: for example, building stronger communications with emergency managers and policy makers so that forecast information can be interpreted correctly and used to maximum advantage. Forward-thinking students of the atmospheric sciences would do well to prepare themselves for this time by building diverse and versatile skill sets, and their skills need not be limited to the physical sciences. Financial, economic, and societal impacts of weather are appreciated far more now than they were a decade ago, and understanding and exploration of the linkages between the physical sciences and society represents a growing opportunity.

## 11.5. COMMENTS ON NOWCASTING, MESOSCALE, AND MEDIUM-RANGE PREDICTION

In keeping with our focus on synoptic-scale prediction and processes, detailed information on the forecasting of mesoscale phenomena will not be provided here. The field of *nowcasting*, which involves prediction with very short lead times of an hour or two, often is utilized for forecasting severe weather. Forecasting mesoscale phenomena also typically involves shorter lead times, consistent with the smaller spatial and temporal scales characterizing these systems. However, an understanding of the linkage between mesoscale phenomena and their larger-scale synoptic environment allows one to make synoptic-scale predictions of regions that are likely to experience a given type of mesoscale system. For example, the Storm Prediction Center (SPC) now issues an experimental 4–8-day convective outlook, in addition to more detailed outlooks for days 1–3. When severe or threatening weather is imminent, mesoscale discussions are provided by the SPC to present the evolving near-term meteorological situation.

The use of mesoscale ensembles of very high-resolution numerical forecasts in the SPC/National Severe Storms Lab (NSSL) Spring Experiment has demonstrated that new metrics for NWP verification and new types of diagnostic analysis are needed to fully exploit the information contained in such systems. Use of the NCEP SREF to formulate probabilistic forecasts of different severe weather threats indicates that ensemble techniques are extremely useful for mesoscale prediction as well (see section 10.6.3).

A type of forecasting that is not represented in this book is medium-range weather prediction, with lead times of 1–2 weeks. Techniques involving teleconnection indices, such as the Pacific–North America (PNA) pattern, the Arctic Oscillation (AO), and the North Atlantic Oscillation (NAO) are useful in understanding the favored modes of atmospheric fluctuations. Diagnosis of Rossby wave packets and downstream development is also helpful in making predictions in this time range. However, deterministic prediction in this range of lead time is of limited value; more typically, forecasts of anomalies are developed (e.g., Fig. 11.6).

The occurrence of certain atmospheric fluctuations and oscillations on the planetary scale has long been recognized, and measures of these oscillations are useful in assessing the state of the seasonal-scale flow regime. For example, the NAO is based on differences in pressure

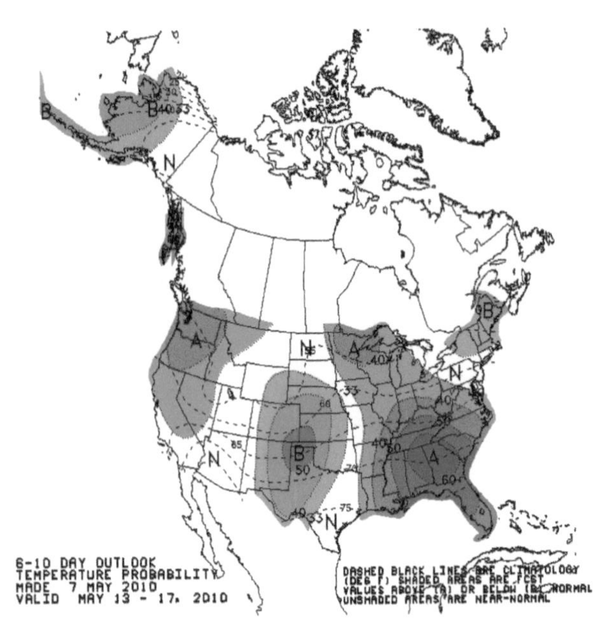


**Figure 11.6.** Climate Prediction Center 6–10-day U.S. temperature probability forecast valid 13–17 May 2010. Shaded regions indicate probability of above or below-normal temperatures, and dashed lines represent predicted average isotherms (°F).

between the Icelandic low and the subtropical anticyclone and can be quantified by sea level pressure differences between a location in Iceland and another to the south in the Azores. A stronger than average pressure difference or positive NAO is consistent with enhanced westerly flow and generally rainier and less extreme weather conditions in Europe. Because the positive NAO is accompanied by an anomalously strong subtropical high, southerly flow is favored over the eastern United States as well, with typically mild wintertime conditions. Conversely, a negative NAO implies anomalously high pressure over the Arctic and a weaker than normal subtropical high. In this state, more extreme conditions can be expected, and high-amplitude blocking patterns can become established.

As with the NAO, AO, and PNA patterns, care must be taken in discussion of cause and effect. During the cold eastern U.S. winter of 2009/10, the explanation of the cause of the anomalous cold should not be "because the NAO was negative." These climate fluctuations are diagnostic, not prognostic. The factors that drive the oscillation are complex, and a given planetary-scale wave pattern could be due to interactions between synoptic-scale transients and persistent large-scale flow anomalies. Thus, a given NAO index is not a cause, but a measure of the observed state of the planetary-scale flow regime. Until predictive skill can be demonstrated, these indices will remain useful as diagnostics; however, they do not represent *physical explanations* for the *cause* of a given winter storm or anomalous stretch of winter weather.

## 11.6. DEFENDING THE FORECAST

In many weather forecasting venues, the forecast and supporting context are provided in the form of a *weather briefing*. Most frequently, the weather briefing consists of a 10–30-min presentation of the past, present, and expected future weather conditions with an emphasis on the justification for the forecast in terms of physical processes and phenomena. Ideally, the weather briefing includes discussion and lively audience participation; differences of opinion frequently lead to improved understanding as well as to more accurate predictions.

Having taught weather briefings for more than a decade and also from observing shift-change briefings in NWS offices and during experimental field programs, I have seen a wide variety of depth and quality of justification for weather forecasts. The following section outlines suggestions for justification of a weather forecast in the form of a weather briefing.

### 11.6.1. The weather briefing

Following Bosart (2003), in the simplest terms, a weather briefing should ask and answer the following questions: *What has been happening, and why? What is happening now, and why? What will happen in the future, and why will it happen?* The point here is that, when developing a weather forecast, one should not immediately jump into the numerical model–forecast output without building proper context from past and present conditions as outlined earlier in this chapter. The above questions are strongly linked, and one must be selective in answering them, because there is not time to treat all meteorological features equally; even if there was, not all features are important to the forecast. The briefing team must ascertain which phenomena or processes are most important to the forecast and build the briefing around those features. For example, if a major coastal cyclone is expected to provide the most significant challenge during a given forecast period for a certain location, questions regarding the past and present evolution of that cyclone and the associated upper-level disturbance(s) should be a focus for the past and present weather. That is not to say that a single weather system or process should be the exclusive focus. If a different system had been affecting the forecast location during the previous day, dropping significant snow cover, this is obviously relevant to the forecast even though other systems may affect the area during the forecast period.

The briefing should begin with larger scales and past weather conditions, telescoping down to synoptic then mesoscale aspects of the forecast in the present and future. The briefing should provide in-depth meteorological discussion of the key weather systems, the physical processes controlling their past evolution and expected future behavior, and the processes relating these synoptic and mesoscale weather systems to local weather conditions.

At the end of a well-constructed weather briefing, the audience should have a clear picture of the important systems, as well as a thorough understanding of where the uncertainty lies in the forecast. Discussion and questions should naturally arise. For the previous coastal cyclone example, if there are large variations in the solutions provided by different model guidance, the presentation would benefit from inclusion of ensemble information, in addition to process-based reasoning as to the cause of the discrepancy. In some situations, current or recent-past observational data can clarify which, if any, of the models are most likely to represent the correct evolution of the system. At this point, very specific evidence and careful analysis are required. Too often, student weather briefers will make statements similar to the following: "I went with model A for this forecast because it seemed like it was verifying the best" or "In these situations, model A always does better." Statements of that type should be followed by immediate demands from the audience for supporting evidence! A more insightful commentary would be "The 6-h QPF from model A valid 1800 UTC today exhibits a region of significant precipitation over southern Georgia, which matches radar-estimated precipitation totals rather well, whereas the QPF from models B and C indicate no precipitation there during that time, as these graphics demonstrate."

The briefing must *not* present *all* of the information the briefing team has examined. On the contrary, the briefers have the task of distilling a huge volume of information to extract only the most relevant and compelling evidence to support their forecast reasoning. Rather than dutifully showing a "laundry list" of graphics, the briefing should tell a cohesive story, building evidence to justify the forecast but also acknowledging sources of uncertainty and potential error for further discussion, if the situation warrants. When done properly, weather briefings are enjoyable, informative, and insightful opportunities to learn about how the atmosphere works. The briefing team is making one or more hypotheses as to what the weather will do, and they must explain the science behind the hypothesis. Nature will test the hypotheses for us, and subsequent briefings should retrospectively analyze what happened, why it happened, and how it related to the expectations presented by the previous forecast team.

### 11.6.2. The postmortem

One of the most valuable activities that a forecaster can undertake is a careful examination of what happened relative to what was predicted. When making a series of forecasts for the same location, this is a natural process, and forecasters often pay careful attention to the fate of their forecast as the weather unfolds. However, if the actual weather diverged strongly from the forecast, the forecaster should make every effort to find out why. This activity may serve to eliminate or reduce similar errors in the future while at the same time increasing the forecaster's meteorological understanding. Experience improves forecast accuracy, but only if the forecaster makes an effort to learn from forecasts that did not verify well.

The postmortem analysis has the advantage of observational data with which to carefully track the evolution of the real atmosphere. Comparisons of past model forecasts with operational analyses, such as those available from the same or different models, are a valuable source of information concerning the model forecast. However, comparisons with radar, satellite, soundings, and surface observations are also highly beneficial. When there are major errors in the model forecast, one should ask if the observations fell within the envelope of uncertainty available from ensemble predictions. If not, what was the source of the model error? On occasion, further investigation or even retrospective model experiments can pinpoint the source of error, be it in the model physics or in the initial conditions.

In some environments, operational forecasters do not have the time to undertake such postmortem case studies. The task of collecting and archiving forecast data, as well as verifying observations, requires diligence, time, and resources. A team of forecasters, sometimes in collaboration with researchers or students, may be best able to collect and analyze the needed information. Postmortem results should be shared among the relevant forecasting staff, perhaps including those who did not work the forecast shift in question. Venues such as the American Meteorological Society (AMS) *Weather Analysis & Forecasting* conference series, the National Weather Association (NWA) annual meeting, the AMS *Severe Local Storms* conference series, and several others are ideally suited for presentation, sharing, and discussion of findings. The AMS journal *Weather and Forecasting* and the NWA *Digest* are designed to help forecasters and researchers communicate results from their investigations more broadly in the forecasting community.

### 11.6.3. The attributes of a good forecaster

When teaching classes that include weather forecasting exercises, students that are new to forecasting often express feelings of apprehension or even anxiety about forecasting. Even the strongest students are often concerned about how their forecasting skills will stack up. Data show that the correlation between student grade point average (GPA) and forecasting skill is not as strong as one might think a priori. It is not uncommon for a very strong student to struggle with forecasting, and occasionally students with relatively low GPAs prove to be excellent forecasters. What qualities distinguish those who enjoy and excel at forecasting? A thought-provoking discussion of the qualities that characterize excellent forecasters is provided by Doswell (2003a), as excerpted below with some modification by the author:

*Decisiveness:* Forecasters often have to make decisions quickly and sometimes without sufficient time to analyze available data.

*Ability to deal with pressure:* Forecasting can carry a large burden of responsibility, especially during severe or high-impact weather. During active weather situations, high-stakes forecasts involving lives and property must be made without hesitation and in a calm, professional manner.

*Strong visualization and conceptualization skills:* The quantity of meteorological data available to forecasters is staggering and will continue to grow. A good forecaster must be able to recognize patterns in the data, link them to conceptual and physical understanding, and translate that into a forecast of the evolution of these features with time. Human abilities are put to the test by the volume and four-dimensional character of numerical model and observational data.

*Passion for meteorology:* Without a passion for weather forecasting, it is very difficult to achieve excellence. If the joy of forecasting has diminished with time, forecasters must find ways to revive to the aspects that sustain the passion they once felt.

*Commitment to learning:* Every forecast contains a lesson, and forecasters must make the effort to learn from poorly verified forecasts. No matter how extensive a forecaster's experience, there will always remain much to learn. A career-long commitment to education, curiosity, and postmortem analysis is critical to improving as a forecaster.

*People skills:* Nearly all forecasters are required to interact with other forecasters as well as those using the forecasts as a part of their job. This may require leadership, a willingness to follow another's lead, and explaining complicated situations to lay persons with little background in meteorology.

*Ability to multitask:* With a plethora of data, deadlines, and responsibilities, strong organizational skills and the ability to manage multiple responsibilities in a timely fashion are required. These necessitate outstanding time-management skills and the ability to prioritize under severe time constraints.

*Able to deal with failure:* Few, if any, forecasts are perfect. Even experienced, highly skilled forecasters make mistakes. Given that forecasters are passionate about their work, it can be difficult to keep such failures in context. Good forecasters must learn to avoid extreme reactions to forecast failures. Rather than just brushing a blown forecast aside or becoming disheartened, a *good forecaster accepts failure but is never satisfied with it.*

*Situational awareness:* Even during benign weather conditions, forecasters should remain focused and prepared. During active weather, keeping an eye on all aspects of the meteorological situation, rather than becoming myopic, is important.

*Communication skills:* Along with people skills, good forecasters are able to communicate complex information to a wide variety of audiences. The ability to remain concise — filtering out superfluous information — is a valuable attribute. Forecast information can be communicated using a variety of media, including the written word, orally, numerically, and graphically.

*Technical skills:* The ability to thrive in a rapidly evolving computer- and technology-intensive environment is a tremendous advantage for modern forecasters. Specific skills include the ability to work with various display, communication, and analysis software, write scripts, manipulate data in a variety of formats, and set up and run numerical models.

*Flexibility:* Good forecasters can anticipate and adapt to rapidly evolving situations. Forecasters are often confronted with unique situations and should not necessarily use a "cookie cutter" approach and treat all situations the same way. Systematic approaches are useful, provided that flexibility is maintained.

*Possess the physical ability to do shift work:* Many forecast offices employ rotating work schedules, and physiologically some individuals are not able to adapt to this without suffering mental or physical impairment. Weather forecasting is a 24/7 endeavor, and critical situations can and will develop at inconvenient times of day or night.

In light of the rapid evolution of the technology and science of weather forecasting, students and future atmospheric scientists would do well to keep a flexible, mobile view of their profession. Development of a versatile set of skills, understanding of the impacts and linkages of weather to society, and retention of fundamental knowledge are challenges that modern students must meet. Participation in internships, broadening knowledge through coursework outside of the traditional meteorology curriculum, and diligent attention to professionalism are critical in today's competitive job market. Students should join professional societies, such as the AMS, NWA, and American Geophysical Union (AGU) to best stay in touch with the changing field of geosciences.

## REVIEW AND STUDY QUESTIONS

1. Every day, spend a few minutes visiting to the Web site of a local NWS forecast office and carefully read the most recent Area Forecast Discussion. For each discussion, identify any portions that you do not understand, and seek additional information concerning these topics.
2. Select a forecast location, and look up the background information suggested in section 11.2.1. Find a way to organize this information in a spreadsheet.
3. For the forecast location you identified in item 2, step through the forecast process outlined in this chapter using the Snellman forecast funnel approach. Make a prediction of maximum and minimum temperature, quantitative precipitation amount, and maximum sustained wind speed for a 24-h period beginning at 1200 UTC the following day.
4. Based on your forecast location from items 2 and 3, write a 2- or 3-paragraph Area Forecast Discussion (AFD) justifying your forecast.
5. Redo the forecast you made in item 3 in *probabilistic* terms. What are expected ranges of temperature and precipitation, and what are the most probable ranges of maximum and minimum temperatures?
6. Familiarize yourself with the NCEP ensemble pages, including both the GEFS and SREF systems. View the experimental ensemble page at the SPC, and also examine output from the NAEFS ensemble system. Viewing these products daily will provide context for future ensemble interpretation.

## REFERENCES


Bjerknes, V., 1919: Weather forecasting. *Mon. Wea. Rev.*, **47**, 90–95.

Bond, N. A. and C. F. Mass, 2009: Development of skill by students enrolled in a weather forecasting laboratory. *Wea. Forecasting,* **24,** 1141–1148.

Bosart, L. F., 2003: Wither the weather analysis and forecasting process? *Wea. Forecasting,* **18,** 520–529.

Bosart, L. F., and F. Sanders, 1981: The Johnstown flood of July 1977: A long-lived convective system. *J. Atmos. Sci.,* **38,** 1616–1642.

Broad, K., A. Leiserowitz, J. Weinkle, and M. Steketee, 2007: Misinterpretations of the "cone of uncertainty" in Florida during the 2004 hurricane season. *Bull. Amer. Meteor. Soc.,* **88,** 651–667.

Carroll, E. B., and T. D. Hewson, 2005: NWP grid editing at the Met Office. *Wea. Forecasting,* **20,** 1021–1033.

Davis, C. A., and S. B. Trier, 2002: Cloud-resolving simulations of mesoscale vortex intensification and its effect on a serial mesoscale convective system. *Mon. Wea. Rev.,* **130,** 2839–2858.

Doswell, C. A., III, 1986: The human element in weather forecasting. *Nat. Wea. Dig.,* **11,** 6–17.

——, 2003a: What does it take to be a good forecaster? [Available online at http://www.flame.org/~cdoswell/forecasting/Forecaster_Qualities.html.]

——, 2003b: Is there a role for humans in the NWS of the future? (Is there an NWS in our future at all?) [Available online at http://www.flame.org/~cdoswell/forecasting/human_role/future_forecasters.html.]

——, 2004: Weather forecasting by humans: Heuristics and decision making. *Wea. Forecasting,* **19,** 1115–1126.

Glahn, H. R., and D. P. Ruth, 2003: The new digital forecast database of the National Weather Service. *Bull. Amer. Meteor. Soc.,* **84,** 195–201.

Haugland, M. J., and K. C. Crawford, 2005: The diurnal cycle of land–atmosphere interactions across Oklahoma's winter wheat belt. *Mon. Wea. Rev.,* **133,** 120–130.

Hawblitzel, D. P., F. Zhang, Z. Meng, and C. A. Davis, 2007: Probabilistic evaluation of the dynamics and predictability of the mesoscale convective vortex of 10-13 June 2003. *Mon. Wea. Rev.,* **135**, 1544–1563.

Jacks, E., J. B. Bower, D. J. Dagostaro, J. P. Dallavalle, M. C. Erickson, and J. C. Su, 1990: New NGM-based MOS guidance for maximum/minimum temperature, probability of precipitation, cloud amount, and surface wind. *Wea. Forecasting,* **5,** 128–138.

Jolliffe, I. T., and D. B. Stephenson, Eds., 2003: *Forecast Verification: A Practitioner's Guide in Atmospheric Science.* Wiley, 240 pp.

Mass, C. F., 2003: IFPS and the future of the National Weather Service. *Wea. Forecasting,* **18,** 75–79.

Morss, R. E., J. Demuth, and J. K., Lazo, 2008: Communicating uncertainty in weather forecasts: A survey of the U.S. public. *Wea. Forecasting,* **23,** 974–991.

Murphy, A. H., 1978: On the evaluation of point precipitation probability forecasts in terms of areal coverage. *Mon. Wea. Rev.,* **106,** 1680–1686.

Novak, D. R., D. R. Bright, and M. J. Brennan, 2008: Operational forecaster uncertainty needs and future roles. *Wea. Forecasting,* **23,** 1069–1084.

Raftery, A. E., T. Gneiting, F. Balabdaoui, and M. Polakowski, 2005: Using Bayesian model averaging to calibrate forecast ensembles. *Mon. Wea. Rev.,* **133,** 1155–1174.

Roebber, P. J., and L. F. Bosart, 1998: The sensitivity of precipitation to circulation details. Part I: An analysis of regional analogs. *Mon. Wea. Rev.,* **126,** 437–455.

Snellman, L. W., 1977: Operational forecasting using automated guidance. *Bull. Amer. Meteor. Soc.,* **58,** 1036–1044.

——, and E. Thaler, 1993: Forecast process. COMET, CD-ROM version 1.0. [Available from COMET, UCAR, P.O. Box 3000, Boulder, CO 80307.].

Stuart, N. A., and Coauthors 2006: The future of humans in an increasingly automated forecast process. *Bull. Amer. Meteor. Soc.,* **87,** 1497–1502.

Thompson, R. L., and R. Edwards, 2000: An overview of environmental conditions and forecast implications of the 3 May 1999 tornado outbreak. *Wea. Forecasting,* **15,** 682–699.

Winkler, R. L. and A. H. Murphy, 1976: Point and area precipitation probability forecasts: Some experimental results. *Mon. Wea. Rev.,* **104,** 86–95.

# CHAPTER 12

# Manual Analysis

*If the data were sufficiently dense that each line could be drawn merely by connecting a series of data points, then there would be no problem at all, and the average person might do as good a job of drawing lines as the one with years of training and experience in analysis. But, owing to limited data, analysis is knowledge, attitude, and application above the mechanical art of line drawing.*

—W. J. Saucier, "Principles of Meteorological Analysis" (1989)

Jacob Bjerknes undertaking a map analysis, Bergen, Norway (from Shapiro and Grønås, *The Life Cycles of Extratropical Cyclones*, Boston, MA: American Meteorological Society, 1999).

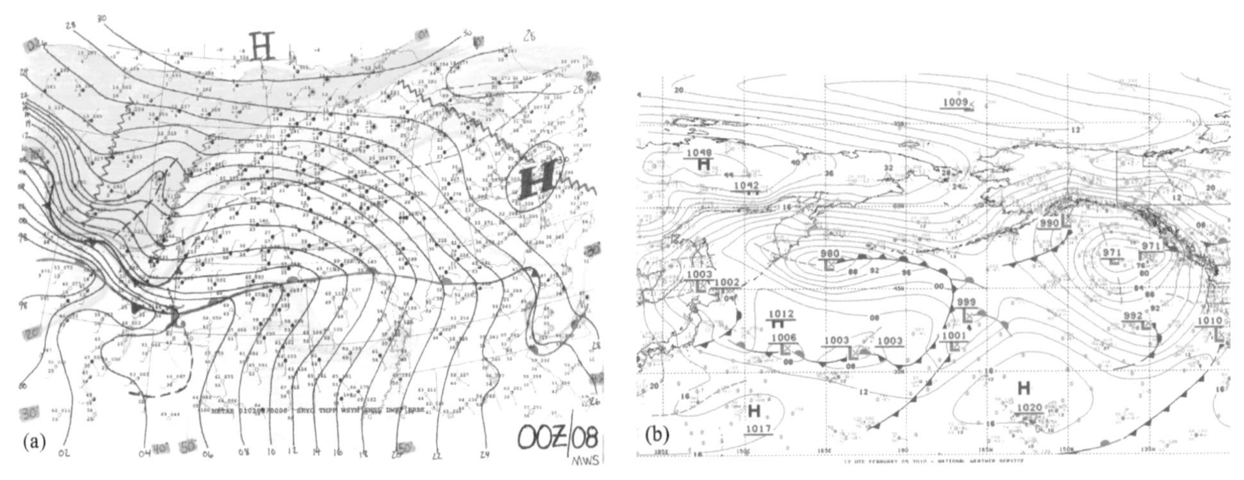


**Figure 12.1.** (a) Manual surface analysis constructed by NWS forecaster Mike Strickler, Raleigh, NC, forecast office, valid 0000 UTC 8 Feb 2001. (b) Unified analysis for Pacific region, valid 1200 UTC 5 Feb 2010, constructed by analysts at NCEP operational centers.

Chapters 2, 3, and 4 presented frameworks for diagnosis of synoptic-scale atmospheric processes. The tools outlined in these earlier chapters are useful, complete with mathematical and physical bases. While these frameworks provide a foundation for problem solving in a variety of applications, it is also important to emphasize a somewhat different kind of analysis tool: *manual analysis* of meteorological data.

With the growing power of computers, along with ever-increasing volumes of data and access to digital data via the Internet, we live in a time when too much data, rather than too little, is often the fundamental challenge. How can one integrate and synthesize so much information into a useful hypothesis or forecast? In light of the advantages provided by high-speed computing, is the time and effort required to construct manual analyses of surface or upper-air maps worthwhile? Automated data assimilation systems can do this much faster, incorporate much more data, and can do so in a manner that is physically consistent.

Nevertheless, there is tremendous value in thinking about observational data and in interpreting individual weather observations. Without questioning the value of modern data assimilation or numerical weather prediction, there remains a place for manual map analysis. Rather than using a pencil and paper, many modern analysts use software to construct manual analyses, such as the NMAP2 program.[1] There is also a contingent of meteorologists who maintain the value of pencil and paper in analysis, and in the synoptic classes I teach, both analysis software and pencil and paper are used for some laboratory assignments. If too many layers of technology come between the user and the original data measurement, the risk of losing touch with the basic physics of the atmosphere increases. Whether sketching on scratch paper or constructing a complete analysis of several meteorological variables, the analyst considers individual observations in this process. For this reason, at many forecast centers, routine manual analyses are still constructed for the purpose of helping forecasters stay in touch with the observational data and identifying important features that can affect the forecast. An example of a surface analysis, drawn on-shift by an operational forecaster, illustrates this point (Fig. 12.1a).

Routine surface analyses are carried out at by the U.S. National Weather Service (NWS) and other global weather services. In the United States, the Ocean Prediction Center, Hydrometeorological Prediction Center, and Tropical Prediction Center all generate and deliver manual analyses. One of the resulting products of analysis coordination between the U.S. centers is the *unified surface analysis* product, an example of which is shown in Fig. 12.1b for the Pacific Ocean region.

The quality and utility of routine manual analyses has been the subject of debate in the scientific community (e.g., Mass 1991; Uccellini et al. 1992; Sanders and Doswell 1995; Sanders 1999; Sanders and Hoffman 2002). Most of the criticism stems from difficulties in the analysis of surface frontal zones, and a lack of consistency between routinely analyzed fronts and the surface temperature or

[1]The use of the National Centers for Environmental Prediction (NCEP)'s Advanced Weather Interactive Processing System (N-AWIPS) software will soon transition to the use of a comparable software package called AWIPS version II (AWIPS-II).

potential temperature field. It is notable that surface isotherms or isentropes are absent in the analyses shown in Fig. 12.1. Despite compelling arguments in support of analyzing surface thermodynamic fields by Sanders and collaborators, these fields are still not included in the United States for routine surface analyses. The reason for the omission of the surface temperature field has been attributed to the fact that surface temperature is sensitive to local effects and elevation differences, which can obscure more useful synoptic signals (Sanders and Doswell 1995). In an effort to minimize these difficulties, analysis of surface *potential temperature* is a useful alternative, despite being less familiar to the general public.

While Sanders and Doswell (1995) concede that detailed surface mesoanalysis can be challenging in a time-constrained operational environment, Sanders and Hoffman (2002) and Bosart (2003) demonstrate that automated techniques can be used to construct credible analyses of the surface potential temperature field, which can then be subjectively modified with other data to locate frontal systems.

At the Storm Prediction Center (SPC), automated and detailed mesoanalyses are available online, and these include a wide variety of options for plotting thermodynamic and kinematic quantities in versatile formats. Additionally, a real-time mesoscale analysis (RTMA) product has been developed, which includes high-resolution surface analyses of thermodynamic variables. The unification of a large variety of surface observations, including nontraditional surface-observing networks, has been accomplished by the MesoWest project at the University of Utah. Access to surface observations from this network has never been better, thanks to a convenient Web page interface. Clearly, efforts are underway to address the concerns expressed in the past.

What is the optimum role for manual analysis in modern weather analysis and forecasting? Given the time pressures facing operational forecasters, is manual analysis the best use of a *forecaster's* time? Is there a good way for forecasters to retain the insights afforded by manual analysis while utilizing efficient automated surface-analysis products?

It is true that there are countless examples of excellent forecasts having been made by meteorologists who did not first construct a manual surface analysis, or even look at one. In the forecast process outlined in the previous chapter, I did not advocate that a manual analysis be drawn prior to making a forecast. The recommended role of surface analysis will depend on the personal preferences and time constraints facing a given forecaster. If, in assessing the critical weather systems affecting the forecast site during a given forecast period it is clear that a manual analysis would be helpful, every effort should be made to construct one. The need for such an analysis is also related to the type of forecast problem; for the prediction of severe convection, knowledge of the presence and strength of surface features, such as outflow boundaries or weak frontal zones, can be of the utmost importance (along with having a clear picture of the synoptic-scale weather pattern). Finally, in postmortem analysis of retrospective events, manual analysis can also be highly beneficial to find out what really happened based on observations. Clearly, an analysis constructed for the purposes of rapid feature location or identification during an operational forecasting shift will be constructed more hastily than an analysis being done retrospectively for research purposes.

## 12.1. MANUAL SURFACE ANALYSIS

As with weather forecasting, an important concept in constructing a surface analysis is *physical consistency* between the observational data and the different analyzed fields. The basic technique of constructing isopleths of a given quantity involves interpolation between adjacent points, smoothing, and consideration of a variety of data sources. For instance, the use of satellite imagery is helpful for analysis of surface features, especially over the ocean.

It is beyond the scope of this text to cover manual analysis in great depth. Saucier (1989) includes much more comprehensive information. Laboratory exercises are best for the presentation of analysis techniques for actual case studies. Here the discussion will be limited to some basic analysis principles.

### 12.1.1. Isopleths 101

Recall that an *isopleth* is a line along which the value of the analyzed quantity is constant, with readings greater than the isopleth value consistently located on one side and lower values on the other. When positioning isopleths, one may assume a linear change between neighboring stations unless data are available to suggest otherwise. For example, a 1012-mb isobar would be located one-third of the way between a station reporting 1013 mb and another reporting 1010 mb (linear interpolation). However, if one station is reporting a calm wind and the other a 20-kt wind, then we might expect the pressure gradient to be concentrated closer to the station with the higher wind speed and the isobar shifted accordingly.

It should go without saying that isopleths should never suddenly end within a region of available data, they must never branch or split, and two different isopleths must never touch or cross! Isopleths must progress sequentially; it is not possible to pass from a large-valued isopleths to a small-valued one without intermediate values being observed (and intermediate isopleths analyzed) in between. Despite the very basic level of information in this paragraph, I am often amazed at the extent to which students in my senior synoptic course have lost sight of these principles!

The lateral spatial extent of the isopleth analysis should be extended until one reaches either the boundary of the plotting domain or the edge of sufficient data. The analysis should not be drawn into regions that are completely devoid of data. When analyzing a region of sparse data, the isopleths may be drawn with dashed lines to indicate a low degree of confidence. Alternate data sources, such as satellite or radar imagery, may be helpful to supplement surface observations, especially over the oceans.

Data values reported by surface stations are characterized by some uncertainty. Some exhibit biases and occasionally unrealistic reports. Even when the data are correct, there is sufficient uncertainty in the observation that the manual analysis does not need to strictly adhere to every single data point. For synoptic analysis, single-station outliers can be ignored or de-emphasized to focus on more credible synoptic features.

### 12.1.2. Analyzing the pressure field

Discontinuities in the pressure field do not exist in the atmosphere. On the synoptic scale, sea level isobars are generally smooth and parallel to one another. To the extent that consistency with data allows, isobars should be drawn so that they are evenly spaced for a synoptic-scale analysis, reflecting a uniform gradient.

Attention to the wind field and recognition of the character of the underlying surface are important consistency checks—regions of light wind should be collocated with weaker pressure gradients and vice versa. In regions of rough terrain, the relation between the wind and sea level pressure field may be complex and weakly evident. In regions of high terrain, the sea level pressure corresponds to a level that is physically far underground (and that does not really exist) and the value may be contaminated by the computation of the sea level pressure (which is based on the assumption of a standard atmosphere lapse rate below ground).

Standard 10-m wind observations often fall within the surface layer, which is the lower portion of the planetary boundary layer; the influence of friction will generally be evident on the wind field. Accordingly, the analyst should expect a larger cross-isobar angle between the winds and isobars in regions of rough terrain or shallow boundary layer depth. Over water or smooth land surfaces, a smaller cross-isobar wind angle is anticipated, perhaps as small as 10°–20°. Regions of strong wind speed should be characterized by closely spaced isobars, and regions of calm or light winds should exhibit a much weaker pressure gradient. If you have drawn closely spaced isobars in a region with calm winds, think again! In regions of complex terrain, winds are often channeled through valleys and may blow at very large angles to the isobars. In such regions, very strong sea level pressure gradients may be a reflection of the pressure correction to sea level, and they are not "real" in the sense that any air parcel is actually experiencing a corresponding pressure gradient force. In some situations, very strong pressure gradients are accompanied by light winds; for example, if the gradient is not realized because of the presence of a mountain range. However, with strong cross-barrier pressure gradients, gaps in the terrain must be monitored for possibly very strong winds.

Given the uncertainty in reported sea level pressure values, partly because of the correction to sea level in elevated terrain and partly because of instrument calibration or error, the analyst should not *force* isobars to follow *all* reported data unless consistency can be maintained. Especially over the ocean, some reports are outright erroneous, as discussed by Sanders (1990). Other stations exhibit consistent biases that can be corrected by an experienced analyst. When should an analyst ignore a given observation? One consideration is the degree of consistency with surrounding observations and between different analyzed fields. If adjusting the analysis to account for a single observation imparts a seemingly unphysical perturbation, then consider ignoring it. However, if there is more than one station reporting consistent conditions, then chances are it is a real (but perhaps mesoscale) feature that should be considered for inclusion in the analysis. If an observation

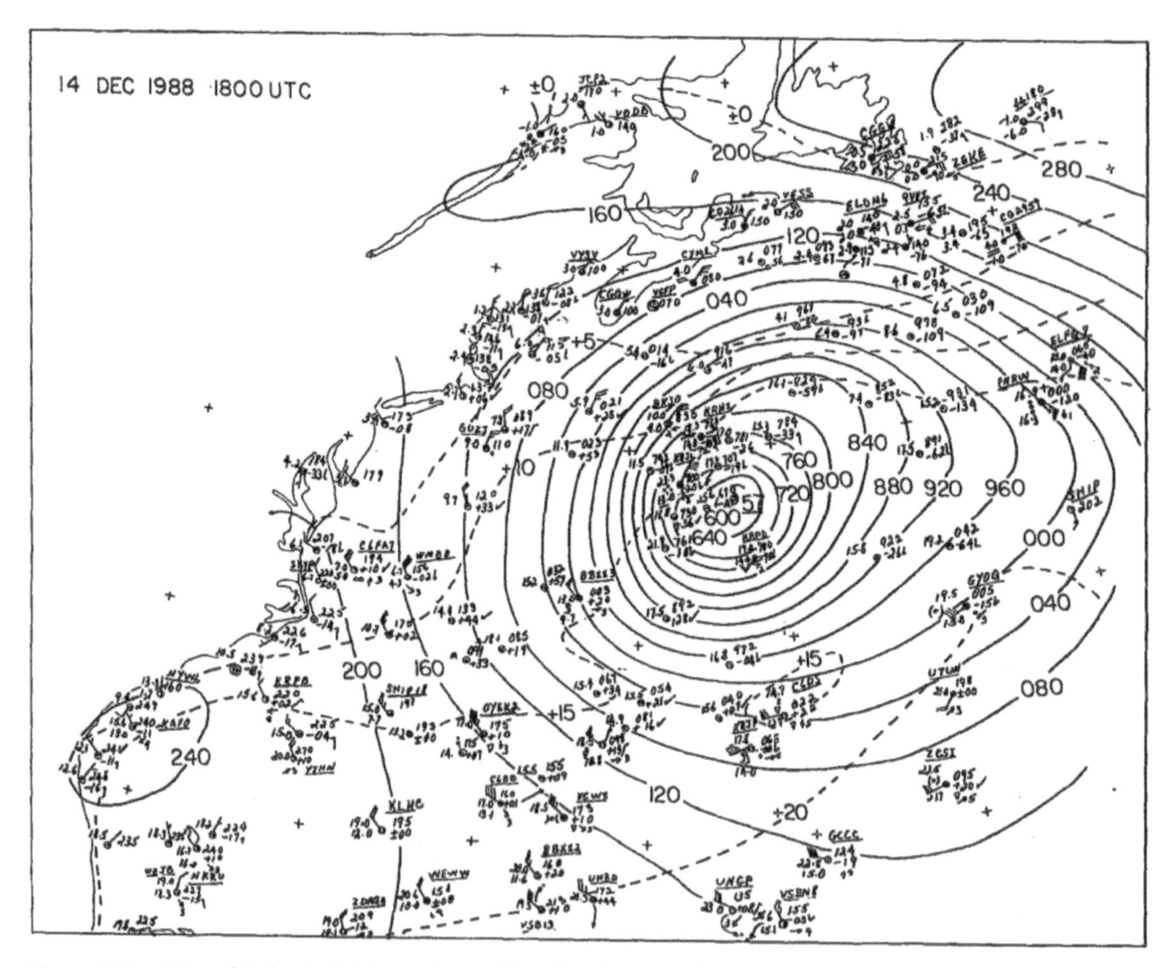


**Figure 12.2.** Manual isobar (solid, interval is 4 mb) and isotherm (dashed, interval is 5°C) analysis by Sanders, valid 1800 UTC 14 Dec 1988, during the Experiment on Rapidly Intensifying Cyclones over the Atlantic (ERICA) field program (from Sanders 1990).

is discounted in the construction of your analysis, it should be flagged to indicate that it was not considered in constructing the analysis.

As an example, consider the manual analysis constructed by the late professor Fred Sanders (Fig. 12.2) for the powerful second ERICA intensive observation period (IOP2) cyclone over the western North Atlantic. Note the even spacing of isobars and the consistency between isobar spacing and wind speed observations. The omission of fronts from the analysis is intentional and was explained by Sanders as a way to "minimize obscuration of data." This is also consistent with his view that the surface isotherm field, in conjunction with the pressure field, is adequate to identify important baroclinic zones: "where fronts were strong and unambiguous, however, the isobars and isotherms leave little doubt where they were" (Sanders 1990).

### 12.1.3. Analysis of surface thermodynamic variables

**12.1.3.1. Temperature and potential temperature.** Surface temperature observations can be relatively noisy because of local land surface effects and elevation differences. The analysis of surface potential temperature has the advantage of reducing differences because of elevation and is suggested as the analysis variable of choice. Many of the same comments made for isobaric analysis apply to isentrope analysis; for instance, one need not draw to individual observations that appear to be outliers from other nearby observations. As data are plotted to the nearest degree Celsius, some flexibility is also available to the analyst on the basis of round off.

Strong thermal gradients can be present at coastlines because of land–sea or land–lake temperature differences.

As shown by Sanders and Hoffman (2002), semipermanent coastal potential temperature gradients are present, for example, along the California coast. This gradient is geographically bound and would not generally be analyzed as a frontal system. More mobile coastal fronts, common along the southeastern coast of the United States, are typically analyzed because of the meteorological importance of their movement (see chapter 6).

The purpose of the analysis is to extract the synoptic-scale and large mesoscale signals; therefore, try to remove high-frequency noise from your analysis by smoothing the fields accordingly. When strong mesoscale features are present, the analyst should consider a smaller-scale domain and the construction of a mesoscale analysis, rather than trying to analyze mesoscale features over a synoptic-scale domain.

**12.1.3.2. Dewpoint analysis.** Because the dewpoint temperature is an absolute measure of water vapor content, surface *isodrosotherm* (dewpoint) analysis can be valuable in the prediction of convective storms and heavy precipitation, as it offers a means of assessing the availability of moisture and moisture transport. Sometimes analysts will simply shade regions in which dewpoint temperature exceeds a threshold value (e.g., Fig. 12.1a), or they will sketch a few key contours, to give a sense of the general regions where moist air is located. Even more than with temperature, dewpoint observations can be noisy, as readings can be strongly influenced by local surface characteristics. In light of this, isodrosotherms should be drawn to smoothly reflect the synoptic character of the moisture field.

**12.1.3.3. Equivalent potential temperature and wet-bulb temperature.** A compact representation of the surface thermodynamic character can be obtained through analysis of equivalent potential temperature, $\theta_e$. Recall from chapter 1 that $\theta_e$ is given by

$$\theta_e = \theta \exp\left(\frac{L_v r}{C_p T_{LCL}}\right). \qquad (12.1)$$

Thus, analysis of $\theta_e$ provides information about both the surface potential temperature and moisture fields. A disadvantage of this quantity is the inability to distinguish gradients of temperature from those of moisture. Furthermore, because $\theta_e$ is not a density variable, it is not as closely linked to frontal circulations. However, because fronts are often characterized by gradients of both temperature and moisture, analysis of $\theta_e$ more readily allows the frontal location to be determined because of the very sharp gradient found there. In situations when the temperature gradient across a front undergoes diurnal weakening, the moisture contrast often remains strong and $\theta_e$ analysis will continue to reveal the frontal location.

The temperature to which air will cool when enough moisture is evaporated to produce saturation is the *wet-bulb temperature*. When winter weather threatens, analysis of the surface wet-bulb temperature can aid forecasters in anticipating the likely location of the surface freezing line once precipitation begins. Outside of winter weather forecasting situations, surface wet-bulb temperature is not typically analyzed, however.

## 12.2. FRONTAL ANALYSIS

That operational frontal analyses have suffered criticism says much about how difficult they are to do properly. As discussed by Sanders and Doswell (1995), Sanders (1999), and Sanders and Hoffman (2002), the lack of operational analysis of surface thermodynamic fields has likely contributed to the perceived inconsistency in operational frontal analysis. In chapter 6, this inconsistency in subjectively analyzed frontal placement was evident even from a group of experienced synoptic analysts, demonstrating subjectivity in the frontal analysis process (Fig. 6.2a). Different analysts weight different factors more or less heavily in their analysis. Despite there being no universally agreed-upon method for frontal analysis, general guidelines for consistent frontal analysis can be drawn:

- It is advantageous to plot fields such as potential or equivalent potential temperature, to best analyze fronts while minimizing elevation dependence.
- A final frontal analysis should await completion of the potential temperature and pressure analysis, with iterative adjustment of these fields once the location of the front has been established.
- It is often useful to emphasize the *wind shift* and *pressure trough*, which are generally (but not always) found near the warm edge of the zone of strongest temperature contrast (e.g., Fig. 6.2b). Given what we know about the relation between wind and pressure, the presence of a pressure trough requires a wind shift.

In some cases prefrontal troughs can develop, but this should be evident when considering consistency among fields (see Schultz 2005).

- When analyzing a front, be sure to seek consistent signals in several meteorological variables and be ready to defend your analysis with specific information.
- Consider previous analyses to ensure continuity, bearing in mind that fronts do change with time and that an analyzed front must be supported by the data in any given analysis.
- Dewpoint temperatures are usually considerably lower on the cold side of a front.
- For occluded fronts, temperatures reach a maximum right at the location of the surface front.
- For warm fronts, a useful rule is that easterly winds are often found north (ahead) of the front. Use rules of this type with caution, however, as fronts can exhibit any orientation, as shown in chapter 6.
- It is essential that consistency be maintained between the different analyzed and plotted fields. In other words, fronts should exhibit wind shifts, changes in isobar orientation, and temperature contrasts in a consistent fashion.

Warm, cold, and stationary fronts are found at the warm edge of regions of strong temperature differences. That is, they mark the boundary of warm air masses. Fronts that are accompanied by a significant wind shift are almost invariably characterized by a trough in the pressure field as well (Fig. 12.3). In fact, the presence of a sharp frontal pressure trough requires that fronts are marked by a wind shift. Examples of subjectively analyzed fronts are provided in Figs. 12.3, 12.4.

At 0000 UTC 11 February 1997, a modest cyclone was centered near the Arkansas–Missouri border, with a cold front extending southward from the center (Fig. 12.3). The temperature gradient is more pronounced along the more southerly portions of the cold front within the analysis area, with a weakening of the thermal gradient near the cyclone center. This weakening of the cold front near the cyclone center, known as "frontal fracture" as discussed in chapter 6, is a common occurrence. A sharp pressure trough and unambiguous wind shift accompany the primary cold front. A secondary cold front is located to the northwest of the cyclone center, although this system is more weakly defined in terms of thermal contrast, wind shift, and pressure trough.

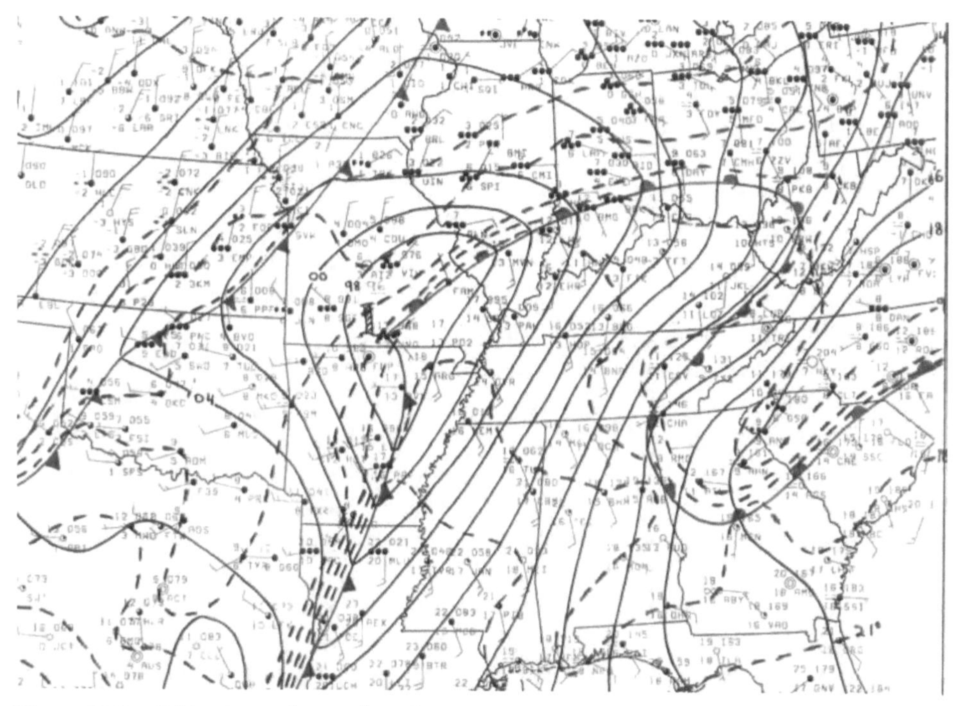

**Figure 12.3.** Subjective surface analysis for 0000 UTC 27 Feb 1997: surface isotherms (dashed red, interval is 3°C), sea level isobars (solid black contours, interval is 2 mb), and fronts (standard plotting convention).

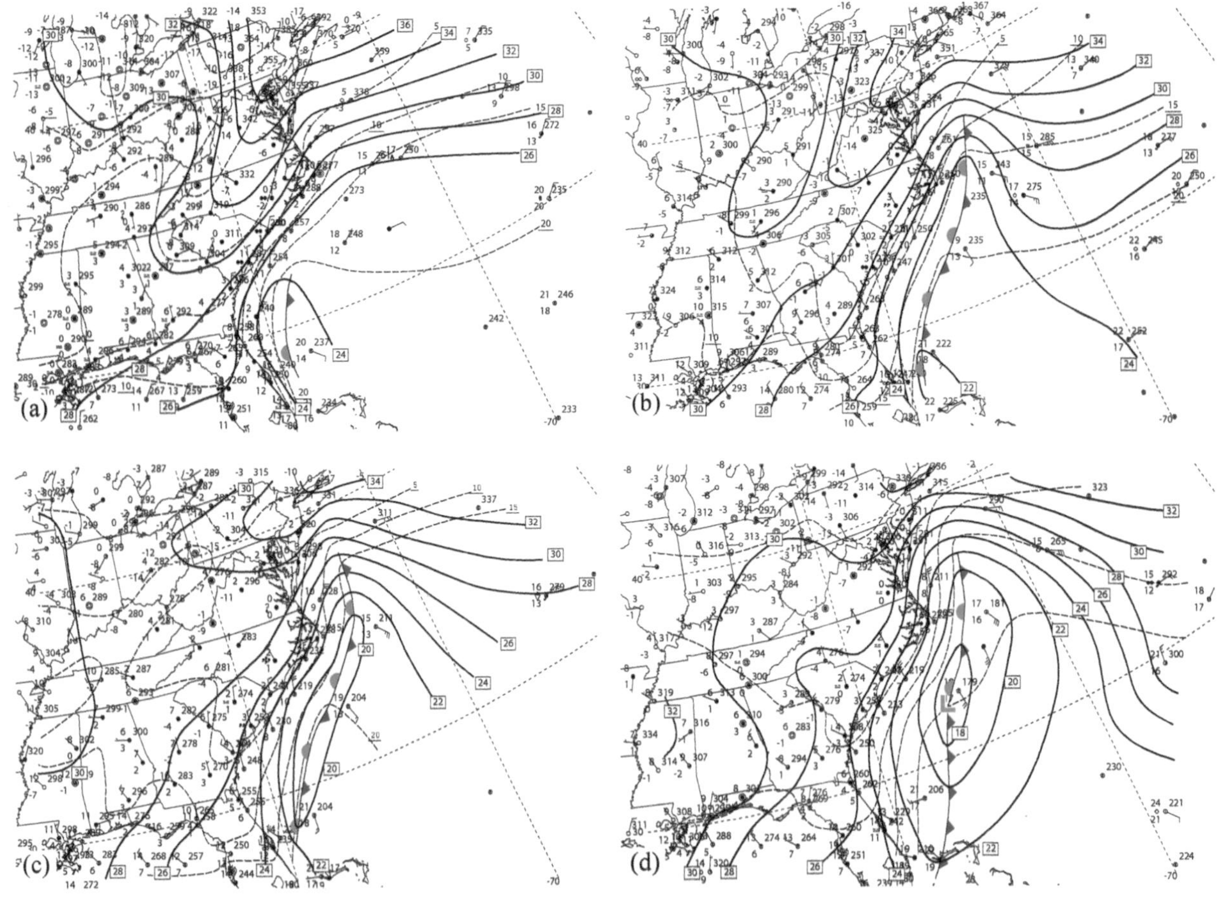


**Figure 12.4.** Sequence of manual surface analyses for a coastal cyclone event dated (a) 1200, (b) 1800, and (c) 2100 UTC 17 Feb 2004; and (d) 0000 UTC 18 Feb 2004 (from Mahoney and Lackmann 2006).

Unlike the cold front, the warm front that extends east from the cyclone center is strongest immediately east of the cyclone and becomes more poorly defined farther east. The remnants of a cold-air damming event are evident in both the southward extension of the warm-frontal zone and in the isotherm field over the Carolinas and Virginia. Much subjectivity is involved in placing the front to the west of the damming region—over Ohio, for instance.

A coastal front (see section 6.4) formed and strengthened along the coast of the southeastern United States in February 2004 (Fig. 12.4). This case was also presented in chapter 10 in conjunction with model sensitivity tests. As the coastal front extended northward along the coast, it remained nearly stationary and is analyzed as such until a cyclone forming along it swept the southern portion of the frontal zone eastward as a cold front (Fig. 12.4d).

## 12.3. UPPER-AIR AND CROSS-SECTIONAL ANALYSES

There are reasons why manual analyses of upper-level isobaric maps may be inferior to those produced by operational model data assimilation (DA) systems; most importantly, there are many upper-air data sources that are available to DA systems that are not easily plotted or available for the purpose of manual analysis. While a skilled analyst has the ability to physically relate different analysis variables, such as wind speed and geopotential height, modern DA systems are able to integrate many more variables, including such nontraditional quantities as satellite radiance or radar-derived Doppler radial velocity (see section 10.5). Despite the probable superiority of automated analyses in terms of accuracy and completeness, manual upper-air analysis remains a worthwhile endeavor by allowing the analyst to think about the patterns and processes taking place aloft. However, only limited attention will be devoted to the discussion of upper-air analysis here.

At and above the 850-mb level, the wind is generally close to a state of gradient balance, and if the flow is relatively straight, it will likely be close to a state of geostrophic balance as well. This provides a very powerful constraint on the analysis of the geopotential height field—the wind will frequently be oriented nearly parallel to the geopotential height contours. Analyses of isotachs, isotherms, and in some instances moisture or dewpoint depression are components of traditional upper-air plots. However, I am not aware of any routinely available manual upper-air analyses.

## 12.4. EVALUATION CRITERIA

Grading or evaluating manual analyses is time consuming and challenging and has a certain element of subjectivity. There are three suggested criteria for the evaluation of manual analyses, as discussed below:

- *Accuracy and consistency:* Skill in interpolation, adherence to the principles outlined above, and, most importantly, *meteorological consistency* are essential to the veracity of an analysis. The extent to which isobars are smooth and parallel is considered, but there also must be consistency between gradients and reported wind speeds, and uniformity in gradients. The winds, isotherms, isobars, and fronts must all be drawn in a manner that is self-consistent.
- *Completeness:* Attention to detail is critical. The analyst must draw for all data, or indicate otherwise; label all isobars (draw all isobars!); and generally follow directions carefully.
- *Neatness:* The final analysis must contain smooth, dark contours without kinks or wiggles, and all isopleths must be neatly and clearly labeled.

## REFERENCES

Bosart, L. F., 2003: Wither the weather analysis and forecasting process? *Wea. Forecasting,* **18,** 520–529.

Mahoney, K. M., and G. M. Lackmann, 2006: The sensitivity of numerical forecasts to convective parameterization: A case study of the 17 February 2004 East Coast cyclone. *Wea. Forecasting,* **21,** 465–488.

Mass, C. F., 1991: Synoptic frontal analysis: Time for a reassessment? *Bull. Amer. Meteor. Soc.,* **72,** 348–363.

Sanders, F., 1990: Surface analysis over the oceans—Searching for sea truth. *Wea. Forecasting,* **5,** 596–612.

——, 1999: A proposed method of surface map analysis. *Mon. Wea. Rev.,* **127,** 945–955.

——, and C. A. Doswell III, 1995: A case for detailed surface analysis. *Bull. Amer. Meteor. Soc.,* **76,** 505–521.

——, and E. G. Hoffman, 2002: A climatology of surface baroclinic zones. *Wea. Forecasting,* **17,** 774–782.

Saucier, W. J., 1989: *Principles of Meteorological Analysis.* 2nd ed. Dover, 438 pp.

Schultz, D. M., 2005: A review of cold fronts with prefrontal troughs and wind shifts. *Mon. Wea. Rev.,* **133,** 2449–2472.

Uccellini, L. W., S. F. Corfidi, N. W. Junker, P. J. Kocin, and D. A. Olson, 1992: Report on the Surface Analysis Workshop held at the National Meteorological Center 25–28 March 1991. *Bull. Amer. Meteor. Soc.,* **73,** 459–472.

# INDEX

# D

# E

## F



## G



## H

# I



# J



# K



# L



# M



# N

# O



# P

## Q



## R



## S

## T



## U



## V



## W

## Z